Applied Algebra and Functional Analysis

ANTHONY N. MICHEL
University of Notre Dame

CHARLES J. HERGET
Lawrence Livermore National Laboratory

DOVER PUBLICATIONS, INC.
New York

Published in Canada by General Publishing Company, Ltd., 30 Lesmill Road, Don Mills, Toronto, Ontario.
Published in the United Kingdom by Constable and Company, Ltd., 3 The Lanchesters, 162–164 Fulham Palace Road, London W6 9ER.

This Dover edition, first published in 1993, is an unabridged, corrected republication of the work first published by Prentice-Hall, Englewood Cliffs, N.J., 1981, under the title *Mathematical Foundations in Engineering and Science: Algebra and Analysis*. For the Dover edition the authors, in addition to correcting a number of errors, have slightly revised the original Preface.

Manufactured in the United States of America
Dover Publications, Inc., 31 East 2nd Street, Mineola, N.Y. 11501

Library of Congress Cataloging in Publication Data

Michel, Anthony N.
[Mathematical foundations in engineering and science]
Applied algebra and functional analysis / Anthony N. Michel, Charles J. Herget.
p. cm.
Originally published: Mathematical foundations in engineering and science. Englewood Cliffs, N.J. : Prentice-Hall, c1981.
Includes bibliographical references and index.
ISBN 0-486-67598-X (pbk.)
1. Algebra. 2. Mathematical analysis. I. Herget, Charles J. II. Title.
QA154.2.M52 1993
512′.15—dc20 93-15006
CIP

CONTENTS

PREFACE

This book has evolved from a one-year sequence of courses offered by the authors at Iowa State University. The audience for this book typically included first- or second-year graduate students in various areas in engineering and science. We believe that it is also suitable for self-study or as a reference book. The prerequisites include the usual background in undergraduate mathematics offered to students majoring in engineering or in the sciences. Thus, this book is suitable for advanced senior undergraduate students as well.

The objectives of this book are to (1) provide the reader with appropriate mathematical background for graduate study in engineering or science, (2) provide the reader with appropriate prerequisites for more advanced subjects in mathematics, (3) allow the student in engineering or science to become familiar with essential background in modern mathematics in an efficient manner, (4) give the reader a unified overview of modern mathematics, thus enabling him or her to choose additional courses in mathematics more intelligently, and, most importantly, (5) make it possible for the student to understand at an early stage of his or her graduate study the mathematics used in the current literature (journal articles and the like).

The book may be viewed as consisting essentially of three parts: set theory (Chapter 1), algebra (Chapters 2–4), and analysis (Chapters 5–7). Chapter 1 is a prerequisite for all subsequent chapters. Chapter 2 emphasizes abstract algebra (semigroups, groups, rings, etc.) and may essentially be skipped by those who are not interested in this topic. Chapter 3, which addresses linear spaces and linear transformations, is a prerequisite for Chapters 4, 6, and 7. Chapter 4, which treats finite-dimensional vector spaces and linear transformations on such spaces (matrices) is required for Chapters 6 and 7. In Chapter 5, metric spaces are treated. This chapter is a prerequisite for the subsequent chapters. Chapters 6 and 7 consider Banach and Hilbert spaces and linear operators on such spaces, respectively. Selected topics in applications, which may be omitted without loss of continuity, are presented at the ends of Chapters 2, 4, 5, 6, and 7 and include topics dealing with ordinary differential equations, integral equations, applications of the contraction mapping principle, minimization of functionals, an example from optimal control, estimation of random variables, and the like. Because of this flexibility, this book can be used as a two-semester course. A comfortable pace can be established by deleting appropriate parts, taking into account the students' backgrounds and interests.

All exercises are an integral part of the text and are given when they arise rather than at the end of each chapter. Their intent is to further the reader's understanding of the subject matter.

Concerning the labeling of items, some comments are in order. Sections are assigned numerals that reflect the chapter and section numbers. For example, Section 2.3 signifies the third section in the second chapter. Extensive sections are usually divided into subsections identified by uppercase common letters A, B, C, etc. Equations, definitions, theorems, corollaries, lemmas, examples, exercises, figures, and special remarks are assigned monotonically increasing numerals which identify the chapter, section, and item number. For example, Theorem 4.4.7 denotes the seventh identified item in the fourth section of Chapter 4. This theorem is followed by Eq. (4.4.8), the eighth identified item in the same section. Within a given chapter, figures are identified by uppercase letters, A, B, C, etc., while outside of the chapter the same figure is identified by the above numbering scheme. Finally, the end of a proof or an example is signified by the symbol ■.

We acknowledge the contributions of the students who used the class notes that served as precursors to this book. We would like to thank Professor J. A. Heinen from Marquette University, and Professors S. E. Dickson, W. R. Madych, R. K. Miller and D. L. Isaacson from Iowa State University for their valuable advice during the preparation of this manuscript. We are particularly appreciative of the efforts of Mrs. Betty A. Carter and Miss Shellie Siders in typing the manuscript.

A. N. Michel
C. J. Herget

1

FUNDAMENTAL CONCEPTS

In this chapter we present fundamental concepts required throughout the remainder of this book. We begin by considering sets in Section 1.1. In Section 1.2 we discuss functions; in Section 1.3 we introduce relations and equivalence relations; and in Section 1.4 we concern ourselves with operations on sets. In Section 1.5 we give a brief indication of the types of mathematical systems which we will consider in this book. The chapter concludes with a brief discussion of references.

1.1. SETS

Virtually every area of modern mathematics is developed by starting from an undefined object called a **set.** There are several reasons for doing this. One of these is to develop a mathematical discipline in a completely axiomatic and totally abstract manner. Another reason is to present a unified approach to what may seem to be highly diverse topics in mathematics. Our reason is the latter, for our interest is not in abstract mathematics for its own sake. However, by using abstraction, many of the underlying principles of modern mathematics are more clearly understood.

Thus, we begin by assuming that a **set** is a well defined collection of

elements or **objects.** We denote sets by common capital letters A, B, C, etc., and elements or objects of sets by lower case letters a, b, c, etc. For example, we write

$$A = \{a, b, c\}$$

to indicate that A is the collection of elements a, b, c. If an element x belongs to a set A, we write

$$x \in A.$$

In this case we say that "x belongs to A," or "x is contained in A," or "x is a member of A," etc. If x is any element and if A is a set, then we assume that one knows whether x belongs to A or whether x does not belong to A. If x does not belong to A we write

$$x \notin A.$$

To illustrate some of the concepts, we assume that the reader is familiar with the set of real numbers. Thus, if we say

R is the set of all real numbers,

then this is a well defined collection of objects. We point out that it is possible to characterize the set of real numbers in a purely abstract manner based on an axiomatic approach. We shall not do so here.

To illustrate a non-well defined collection of objects, consider the statement "the set of all tall people in Ames, Iowa." This is clearly not precise enough to be considered here.

We will agree that any set A may not contain any given element x more than once unless we explicitly say so. Moreover, we assume that the concept of "order" will play no role when representing elements of a set, unless we say so. Thus, the sets $A = \{a, b, c\}$ and $B = \{c, b, a\}$ are to be viewed as being exactly the same set.

We usually do not describe a set by listing every element between the curly brackets $\{\quad\}$ as we did for set A above. A convenient method of characterizing sets is as follows. Suppose that for each element x of a set A there is a statement $P(x)$ which is either true or false. We may then define a set B which consists of all elements $x \in A$ such that $P(x)$ is true, and we may write

$$B = \{x \in A : P(x) \text{ is true}\}.$$

For example, let A denote the set of all people who live in Ames, Iowa, and let B denote the set of all males who live in Ames. We can write, then,

$$B = \{x \in A : x \text{ is a male}\}.$$

When it is clear which set x belongs to, we sometimes write $\{x : P(x) \text{ is true}\}$ (instead of, say, $\{x \in A : P(x) \text{ is true}\}$).

It is also necessary to consider a set which has no members. Since a set is determined by its elements, there is only one such set which is called the

empty set, or the **vacuous set,** or the **null set,** or the **void set** and which is denoted by $\varnothing$. Any set, A, consisting of one or more elements is said to be **non-empty** or **non-void.** If A is non-void we write $A \neq \varnothing$.

If A and B are sets and if every element of B also belongs to A, then we say that B is a **subset** of A or A **includes** B, and we write $B \subset A$ or $A \supset B$. Furthermore, if $B \subset A$ and if there is an $x \in A$ such that $x \notin B$, then we say that B is a **proper subset** of A. Some texts make a distinction between proper subset and any subset by using the notation $\subset$ and $\subseteq$, respectively. We shall not use the symbol $\subseteq$ in this book. We note that if A is any set, then $\varnothing \subset A$. Also, $\varnothing \subset \varnothing$. If B is not a subset of A, we write $B \not\subset A$ or $A \not\supset B$.

1.1.1. Example. Let R denote the set of all real numbers, let Z denote the set of all integers, let J denote the set of all positive integers, and let Q denote the set of all rational numbers. We could alternately describe the set Z as

$$Z = \{x \in R: x \text{ is an integer}\}.$$

Thus, for every $x \in R$, the statement x is an integer is either true or false. We frequently also specify sets such as J in the following obvious manner,

$$J = \{x \in Z: x = 1, 2, \ldots\}.$$

We can specify the set Q as

$$Q = \left\{x \in R: x = \frac{p}{q}, p, q \in Z, q \neq 0\right\}.$$

It is clear that $\varnothing \subset J \subset Z \subset Q \subset R$, and that each of these subsets are proper subsets. We note that $0 \notin J$. ■

We now wish to state what is meant by equality of sets.

1.1.2. Definition. Two sets, A and B, are said to be **equal** if $A \subset B$ and $B \subset A$. In this case we write $A = B$. If two sets, A and B, are not equal, we write $A \neq B$. If x and y denote the same element of a set, we say that they are equal and we write $x = y$. If x and y denote distinct elements of a set, we write $x \neq y$.

We emphasize that all definitions are "if and only if" statements. Thus, in the above definition we should actually have said: A and B are equal if and only if $A \subset B$ and $B \subset A$. Since this is always understood, *hereafter all definitions will imply the "only if" portion*. Thus, we simply say: two sets A and B are said to be equal if $A \subset B$ and $B \subset A$.

In Definition 1.1.2 we introduced two concepts of equality, one of equality of sets and one of equality of elements. We shall encounter many forms of equality throughout this book.

Now let X be a set and let $A \subset X$. The **complement of subset A with respect to X** is the set of elements of X which do not belong to A. We denote the complement of A with respect to X by $C_X A$. When it is clear that the complement is with respect to X, we simply say the **complement of** A (instead of the complement of A with respect to X), and simply write $A^{\sim}$. Thus, we have

$$A^{\sim} = \{x \in X: x \notin A\}. \tag{1.1.3}$$

In every discussion involving sets, we will always have a given fixed set in mind from which we take elements and subsets. We will call this set the **universal set**, and we will usually denote this set by X.

Throughout the remainder of the present section, X denotes always an arbitrary non-void fixed set.

We now establish some properties of sets.

1.1.4. Theorem. Let A, B, and C be subsets of X. Then

(i) if $A \subset B$ and $B \subset C$, then $A \subset C$;
(ii) $X^{\sim} = \varnothing$;
(iii) $\varnothing^{\sim} = X$;
(iv) $(A^{\sim})^{\sim} = A$;
(v) $A \subset B$ if and only if $A^{\sim} \supset B^{\sim}$; and
(vi) $A = B$ if and only if $A^{\sim} = B^{\sim}$.

Proof. To prove (i), first assume that A is non-void and let $x \in A$. Since $A \subset B$, $x \in B$, and since $B \subset C$, $x \in C$. Since x is arbitrary, every element of A is also an element of C and so $A \subset C$. Finally, if $A = \varnothing$, then $A \subset C$ follows trivially.

The proofs of parts (ii) and (iii) follow immediately from (1.1.3).

To prove (iv), we must show that $A \subset (A^{\sim})^{\sim}$ and $(A^{\sim})^{\sim} \subset A$. If $A = \varnothing$, then clearly $A \subset (A^{\sim})^{\sim}$. Now suppose that A is non-void. We note from (1.1.3) that

$$(A^{\sim})^{\sim} = \{x \in X: x \notin A^{\sim}\}. \tag{1.1.5}$$

If $x \in A$, it follows from (1.1.3) that $x \notin A^{\sim}$, and hence we have from (1.1.5) that $x \in (A^{\sim})^{\sim}$. This proves that $A \subset (A^{\sim})^{\sim}$.

If $(A^{\sim})^{\sim} = \varnothing$, then $A = \varnothing$; otherwise we would have a contradiction by what we have already shown; i.e., $A \subset (A^{\sim})^{\sim}$. So let us assume that $(A^{\sim})^{\sim} \neq \varnothing$. If $x \in (A^{\sim})^{\sim}$ it follows from (1.1.5) that $x \notin A^{\sim}$, and thus we have $x \in A$ in view of (1.1.3). Hence, $(A^{\sim})^{\sim} \subset A$.

We leave the proofs of parts (v) and (vi) as an exercise. ■

1.1.6. Exercise. Prove parts (v) and (vi) of Theorem 1.1.4.

The proofs given in parts (i) and (iv) of Theorem 1.1.4 are intentionally quite detailed in order to demonstrate the exact procedure required to prove containment and equality of sets. Frequently, the manipulations required to prove some seemingly obvious statements are quite long. It is suggested that the reader carry out all the details in the manipulations of the above exercise and the exercises that follow.

Next, let A and B be subsets of X. We define the **union** of sets A and B, denoted by $A \cup B$, as the set of all elements that are in A or B; i.e.,

$$A \cup B = \{x \in X : x \in A \text{ or } x \in B\}.$$

When we say $x \in A$ or $x \in B$, we mean x is in either A or in B or in both A and B. This *inclusive* use of "or" is standard in mathematics and logic.

If A and B are subsets of X, we define their **intersection** to be the set of all elements which belong to both A and B and denote the intersection by $A \cap B$. Specifically,

$$A \cap B = \{x \in X : x \in A \text{ and } x \in B\}.$$

If the intersection of two sets A and B is empty, i.e., if $A \cap B = \varnothing$, we say that A and B are **disjoint**.

For example, let $X = \{1, 2, 3, 4, 5\}$, let $A = \{1, 2\}$, let $B = \{3, 4, 5\}$, let $C = \{2, 3\}$, and let $D = \{4, 5\}$. Then $A^{\sim} = B$, $B^{\sim} = A$, $D \subset B$, $A \cup B = X$, $A \cap B = \varnothing$, $A \cup C = \{1, 2, 3\}$, $B \cap D = D$, $A \cap C = \{2\}$, etc.

In the next result we summarize some of the important properties of union and intersection of sets.

1.1.7. Theorem. Let A, B, and C be subsets of X. Then

(i) $A \cap B = B \cap A$;
(ii) $A \cup B = B \cup A$;
(iii) $A \cap \varnothing = \varnothing$;
(iv) $A \cup \varnothing = A$;
(v) $A \cap X = A$;
(vi) $A \cup X = X$;
(vii) $A \cap A = A$;
(viii) $A \cup A = A$;
(ix) $A \cup A^{\sim} = X$;
(x) $A \cap A^{\sim} = \varnothing$;
(xi) $A \cap B \subset A$;
(xii) $A \cap B = A$ if and only if $A \subset B$;
(xiii) $A \subset A \cup B$;
(xiv) $A = A \cup B$ if and only if $B \subset A$;
(xv) $(A \cap B) \cap C = A \cap (B \cap C)$;
(xvi) $(A \cup B) \cup C = A \cup (B \cup C)$;
(xvii) $A \cap (B \cup C) = (A \cap B) \cup (A \cap C)$;

(xviii) $(A \cap B) \cup C = (A \cup C) \cap (B \cup C)$;

(xix) $(A \cup B)^\sim = A^\sim \cap B^\sim$; and

(xx) $(A \cap B)^\sim = A^\sim \cup B^\sim$.

Proof. We only prove part (xviii) of this theorem, again as an illustration of the manipulations involved. We will first show that $(A \cap B) \cup C \subset (A \cup C) \cap (B \cup C)$, and then we show that $(A \cap B) \cup C \supset (A \cup C) \cap (B \cup C)$.

Clearly, if $(A \cap B) \cup C = \varnothing$, the assertion is true. So let us assume that $(A \cap B) \cup C \neq \varnothing$, and let x be any element of $(A \cap B) \cup C$. Then $x \in A \cap B$ or $x \in C$. Suppose $x \in A \cap B$. Then x belongs to both A and B, and hence $x \in A \cup C$ and $x \in B \cup C$. From this it follows that $x \in (A \cup C) \cap (B \cup C)$. On the other hand, let $x \in C$. Then $x \in A \cup C$ and $x \in B \cup C$, and hence $x \in (A \cup C) \cap (B \cup C)$. Thus, if $x \in (A \cap B) \cup C$, then $x \in (A \cup C) \cap (B \cup C)$, and we have

$$(A \cap B) \cup C \subset (A \cup C) \cap (B \cup C). \tag{1.1.8}$$

To show that $(A \cap B) \cup C \supset (A \cup C) \cap (B \cup C)$ we need to prove the assertion only when $(A \cup C) \cap (B \cup C) \neq \varnothing$. So let x be any element of $(A \cup C) \cap (B \cup C)$. Then $x \in A \cup C$ and $x \in B \cup C$. Since $x \in A \cup C$, then $x \in A$ or $x \in C$. Furthermore, $x \in B \cup C$ implies that $x \in B$ or $x \in C$. We know that either $x \in C$ or $x \notin C$. If $x \in C$, then $x \in (A \cap B) \cup C$. If $x \notin C$, then it follows from the above comments that $x \in A$ and also $x \in B$. Then $x \in A \cap B$, and hence $x \in (A \cap B) \cup C$. Thus, if $x \notin C$, then $x \in (A \cap B) \cup C$. Since this exhausts all the possibilities, we conclude that

$$(A \cup C) \cap (B \cup C) \subset (A \cap B) \cup C. \tag{1.1.9}$$

From (1.1.8) and (1.1.9) it follows that $(A \cup C) \cap (B \cup C) = (A \cap B) \cup C$. ■

1.1.10. Exercise. Prove parts (i) through (xvii) and parts (xix) and (xx) of Theorem 1.1.7.

In view of part (xvi) of Theorem 1.1.7, there is no ambiguity in writing $A \cup B \cup C$. Extending this concept, let n be any positive integer and let $A_1, A_2, \ldots, A_n$ denote subsets of X. The set $A_1 \cup A_2 \cup \ldots \cup A_n$ is defined to be the set of all $x \in X$ which belong to at least one of the subsets A_i, and we write

$$\bigcup_{i=1}^{n} A_i = A_1 \cup A_2 \cup \ldots \cup A_n = \{x \in X : x \in A_i \text{ for some } i = 1, \ldots, n\}.$$

Similarly, by part (xv) of Theorem 1.1.7, there is no ambiguity in writing $A \cap B \cap C$. We define

$$\bigcap_{i=1}^{n} A_i = A_1 \cap A_2 \cap \ldots \cap A_n = \{x \in X : x \in A_i \text{ for all } i = 1, \ldots, n\}.$$

That is, $\bigcap_{i=1}^{n} A_i$ consists of those members of X which belong to all the subsets $A_1, A_2, \ldots, A_n$.

We will consider the union and the intersection of an infinite number of subsets A_i at a later point in the present section.

The following is a generalization of parts (xix) and (xx) of Theorem 1.1.7.

1.1.11. Theorem. Let $A_1, \ldots, A_n$ be subsets of X. Then

(i) $\left[\bigcup_{i=1}^{n} A_i\right]^{\sim} = \bigcap_{i=1}^{n} A_i^{\sim}$, and (1.1.12)

(ii) $\left[\bigcap_{i=1}^{n} A_i\right]^{\sim} = \bigcup_{i=1}^{n} A_i^{\sim}$. (1.1.13)

1.1.14. Exercise. Prove Theorem 1.1.11.

The results expressed in Eqs. (1.1.12) and (1.1.13) are usually referred to as **De Morgan's laws**. We will see later in this section that these laws hold under more general conditions.

Next, let A and B be two subsets of X. We define the **difference** of B and A, denoted $(B - A)$, as the set of elements in B which are not in A, i.e.,

$$B - A = \{x \in X : x \in B \text{ and } x \notin A\}.$$

We note here that A is not required to be a subset of B. It is clear that

$$B - A = B \cap A^{\sim}.$$

Now let A and B be again subsets of the set X. The **symmetric difference** of A and B is denoted by $A \,\Delta\, B$ and is defined as

$$A \,\Delta\, B = (A - B) \cup (B - A).$$

The following properties follow immediately.

1.1.15. Theorem. Let A, B, and C denote subsets of X. Then

(i) $A \,\Delta\, B = B \,\Delta\, A$;
(ii) $A \,\Delta\, B = (A \cup B) - (A \cap B)$;
(iii) $A \,\Delta\, A = \varnothing$;
(iv) $A \,\Delta\, \varnothing = A$;
(v) $A \,\Delta\, (B \,\Delta\, C) = (A \,\Delta\, B) \,\Delta\, C$;
(vi) $A \cap (B \,\Delta\, C) = (A \cap B) \,\Delta\, (A \cap C)$; and
(vii) $A \,\Delta\, B \subset (A \,\Delta\, C) \cup (C \,\Delta\, B)$.

1.1.16. Exercise. Prove Theorem 1.1.15.

In passing, we point out that the use of **Venn diagrams** is highly useful in visualizing properties of sets; however, under no circumstances should such diagrams take the place of a proof. In Figure A we illustrate the concepts of union, intersection, difference, and symmetric difference of two sets, and the complement of a set, by making use of Venn diagrams. Here, the shaded regions represent the indicated sets.

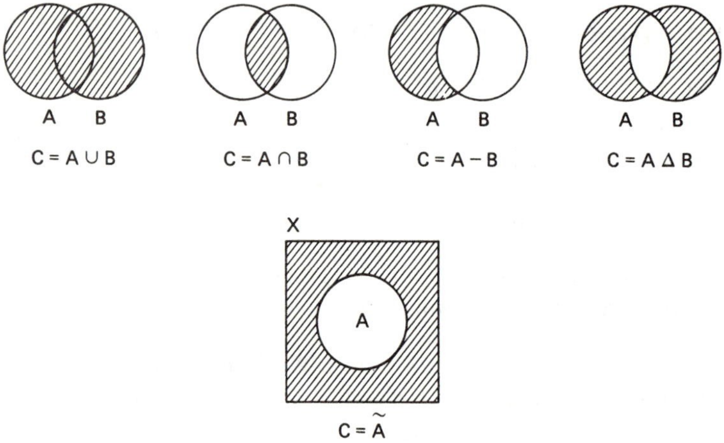

1.1.17. **Figure A.** Venn diagrams.

1.1.18. Definition. A non-void set A is said to be **finite** if A contains n distinct elements, where n is some positive integer; such a set A is said to be of **order** n. The null set is defined to be finite with **order zero**. A set consisting of exactly one element, say $A = \{a\}$, is called a **singleton** or the **singleton of** a. If a set A is not finite, then we say that A is **infinite**.

In Section 1.2 we will further categorize infinite sets as being countable or uncountable.

Next, we need to consider sets whose elements are sets themselves. For example, if A, B, and C are subsets of X, then the collection $\mathcal{A} = \{A, B, C\}$ is a set whose elements are A, B, and C. We usually call a set whose elements are subsets of X a **family of subsets of** X or a **collection of subsets of** X.

We will usually employ a hierarchical system of notation where lower case letters, e.g., a, b, c, are elements of X, upper case letters, e.g., A, B, C, are subsets of X, and script letters, e.g., $\mathcal{A}$, $\mathcal{B}$, $\mathcal{C}$, are families of subsets of X. We could, of course, continue this process and consider a set whose elements are families of subsets, e.g., $\{\mathcal{A}, \mathcal{B}, \mathcal{C}\}$.

In connection with the above comments, we point out that the empty set, $\varnothing$, is a subset of X. It is possible to form a non-empty set whose only element is the empty set, i.e., $\{\varnothing\}$. In this case, $\{\varnothing\}$ is a singleton. We see that $\varnothing \in \{\varnothing\}$ and $\varnothing \subset \{\varnothing\}$.

In principle, we could also consider sets made up of both elements of X and subsets of X. For example, if $x \in X$ and $A \subset X$, then $\{x, A\}$ is a valid set. However, we shall not make use of sets of this nature in this book.

There is a special family of subsets of X to which we given a special name.

1.1.19. Definition. Let A be any subset of X. We define the **power class** of A or the **power set** of A to be the family of all subsets of A. We denote the power class of A by $\mathcal{P}(A)$. Specifically,

$$\mathcal{P}(A) = \{B: B \subset A\}.$$

1.1.20. Example. The power class of the empty set, $\mathcal{P}(\varnothing) = \{\varnothing\}$, i.e., the singleton of $\varnothing$. The power class of a singleton, $\mathcal{P}(\{a\}) = \{\varnothing, \{a\}\}$. For the set $A = \{a, b\}$, $\mathcal{P}(A) = \{\varnothing, \{a\}, \{b\}, \{a, b\}\}$. In general, if A is a finite set with n elements, then $\mathcal{P}(A)$ contains 2^n elements. ■

Before proceeding further, it should be pointed out that a free and uncritical use of a set theory can lead to contradictions and that set theory has had a careful development with various devices used to exclude the contradictions. Roughly speaking, contradictions arise when one uses sets which are "too big," such as trying to speak of a set which contains everything. In all of our subsequent discussions we will keep away from these contradictions by always having some set or space X fixed for a given discussion and by considering only sets whose elements are elements of X, or sets (collections) whose elements are subsets of X, or sets (families) whose elements are collections of subsets of X, etc.

Let us next consider ordered sets. Above, we defined set in such a manner that the ordering of the elements is immaterial, and furthermore that each element is distinct. Thus, if a and b are elements of X, then $\{a, b\} = \{b, a\}$; i.e., there is no preference given to a or b. Furthermore, we have $\{a, a, b\} = \{a, b\}$. In this case we sometimes speak of an **unordered pair** $\{a, b\}$.

Frequently, we will need to consider the **ordered pair** (a, b), (a and b need not belong to the same set) where we distinguish between the *first element* a and the *second element* b. In this case $(a, b) = (u, v)$ if and only if $u = a$ and $v = b$. Thus, $(a, b) \neq (b, a)$ if $a \neq b$. Also, we will consider **ordered triplets** (a, b, c), **ordered quadruplets** (a, b, c, d), etc., where we need to distinguish between the first element, second element, third element, fourth element, etc. Ordered pairs, ordered triplets, ordered quadruplets, etc., are examples of **ordered sets**.

We point out here that our characterization of ordered sets is not axiomatic, since we are assuming that the reader knows what is meant by the first element, second element, third element, etc. (However, it is possible to define ordered sets in a totally abstract fashion without assuming this simple fact. We shall forego these subtle distinctions and accept the preceding as a definition.)

Now let X and Y be two non-void sets. We define the **Cartesian** or **direct product** of X and Y, denoted by $X \times Y$, as the set of all ordered pairs whose first element belongs to X and whose second element belongs to Y. Thus,

$$X \times Y = \{(x, y): x \in X, y \in Y\}. \tag{1.1.21}$$

Next, let $X_1, \ldots, X_n$ denote n arbitrary non-void sets. We similarly define the (n-fold) **Cartesian product** of $X_1, \ldots, X_n$, denoted by $X_1 \times X_2 \times \ldots \times X_n$, as

$$X_1 \times X_2 \times \ldots \times X_n = \{(x_1, x_2, \ldots, x_n): x_1 \in X_1, x_2 \in X_2, \ldots, x_n \in X_n\}. \tag{1.1.22}$$

We call x_i the ith **element of the ordered set** $(x_1, \ldots, x_n) \in X_1 \times X_2 \times \ldots \times X_n$, $i = 1, \ldots, n$. Here again, two ordered sets $(x_1, \ldots, x_n)$ and $(y_1, \ldots, y_n)$ are said to be equal if and only if $x_i = y_i$, $i = 1, \ldots, n$.

In the following example, the symbol $\triangleq$ means **equal by definition.**

1.1.23. Example. Let R be the set of all real numbers. We denote the Cartesian product, $R \times R$, by $R^2 \triangleq R \times R$. Thus, if $x, y \in R$, the ordered pair $(x, y) \in R \times R$. We may interpret (x, y) geometrically as being the coordinates of a point in the plane, x being the first coordinate and y the second coordinate. ■

1.1.24. Example. Let $A = \{0, 1\}$, and let $B = \{a, b, c\}$. Then

$$A \times B = \{(0, a), (0, b), (0, c), (1, a), (1, b), (1, c)\}$$

and

$$B \times A = \{(a, 0), (a, 1), (b, 0), (b, 1), (c, 0), (c, 1)\}.$$

From this example it follows that, in general, if A and B are distinct sets, then

$$A \times B \neq B \times A. \quad ■$$

Next, we consider some generalizations to an ordered set. To this end, let I denote any non-void set which we call **index set**. Now for each $\alpha \in I$, suppose there is a unique $A_\alpha \subset X$. We call $\{A_\alpha: \alpha \in I\}$ an **indexed family of sets**. This notation requires some clarification. Strictly speaking, the set notation $\{A_\alpha: \alpha \in I\}$ would normally indicate that none of the sets A_α, $\alpha \in I$ may be repeated. However, in the case of indexed family we agree to permit the possibility that the sets A_α, $\alpha \in I$ need not be distinct.

We define an indexed set in a similar manner. Let I be an index set, and for each $\alpha \in I$ let there be a unique element $x_\alpha \in X$. Then the set $\{x_\alpha: \alpha \in I\}$

is called an **indexed set**. Here again, we agree to permit the possibility that the elements x_α, $\alpha \in I$ need not be distinct. Clearly, if I is a finite non-void set, then an indexed set is simply an ordered set.

In the next definition, and *throughout the remainder of this section*, J denotes the set of positive integers.

1.1.25. Definition. A **sequence** is an indexed set whose index set is J. A **sequence of sets** is an indexed family of sets whose index set is J.

We usually abbreviate the sequence $\{x_n \in X: n \in J\}$ by $\{x_n\}$, when no possibility for confusion exists. (Even though the same notation is used for the sequence $\{x_n\}$ and the singleton of x_n, the meaning as to which is meant will always be clear from context.) Some authors write $\{x_n\}_{n=1}^{\infty}$ to indicate that the index set of the sequence is J. Also, some authors allow the index set of a sequence to be finite.

We are now in a position to consider the following additional generalizations.

1.1.26. Definition. Let $\{A_\alpha : \alpha \in I\}$ be an indexed family of sets, and let K be any subset of I. If K is non-void, we define

$$\bigcup_{\alpha \in K} A_\alpha = \{x \in X: x \in A_\alpha \text{ for some } \alpha \in K\}$$

and

$$\bigcap_{\alpha \in K} A_\alpha = \{x \in X: x \in A_\alpha \text{ for all } \alpha \in K\}.$$

If $K = \varnothing$, we define $\bigcup_{\alpha \in \varnothing} A_\alpha = \varnothing$ and $\bigcap_{\alpha \in \varnothing} A_\alpha = X$.

The union and intersection of families of sets which are not necessarily indexed is defined in a similar fashion. Thus, if $\mathfrak{F}$ is any non-void family of subsets of X, then we define

$$\bigcup_{F \in \mathfrak{F}} F = \{x \in X: x \in F \text{ for some } F \in \mathfrak{F}\}$$

and

$$\bigcap_{F \in \mathfrak{F}} F = \{x \in X: x \in F \text{ for all } F \in \mathfrak{F}\}.$$

When, in Definition 1.1.26, K is of the form $K = \{k, k+1, k+2, \ldots\}$, where k is an integer, we sometimes write $\bigcup_{n=k}^{\infty} A_n$ and $\bigcap_{n=k}^{\infty} A_n$.

1.1.27. Example. Let $X = R$, the set of real numbers, and let $I = \{x \in R: 0 \leq x < 1\}$. Let $A_\alpha = \{x \in R: 0 \leq x < \alpha\}$ for all $\alpha \in I$. Then, $\bigcup_{\alpha \in I} A_\alpha = I$ and $\bigcap_{\alpha \in I} A_\alpha = \{0\}$, i.e., the singleton containing only the element 0. ■

1.1.28. Example. Let $X = R$, the set of real numbers, and let $I = J$. Let $A_n = \{x: -n < x < +n\}$. Then, $\bigcup_{n=1}^{\infty} A_n = R$ and $\bigcap_{n=1}^{\infty} A_n = \{x: -1 < x < 1\}$. Let $B_n = \left\{x \in R: -\frac{1}{n} < x < 1 + \frac{1}{n}\right\}$. Then, $\bigcup_{n=1}^{\infty} B_n = \{x: -1 < x < 2\}$ and $\bigcap_{n=1}^{\infty} B_n = \{x: 0 \leq x \leq 1\}$. ■

The reader is now in a position to prove the following results.

1.1.29. Theorem. Let $\{A_\alpha: \alpha \in I\}$ be an indexed family of sets. Let B be any subset of X, and let K be any subset of I. Then

(i) $B \cap [\bigcup_{\alpha \in K} A_\alpha] = \bigcup_{\alpha \in K} [B \cap A_\alpha]$;

(ii) $B \cup [\bigcap_{\alpha \in K} A_\alpha] = \bigcap_{\alpha \in K} [B \cup A_\alpha]$;

(iii) $B - \bigcup_{\alpha \in K} A_\alpha = \bigcap_{\alpha \in K} (B - A_\alpha)$;

(iv) $B - \bigcap_{\alpha \in K} A_\alpha = \bigcup_{\alpha \in K} (B - A_\alpha)$;

(v) $[\bigcup_{\alpha \in K} A_\alpha]^{\sim} = \bigcap_{\alpha \in K} A_\alpha^{\sim}$; and

(vi) $[\bigcap_{\alpha \in K} A_\alpha]^{\sim} = \bigcup_{\alpha \in K} A_\alpha^{\sim}$.

1.1.30. Exercise. Prove Theorem 1.1.29.

Parts (v) and (vi) of Theorem 1.1.29 are called **De Morgan's laws.**

We conclude the present section with the following:

1.1.31. Definition. Let $\mathfrak{F}$ be any family of subsets of X. $\mathfrak{F}$ is said to be a **family of disjoint sets** if for all $A, B \in \mathfrak{F}$ such that $A \neq B$, then $A \cap B = \varnothing$. A sequence of sets $\{E_n\}$ is said to be a **sequence of disjoint sets** if for every m, $n \in J$ such that $m \neq n$, $E_m \cap E_n = \varnothing$.

1.2. FUNCTIONS

We first give the definition of a function in a set theoretic manner. Then we discuss the meaning of function in more intuitive terms.

1.2.1. Definition. Let X and Y be non-void sets. A **function f from X into Y** is a subset of $X \times Y$ such that for every $x \in X$ there is one and only one $y \in Y$ (i.e., there is a unique $y \in Y$) such that $(x, y) \in f$. The set X is called the **domain** of f (or the **domain of definition** of f), and we say that **f is defined on X**. The set $\{y \in Y: (x, y) \in f$ for some $x \in X\}$ is called the **range** of f and is denoted by $\mathfrak{R}(f)$. For each $(x, y) \in f$, we call y the **value of f at x**

and denote it by $f(x)$. We sometimes write $f: X \longrightarrow Y$ to denote the function f from X into Y.

The terms *mapping*, *map*, *operator*, *transformation*, and *function* are used interchangeably. When using the term *mapping*, we usually say "a mapping of X into Y." Although the distinction between the words "of X" and "from X" is immaterial, as we shall see, the wording "into Y" becomes important as opposed to the wording "onto Y," which we will encounter later.

Sometimes it is convenient not to insist that the domain of definition of f be all of X; i.e., a function is sometimes defined on a subset of X rather than on all of X. In any case, the domain of definition of f is denoted by $\mathfrak{D}(f) \subset X$. Unless specified otherwise, we shall always assume that $\mathfrak{D}(f) = X$.

Intuitively, a function f is a "rule" whereby for each $x \in X$ a unique $y \in Y$ is assigned to x. When viewed in this manner, the term *mapping* is quite descriptive. However, defining a function as a "rule" involves usage of yet another undefined term.

Concerning functions, some additional comments are in order.

1. So-called "multivalued functions" are not allowed by the above definition. They will be treated later under the topic of relations (Section 1.3).
2. The set X (or Y) may be the Cartesian product of sets, e.g., $X = X_1 \times X_2 \times \ldots \times X_n$. In this case we think of f as being a function of n variables. We write $f(x_1, \ldots, x_n)$ to denote the value of f at $(x_1, \ldots, x_n) \in X = X_1 \times \ldots \times X_n$.
3. It is important that the distinction between a function and the value of a function be clearly understood. The value of a function, $f(x)$, is an element of Y. The function f is a much larger entity, and it is to be thought of as a single object. Note that $f \in \mathcal{P}(X \times Y)$ (the power set of $X \times Y$), but not every element of $\mathcal{P}(X \times Y)$ is a function. The set of all functions from X into Y is a subset of $\mathcal{P}(X \times Y)$ and is sometimes denoted by Y^X.

1.2.2. Example. Let A and B be the sets defined in Example 1.1.24. Let f be the subset of $A \times B$ given by $f = \{(0, a), (1, b)\}$. Then f is a function from A into B. We see that $f(0) = a$ and $f(1) = b$. The range of f is the set $\{a, b\}$ which is a proper subset of B. ■

Although we have defined a function as being a set, we usually characterize a function according to a rule as shown, for example, in the following.

1.2.3. Example. Let R denote the real numbers, and let f be a function from R into R whose value at each $x \in R$ is given by $f(x) = \sin x$. The function f is the sine function. Expressed explicitly as a set, we see that $f = \{(x, y):$

$y = \sin x\}$. Note that the subset $\{(x, y): x = \sin y\} \subset R \times R$ is not a function. ■

The preceding example also illustrates the notion of the graph of a function. Let X and Y denote the set of real numbers, let $X \times Y$ denote their Cartesian product, and let f be a function from X into Y. The collection of ordered pairs $(x, f(x))$ in $X \times Y$ is called the **graph** of the function f. Thus, a subset G of $X \times Y$ is the graph of a function defined on X if and only if for each $x \in X$ there is a unique ordered pair in G whose first element is x. In fact, the graph of a function and the function itself are one and the same thing.

Since functions are defined as sets, **equality of functions** is to be interpreted in the sense as equality of sets. With this in mind, the reader will have no difficulty in proving the following.

1.2.4. Theorem. Two mappings f and g of X into Y are equal if and only if $f(x) = g(x)$ for every $x \in X$.

1.2.5. Exercise. Prove Theorem 1.2.4.

We now wish to further characterize and classify functions. If f is a function from X into Y, we denote the range of f by $\mathcal{R}(f)$. In general, $\mathcal{R}(f) \subset Y$ may or may not be a proper subset of Y. Thus, we have the following definition.

1.2.6. Definition. Let f be a function from X into Y. If $\mathcal{R}(f) = Y$, then f is said to be **surjective** or a **surjection**, and we say that f **maps** X **onto** Y. If f is a function such that for every $x_1, x_2 \in X, f(x_1) = f(x_2)$ implies that $x_1 = x_2$, then f is said to be **injective** or a **one-to-one mapping**, or an **injection**. If f is both injective and surjective, we say that f is **bijective** or **one-to-one and onto**, or a **bijection**.

Let's go over this again. Every function $f: X \longrightarrow Y$ is a mapping of X *into* Y. If the range of f happens to be all of Y, then we say f maps X *onto* Y. For each $x \in X$, there is always a *unique* $y \in Y$ such that $y = f(x)$. However, there may be distinct elements x_1 and x_2 in X such that $f(x_1) = f(x_2)$. If there is a *unique* $x \in X$ such that $f(x) = y$ for each $y \in \mathcal{R}(f)$, then we say that f is a one-to-one mapping. If f maps X onto Y and is one-to-one, we say that f is one-to-one and onto. In Figure B an attempt is made to illustrate these concepts pictorially. In this figure the dots denote elements of sets and the arrows indicate the rules of the various functions.

The reader should commit to memory the following associations: surjective $\longleftrightarrow$ onto; injective $\longleftrightarrow$ one-to-one; bijective $\longleftrightarrow$ one-to-one and onto. Frequently, the term *one-to-one* is abbreviated as (1-1).

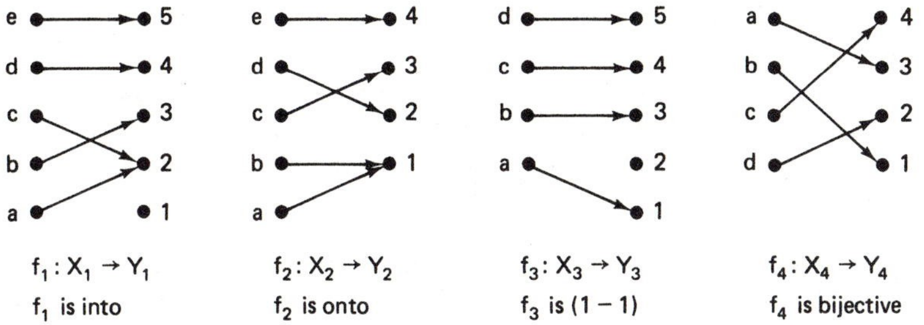

1.2.7. Figure B. Illustration of different types of mappings.

We now prove the following important but obvious result.

1.2.8. Theorem. Let f be a function from X into Y, and let $Z = \mathcal{R}(f)$, the range of f. Let g denote the set $\{(y, x) \in Z \times X: (x, y) \in f\}$. Then, clearly, g is a subset of $Z \times X$ and f is injective if and only if g is a function from Z into X.

Proof. Let f be injective, and let $y \in Z$. Since $y \in \mathcal{R}(f)$, there is an $x \in X$ such that $(x, y) \in f$ and hence $(y, x) \in g$. Now suppose there is another $x_1 \in X$ such that $(y, x_1) \in g$. Then $(x_1, y) \in f$. Since f is injective and $y = f(x) = f(x_1)$, this implies that $x = x_1$ and so x is unique. This means that g is a function from Z into X.

Conversely, suppose g is a function from Z into X. Let $x_1, x_2 \in X$ be such that $f(x_1) = f(x_2)$. This implies that $(x_1, f(x_1))$ and $(x_2, f(x_2)) \in f$ and so $(f(x_1), x_1)$ and $(f(x_2), x_2) \in g$. Since $f(x_1) = f(x_2)$ and g is a function, we must have $x_1 = x_2$. Therefore, f is injective. ■

The above result motivates the following definition.

1.2.9. Definition. Let f be an injective mapping of X into Y. Then we say that f **has an inverse**, and we call the mapping g defined in Theorem 1.2.8 the **inverse of** f. Hereafter, we will denote the inverse of f by f^{-1}.

Clearly, if f has an inverse, then f^{-1} is a mapping from $\mathcal{R}(f)$ onto X.

1.2.10. Theorem. Let f be an injective mapping of X into Y. Then

(i) f is a one-to-one mapping of X *onto* $\mathcal{R}(f)$;
(ii) f^{-1} is a one-to-one mapping of $\mathcal{R}(f)$ *onto* X;
(iii) for every $x \in X$, $f^{-1}(f(x)) = x$; and
(iv) for every $y \in \mathcal{R}(f)$, $f(f^{-1}(y)) = y$.

1.2.11. Exercise. Prove Theorem 1.2.10.

Note that in the above definition, the domain of f^{-1} is $\mathcal{R}(f)$,which need not be all of Y.

Some texts insist that in order for a function f to have an inverse, it must be bijective. Thus, when reading the literature it is important to note which definition of f^{-1} the author has in mind. (Note that an injective function $f: X \longrightarrow Y$ is a bijective function from X onto $\mathcal{R}(f)$.)

1.2.12. Example. Let $X = Y = R$, the set of real numbers. Let $f : X \longrightarrow Y$ be given by $f(x) = x^3$ for every $x \in R$. Then f is a (1-1) mapping of X onto Y and $f^{-1}(y) = (y)^{1/3}$ for all y. ■

1.2.13. Example. Let $X = Y = J$, the set of positive integers. Let $f : X \longrightarrow Y$ be given by $f(n) = n + 3$ for all $n \in J$. Then f is a (1-1) mapping of X into Y. However, the range of f, $\mathcal{R}(f) = \{y \in Y: y \geq 4\} = \{4, 5, \ldots\} \neq Y$. Therefore, f has an inverse, f^{-1}, which is defined only on $\mathcal{R}(f)$ and not on all of Y. In this case we have $f^{-1}(y) = y - 3$ for all $y \in \mathcal{R}(f)$. ■

1.2.14. Example. Let $X = Y = R$, the set of all real numbers. Let $f : X \longrightarrow Y$ be given by $f(x) = \dfrac{x}{1 + |x|}$ for all $x \in R$. Then f is an injective mapping and $\mathcal{R}(f) = \{y \in Y: -1 < y < +1\}$. Also, f^{-1} is a mapping from $\mathcal{R}(f)$ into R given by $f^{-1}(y) = \dfrac{y}{1 - |y|}$ for all $y \in \mathcal{R}(f)$. ■

Next, let X, Y, and Z be non-void sets. Suppose that $f : X \longrightarrow Y$ and $g : Y \longrightarrow Z$. For each $x \in X$, we have $f(x) \in Y$ and $g(f(x)) \in Z$. Since f and g are mappings from X into Y and from Y into Z, respectively, it follows that for each $x \in X$ there is one and only one element $g(f(x)) \in Z$. Hence, the set

$$\{(x, z) \in X \times Z: z = g(f(x)), x \in X\} \tag{1.2.15}$$

is a function from X into Z. We call this function the **composite function of** g **and** f and denote it by $g \circ f$. The value of $g \circ f$ at x is given by

$$(g \circ f)(x) = g \circ f(x) \triangleq g(f(x)).$$

In Figure C, a pictorial interpretation of a composite function is given.

1.2.17. Theorem. If f is a mapping of a set X *onto* a set Y and g is a mapping of the set Y *onto* a set Z, then $g \circ f$ is a mapping of X *onto* Z.

Proof. In order to show that $g \circ f$ is an onto mapping we must show that for any $z \in Z$ there exists an $x \in X$ such that $g(f(x)) = z$. If $z \in Z$ then since g is a mapping of Y onto Z, there is an element $y \in Y$ such that $g(y) = z$. Furthermore, since f is a mapping of X onto Y, there is an $x \in X$ such that $f(x) = y$. Since $g \circ f(x) = g(f(x)) = g(y) = z$, it readily follows that $g \circ f$ is a mapping of X onto Z, which proves the theorem. ■

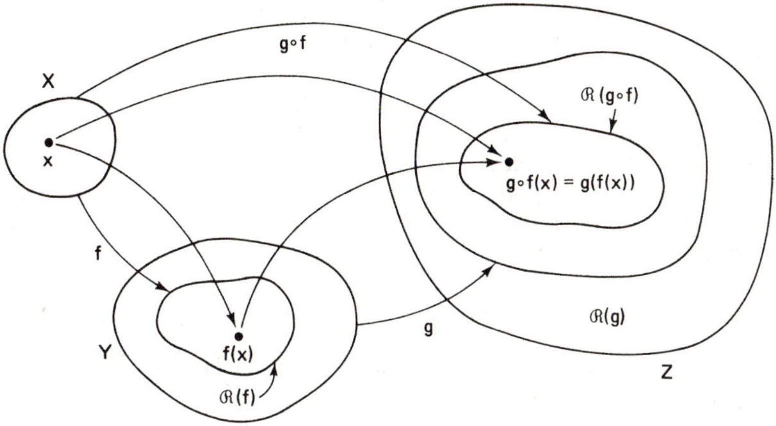

1.2.16. **Figure C.** Illustration of a composite function.

We also have

1.2.18. Theorem. If f is a (1-1) mapping of a set X onto a set Y, and if g is a (1-1) mapping of the set Y onto a set Z, then $g \circ f$ is a (1-1) mapping of X onto Z.

1.2.19. Exercise. Prove Theorem 1.2.18.

Next we prove:

1.2.20. Theorem. If f is a (1-1) mapping of a set X onto a set Y, and if g is a (1-1) mapping of Y onto a set Z, then $(g \circ f)^{-1} = (f^{-1}) \circ (g^{-1})$.

Proof. Let $z \in Z$. Then there exists an $x \in X$ such that $g \circ f(x) = z$, and hence $(g \circ f)^{-1}(z) = x$. Also, since $g \circ f(x) = g(f(x)) = z$, it follows that $g^{-1}(z) = f(x)$, from which we have $f^{-1}(g^{-1}(z)) = x$. But $f^{-1}(g^{-1}(z)) = f^{-1} \circ g^{-1}(z)$ and since this is equal to x, we have $f^{-1} \circ g^{-1}(z) = (g \circ f)^{-1}(z)$. Since z is arbitrary, the theorem is proved. ■

Note carefully that in Theorem 1.2.20 f is a mapping of X *onto* Y. If it had simply been an injective mapping, the composite function $(f^{-1}) \circ (g^{-1})$ may not be defined. That is, the range of g^{-1} is Y; however, the domain of f^{-1} is $\mathcal{R}(f)$. Clearly, the domain of f^{-1} must include the range of g^{-1} in order that the composition $(f^{-1}) \circ (g^{-1})$ be defined.

1.2.21. Example. Let $A = \{r, s, t, u\}$, $B = \{u, v, w, x\}$, and $C = \{w, x, y, z\}$. Let the function $f: A \longrightarrow B$ be defined as

$$f = \{(r, u), (s, w), (t, v), (u, x)\}.$$

We find it convenient to represent this function in the following way:

$$f = \begin{pmatrix} r & s & t & u \\ u & w & v & x \end{pmatrix}.$$

That is, the top row identifies the domain of f and the bottom row contains each unique element in the range of f directly below the appropriate element in the domain. Clearly, this representation can be used for any function defined on a finite set. In a similar fashion, let the function $g : B \longrightarrow C$ be defined as

$$g = \begin{pmatrix} u & v & w & x \\ x & w & z & y \end{pmatrix}.$$

Clearly, both f and g are bijective. Also, $g \circ f$ is the (1-1) mapping of A onto C given by

$$g \circ f = \begin{pmatrix} r & s & t & u \\ x & z & w & y \end{pmatrix}.$$

Furthermore,

$$f^{-1} = \begin{pmatrix} u & w & v & x \\ r & s & t & u \end{pmatrix}, \quad g^{-1} = \begin{pmatrix} x & w & z & y \\ u & v & w & x \end{pmatrix}, \quad (g \circ f)^{-1} = \begin{pmatrix} x & z & w & y \\ r & s & t & u \end{pmatrix}.$$

Now

$$f^{-1} \circ g^{-1} = \begin{pmatrix} x & w & z & y \\ r & t & s & u \end{pmatrix},$$

i.e., $f^{-1} \circ g^{-1} = (g \circ f)^{-1}$. ■

The reader can prove the next result readily.

1.2.22. Theorem. Let W, X, Y, and Z be non-void sets. If f is a mapping of set W into set X, if g is a mapping of X into set Y, and if h is a mapping of Y into set Z (sets W, X, Y, Z are not necessarily distinct), then $h \circ (g \circ f) = (h \circ g) \circ f$.

1.2.23. Exercise. Prove Theorem 1.2.22.

1.2.24. Example. Let $A = \{m, n, p, q\}$, $B = \{m, r, s\}$, $C = \{r, t, u, v\}$, $D = \{w, x, y, z\}$, and define $f : A \longrightarrow B$, $g : B \longrightarrow C$, and $h : C \longrightarrow D$ as

$$f = \begin{pmatrix} m & n & p & q \\ m & r & s & m \end{pmatrix}, \quad g = \begin{pmatrix} m & r & s \\ r & t & v \end{pmatrix}, \quad h = \begin{pmatrix} r & t & u & v \\ z & w & y & x \end{pmatrix}.$$

Then

$$g \circ f = \begin{pmatrix} m & n & p & q \\ r & t & v & r \end{pmatrix} \quad \text{and} \quad h \circ g = \begin{pmatrix} m & r & s \\ z & w & x \end{pmatrix}.$$

Thus,

$$h \circ (g \circ f) = \begin{pmatrix} m & n & p & q \\ z & w & x & z \end{pmatrix} \quad \text{and} \quad (h \circ g) \circ f = \begin{pmatrix} m & n & p & q \\ z & w & x & z \end{pmatrix},$$

i.e., $h \circ (g \circ f) = (h \circ g) \circ f$. ■

There is a special mapping which is so important that we give it a special name. We have:

1.2.25. Definition. Let X be a non-void set. Let $e : X \longrightarrow X$ be defined by $e(x) = x$ for all $x \in X$. We call e the **identity function** on X.

It is clear that the identity function is bijective.

1.2.26. Theorem. Let X and Y be non-void sets, and left $f : X \longrightarrow Y$. Let e_X, e_Y, and e_1 be the identity functions on X, Y, and $\mathcal{R}(f)$, respectively. Then

(i) if f is injective, then $f^{-1} \circ f = e_X$ and $f \circ f^{-1}; = e_1$; and

(ii) f is bijective if and only if there is a $g : Y \longrightarrow X$ such that $g \circ f = e_X$ and $f \circ g = e_Y$.

Proof. Part (i) follows immediately from parts (iii) and (iv) of Theorem 1.2.10.

The proof of part (ii) is left as an exercise. ■

1.2.27. Exercise. Prove part (ii) of Theorem 1.2.26.

Another special class of important functions are permutations.

1.2.28. Definition. A **permutation on a set** X is a (1-1) mapping of X onto X.

It is clear that the identity mapping on X is a permutation on X. For this reason it is sometimes called the **identity permutation on** X. It is also clear that the inverse of a permutation is also a permutation.

1.2.29. Exercise. Let $X = \{a, b, c\}$, and define $f : X \longrightarrow X$ and $g : X \longrightarrow X$ as

$$f = \begin{pmatrix} a & b & c \\ c & b & a \end{pmatrix}, \quad g = \begin{pmatrix} a & b & c \\ b & c & a \end{pmatrix}.$$

Show that f, g, f^{-1}, and g^{-1} are permutations on X.

1.2.30. Exercise. Let Z denote the set of integers, and let $f : Z \longrightarrow Z$ be defined by $f(n) = n + 3$ for all $n \in Z$. Show that f and f^{-1} are permutations on Z and that $f^{-1} \circ f = f \circ f^{-1}$.

The reader can readily prove the following results.

1.2.31. Theorem. If f is a (1-1) mapping of a set A onto a set B and if g is a (1-1) mapping of the set B onto the set A, then $g \circ f$ is a permutation on A.

1.2.32. Corollary. If f and g are both permutations on a set A, then $g \circ f$ is a permutation on A.

1.2.33. Exercise. Prove Theorem 1.2.31 and Corollary 1.2.32.

1.2.34. Exercise. Show that if a set A consists of n elements, then there are exactly $n!$ (n factorial) distinct permutations on A.

Now let f be a mapping of a set X into a set Y. If X_1 is a subset of X, then for each element $x' \in X_1$ there is a unique element $f(x') \in Y$. Thus, f may be used to define a mapping f' of X_1 into Y defined by

$$f'(x') = f(x') \tag{1.2.35}$$

for all $x' \in X_1$. This motivates the following definition.

1.2.36. Definition. The mapping f' of subset $X_1 \subset X$ into Y of Eq. (1.2.35) is called the mapping of X_1 into Y **induced** by the mapping $f: X \longrightarrow Y$. In this case f' is called the **restriction** of f to the set X_1.

We also have:

1.2.37. Definition. If f is a mapping of X_1 into Y and if $X_1 \subset X$, then any mapping $\bar{f}$ of X into Y is said to be an **extension** of f if

$$\bar{f}(x) = f(x) \tag{1.2.38}$$

for every $x \in X_1$.

Thus, if $\bar{f}$ is an extension of f, then f is a mapping of a set $X_1 \subset X$ into Y which is induced by the mapping $\bar{f}$ of X into Y.

1.2.39. Example. Let $X_1 = \{u, v, x\}$, $X = \{u, v, x, y, z\}$, and $Y = \{n, p, q, r, s, t\}$. Clearly $X_1 \subset X$. Define $f: X_1 \longrightarrow Y$ as

$$f = \begin{pmatrix} u & v & x \\ n & p & q \end{pmatrix}.$$

Also, define $\bar{f}, \bar{\bar{f}} : X \longrightarrow Y$ as

$$\bar{f} = \begin{pmatrix} u & v & x & y & z \\ n & p & q & r & s \end{pmatrix}, \quad \bar{\bar{f}} = \begin{pmatrix} u & v & x & y & z \\ n & p & q & n & t \end{pmatrix}.$$

Then $\bar{f}$ and $\bar{\bar{f}}$ are two different extensions of f. Moreover, f is the mapping of X_1 into Y induced either by $\bar{f}$ or $\tilde{f}$. In general, two distinct mappings may induce the same mapping on a subset. ■

Let us next consider the image and the inverse image of sets under mappings. Specifically, we have

1.2.40. Definition. Let f be a function from a set X into a set Y. Let $A \subset X$, and let $B \subset Y$. We define the **image of A under** f, denoted by $f(A)$, to be the set

$$f(A) = \{y \in Y: y = f(x), x \in A\}.$$

We define the **inverse image of B under** f, denoted by $f^{-1}(B)$, to be the set

$$f^{-1}(B) = \{x \in X: f(x) \in B\}.$$

Note that $f^{-1}(B)$ is always defined for any $f: X \longrightarrow Y$. That is, there is no implication here that f has an inverse. The notation is somewhat unfortunate in this respect. Note also that the range of f is $f(X)$.

In the next result, some of the important properties of images and inverse images of functions are summarized.

1.2.41. Theorem. Let f be a function from X into Y, let A, A_1, and A_2 be subsets of X, and let B, B_1, and B_2 be subsets of Y. Then

(i) if $A_1 \subset A$, then $f(A_1) \subset f(A)$;
(ii) $f(A_1 \cup A_2) = f(A_1) \cup f(A_2)$;
(iii) $f(A_1 \cap A_2) \subset f(A_1) \cap f(A_2)$;
(iv) $f^{-1}(B_1 \cup B_2) = f^{-1}(B_1) \cup f^{-1}(B_2)$;
(v) $f^{-1}(B_1 \cap B_2) = f^{-1}(B_1) \cap f^{-1}(B_2)$;
(vi) $f^{-1}(B^{\sim}) = [f^{-1}(B)]^{\sim}$;
(vii) $f^{-1}[f(A)] \supset A$; and
(viii) $f[f^{-1}(B)] \subset B$.

Proof. We prove parts (i) and (ii) to demonstrate the method of proof. The remaining parts are left as an exercise.

To prove part (i), let $y \in f(A_1)$. Then there is an $x \in A_1$ such that $y = f(x)$. But $A_1 \subset A$ and so $x \in A$. Hence, $f(x) = y \in f(A)$. This proves that $f(A_1) \subset f(A)$.

To prove part (ii), let $y \in f(A_1 \cup A_2)$. Then there is an $x \in A_1 \cup A_2$ such that $y = f(x)$. If $x \in A_1$, then $f(x) = y \in f(A_1)$. If $x \in A_2$, then $f(x) = y \in f(A_2)$. Since x is in A_1 or in A_2, $f(x)$ must be in $f(A_1)$ or $f(A_2)$. Therefore, $f(A_1 \cup A_2) \subset f(A_1) \cup f(A_2)$. To prove that $f(A_1) \cup f(A_2) \subset f(A_1 \cup A_2)$, we note that $A_1 \subset A_1 \cup A_2$. So by part (i), $f(A_1) \subset f(A_1$

$\cup A_2)$. Similarly, $f(A_2) \subset f(A_1 \cup A_2)$. From this it follows that $f(A_1) \cup f(A_2) \subset f(A_1 \cup A_2)$. We conclude that $f(A_1 \cup A_2) = f(A_1) \cup f(A_2)$. ■

1.2.42. Exercise. Prove parts (iii) through (viii) of Theorem 1.2.41.

We note that, in general, equality is not attained in parts (iii), (vii), and (viii) of Theorem 1.2.41. However, by considering special types of mappings we can obtain the following results for these cases.

1.2.43. Theorem. Let f be a function from X into Y, let A, A_1, and A_2 be subsets of X, and let B be a subset of Y. Then

(i) $f(A_1 \cap A_2) = f(A_1) \cap f(A_2)$ for all pairs of subsets A_1, A_2 of X if and only if f is injective;

(ii) $f^{-1}[f(A)] = A$ for all $A \subset X$ if and only if f is injective; and

(iii) $f[f^{-1}(B)] = B$ for all $B \subset Y$ if and only if f is surjective.

Proof. We will prove only part (i) and leave the proofs of parts (ii) and (iii) as an exercise.

To prove sufficiency, let f be injective and let A_1 and A_2 be subsets of X. In view of part (iii) of Theorem 1.2.41, we need only show that $f(A_1) \cap f(A_2) \subset f(A_1 \cap A_2)$. In doing so, let $y \in f(A_1) \cap f(A_2)$. Then $y \in f(A_1)$ and $y \in f(A_2)$. This means there is an $x_1 \in A_1$ and an $x_2 \in A_2$ such that $y = f(x_1) = f(x_2)$. Since f is injective, $x_1 = x_2$. Hence, $x_1 \in A_1 \cap A_2$. This implies that $y \in f(A_1 \cap A_2)$; i.e., $f(A_1) \cap f(A_2) \subset f(A_1 \cap A_2)$.

To prove necessity, assume that $f(A_1 \cap A_2) = f(A_1) \cap f(A_2)$ for all subsets A_1 and A_2 of X. For purposes of contradiction, suppose there are $x_1, x_2 \in X$ such that $x_1 \neq x_2$ and $f(x_1) = f(x_2)$. Let $A_1 = \{x_1\}$ and $A_2 = \{x_2\}$; i.e., A_1 and A_2 are singletons of x_1 and x_2, respectively. Then $A_1 \cap A_2 = \varnothing$, and so $f(A_1 \cap A_2) = \varnothing$. However, $f(A_1) = \{y\}$ and $f(A_2) = \{y\}$, and thus $f(A_1) \cap f(A_2) = \{y\} \neq \varnothing$. This contradicts the fact that $f(A_1) \cap f(A_2) = f(A_1 \cap A_2)$ for all subsets A_1 and A_2 of X. Thus, f is injective. ■

1.2.44. Exercise. Prove parts (ii) and (iii) of Theorem 1.2.43.

Some of the preceding results can be extended to families of sets. For example, we have:

1.2.45. Theorem. Let f be a function from X into Y, let $\{A_\alpha : \alpha \in I\}$ be an indexed family of sets in X, and let $\{B_\alpha : \alpha \in K\}$ be an indexed family of sets in Y. Then

(i) $f(\bigcup_{\alpha \in I} A_\alpha) = \bigcup_{\alpha \in I} f(A_\alpha)$;

(ii) $f(\bigcap_{\alpha \in I} A_\alpha) \subset \bigcap_{\alpha \in I} f(A_\alpha)$;

(iii) $f^{-1}(\bigcup_{\alpha \in K} B_\alpha) = \bigcup_{\alpha \in K} f^{-1}(B_\alpha)$;

(iv) $f^{-1}(\bigcap_{\alpha \in K} B_\alpha) = \bigcap_{\alpha \in K} f^{-1}(B_\alpha)$; and

(v) if $B \subset Y$, $f^{-1}(B^\sim) = [f^{-1}(B)]^\sim$.

Proof. We prove parts (i) and (iii) and leave the proofs of the remaining parts as an exercise.

To prove part (i), let $y \in f(\bigcup_{\alpha \in I} A_\alpha)$. This means that there is an $x \in \bigcup_{\alpha \in I} A_\alpha$ such that $y = f(x)$. Thus, for some $\alpha \in I$, $x \in A_\alpha$. This implies that $f(x) \in f(A_\alpha)$ and so $y \in f(A_\alpha)$. Hence, $y \in \bigcup_{\alpha \in I} f(A_\alpha)$. This shows that $f(\bigcup_{\alpha \in I} A_\alpha) \subset \bigcup_{\alpha \in I} f(A_\alpha)$.

To prove the converse, let $y \in \bigcup_{\alpha \in I} f(A_\alpha)$. Then $y \in f(A_\alpha)$ for some $\alpha \in I$. This means there is an $x \in A_\alpha$ such that $f(x) = y$. Now $x \in \bigcup_{\alpha \in I} A_\alpha$, and so $f(x) = y \in f(\bigcup_{\alpha \in I} A_\alpha)$. Therefore, $\bigcup_{\alpha \in I} f(A_\alpha) \subset f(\bigcup_{\alpha \in I} A_\alpha)$. This completes the proof of part (i).

To prove part (iii), let $x \in f^{-1}(\bigcup_{\alpha \in K} B_\alpha)$. This means that $f(x) \in \bigcup_{\alpha \in K} B_\alpha$. Hence, $f(x) \in B_\alpha$ for some $\alpha \in K$. Thus, $x \in f^{-1}(B_\alpha)$, and so $x \in \bigcup_{\alpha \in K} f^{-1}(B_\alpha)$. Therefore, $f^{-1}(\bigcup_{\alpha \in K} B_\alpha) \subset \bigcup_{\alpha \in K} f^{-1}(B_\alpha)$.

Conversely, let $x \in \bigcup_{\alpha \in K} f^{-1}(B_\alpha)$. Then $x \in f^{-1}(B_\alpha)$ for some $\alpha \in K$. Thus, $f(x) \in B_\alpha$. Hence, $f(x) \in \bigcup_{\alpha \in K} B_\alpha$, and so $x \in f^{-1}(\bigcup_{\alpha \in K} B_\alpha)$. This means that $\bigcup_{\alpha \in K} f^{-1}(B_\alpha) \subset f^{-1}(\bigcup_{\alpha \in K} B_\alpha)$, which completes the proof of part (iii). ■

1.2.46. Exercise. Prove parts (ii), (iv), and (v) of Theorem 1.2.45.

Having introduced the concept of mapping, we are in a position to consider an important classification of infinite sets. We first consider the following definition.

1.2.47. Definition. Let A and B be any two sets. The set A is said to be **equivalent** to set B if there exists a bijective mapping of A onto B.

Clearly, if A is equivalent to B, then B is equivalent to A.

1.2.48. Definition. Let J be the set of positive integers, and let A be any set. Then A is said to be **countably infinite** if A is equivalent to J. A set is said to be **countable** or **denumerable** if it is either finite or countably infinite. If a set is not countable, it is said to be **uncountable**.

We have:

1.2.49. Theorem. Let J be the set of positive integers, and let $I \subset J$. If I is infinite, then I is equivalent to J.

Proof. We shall construct a bijective mapping, f, from J onto I. Let $\{J_n: n \in J\}$ be the family of sets given by $J_n = \{1, 2, \ldots, n\}$ for $n = 1, 2, \ldots$. Clearly, each J_n is finite and of order n. Therefore, $J_n \cap I$ is finite. Since I is infinite, $I - J_n \neq \varnothing$ for all n. Let us now define $f: J \longrightarrow I$ as follows. Let $f(1)$ be the smallest integer in I. We now proceed inductively. Assume $f(n) \in I$ has been defined and let $f(n+1)$ be the smallest integer in I which is greater than $f(n)$. Now $f(n+1) > f(n)$, and so $f(n_1) > f(n_2)$ for any $n_1 > n_2$. This implies that f is injective.

Next, we want to show that f is surjective. We do so by contradiction. Suppose that $f(J) \neq I$. Since $f(J) \subset I$, this implies that $I - f(J) \neq \varnothing$. Let q be the smallest integer in $I - f(J)$. Then $q \neq f(1)$ because $f(1) \in f(J)$, and so $q > f(1)$. This implies that $I \cap J_{q-1} \neq \varnothing$. Since $I \cap J_{q-1}$ is non-void and finite, we may find the largest integer in this set, say r. It follows that $r \leq q - 1 < q$. Now r is the largest integer in I which is less than q. But $r < q$ implies that $r \in f(J)$. This means there is an $s \in J$ such that $r = f(s)$. By definition of f, $f(s+1) = q$. Hence, $q \in f(J)$ and we have arrived at a contradition. Thus, f is surjective. This completes the proof. ∎

We now have the following corollary.

1.2.50. Corollary. Let $A \subset B \subset X$. If B is a countable set, then A is countable.

Proof. If A is finite, then there is nothing to prove. So let us assume that A is infinite. This means that B is countably infinite, and so there exists a bijective mapping $f: B \longrightarrow J$. Let g be the restriction of f to A. Then for all $x_1, x_2 \in A$ such that $x_1 \neq x_2$, $g(x_1) = f(x_1) \neq f(x_2) = g(x_2)$. Thus, g is an injective mapping of A into J. By part (i) of Theorem 1.2.10, g is a bijective mapping of A onto $g(A)$. This means A is equivalent to $g(A)$, and thus $g(A)$ is an infinite set. Since $g(A) \subset J$, $g(A)$ is equivalent to J. Hence, there is a bijective mapping of $g(A)$ onto J, which we call h. By Theorem 1.2.18, the composite mapping $h \circ g$ is a bijective mapping of A onto J. This means that J is equivalent to A. Therefore, A is countable. ∎

We conclude the present section by considering the cardinality of sets. Specifically, if a set is finite, we say the **cardinal number** of the set is equal to the number of elements of the set. If two sets are countably infinite, then we say they have the same cardinal number, which we can define to be the cardinal number of the positive integers. More generally, two arbitrary sets are said to have the same cardinal number if we can establish a bijective mapping between the two sets (i.e., the sets are equivalent).

1.3. RELATIONS AND EQUIVALENCE RELATIONS

Throughout the present section, X denotes a non-void set.

We begin by introducing the notion of relation, which is a generalization of the concept of function.

1.3.1 Definition. Let X and Y be non-void sets. Any subset of $X \times Y$ is called a **relation from X to Y**. Any subset of $X \times X$ is called a **relation in X**.

1.3.2. Example. Let $A = \{u, v, x, y\}$ and $B = \{a, b, c, d\}$. Let $\alpha = \{(u, a), (v, b), (u, c), (x, a)\}$. Then α is a relation from A into B. It is clearly not a function from A into B (why?). ■

1.3.3. Example. Let $X = Y = R$, the set of real numbers. The set $\{(x, y) \in R \times R: x \leq y\}$ is a relation in R. Also, the set $\{(x, y) \in R \times R: x = \sin y\}$ is a relation in R. This shows that so-called *multivalued functions* are actually relations rather than mappings. ■

As in the case of mappings, it makes sense to speak of the domain and the range of a relation. We have:

1.3.4. Definition. Let ρ be a relation from X to Y. The subset of X,

$$\{x \in X: (x, y) \in \rho, y \in Y\},$$

is called the **domain of** ρ. The subset of Y

$$\{y \in Y: (x, y) \in \rho, x \in X\},$$

is called the **range** of ρ.

Now let ρ be a relation from X to Y. Then, clearly, the set $\rho^{-1} \subset Y \times X$, defined by

$$\rho^{-1} = \{(y, x) \in Y \times X: (x, y) \in \rho \subset X \times Y\},$$

is a relation from Y to X. The relation ρ^{-1} is called the **inverse relation** of ρ. Note that whereas the inverse of a function does not always exist, the inverse of a relation does always exist.

Next, we consider equivalence relations. Let ρ denote a relation in X; i.e., $\rho \subset X \times X$. Then for any $x, y \in X$, either $(x, y) \in \rho$ or $(x, y) \notin \rho$, but not both. If $(x, y) \in \rho$, then we write $x \rho y$ and if $(x, y) \notin \rho$, we write $x \not\rho y$.

1.3.5. Definition. Let ρ be a relation in X.

(i) If $x \rho x$ for all $x \in X$, then ρ is said to be **reflexive**;

(ii) if $x \rho y$ implies $y \rho x$ for all $x, y \in \rho$, then ρ is said to be **symmetric**; and

(iii) if for all $x, y, z \in X$, $x \rho y$ and $y \rho z$ implies $x \rho z$, then ρ is said to be **transitive**.

1.3.6. Example. Let R denote the set of real numbers. The relation in R given by $\{(x, y): x < y\}$ is transitive but not reflexive and not symmetric. The relation in R given by $\{(x, y): x \neq y\}$ is symmetric but not reflexive and not transitive. ■

1.3.7. Example. Let ρ be the relation in $\mathcal{P}(X)$ defined by $\rho = \{(A \times B): A \subset B\}$. That is, $A \rho B$ if and only if $A \subset B$. Then ρ is reflexive and transitive but not symmetric. ■

In the following, we use the symbol $\sim$ to denote a relation in X. If $(x, y) \in \sim$, then we write, as before, $x \sim y$.

1.3.8. Definition. Let $\sim$ be a relation in X. Then $\sim$ is said to be an **equivalence relation in** X if $\sim$ is reflexive, symmetric, and transitive. If $\sim$ is an equivalence relation and if $x \sim y$, we say that x is **equivalent** to y.

In particular, the equivalence relation in X characterized by the statement "$x \sim y$ if and only if $x = y$" is called the **equals relation in** X or the **identity relation in** X.

1.3.9. Example. Let X be a finite set, and let $A, B, C \in \mathcal{P}(X)$. Let $\sim$ on $\mathcal{P}(X)$ be defined by saying that $A \sim B$ if and only if A and B have the same number of elements. Clearly $A \sim A$. Also, if $A \sim B$ then $B \sim A$. Furthermore, if $A \sim B$ and $B \sim C$, then $A \sim C$. Hence, $\sim$ is reflexive, symmetric, and transitive. Therefore, $\sim$ is an equivalence relation in $\mathcal{P}(X)$. ■

1.3.10. Example. Let $R^2 = R \times R$, the real plane. Let X be the family of all triangles in R^2. Then each of the following statements can be used to define an equivalence relation in X: "is similar to," "is congruent to," "has the same area as," and "has the same perimeter as." ■

1.4. OPERATIONS ON SETS

In the present section we introduce the concept of operation on set, and we consider some of the properties of operations. *Throughout this section, X denotes a non-void set.*

1.4.1. Definition. A **binary operation** on X is a mapping of $X \times X$ into X. A **ternary operation** on X is a mapping of $X \times X \times X$ into X.

We could proceed in an obvious manner and define an n-ary operation on X. Since our primary concern in this book will be with binary operations, *we will henceforth simply say "an operation on X" when we actually mean a binary operation on X.*

If $\alpha: X \times X \longrightarrow X$ is an operation, then we usually use the notation $\alpha(x, y) \triangleq x\alpha y$.

1.4.2. Example. Let R denote the real numbers. Let $f: R \times R \longrightarrow R$ be given by $f(x, y) = x + y$ for all $x, y \in R$, where $x + y$ denotes the customary sum of x plus y (i.e., $+$ denotes the usual operation of addition of real numbers). Then f is clearly an operation on R, in the sense of Definition 1.4.1. We could just as well have defined "$+$" as being the operation on R, i.e., $+: R \times R \longrightarrow R$, where $+(x, y) \triangleq x + y$. Similarly, the ordinary rules of subtraction and multiplication on R, "$-$" and "$\cdot$", respectively, are also operations on R. Notice that division, $\div$, is not an operation on R, because $x \div y$ is not defined for all $y \in R$ (i.e., $x \div y$ is not defined for $y = 0$). However, if we let $R^\# = R - \{0\}$, then "$\div$" is an operation on $R^\#$. ■

1.4.3. Exercise. Show that if A is a set consisting of n distinct elements, then there exist exactly $n^{(n^2)}$ distinct operations on A.

1.4.4. Example. Let $A = \{a, b\}$. An example of an operation on A is the mapping $\alpha: A \times A \longrightarrow A$ defined by

$$\alpha(a, a) \triangleq a\,\alpha\,a = a, \quad \alpha(a, b) \triangleq a\,\alpha\,b = b,$$
$$\alpha(b, a) \triangleq b\,\alpha\,a = b, \quad \alpha(b, b) = b\,\alpha\,b = a.$$

It is convenient to utilize the following **operation table** to define α:

α	a	b
a	a	b
b	b	a

(1.4.5)

If, in general, α is an operation on an arbitrary finite set A, or sometimes even on a countably infinite set A, then we can construct an operation table as follows:

α	$\cdots$	y	$\cdots$
$\vdots$		$\vdots$	
x	$\cdots$	$x\,\alpha\,y$	$\cdots$
$\vdots$		$\vdots$	

If $A = \{a, b\}$, as at the beginning of this example, then in addition to α

given in (1.4.5), we can define, for example, the operations β, γ, and δ on A as

β	a	b
a	a	a
b	b	a

γ	a	b
a	a	b
b	a	b

δ	a	b
a	a	a
b	b	b

■

We now consider operations with important special properties.

1.4.6. Definition. An operation α on X is said to be **commutative** if $x \alpha y = y \alpha x$ for all $x, y \in X$.

1.4.7. Definition. An operation α on X is said to be **associative** if $(x \alpha y) \alpha z = x \alpha (y \alpha z)$ for $x, y, z \in X$.

In the case of the real numbers R, the operations of addition and multiplication are both associative and commutative. The operation of subtraction is neither associative nor commutative.

1.4.8. Definition. If α and β are operations on X (not necessarily distinct), then

(i) α is said to be **left distributive over** β if
$$x \alpha (y \beta z) = (x \alpha y) \beta (x \alpha z)$$
for every $x, y, z \in X$;

(ii) α is said to be **right distributive over** β if
$$(x \beta y) \alpha z = (x \alpha z) \beta (y \alpha z)$$
for every $x, y, z \in X$; and

(iii) α is said to be **distributive over** β if α is both left and right distributive over β.

In Example 1.4.4, α is the only commutative operation. The operation β of Example 1.4.4 is not associative. The operations α, γ, and δ of this example are associative. In this example, γ is distributive over δ and δ is distributive over γ.

In the case of the real numbers R, multiplication, "$\cdot$", is distributive over addition, "$+$". The converse is not true.

1.4.9. Definition. If α is an operation on X, and if X_1 is a subset of X, then X_1 is said to be **closed** relative to α if for every $x, y \in X_1$, $x \alpha y \in X_1$.

Clearly, every set is closed with respect to an operation on it.

The set of all integers Z, which is a subset of the real numbers R, is closed with respect to the operations of addition and multiplication defined on R. The even integers are also closed with respect to both of these operations, whereas the odd integers are not a closed set relative to addition.

1.4.10. Definition. If a subset X_1 of X is closed relative to an operation α on X, then the operation α' on X_1 defined by

$$\alpha'(x, y) = x\,\alpha'\,y = x\,\alpha\,y$$

for all $x, y \in X_1$, is called the operation on X_1 **induced** by α.

If $X_1 = X$, then $\alpha' = \alpha$. If $X_1 \subset X$ but $X_1 \neq X$, then $\alpha' \neq \alpha$ since α' and α are operations on different sets, namely X_1 and X, respectively. In general, an induced operation α' differs from its **predecessor** α; however, it does inherit the essential properties which α possesses, as shown in the following result.

1.4.11. Theorem. Let α be an operation on X, let $X_1 \subset X$, where X_1 is closed relative to α, and let α' be the operation on X_1 induced by α. Then

(i) if α is commutative, then α' is commutative;
(ii) if α is associative, then α' is associative; and
(iii) if β is an operation on X and X_1 is closed relative to β, and if α is left (right) distributive over β, then α' is left (right) distributive over β', where β' is the operation on X_1 induced by β.

1.4.12. Exercise. Prove Theorem 1.4.11.

The operation α' on a subset X_1 induced by an operation α on X will frequently be denoted by α, and we will refer to α as an operation on X_1. In such cases one must keep in mind that we are actually referring to the induced operation α' and not to α.

1.4.13. Definition. Let X_1 be a subset of X. An operation $\bar{\alpha}$ on X is called an **extension** of an operation α on X_1 if X_1 is closed relative to $\bar{\alpha}$ and if α is equal to the operation on X_1 induced by $\bar{\alpha}$.

A given operation α on a subset X_1 of a set X may, in general, have many different extensions.

1.4.14. Example. Let $X_1 = \{a, b, c\}$, and let $X = \{a, b, c, d, e\}$. Define α on X_1 and $\bar{\alpha}$ and $\tilde{\alpha}$ on X as

α	a	b	c
a	a	c	b
b	c	b	a
c	b	a	c

$\bar{\alpha}$	a	b	c	d	e
a	a	c	b	e	d
b	c	b	a	d	e
c	b	a	c	e	d
d	c	d	a	b	e
e	d	c	a	b	e

$\tilde{\alpha}$	a	b	c	d	e
a	a	c	b	d	e
b	c	b	a	e	d
c	b	a	c	d	e
d	d	c	b	a	e
e	d	a	c	b	e

Clearly, α is an operation on X_1 and $\bar{\alpha}$ and $\tilde{\alpha}$ are operations on X. Moreover, both $\bar{\alpha}$ and $\tilde{\alpha}$ ($\bar{\alpha} \neq \tilde{\alpha}$) are extensions of α. Also, α may be viewed as being induced by $\bar{\alpha}$ and $\tilde{\alpha}$. ■

1.5. MATHEMATICAL SYSTEMS CONSIDERED IN THIS BOOK

We will concern ourselves with several different types of mathematical systems in the subsequent chapters. Although it is possible to give an abstract definition of the term *mathematical system*, we will not do so. Instead, we will briefly indicate which *types* of mathematical systems we shall consider in this book.

1. In Chapter 2 we will begin by considering mathematical systems which are made up of an underlying set X and an operation α defined on X. We will identify such systems by writing $\{X; \alpha\}$. We will be able to characterize a system $\{X; \alpha\}$ according to certain properties which X and α possess. Two important cases of such systems that we will consider are **semigroups** and **groups**.

In Chapter 2 we will also consider mathematical systems consisting of a basic set X and two operations, say α and β, defined on X, where a special relation exists between α and β. We will identify such systems by writing $\{X; \alpha, \beta\}$. Included among the mathematical systems of this kind which we will consider are **rings** and **fields**.

In Chapter 2 we will also consider **composite mathematical systems**. Such systems are endowed with two underlying sets, say X and F, and possess a much more complex (algebraic) structure than semigroups, groups, rings, and fields. Composite systems which we will consider include **modules, vector spaces** over a field F which are also called **linear spaces**, and **algebras**.

In Chapter 2 we will also study various types of important mappings (e.g., **homomorphisms** and **isomorphisms**) defined on semigroups, groups, rings, etc.

Mathematical systems of the type considered in Chapter 2 are sometimes called **algebraic systems**.

2. In Chapters 3 and 4 we will study in some detail **vector spaces** and special types of mappings on vector spaces, called **linear transformations**. An important class of linear transformations can be represented by **matrices**, which we will consider in Chapter 4. In this chapter we will also study in some detail important vector spaces, called **Euclidean spaces**.

3. Most of Chapter 5 is devoted to mathematical systems consisting of a basic set X and a function $\rho: X \times X \rightarrow R$ (R denotes the real numbers), where ρ possesses certain properties (namely, the properties of distance

between points or elements in X). The function ρ is called a **metric** (or a distance function) and the pair $\{X; \rho\}$ is called a **metric space**.

In Chapter 5 we will also consider mathematical systems consisting of a basic set X and a family of subsets of X (called open sets) denoted by $\mathfrak{J}$. The pair $\{X; \mathfrak{J}\}$ is called a **topological space**. It turns out that all metric spaces are in a certain sense topological spaces.

We will also study functions and their properties on metric (topological) spaces in Chapter 5.

4. In Chapters 6 and 7 we will consider **normed linear spaces**, **inner product spaces**, and an important class of functions (**linear operators**) defined on such spaces.

A normed linear space is a mathematical system consisting of a vector space X and a real-valued function denoted by $\|\cdot\|$, which takes elements of X into R and which possesses the properties which characterize the "length" of a vector. We will denote normed spaces by $\{X; \|\cdot\|\}$.

An inner product space consists of a vector space X (over the field of real numbers R or over the field of complex numbers C) and a function $(\cdot, \cdot)$, which takes elements from $X \times X$ into R (or into C) and possesses certain properties which allow us to introduce, among other items, the concept of orthogonality. We will identify such mathematical systems by writing $\{X; (\cdot, \cdot)\}$.

It turns out that in a certain sense all inner product spaces are normed linear spaces, that all normed linear spaces are metric spaces, and as indicated before, that all metric spaces are topological spaces. Since normed linear spaces and inner product spaces are also vector spaces, it should be clear that, in the case of such spaces, properties of algebraic systems (called **algebraic structure**) and properties of topological systems (called **topological structure**) are combined.

A class of normed linear spaces which are very important are **Banach spaces**, and among the more important inner product spaces are **Hilbert spaces**. Such spaces will be considered in some detail in Chapter 6. Also, in Chapter 7, linear transformations defined on Banach and Hilbert spaces will be considered.

5. Applications are considered at the ends of Chapters 4, 5, and 7.

1.6. REFERENCES AND NOTES

A classic reference on set theory is the book by Hausdorff [1.5]. The many excellent references on the present topics include the elegant text by Hanneken [1.4], the standard reference by Halmos [1.3] as well as the books by Gleason [1.1] and Goldstein and Rosenbaum [1.2].

REFERENCES

[1.1] A. M. Gleason, *Fundamentals of Abstract Analysis*. Reading, Mass.: Addison-Wesley Publishing Co., Inc., 1966.

[1.2] M. E. Goldstein and B. M. Rosenbaum, "Introduction to Abstract Analysis," National Aeronautics and Space Administration, Report No. SP-203, Washington, D.C., 1969.

[1.3] P. R. Halmos, *Naive Set Theory*. Princeton, N.J.: D. Van Nostrand Company, Inc., 1960.

[1.4] C. B. Hanneken, *Introduction to Abstract Algebra*. Belmont, Calif.: Dickenson Publishing Co., Inc., 1968.

[1.5] F. Hausdorff, *Mengenlehre*. New York: Dover Publications, Inc., 1944.

2

ALGEBRAIC STRUCTURES

The subject matter of the previous chapter is concerned with set theoretic structure. We emphasized essential elements of set theory and introduced related concepts such as mappings, operations, and relations.

In the present chapter we concern ourselves with algebraic structure. The material of this chapter falls usually under the heading of *abstract algebra* or *modern algebra*. In the next two chapters we will continue our investigation of algebraic structure. The topics of those chapters go usually under the heading of *linear algebra*.

This chapter is divided into three parts. The first section is concerned with some basic algebraic structures, including semigroups, groups, rings, fields, modules, vector spaces, and algebras. In the second section we study properties of special important mappings on the above structures, including homomorphisms, isomorphisms, endomorphisms, and automorphisms of semigroups, groups and rings. Because of their importance in many areas of mathematics, as well as in applications, polynomials are considered in the third section. Some appropriate references for further reading are suggested at the end of the chapter.

The subject matter of the present chapter is widely used in pure as well as in applied mathematics, and it has found applications in diverse areas, such as modern physics, automata theory, systems engineering, information theory, graph theory, and the like.

Our presentation of modern algebra is by necessity very brief. However, mastery of the topics covered in the present chapter will provide the reader with the foundation required to make contact with the literature in applications, and it will enable the interested reader to pursue this subject further at a more advanced level.

2.1. SOME BASIC STRUCTURES OF ALGEBRA

We begin by developing some of the more important properties of mathematical systems, $\{X; \alpha\}$, where α is an operation on a non-void set X.

2.1.1. Definition. Let α be an operation on X. If for all $x, y, z \in X$, $x \,\alpha\, y = x \,\alpha\, z$ implies that $y = z$, then we say that $\{X; \alpha\}$ possesses the **left cancellation property**. If $x \,\alpha\, y = z \,\alpha\, y$ implies that $x = z$, then $\{X; \alpha\}$ is said to possess the **right cancellation property**. If $\{X; \alpha\}$ possesses both the left and right cancellation properties, then we say that the **cancellation laws hold** in $\{X; \alpha\}$.

In the following exercise, some specific cases are given.

2.1.2. Exercise. Let $X = \{x, y\}$ and let α, β, γ, and δ be defined as

α	x	y
x	x	y
y	y	x

β	x	y
x	x	x
y	y	x

γ	x	y
x	x	y
y	x	y

δ	x	y
x	x	x
y	y	y

Show that (i) $\{X; \beta\}$ possesses neither the right nor the left cancellation property; (ii) $\{X; \gamma\}$ possesses the left cancellation property but not the right cancellation property; (iii) $\{X; \delta\}$ possesses the right cancellation property but not the left cancellation property; and (iv) $\{X; \alpha\}$ possesses both the left and the right cancellation property.

In an arbitrary mathematical system $\{X; \alpha\}$ there are sometimes special elements in X which possess important properties relative to the operation α. We have:

2.1.3. Definition. Let α be an operation on a set X and let X contain an element e_r such that

$$x \,\alpha\, e_r = x,$$

for all $x \in X$. We call e_r a **right identity element** of X relative to α, or simply a **right identity** of the system $\{X; \alpha\}$. If X contains an element e_l which satisfies the condition

$$e_l \,\alpha\, x = x,$$

for all $x \in X$, then e_ℓ is called a **left identity element** of X relative to α, or simply a **left identity** of the system $\{X; \alpha\}$.

We note that a system $\{X; \alpha\}$ may contain more than one right identity element of X (e.g., system $\{X; \delta\}$ of Exercise 2.1.2) or left identity element of X (e.g., system $\{X; \gamma\}$ of Exercise 2.1.2).

2.1.4. Definition. An element e of a set X is called an **identity element** of X relative to an operation α on X if

$$e \,\alpha\, x = x \,\alpha\, e = x$$

for every $x \in X$.

2.1.5. Exercise. Let $X = \{0, 1\}$ and define the operations "+" and "·" by

+	0	1
0	0	1
1	1	0

·	0	1
0	0	0
1	0	1

Does either $\{X; +\}$ or $\{X; \cdot\}$ have an identity element?

Identity elements have the following properties.

2.1.6. Theorem. Let α be an operation on X.

(i) If $\{X; \alpha\}$ has an identity element e, then e is unique.
(ii) If $\{X; \alpha\}$ has a right identity e_r and a left identity e_ℓ, then $e_r = e_\ell$.
(iii) If α is a commutative operation and if $\{X; \alpha\}$ has a right identity element e_r, then e_r is also a left identity.

Proof. To prove the first part, let e' and e'' be identity elements of $\{X; \alpha\}$. Then $e' \,\alpha\, e'' = e'$ and $e' \,\alpha\, e'' = e''$. Hence, $e' = e''$.

To prove the second part, note that since e_r is a right identity, $e_\ell \,\alpha\, e_r = e_\ell$. Also, since e_ℓ is a left identity, $e_\ell \,\alpha\, e_r = e_r$. Thus, $e_\ell = e_r$.

To prove the last part, note that for all $x \in X$ we have $x = x \,\alpha\, e_r = e_r \,\alpha\, x$. ■

In summary, if $\{X; \alpha\}$ has an identity element, then that element is unique. Furthermore, if $\{X; \alpha\}$ has both a right identity and a left identity element, then these elements are equal, and in fact they are equal to the unique identity element. Also, if $\{X; \alpha\}$ has a right (or left) identity element and α is a commutative operation, then $\{X; \alpha\}$ has an identity element.

2.1.7. Definition. Let α be an operation on X and let e be an identity of X relative to α. If $x \in X$, then $x' \in X$ is called a **right inverse** of x relative to

α provided that

$$x \alpha x' = e.$$

An element $x'' \in X$ is called a **left inverse** of x relative to α if

$$x'' \alpha x = e.$$

The following exercise shows that some elements may not possess any right or left inverses. Some other elements may possess several inverses of one kind and none of the other, and other elements may possess a number of inverses of both kinds.

2.1.8. Exercise. Let $X = \{x, y, u, v\}$ and define α as

α	x	y	u	v
x	x	y	x	y
y	x	y	y	x
u	x	y	u	v
v	x	y	v	u

.

(i) Show that $\{X; \alpha\}$ contains an identity element.
(ii) Which elements possess neither left inverses nor right inverses?
(iii) Which element has a left and a right inverse?

A. Semigroups and Groups

Of crucial importance are mathematical systems called **semigroups**. Such mathematical systems serve as the natural setting for many important results in algebra and are used in several diverse areas of applications (e.g., qualitative analysis of dynamical systems, automata theory, etc.).

2.1.9. Definition. Let α be an operation on X. We call $\{X; \alpha\}$ a **semigroup** if α is an associative operation on X.

Now let $x, y, z \in X$, and let α be an associative operation on X. Then $x \alpha (y \alpha z) = (x \alpha y) \alpha z = u \in X$. Henceforth, we will often simply write $u = x \alpha y \alpha z$. As a result of this convention we see that for $x, y, u, v \in X$,

$$\begin{aligned} x \alpha y \alpha u \alpha v &= x \alpha (y \alpha u) \alpha v = x \alpha y \alpha (u \alpha v) \\ &= (x \alpha y) \alpha (u \alpha v) = (x \alpha y) \alpha u \alpha v. \end{aligned} \tag{2.1.10}$$

As a generalization of the above we have the so-called **generalized associative law**, which asserts that if $x_1, x_2, \ldots, x_n$ are elements of a semigroup $\{X; \alpha\}$, then any two products, each involving these elements in a particular order, are equal. This allows us to simply write $x_1 \alpha x_2 \alpha \ldots \alpha x_n$.

In view of Theorem 2.1.6, part (i), if a semigroup has an identity element, then such an element is unique. We give a special name to such a semigroup.

2.1.11. Definition. A semigroup $\{X; \alpha\}$ is called a **monoid** if X contains an identity element relative to α. Henceforth, the unique identity element of a monoid $\{X; \alpha\}$ will be denoted by e.

Subsequently, we frequently single out elements of monoids which possess inverses.

2.1.12. Definition. Let $\{X; \alpha\}$ be a monoid. If $x \in X$ possesses a right inverse $x' \in X$, then x is called a **right invertible element** in X. If $x \in X$ possesses a left inverse $x'' \in X$, then x is called a **left invertible element** in X. If $x \in X$ is both right invertible and left invertible in X, then we say that x is an **invertible element** or a **unit** of X.

Clearly, if $e \in X$, then e is an invertible element.

2.1.13. Theorem. Let $\{X; \alpha\}$ be a monoid, and let $x \in X$. If there exists a left inverse of x, say x', and a right inverse of x, say x'', then $x' = x''$ and x' is unique.

Proof. Since α is associative, we have $(x' \alpha x) \alpha x'' = x''$ and $x' \alpha (x \alpha x'') = x'$. Thus, $x' = x''$. Now suppose there is another left inverse of x, say x'''. Then $x''' = x''$ and therefore $x''' = x'$. ■

Theorem 2.1.13 does, in general, not hold for arbitrary mathematical systems $\{X; \alpha\}$ with identity, as is evident from the following:

2.1.14. Exercise. Let $X = \{u, v, x, y\}$ and define α as

α	u	v	x	y
u	v	v	u	u
v	u	u	v	x
x	u	v	x	y
y	x	v	y	x

Use this operations table to demonstrate that Theorem 2.1.13 does not, in general, hold if monoid $\{X; \alpha\}$ is replaced by system $\{X; \alpha\}$ with identity.

By Theorem 2.1.13, any invertible element of a monoid possesses a unique right inverse and a unique left inverse, and moreover these inverses are equal. This gives rise to the following.

2.1.15. Definition. Let $\{X; \alpha\}$ be a monoid. If $x \in X$ has a left inverse and a right inverse, x' and x'', respectively, then this unique element $x' = x''$ is called the **inverse** of x and is denoted by x^{-1}.

Concerning inverses we have.

2.1.16. Theorem. Let $\{X; \alpha\}$ be a monoid.

(i) If $x \in X$ has an inverse, x^{-1}, then x^{-1} has an inverse $(x^{-1})^{-1} = x$.

(ii) If $x, y \in X$ have inverses x^{-1}, y^{-1}, respectively, then $x \,\alpha\, y$ has an inverse, and moreover $(x \,\alpha\, y)^{-1} = y^{-1} \,\alpha\, x^{-1}$.

(iii) The identity element $e \in X$ has an inverse e^{-1} and $e^{-1} = e$.

Proof. To prove the first part, note that $x \,\alpha\, x^{-1} = e$ and $x^{-1} \,\alpha\, x = e$. Thus, x is both a left and a right inverse of x^{-1} and $(x^{-1})^{-1} = x$.

To prove the second part, note that

$$(x \,\alpha\, y) \,\alpha\, (y^{-1} \,\alpha\, x^{-1}) = x \,\alpha\, (y \,\alpha\, y^{-1}) \,\alpha\, x^{-1} = e$$

and

$$(y^{-1} \,\alpha\, x^{-1}) \,\alpha\, (x \,\alpha\, y) = y^{-1} \,\alpha\, (x^{-1} \,\alpha\, x) \,\alpha\, y = e.$$

The third part of the theorem follows trivially from $e \,\alpha\, e = e$. ■

In the remainder of the present chapter we will often use the symbols "$+$" and "$\cdot$" to denote operations in place of α, β, etc. We will call these "addition" and "multiplication." However, we strongly emphasize here that "$+$" and "$\cdot$" will, in general, *not* denote addition and multiplication of real numbers but, instead, arbitrary operations. In cases where there exists an identity element relative to "$+$", we will denote this element by "0" and call it "zero." If there exists an identity element relative to "$\cdot$", we will denote this element either by "1" or by e. Our usual notation for representing an identity relative to an arbitrary operation α will still be e. If in a system $\{X; +\}$ an element $x \in X$ possesses an inverse, we will denote this element by $-x$ and we will call it "minus x". For example, if $\{X; +\}$ is a semigroup, then we denote the inverse of an invertible element $x \in X$ by $-x$, and in this case we have $x + (-x) = (-x) + x = 0$, and also, $-(-x) = x$. Furthermore, if $x, y \in X$ are invertible elements, then the "sum" $x + y$ is also invertible, and $-(x + y) = (-y) + (-x)$. Note, however, that unless "$+$" is commutative, $-(x + y) \neq (-x) + (-y)$. Finally, if $x, y \in X$ and if y is an invertible element, then $-y \in X$. In this case we often will simply write $x + (-y) = x - y$.

2.1.17. Example. Let $X = \{0, 1, 2, 3\}$, and let the systems $\{X; +\}$ and $\{X; \cdot\}$ be defined by means of the operation tables

+	0	1	2	3
0	0	1	2	3
1	1	2	3	0
2	2	3	0	1
3	3	0	1	2

·	0	1	2	3
0	0	0	0	0
1	0	1	2	3
2	0	2	0	2
3	0	3	2	1

The reader should readily show that the systems $\{X; +\}$ and $\{X; \cdot\}$ are monoids. In this case the operation "+" is called "addition mod 4" and "·" is called "multiplication mod 4." ■

The most important special type of semigroup that we will encounter in this chapter is the group.

2.1.18. Definition. A **group** is a monoid in which every element is invertible; i.e., a group is a semigroup, $\{X; \alpha\}$, with identity in which every element is invertible.

The set R of real numbers with the operation of addition is an example of a group. The set of real numbers with the operation of multiplication does *not* form a group, since the number zero does not have an inverse relative to multiplication. However, the latter system is a monoid. If we let $R^\# = R - \{0\}$, then $\{R^\#; \cdot\}$ is a group.

Groups possess several important properties. Some of these are summarized in the next result.

2.1.19. Theorem. Let $\{X; \alpha\}$ be a group, and let e denote the identity element of X relative to α. Let x and y be arbitrary elements in X. Then

(i) if $x \alpha x = x$, then $x = e$;
(ii) if $z \in X$ and $x \alpha y = x \alpha z$, then $y = z$;
(iii) if $z \in X$ and $x \alpha y = z \alpha y$, then $x = z$;
(iv) there exists a unique $w \in X$ such that

$$w \alpha x = y; \text{ and} \tag{2.1.20}$$

(v) there exists a unique $z \in X$ such that

$$x \alpha z = y. \tag{2.1.21}$$

Proof. To prove the first part, let $x \alpha x = x$. Then $x^{-1} \alpha (x \alpha x) = x^{-1} \alpha x$, and so $(x^{-1} \alpha x) \alpha x = e$. This implies that $x = e$.

To prove the second part, let $x \alpha y = x \alpha z$. Then $x^{-1} \alpha (x \alpha y) = x^{-1} \alpha (x \alpha z)$, and so $(x^{-1} \alpha x) \alpha y = (x^{-1} \alpha x) \alpha z$. This implies that $y = z$.

The proof of part (iii) is similar to that of part (ii).

To prove part (iv), let $w = y \alpha x^{-1}$. Then $w \alpha x = (y \alpha x^{-1}) \alpha x = y \alpha (x^{-1} \alpha x) = y$. To show that w is unique, suppose there is a $v \in X$ such that $v \alpha x = y$. Then $w \alpha x = v \alpha x$. By part (iii), $w = v$.

The proof of the last part of the theorem is similar to the proof of part (iv). ■

In part (iv) of Theorem 2.1.19 the element w is called **the left solution** of Eq. (2.1.20), and in part (v) of this theorem the element z is called the **right solution** of Eq. (2.1.21).

We can classify groups in a variety of ways. Some of these classifications are as follows. Let $\{X; \alpha\}$ be a group. If the set X possesses a finite number of elements, then we speak of a **finite group**. If the operation α is commutative then we have a **commutative group**, also called an **abelian group**. If α is not commutative, then we speak of a **non-commutative group** or a **non-abelian group**. Also, by the **order of a group** we understand the order of the set X.

Now let $\{X; \alpha\}$ be a semigroup and let X_1 be a non-void subset of X which is closed relative to α. Then by Theorem 1.4.11, the operation α_1 on X_1 induced by the associative operation α is also associative, and thus the mathematical system $\{X_1; \alpha_1\}$ is also a semigroup. The system $\{X_1; \alpha_1\}$ is called a **subsystem** of $\{X; \alpha\}$. This gives rise to the following concept.

2.1.22. Definition. Let $\{X; \alpha\}$ be a semigroup, let X_1 be a non-void subset of X which is closed relative to α, and let α_1 be the operation on X_1 induced by α. The semigroup $\{X_1; \alpha_1\}$ is called a **subsemigroup of** $\{X; \alpha\}$.

In order to simplify our notation, we will henceforth use the notation $\{X_1; \alpha\}$ to denote the subsemigroup $\{X_1; \alpha_1\}$ (i.e., we will suppress the subscript of α).

The following result allows us to generate subsemigroups in a variety of ways.

2.1.23. Theorem. Let $\{X; \alpha\}$ be a semigroup and let $X_i \subset X$ for all $i \in I$, where I denotes some index set. Let $Y = \bigcap_{i \in I} X_i$. If $\{X_i; \alpha\}$ is a subsemigroup of $\{X; \alpha\}$ for every $i \in I$, and if Y is not empty, then $\{Y; \alpha\}$ is a subsemigroup of $\{X; \alpha\}$.

Proof. Let $x, y \in Y$. Then $x, y \in X_i$ for all $i \in I$ and so $x \alpha y \in X_i$ for every i, and hence $x \alpha y \in Y$. This implies that $\{Y; \alpha\}$ is a subsemigroup. ■

Now let W be any non-void subset of X, where $\{X; \alpha\}$ is a semigroup, and let

$$\mathcal{Y} = \{Y: W \subset Y \subset X \text{ and } \{Y; \alpha\} \text{ is a subsemigroup of } \{X; \alpha\}\}.$$

Then $\mathcal{Y}$ is non-empty, since $X \in \mathcal{Y}$. Also, let

$$G = \bigcap_{Y \in \mathcal{Y}} Y.$$

Then $W \subset G$, and by Theorem 2.1.23 $\{G; \alpha\}$ is a subsemigroup of $\{X; \alpha\}$. This subsemigroup is called the **subsemigroup generated by** W.

2.1.24. Theorem. Let $\{X; \alpha\}$ be a monoid with e its identity element, and let $\{X_1; \alpha_1\}$ be a subsemigroup of $\{X; \alpha\}$. If $e \in X_1$, then e is an identity element of $\{X_1; \alpha_1\}$ and $\{X_1; \alpha_1\}$ is a monoid.

2.1.25. Exercise. Prove Theorem 2.1.24.

Next we define subgroup.

2.1.26. Definition. Let $\{X; \alpha\}$ be a semigroup, and let $\{X_1; \alpha_1\}$ be a subsemigroup of $\{X; \alpha\}$. If $\{X_1; \alpha_1\}$ is a group, then $\{X_1; \alpha_1\}$ is called a **subgroup** of $\{X; \alpha\}$. We denote this subgroup by $\{X_1; \alpha\}$, and we say the set X_1 determines a subgroup of $\{X; \alpha\}$.

We consider a specific example in the following:

2.1.27. Exercise. Let $Z_6 = \{0, 1, 2, 3, 4, 5\}$ and define the operation $+$ on Z_6 by means of the following operation table:

+	0	1	2	3	4	5
0	0	1	2	3	4	5
1	1	0	4	5	2	3
2	2	5	0	4	3	1
3	3	4	5	0	1	2
4	4	3	1	2	5	0
5	5	2	3	1	0	4

(a) Show that $\{Z_6; +\}$ is a group.
(b) Let $K = \{0, 1\}$. Show that $\{K; +\}$ is a subgroup of $\{Z_6; +\}$.
(c) Are there any other subgroups of $\{Z_6; +\}$?

We have seen in Theorem 2.1.24 that if $e \in X_1 \subset X$, then it is also an identity of the subsemigroup $\{X_1; \alpha\}$. We can state something further.

2.1.28. Theorem. Let $\{X; \alpha\}$ be a group with identity element e, and let $\{X_1; \alpha\}$ be a subgroup of $\{X; \alpha\}$. Then e_1 is the identity element of $\{X_1; \alpha\}$ if and only if $e_1 = e$.

2.1.29. Exercise. Prove Theorem 2.1.28.

It should be noted that a semigroup $\{X; \alpha\}$ which has no identity element may contain a subgroup $\{X_1; \alpha\}$, since it is possible for a subsystem to possess an identity element while the original system may not possess an identity. If $\{X; \alpha\}$ is a semigroup with an identity element and if $\{X_1; \alpha\}$ is a subgroup, then the identity element of X may or may not be the identity element of X_1. However, if $\{X; \alpha\}$ is a group, then the subgroup must satisfy the conditions given in the following:

2.1.30. Theorem. Let $\{X; \alpha\}$ be a group, and let X_1 be a non-empty subset of X. Then $\{X_1; \alpha\}$ is a subgroup if and only if

(i) $e \in X_1$;
(ii) for every $x \in X_1$, $x^{-1} \in X_1$; and
(iii) for every $x, y \in X_1$, $x \,\alpha\, y \in X_1$.

Proof. Assume that $\{X_1; \alpha\}$ is a subgroup. Then (i) follows from Theorem 2.1.28, and (ii) and (iii) follow from the definition of a group.

Conversely, assume that hypotheses (i), (ii), and (iii) hold. Condition (iii) implies that X_1 is closed relative to α, and therefore $\{X_1; \alpha\}$ is a subsemigroup. Condition (i) along with Theorem 2.1.24 imply that $\{X_1; \alpha\}$ is a monoid, and condition (ii) implies that $\{X_1; \alpha\}$ is a group. ■

Analogous to Theorem 2.1.23 we have:

2.1.31. Theorem. Let $\{X; \alpha\}$ be a group, and let $X_i \subset X$ for all $i \in I$, where I is some index set. Let $Y = \bigcap_{i \in I} X_i$. If $\{X_i; \alpha\}$ is a subgroup of $\{X; \alpha\}$ for every $i \in I$, then $\{Y; \alpha\}$ is a subgroup of $\{X; \alpha\}$.

Proof. Since $e \in X_i$ for every $i \in I$ it follows that $e \in Y$. Therefore, Y is non-empty. Now let $y \in Y$. Then $y \in X_i$ for all $i \in I$, and thus $y^{-1} \in X_i$ so that $y^{-1} \in Y$. Since $y \in X$, it follows that $Y \subset X$. Also, for every $x, y \in Y$, $x, y \in X_i$ for every $i \in I$, and thus $x \,\alpha\, y \in X_i$ for every i and hence $x \,\alpha\, y \in Y$. Therefore, we conclude from Theorem 2.1.30 that $\{Y; \alpha\}$ is a subgroup of $\{X; \alpha\}$. ■

A direct consequence of the above result is the following:

2.1.32. Corollary. Let $\{X; \alpha\}$ be a group, and let $\{X_1; \alpha\}$ and $\{X_2; \alpha\}$ be subgroups of $\{X; \alpha\}$. Let $X_3 = X_1 \cap X_2$. Then $\{X_3; \alpha\}$ is a subgroup of $\{X_1; \alpha\}$ and $\{X_2; \alpha\}$.

2.1.33. Exercise. Prove Corollary 2.1.32.

We can define a generated subgroup in a similar manner as was done in the case of semigroups. To this end let W be any subset of X, where $\{X; \alpha\}$ is a group, and let

$$\mathcal{Y} = \{Y: W \subset Y \subset X \text{ and } \{Y; \alpha\} \text{ is a subgroup of } \{X; \alpha\}\}.$$

The set $\mathcal{Y}$ is clearly non-empty because $X \in \mathcal{Y}$. Now let

$$G = \bigcap_{Y \in \mathcal{Y}} Y.$$

Then $W \subset G$, and by Theorem 2.1.31 $\{G; \alpha\}$ is a subgroup of $\{X; \alpha\}$. This subgroup is called the **subgroup generated by** W.

2.1.34. Exercise. Let W be defined as above. Show that if $\{W; \alpha\}$ is a subgroup of $\{X; \alpha\}$, then it is the subgroup generated by W.

Let us now consider the following:

2.1.35. Example. Let Z denote the set of integers, and let "$+$" denote the usual operation of addition of integers. Let $W = \{1\}$. If Y is any subset of Z such that $\{Y; +\}$ is a subgroup of $\{Z; +\}$ and $W \subset Y$, then $Y = Z$. To prove this statement, let n be any positive integer. Since Y is closed with respect to $+$, we must have $1 + 1 = 2 \in Y$. Similarly, we must have $1 + 1 + \ldots + 1 = n \in Y$. Also, $n^{-1} = -n$, and therefore all the negative integers are in Y. Also, $n - n = 0 \in Y$, i.e., $Y = Z$. Thus, $G = \bigcap_{Y \in \mathcal{Y}} Y = Z$, and so the group $\{Z; +\}$ is the subgroup generated by $\{1\}$. ■

The above is an example of a special class of generated subgroups, the so-called **cyclic groups**, which we will define after our next result.

2.1.36. Theorem. Let Z denote the set of all integers, and let $\{X; \alpha\}$ be a group. Let $x \in X$ and define $x^k = x \,\alpha\, x \,\alpha \ldots \alpha\, x$ (k times), for k a positive integer. Let $x^{-k} = (x^k)^{-1}$, and let $x^0 = e$. Let $Y = \{x^k : k \in Z\}$. Then $\{Y; \alpha\}$ is the subgroup of $\{X; \alpha\}$ generated by $\{x\}$.

Proof. We first show that $\{Y; \alpha\}$ is a subgroup of $\{X; \alpha\}$. Clearly, $Y \subset X$ and $e \in Y$ and for every $y \in Y$ we have $y^{-1} \in Y$. Also, for every $x, y \in Y$ we have $x \,\alpha\, y \in Y$. Thus, by Theorem 2.1.30, $\{Y; \alpha\}$ is a subgroup of $\{X; \alpha\}$. Next, we must show that $\{Y; \alpha\}$ is the subgroup generated by $\{x\}$. To do so, it suffices to show that $Y \subset Y_i$ for every Y_i such that $x \in Y_i$ and such that $\{Y_i; \alpha\}$ is a subgroup of $\{X; \alpha\}$. But this is certainly true, since $y \in Y$ implies $y = x^k$ for some $k \in Z$. Since $x \in Y_i$ it follows that $x^k \in Y_i$ and therefore $y \in Y_i$. ■

The preceding result motivates the following:

2.1.37. Definition. Let $\{X; \alpha\}$ be a group. If there exists an element $x \in X$ such that the subgroup generated by $\{x\}$ is equal to $\{X; \alpha\}$, then $\{X; \alpha\}$ is called the **cyclic group** generated by x.

By Theorem 2.1.36, we see that a cyclic group has elements of such a form that $X = \{\ldots, x^{-3}, x^{-2}, x^{-1}, e, x, x^2, \ldots\}$. Now suppose there is some positive integer n such that $x^n = e$. Then we see that $x^{n+1} = x$. Similarly, $x^{-n} = e$, and $x^{-n+1} = x$. Thus, $X = \{e, x, \ldots, x^{n-1}\}$, and X is a finite set of order n. If there is no n such that $x^n = e$, then X is an infinite set.

We consider next another important class of groups, the so-called **permutation groups**. To this end let X be a non-empty set and let $M(X)$ denote the set of all mappings of X into itself. Now, if $\alpha, \beta \in M(X)$ then it follows from (1.2.15) that the composite mapping $\beta \circ \alpha$ belongs also to $M(X)$, and we can define an operation on $M(X)$ (i.e., a mapping from $M(X) \times M(X)$ into $M(X)$) by associating with each ordered pair (β, α) the element $\beta \circ \alpha$. We denote this operation by "$\cdot$" and write

$$\beta \cdot \alpha = \beta \circ \alpha, \qquad \alpha, \beta \in M(X). \tag{2.1.38}$$

We call this operation "**multiplication**," we refer to $\beta \cdot \alpha$ as the **product** of β and α, and we note that $(\beta \circ \alpha)(x) = (\beta \cdot \alpha)(x)$ for all $x \in X$. We also note that "$\cdot$" is associative, for if $\alpha, \beta, \gamma \in M(X)$, then

$$(\alpha \cdot \beta) \cdot \gamma = (\alpha \circ \beta) \circ \gamma = \alpha \circ (\beta \circ \gamma) = \alpha \cdot (\beta \cdot \gamma).$$

Thus, the system $\{M(X); \cdot\}$ is a semigroup, which we call the **semigroup of transformations** on X.

Next, let us recall that a permutation on X is a one-to-one mapping of X onto X. Clearly, any permutation on X belongs to $M(X)$. In particular, the **identity permutation** $e: X \longrightarrow X$ defined by

$$e(x) = x \text{ for all } x \in X,$$

belongs to $M(X)$. We thus can readily prove the following:

2.1.39. Theorem. $\{M(X); \cdot\}$ is a monoid whose identity element is the identity permutation of $M(X)$.

Proof. Let $\alpha \in M(X)$. The $(e \cdot \alpha)(x) = e(\alpha(x)) = \alpha(x)$ for every $x \in X$, and so $e \cdot \alpha = \alpha$. Similarly, $(\alpha \cdot e)(x) = \alpha(e(x)) = \alpha(x)$ for all $x \in X$, and so $\alpha \cdot e = \alpha$. ■

Next, we prove:

2.1.40. Theorem. Let $\{M(X); \cdot\}$ be the semigroup of transformations on the set X. An element $\alpha \in M(X)$ has an inverse in $M(X)$ if and only if α is a permutation on X. Moreover, the inverse of a unit α is the inverse mapping α^{-1} determined by the permutation α.

Proof. Suppose that $\alpha \in M(X)$ is a permutation on X. Then it follows from Theorem 1.2.10, part (ii), that α^{-1} is a permutation on X and hence $\alpha^{-1} \in M(X)$. Since $\alpha \circ \alpha^{-1} = \alpha^{-1} \circ \alpha = e$, it follows that $\alpha \cdot \alpha^{-1} = \alpha^{-1} \cdot \alpha = e$, and thus α has an inverse.

Next, suppose that α has an inverse in $M(X)$ and let α' denote that inverse relative to "$\cdot$". Then $\alpha' \in M(X)$ and $\alpha \cdot \alpha' = \alpha' \cdot \alpha = e$. To show that α is a permutation on X we must show that α is a one-to-one mapping of X onto X. To prove that α is onto, we must show that for any $x \in X$ there exists a $y \in X$ such that $\alpha(y) = x$. Since $\alpha' \in M(X)$ it follows that $\alpha'(x) \in X$ for every $x \in X$ and $\alpha \circ \alpha'(x) = e(x) = x$. Letting $y = \alpha'(x)$ it follows that α is onto. To show that α is one-to-one we assume that $\alpha(x) = \alpha(y)$. Then, $\alpha'(\alpha(x)) = \alpha'(\alpha(y))$ and since $\alpha \circ \alpha' = e$, we have

$$x = e(x) = \alpha \circ \alpha'(x) = \alpha' \circ \alpha(x) = \alpha' \circ \alpha(y) = e(y) = y.$$

Therefore, α is one-to-one. Hence, if $\alpha \in M(X)$ has an inverse, α^{-1}, it is a permutation on X. ■

Henceforth, we employ the following *notation*: the set of all permutations on a given set X is denoted by $P(X)$. As pointed out in Chapter 1, if a set X has n elements, then there are $n!$ distinct permutations on X.

The reader is now in a position to prove the following result.

2.1.41. Theorem. $\{P(X); \cdot\}$ is a subgroup of $\{M(X); \cdot\}$.

2.1.42. Exercise. Prove Theorem 2.1.41.

The preceding result gives rise to a very important class of groups.

2.1.43. Definition. Any subgroup of the group $\{P(X); \cdot\}$ is called **a permutation group** or **a transformation group** on X, and $\{P(X); \cdot\}$ is called **the permutation group** or **the transformation group** on X.

Occasionally, we speak of a permutation group on X, say $\{Y; \cdot\}$, without making reference to the set X. In such cases it is assumed that $\{Y; \cdot\}$ is a subgroup of the permutation group $P(X)$ for some set X.

2.1.44. Example. Let $X = \{x, y, z\}$. Then $P(X)$ consists of $3! = 6$ permutations, namely,

$$\alpha_1 = \begin{pmatrix} x & y & z \\ x & y & z \end{pmatrix}, \quad \alpha_2 = \begin{pmatrix} x & y & z \\ x & z & y \end{pmatrix}, \quad \alpha_3 = \begin{pmatrix} x & y & z \\ y & x & z \end{pmatrix},$$

$$\alpha_4 = \begin{pmatrix} x & y & z \\ y & z & x \end{pmatrix}, \quad \alpha_5 = \begin{pmatrix} x & y & z \\ z & x & y \end{pmatrix}, \quad \alpha_6 = \begin{pmatrix} x & y & z \\ z & y & x \end{pmatrix}.$$

We can readily verify that $\alpha_1 = e$. If $X_1 = \{e, \alpha_2\}$, then $\{X_1; \cdot\}$ is a subgroup of $P(X)$ and hence a permutation group on X. Let $X_2 = \{e, \alpha_4, \alpha_5\}$. Then

$\{X_2; \cdot\}$ is also a permutation group on X. Note that $\{X_1; \cdot\}$ is of order 2 and $\{X_2; \cdot\}$ is of order 3. ■

B. Rings and Fields

Thus far we have concerned ourselves with mathematical systems consisting of a set and an operation on the set. Presently we consider mathematical systems consisting of a basic set X with two operations α and β defined on the set, denoted by $\{X; \alpha, \beta\}$. Associated with such systems there are two mathematical systems (called **subsystems**) $\{X; \alpha\}$ and $\{X; \beta\}$. By insisting that the systems $\{X; \alpha\}$ and $\{X; \beta\}$ possess certain properties and that one of the operations be distributive over the other, we introduce the important mathematical systems known as **rings**. We then concern ourselves with special types of important rings called **integral domains**, **division rings**, and **fields**.

2.1.45. Definition. Let X be a non-empty set, and let α and β be operations on X. The set X together with the operations α and β on X, denoted by $\{X; \alpha, \beta\}$, is called a **ring** if

(i) $\{X; \alpha\}$ is an abelian group;
(ii) $\{X; \beta\}$ is a semigroup; and
(iii) β is distributive over α.

We refer to $\{X; \alpha\}$ as the **group component** of the ring, to $\{X; \beta\}$ as the **semigroup component** of the ring, to α as the **group operation** of the ring, and to β as the **semigroup operation** of the ring. For convenience we often denote a ring $\{X; \alpha, \beta\}$ by X and simply refer to "ring X". For obvious reasons, we often use the symbols "$+$" and "$\cdot$" ("addition" and "multiplication") in place of α and β, respectively. Thus, if X is a ring we may write $\{X; +, \cdot\}$ and assume that $\{X; +\}$ is the group component of X, and $\{X; \cdot\}$ is the semigroup component of X. We call $\{X; +\}$ the **additive group** of ring X, $\{X; \cdot\}$ the **multiplicative semigroup** of ring X, $x + y$ the **sum** of x and y, and $x \cdot y$ the **product** of x and y.

We use 0 ("zero") to denote the identity element of $\{X; +\}$. If $\{X; \cdot\}$ has an identity element, we denote that identity by e.

The inverse of an element x relative to $+$ is denoted by $-x$. If x has an inverse relative to "$\cdot$", we denote it by x^{-1}. Furthermore, we denote $x + (-y)$ by $x - y$ (the "difference of x and y") and $(-x) + y$ by $-x + y$. Note that the elements 0, e, $-x$, and x^{-1} are unique.

Subsequently, we adopt the convention that when operations "$+$" and "$\cdot$" appear mixed without parentheses to clarify the order of operation, the operation should be taken with respect to "$\cdot$" first and then with respect to $+$. For example,

$$x \cdot y + z = (x \cdot y) + z$$

and *not* $x \cdot (y + z)$. The latter would have to be written with parentheses. Thus, we have

$$x \cdot (y + z) = (x \cdot y) + (x \cdot z) = x \cdot y + x \cdot z.$$

In general, the semigroup $\{X; \cdot\}$ does not contain an identity. However, if it does we have:

2.1.46. Definition. Let $\{X; +, \cdot\}$ be a ring. If the semigroup $\{X; \cdot\}$ has an identity element, we say that X is a **ring with identity**.

There should be no ambiguity concerning the above statement. The group $\{X; +\}$ always has an identity, so if we say "ring with identity," we must refer to $\{X; \cdot\}$.

We note that it is always true that the operation "$+$" is commutative for a given ring. If in addition the operation "$\cdot$" is also commutative, we have

2.1.47. Definition. Let $\{X; +, \cdot\}$ be a ring. If the operation "$\cdot$" is commutative on the set X then the ring X is called a **commutative ring**.

For rings we also have:

2.1.48. Definition. Let $\{X; +, \cdot\}$ be a ring with identity. An element $x \in X$ is called a **unit** of X if x has an inverse as an element of the semigroup $\{X; \cdot\}$. We denote this inverse of x by x^{-1}.

The reader can readily verify that the following examples are rings.

2.1.49. Exercise. Letting "$+$" and "$\cdot$" denote the usual operations of addition and multiplication, show that $\{X; +, \cdot\}$ is a commutative ring with identity if

(i) X is the set of integers;
(ii) X is the set of rational numbers; and
(iii) X is the set of real numbers.

2.1.50. Exercise. Let $X = \{0, 1\}$ and define "$+$" and "$\cdot$" by the following operation tables:

$+$	0	1
0	0	1
1	1	0

$\cdot$	0	1
0	0	0
1	0	1

Show that $\{X; +, \cdot\}$ is a commutative ring with identity.

2.1.51. Exercise. Let $\{X; \alpha\}$ be an abelian group with identity element e. Define the operation β on X as $x \,\beta\, y = e$ for every $x, y \in X$. Show that $\{X; \alpha, \beta\}$ is a ring.

For rings we have:

2.1.52. Theorem. If $\{X; +, \cdot\}$ is a ring then for every $x, y \in X$ we have

(i) $x + 0 = 0 + x = x$;
(ii) $-(x + y) = (-x) + (-y) = (-x) - y = -x - y$;
(iii) if $x + y = 0$, then $x = -y$;
(iv) $-(-x) = x$;
(v) $0 = x \cdot 0 = 0 \cdot x$;
(vi) $(-x) \cdot y = -(x \cdot y) = x \cdot (-y)$; and
(vii) $(-x) \cdot (-y) = x \cdot y$.

Proof. Parts (i)–(iv) follow from the fact that $\{X; +\}$ is an abelian group and from our notation convention.

To prove part (v) we note that since $z + 0 = z$ for every $z \in X$ we have for every $x \in X$, $0 \cdot x + 0 = 0 \cdot x = (0 + 0) \cdot x = 0 \cdot x + 0 \cdot x$, and thus $0 = 0 \cdot x$. Also, $x \cdot 0 + 0 = x \cdot 0 = x \cdot (0 + 0) = x \cdot 0 + x \cdot 0$, so that $0 = x \cdot 0$. Hence, $0 = x \cdot 0 = 0 \cdot x$ for every $x \in X$.

To prove part (vi), note that $0 \cdot y = 0$ for every $y \in X$ and since $x + (-x) = 0$ we have $0 = 0 \cdot y = [x + (-x)] \cdot y = x \cdot y + (-x) \cdot y$. This implies that $-(x \cdot y) = (-x) \cdot y$ since $-(x \cdot y)$ is the additive inverse of $x \cdot y$. Similarly, $0 = x \cdot 0 = x \cdot [y + (-y)] = x \cdot y + x \cdot (-y)$. This implies that $x \cdot (-y) = -(x \cdot y)$. Thus, $(-x) \cdot y = -(x \cdot y) = x \cdot (-y)$.

Finally, to prove part (vii), we note that since $-(-z) = z$ for every $z \in X$ and since part (vi) holds for any $x \in X$, we obtain, replacing x by $-x$, $(-x) \cdot (-y) = -[(-x) \cdot y] = -[-(x \cdot y)] = x \cdot y$. ■

Now let $\{X; +, \cdot\}$ denote a ring for which the two operations are equal, i.e., "$+$" = "$\cdot$". Then $x + y = x \cdot y$ for all $x, y \in X$. In particular, if $y = 0$, then $x + 0 = x \cdot 0 = 0$ for all $x \in X$ and we conclude that 0 is the only element of the set X. This gives rise to:

2.1.53. Definition. A ring $\{X; +, \cdot\}$ is called a **trivial ring** if $X = \{0\}$.

We next introduce:

2.1.54. Definition. Let $\{X; +, \cdot\}$ be a ring. If there exist non-zero elements $x, y \in X$ (not necessarily distinct) such that $x \cdot y = 0$, then x and y are both called **divisors of zero**.

We have:

2.1.55. Theorem. Let $\{X; +, \cdot\}$ be a ring, and let $X^{\#} = X - \{0\}$. Then X has no divisors of zero if and only if $\{X^{\#}; \cdot\}$ is a subsemigroup of $\{X; \cdot\}$.

Proof. Assume that X has no divisors of zero. Then $x, y \in X^{\#}$ implies $x \cdot y \neq 0$, so $x \cdot y \in X^{\#}$ and $X^{\#}$ is a subsemigroup.

Conversely, if $x, y \in X^{\#}$ implies $x \cdot y \in X^{\#}$, then $x \cdot y \neq 0$ if $x \neq 0$ and $y \neq 0$. ■

We now consider special types of rings called **integral domains.**

2.1.56. Definition. A ring $\{X; +, \cdot\}$ is called an **integral domain** if it has no divisors of zero.

Our next result enables us to characterize integral domains in another equivalent fashion.

2.1.57. Theorem. A ring X is an integral domain if and only if for every $x \neq 0$, the following three statements are equivalent for every $y, z \in X$:

(i) $y = z$;

(ii) $x \cdot y = x \cdot z$; and

(iii) $y \cdot x = z \cdot x$.

Proof. Assume that X is an integral domain. Clearly (i) implies (ii) and (iii). To show that (ii) implies (i), let $x \cdot y = x \cdot z$. Then $x \cdot (y - z) = 0$. Since $x \neq 0$ and X has no zero divisors, $y - z = 0$ or $y = z$. Thus, (ii) implies (i). Similarly, it follows that (iii) implies (i). This proves that (i), (ii), and (iii) are equivalent.

Conversely, assume that $x \neq 0$ and that (i), (ii), and (iii) are equivalent. Let $x \cdot y = 0$. Then $x \cdot 0 = x \cdot y$, and it follows that y must be zero since (ii) implies (i). Thus, $x \cdot y \neq 0$ for $y \neq 0$, and X has no zero divisors. ■

We now introduce divisors of elements.

2.1.58. Definition. Let $\{X; +, \cdot\}$ be a commutative integral domain with identity, and let $x, y \in X$. We say y is a **divisor** of x if there exists an element $z \in X$ such that $x = y \cdot z$. If y is a divisor of x, we write $y | x$.

If $y | x$, it is customary to say that y **divides** x.

2.1.59. Theorem. Let $\{X; +, \cdot\}$ be a commutative integral domain with identity, and let $x \in X$. Then x is a unit of X if and only if $x | e$.

Proof. Let $x | e$. Then there is a $z \in X$ such that $e = x \cdot z = z \cdot x$. Thus, z is an inverse of x, i.e., $z = x^{-1}$.

Conversely, let x be a unit of X. Then there exists $x^{-1} \in X$ such that $e = x \cdot x^{-1}$, and thus $x | e$. ■

We notice that if in an integral domain $x \cdot y = 0$, then either $x = 0$ or

$y = 0$. Now a divisor of zero cannot have an inverse. To show this, we let x and y be divisors of zero, i.e., $x \cdot y = 0$. Suppose that y has an inverse. Then $x \cdot y \cdot y^{-1} = 0 \cdot y^{-1}$, or $x = 0$, which contradicts the fact that x and y are zero divisors. However, the fact that an element is not a zero divisor does not imply it has an inverse. If all of the elements except zero have an inverse, we have yet another special type of ring.

2.1.60. Definition. Let $\{X; +, \cdot\}$ be a non-trivial ring, and let $X^\# = X - \{0\}$. The ring X is called a **division ring** if $\{X^\#; \cdot\}$ is a subgroup of $\{X; \cdot\}$.

In the case of division rings we have:

2.1.61. Theorem. Let $\{X; +, \cdot\}$ be a division ring. Then X is a ring with identity.

Proof. Let $X^\# = X - \{0\}$. Then $\{X^\#; \cdot\}$ has an identity element e. Let $x \in X$. If $x \in X^\#$, then $e \cdot x = x \cdot e = x$. If $x \notin X^\#$, then $x = 0$ and $0 \cdot e = e \cdot 0 = 0$. Therefore, e is an identity element of X. ■

Of utmost importance is the following special type of ring.

2.1.62. Definition. Let $\{X; +, \cdot\}$ be a division ring. Then X is called a **field** if the operation "$\cdot$" is commutative.

Because of the prominence of fields in mathematics as well as in applications, and because we will have occasion to make repeated use of fields, it may be worthwhile to restate the above definition, by listing all the properties of fields.

2.1.63. Definition. Let X be a set containing more than one element, and let there be two operations "$+$" and "$\cdot$" defined on X. Then $\{X; +, \cdot\}$ is a **field** provided that:

(i) $x + (y + z) = (x + y) + z$ and $x \cdot (y \cdot z) = (x \cdot y) \cdot z$ for all $x, y, z \in X$ (i.e., "$+$" and "$\cdot$" are associative operations);

(ii) $x + y = y + x$ and $x \cdot y = y \cdot x$ for all $x, y \in X$ (i.e., "$+$" and "$\cdot$" are commutative operations);

(iii) there exists an element $0 \in X$ such that $0 + x = x$ for all $x \in X$;

(iv) for every $x \in X$ there exists an element $-x \in X$ such that $x + (-x) = 0$;

(v) $x \cdot (y + z) = x \cdot y + x \cdot z$ for all $x, y, z \in X$ (i.e., "$\cdot$" is distributive over "$+$");

(vi) there exists an element $e \neq 0$ such that $e \cdot x = x$ for all $x \in X$; and

(vii) for any $x \neq 0$, there exists an $x^{-1} \in X$ such that $x \cdot (x^{-1}) = e$.

2.1.64. Example. Perhaps the most widely known field is the set of real numbers with the usual rules for addition and multiplication. ■

2.1.65. Exercise. Let Z denote the set of all integers and "$+$" and "$\cdot$" denote the usual operations of addition and multiplication on Z. Show that $\{Z; +, \cdot\}$ is an integral domain, but not a division ring, and hence not a field.

The above example and exercise yield:

2.1.66. Definition. Let R denote the set of all real numbers, let Z denote the set of all integers, and let "$+$" and "$\cdot$" denote the usual operations of addition and multiplication, respectively. We call $\{R; +, \cdot\}$ the **field of real numbers** and $\{Z; +, \cdot\}$ the **ring of integers**.

Another very important field is considered in the following:

2.1.67. Exercise. Let $C = R \times R$, where R is given in Definition 2.1.66. For any $x, y \in C$, let $x = (a, b)$ and $y = (c, d)$, where $a, b, c, d \in R$. We define $x = y$ if and only if $a = c$ and $b = d$. Also, we define the operations "$+$" and "$\cdot$" on C by

$$x + y = (a + c, b + d)$$

and

$$x \cdot y = (ac - bd, ad + bc).$$

Show that $\{C; +, \cdot\}$ is a field.

In view of the last exercise we have:

2.1.68. Definition. The field $\{C; +, \cdot\}$ defined in Exercise 2.1.67 is called the **field of complex numbers**.

2.1.69. Exercise. Let Q denote the set of rational numbers, let P denote the set of irrational numbers, and let "$+$" and "$\cdot$" denote the usual operations of addition and multiplication on P and Q.

(a) Discuss the system $\{Q; +, \cdot\}$.
(b) Discuss the system $\{P; +, \cdot\}$.

2.1.70. Exercise. (This exercise shows that the family of 2×2 matrices forms a ring but not a field.) Let $\{R; +, \cdot\}$ denote the field of real numbers. Define M to be the set characterized as follows. If $u, v \in M$, then u and v are of the form

$$u = \begin{bmatrix} a & b \\ c & d \end{bmatrix}, \quad v = \begin{bmatrix} m & n \\ p & q \end{bmatrix},$$

where a, b, c, d and $m, n, p, q \in R$. Define the operations "+" and "·" on M by

$$u + v = \begin{bmatrix} a & b \\ c & d \end{bmatrix} + \begin{bmatrix} m & n \\ p & q \end{bmatrix} = \begin{bmatrix} a+m & b+n \\ c+p & d+q \end{bmatrix}$$

and

$$u \cdot v = \begin{bmatrix} a & b \\ c & d \end{bmatrix} \cdot \begin{bmatrix} m & n \\ p & q \end{bmatrix} = \begin{bmatrix} a \cdot m + b \cdot p & a \cdot n + b \cdot q \\ c \cdot m + d \cdot p & c \cdot n + d \cdot q \end{bmatrix}.$$

(Note that in the preceding the operations + and · defined on M are entirely different from the operations + and · for the field R.)

(a) Show that $\{M; +\}$ is a monoid.
(b) Show that $\{M; +\}$ is an abelian group.
(c) Show that $\{M; +, \cdot\}$ is a ring.
(d) Show that $\{M; +, \cdot\}$ has divisors of zero.

Next, we introduce the concept of subring.

2.1.71. Definition. Let X be a ring, and let Y be a non-void subset of X which is closed relative to both operations "+" and "·" of the ring X. The set Y, together with the (induced) operations "+" and "·", $\{Y; +, \cdot\}$, is called a **subring** of the ring X provided that $\{Y; +, \cdot\}$ is itself a ring.

In connection with the above definition we say that **subset Y determines the subring** $\{Y; +, \cdot\}$. We have:

2.1.72. Theorem. If X is a ring then a non-void subset Y of X determines a subring of the ring X if and only if

(i) Y is closed with respect to both operations "+" and "·"; and
(ii) $-x \in Y$ whenever $x \in Y$.

2.1.73. Exercise. Prove Theorem 2.1.72.

Using the concept of subring, we now introduce subdomains.

2.1.74. Definition. Let X be a ring, and let Y be a subring of X. If Y is an integral domain, then it is called a **subdomain** of X.

We also define subfield in a natural way.

2.1.75. Definition. Let X be a ring, and let Y be a subring of X. If Y is a field, then it is called a **subfield** of X.

Before, we characterized a trivial ring as a ring for which the set X consists only of the 0 element. In the case of subrings we have:

2.1.76. Definition. Let $\{X; +, \cdot\}$ be a ring, and let $\{Y; +, \cdot\}$ be a subring. Then subring Y is called a **trivial subring** if either

(i) $Y = \{0\}$ or
(ii) $Y = X$.

For subdomains we have:

2.1.77. Theorem. Let X be an integral domain, and let Y be a non-trivial subring of X. Then Y is a subdomain of X.

Proof. Let $x, y \in Y$, and let $x \cdot y = 0$. Since $x, y \in X$, x and y cannot be zero divisors. Thus, Y has no zero divisors. ■

For subfields we have:

2.1.78. Theorem. Let X be a field, and let Y be a subring of X. Then Y is a subfield of X if and only if for every $x \in Y$, $x \neq 0$, $x^{-1} \in Y$.

2.1.79. Exercise. Prove Theorem 2.1.78.

For the intersection of arbitrary subrings we have the following:

2.1.80. Theorem. Let X be a ring, and let X_i be a subring of X for each $i \in I$, where I is some index set. Let $Y = \bigcap_{i \in I} X_i$. Then $\{Y; +, \cdot\}$ is a subring of $\{X; +, \cdot\}$.

Proof. Since $0 \in X_i$ for all $i \in I$, it follows that $0 \in Y$ and Y is non-empty. Let $x, y \in Y$. Then $x, y \in X_i$ for all $i \in I$. Hence, $x + y \in X_i$ and $x \cdot y \in X_i$ for all $i \in I$ so that Y is closed with respect to "+" and "$\cdot$" Also, $-x \in X_i$ for every $i \in I$. Thus, by Theorem 2.1.72, Y is a subring of X. ■

Now let $\{X; +, \cdot\}$ be a ring and let W be any subset of X. Also, let

$$\mathcal{Y} = \{Y: W \subset Y \subset X \text{ and } Y \text{ is a subring of } X\}.$$

Then $\mathcal{Y}$ is non-empty because $X \in \mathcal{Y}$. Now let $R = \bigcap_{Y \in \mathcal{Y}} Y$. Then $W \subset R$ and, by Theorem 2.1.80, $\{R; +, \cdot\}$ is a subring of $\{X; +, \cdot\}$. This subring is called the **subring generated by** W.

C. Modules, Vector Spaces, and Algebras

Thus far we have considered mathematical systems consisting of a set X of elements and of mappings from $X \times X$ into X called **operations on** X. Since a mapping may be regarded as a set and since an operation is a mapping (see Chapter 1), the various components of the mathematical systems considered up to this point may be thought as being derived from *one set* X.

Next, we concern ourselves with mathematical systems which are not restricted to possessing one single fundamental set. We have seen that a single set admits a number of basic derived sets. Clearly, the number of sets that may be derived from two sets, say X and Y, will increase considerably. For example, there are sets which may be generated by utilizing operations on X and Y, and then there are sets which may be derived from mappings of $X \times Y$ into X or into Y.

Mathematical systems which possess several fundamental sets and operations on at least one of these sets may, at least in part, be analyzed by making use of the development given thus far in the present section. Indeed, one may view many such complex systems as a composite of simpler mathematical systems and refer to such systems simply as **composite mathematical systems**. Important examples of such systems include vector spaces, algebras, and modules.

2.1.81. Definition. Let $\{R; +, \cdot\}$ be a ring with identity, e, and let $\{X; +\}$ be an abelian group. Let $\mu: R \times X \to X$ be any function satisfying the following four conditions for all $\alpha, \beta \in R$ and for all $x, y \in X$:

(i) $\mu(\alpha + \beta, x) = \mu(\alpha, x) + \mu(\beta, x)$;
(ii) $\mu(\alpha, x + y) = \mu(\alpha, x) + \mu(\alpha, y)$;
(iii) $\mu(\alpha, \mu(\beta, x)) = \mu(\alpha \cdot \beta, x)$; and
(iv) $\mu(e, x) = x$.

Then the composite system $\{R, X, \mu\}$ is called a **module**.

Since the function μ is defined on $R \times X$, the module defined above is sometimes called a **left R-module**. A **right R-module** is defined in an analogous manner. We will consider only left R-modules and simply refer to them as **modules**, or **R-modules**.

The mapping $\mu: R \times X \to X$ is usually abbreviated by writing $\mu(\alpha, x) = \alpha x$, i.e., in the same manner as "multiplication of α times x." Using this notation, conditions (i) to (iv) above become

(i) $(\alpha + \beta)x = \alpha x + \beta x$;
(ii) $\alpha(x + y) = \alpha x + \alpha y$;
(iii) $\alpha(\beta x) = (\alpha \cdot \beta)x$; and
(iv) $ex = x$;

respectively. We usually refer to the module $\{R, X, \mu\}$ by simply referring to X and calling it an R-module or a module over R.

To simplify notation, we used in the preceding the same operation symbol, $+$, for ring R as well as for group X. However, this should cause no confusion, since it will always be clear from context which operation is used. We will follow similar practices on numerous other occasions in this book.

2.1.82. Example. Let $\{Z; +, \cdot\}$ denote the ring of integers, and let $\{X; +\}$ be any abelian group. Define $\mu: Z \times X \rightarrow X$ by $\mu(n, x) = x + \ldots + x$, where the summation includes x n times. We abreviate this as $\mu(n, x) = nx$ and think of it as "n times x." The identity element in Z is 1, and we see that the conditions (i) to (iv) in Definition 2.1.81 are satisfied. Thus, any abelian group may be viewed as a module over the ring of integers. ■

2.1.83. Example. Let $\{X; +, \cdot\}$ be a ring with identity, and let R be a subring of X with $e \in R$. By defining $\mu: R \times X \rightarrow X$ as $\mu(\alpha, x) = \alpha \cdot x$, it is clear that X is an R-module. In particular, if $R = X$, we see that any ring with identity can be made into a module over itself. ■

For modules we have:

2.1.84. Theorem. Let X be an R-module. Then for all $\alpha \in R$ and $x \in X$ we have

(i) $\alpha 0 = 0$;
(ii) $\alpha(-x) = -(\alpha x)$;
(iii) $0x = 0$; and
(iv) $(-\alpha)x = -(\alpha x)$.

Proof. To prove the first part, we note that for $0 \in X$ we have $0 + 0 = 0$. Thus, $\alpha(0 + 0) = \alpha 0 + \alpha 0 = \alpha 0$, and so $\alpha 0 = 0$.

To prove the second part, note that for any $x \in X$ we have $x + (-x) = 0$, and thus $\alpha(x + (-x)) = \alpha x + \alpha(-x) = \alpha 0 = 0$. Therefore, $\alpha(-x) = -(\alpha x)$.

To prove the third part observe that for $0 \in R$ we have $0 + 0 = 0$. Hence, $(0 + 0)x = 0x + 0x = 0x$, and therefore $0x = 0$.

To prove the last part, note that since $\alpha + (-\alpha) = 0$ it follows that $(\alpha + (-\alpha))x = 0x = 0$. Therefore, $\alpha x + (-\alpha)x = 0$, and $(-\alpha)x = -(\alpha x)$. ■

We next introduce the important concept of vector space.

2.1.85. Definition. Let $\{F; +, \cdot\}$ be a field, and let $\{X; +\}$ be an abelian group. If X is an F-module, then X is called a **vector space over** F.

The notion of vector space, also called **linear space**, is among the most important concepts encountered in mathematics. We will devote the next two chapters and a large portion of the remainder of this book to vector spaces and to mappings on such spaces.

2.1.86. Theorem. Let $\{R; +, \cdot\}$ be a ring, and let $R^n = R \times \ldots \times R$; i.e., R^n denotes the n-fold Cartesian product of R. We denote the element

$x \in R^n$ by $x = (x_1, x_2, \ldots, x_n)$ and define the operation "+" on R^n by

$$x + y = (x_1 + y_1, \ldots, x_n + y_n)$$

for all $x, y \in R^n$. Also, we define $\mu: R \times R^n \longrightarrow R^n$ by

$$\alpha x = (\alpha x_1, \ldots, \alpha x_n)$$

for all $\alpha \in R$ and $x \in R^n$. Then, R^n is an R-module.

2.1.87. Exercise. Prove Theorem 2.1.86.

We also have:

2.1.88. Theorem. Let $\{F; +, \cdot\}$ be a field, and let $F^n = F \times \ldots \times F$ be the n-fold Cartesian product of F. Denote the element $x \in F^n$ by $x = (\xi_1, \xi_2, \ldots, \xi_n)$ and define the operation "+" on F^n by

$$x + y = (\xi_1 + \eta_1, \ldots, \xi_n + \eta_n)$$

for all $x, y \in F^n$. Also, define $\mu: F \times F^n \longrightarrow F^n$ by

$$\alpha x = (\alpha \xi_1, \ldots, \alpha \xi_n)$$

for all $\alpha \in F$ and $x \in F^n$. Then F^n is a vector space over F.

2.1.89. Exercise. Prove Theorem 2.1.88.

In view of Theorem 2.1.88 we have:

2.1.90. Definition. Let $\{F; +, \cdot\}$ be a field. The vector space F^n over F is called the **vector space of n-tuples over F**.

Another very important concept encountered in mathematics is that of an **algebra**. We have:

2.1.91. Definition. Let X be a vector space over a field F. Let a binary operation called "multiplication" and denoted by "$\cdot$" be defined on X, satisfying the following axioms:

(i) $x \cdot (y + z) = x \cdot y + x \cdot z$;
(ii) $(x + y) \cdot z = x \cdot z + y \cdot z$; and
(iii) $(\alpha x) \cdot (\beta y) = (\alpha \cdot \beta)(x \cdot y)$

for all $x, y, z \in X$ and for all $\alpha, \beta \in F$. Then, X is called an **algebra over F**. If, in addition to the above axioms, the binary operation of multiplication is associative, then X is called an **associative algebra**. If the operation is commutative, then X is called a **commutative algebra**. If X has an identity element, then X is called an **algebra with identity**.

Note that in hypothesis (iii) the symbol "$\cdot$" is used to denote two different operations. Thus, in the case of $x \cdot y$ the operation used is defined on X while in the case of $\alpha \cdot \beta$ the operation used is defined on F.

The reader is cautioned that in some texts the term **algebra** means what we defined to be an associative algebra.

2.1.92. Exercise. Let $\{M; +, \cdot\}$ denote the ring of 2×2 matrices defined in Exercise 2.1.70, and let $\{R; +, \cdot\}$ be the field of real numbers. For $u \in M$ given by

$$u = \begin{bmatrix} a & b \\ c & d \end{bmatrix},$$

where $a, b, c, d \in R$, define αu for $\alpha \in R$ by

$$\alpha u = \begin{bmatrix} \alpha a & \alpha b \\ \alpha c & \alpha d \end{bmatrix}.$$

Show that M is an associative algebra over R.

In some areas of application, so-called Lie algebras are of importance. We have:

2.1.93. Definition. A non-associative algebra R is called a **Lie algebra** if $x \cdot x = 0$ for every $x \in R$ and if

$$x \cdot (y \cdot z) + y \cdot (z \cdot x) + z \cdot (x \cdot y) = 0 \qquad (2.1.94)$$

for every $x, y, z \in R$. Equation (2.1.94) is called the **Jacobi identity**.

Let us now consider some specific cases of Lie algebras. Our first exercise shows that any associative algebra can be made into a Lie algebra.

2.1.95. Exercise. Let R be an associative algebra over F, and define the operation "$*$" on R by

$$x * y = x \cdot y - y \cdot x$$

for all $x, y \in R$ (where "$\cdot$" is the operation on the associative algebra R over F). Show that R with "$*$" defined on it is a Lie algebra.

2.1.96. Example. In Exercise 2.1.70 we showed that the set of 2×2 matrices forms a ring but not a field, and in Exercise 2.1.92 we showed that this set forms an algebra over F, the field of real numbers. This set can be made into a Lie algebra by Exercise 2.1.95. ■

2.1.97. Exercise. Let X denote the usual "three-dimensional space," and let i, j, k denote the elements of X depicted in Figure A.

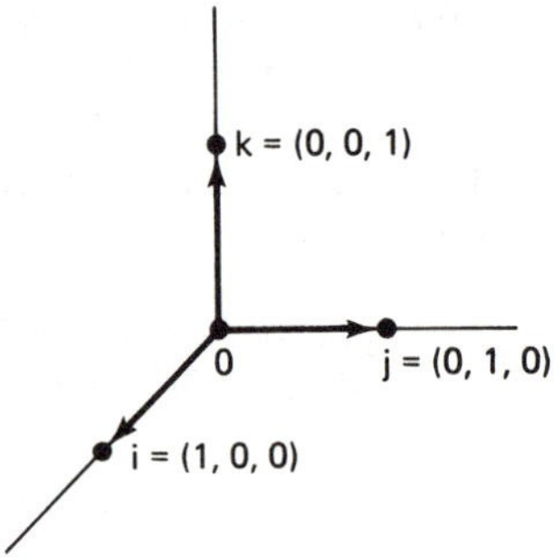

2.1.98. Figure A. Unit vectors i, j, k in three-dimensional space.

Define the operation "$\times$" on X by

$\times$	i	j	k
i	0	k	$-j$
j	$-k$	0	i
k	j	$-i$	0

i.e., "$\times$" denotes the usual "cross product," also called "outer product," encountered in vector analysis. Show that X is a Lie algebra.

Let us next consider submodules.

2.1.99. Definition. Let $\{R; +, \cdot\}$ be a ring with identity, and let $\{X; +\}$ be an abelian group, where X is an R-module. Let $\{Y; +\}$ be a subgroup of $\{X; +\}$. If Y is an R-module, then Y is called an ***R*-submodule of *X*.**

We can characterize submodules by the following:

2.1.100. Theorem. Let X be an R-module, and let Y be a non-empty subset of X. Then, Y is an R-submodule if and only if

(i) $\{Y; +\}$ is a subgroup of $\{X; +\}$; and
(ii) for all $\alpha \in R$ and $x \in Y$, we have $\alpha x \in Y$.

Proof. We give the sufficiency part of the proof and leave the necessity part as an exercise.

Let $\alpha, \beta \in R$ and let $x \in Y$. Then $\alpha x, \beta x, (\alpha + \beta)x \in Y$ by hypothesis (ii). Since Y is a group, it follows that $\alpha x + \beta y \in Y$ and since $x \in X$ we have

$(\alpha + \beta)x = \alpha x + \beta x$. Now let $\alpha \in R$ and let $x, y \in Y$. Then $(x + y) \in Y$ and, also, $\alpha(x + y), \alpha x, \alpha y \in Y$. Thus, $\alpha(x + y) = \alpha x + \alpha y$, since Y is a subgroup of X. Now let $\alpha, \beta \in R$, and let $x \in Y$. Then $\beta x \in Y$, and hence $\alpha(\beta x) \in Y$. We have $(\alpha \cdot \beta)x \in Y$, and so $\alpha(\beta x) = (\alpha \cdot \beta)x$. Also, since $e \in R$, we have $ex \in Y$ for all $x \in Y$ and furthermore, since $x \in X$, we have $ex = x$. This proves that Y is an R-module and hence an R-submodule of X. ■

2.1.101. Exercise. Prove the necessity part of the preceding theorem.

We next introduce the notion of **vector subspace**, also called **linear subspace**.

2.1.102. Definition. Let F be a field, and let X be a vector space over F. Let Y be a subset of X. If Y is an F-submodule of X, then Y is called a **vector subspace**.

Let us consider some specific cases.

2.1.103. Example. Let R be a ring, let X be an R-module, and let $x_i \in X$ for $i = 1, \ldots, n$. Then the subset of X given by $\left\{x \in X : x = \sum_{i=1}^{n} \alpha_i x_i, \alpha_i \in R\right\}$ is an R-submodule of X. ■

2.1.104. Example. Let F be a field, and let F^n be the vector space of n-tuples over F. Let $x_1 = (1, 0, \ldots, 0)$ and $x_2 = (0, 1, 0, \ldots, 0)$. Then $x_1, x_2 \in F^n$. Let $Y = \{x \in F^n : x = \alpha_1 x_1 + \alpha_2 x_2, \alpha_1, \alpha_2 \in F\}$. Then Y is a vector subspace. We see that if $x \in Y$, then x is of the form $x = (\alpha_1, \alpha_2, 0, \ldots, 0)$. ■

We next prove:

2.1.105. Theorem. Let X be an R-module, and let $\mathcal{Y}$ denote a family of R-submodules of X, i.e., Y_i is a submodule of X for every $Y_i \in \mathcal{Y}$, where $i \in I$ and I is some index set. Let $Y = \bigcap_{i \in I} Y_i$. Then Y is an R-submodule of X.

Proof. Since Y_i is a subgroup of X for all $Y_i \in \mathcal{Y}$, it follows that Y is a subgroup of X by Theorem 2.1.31. Now let $\alpha \in R$ and let $y \in Y$. Then $y \in Y_i$ for all $Y_i \in \mathcal{Y}$. Hence, $\alpha y \in Y_i$ for all $Y_i \in \mathcal{Y}$, and so $\alpha y \in Y$. Therefore, by Theorem 2.1.100, Y is an R-submodule of X. ■

The above result gives rise to:

2.1.106. Definition. Let X be an R-module, and let W be a subset of X. Let $\mathcal{Y}$ be the family of subsets of X given by

$$\mathcal{Y} = \{Y\colon W \subset Y \subset X \text{ and } Y \text{ is an } R\text{-submodule of } X\}.$$

Let $G = \bigcap_{Y \in \mathcal{Y}} Y$. Then G is called the **R-submodule of X generated by W.**

Let us next prove:

2.1.107. Theorem. Let X be an R-module, and let $x_1, \ldots, x_n \in X$. Let $Y(x_1, \ldots, x_n)$ denote the subset of X given by

$$Y(x_1, \ldots, x_n) = \{x \in X\colon x = \alpha_1 x_1 + \ldots + \alpha_n x_n, \alpha_1, \ldots, \alpha_n \in R\}.$$

Then $Y(x_1, \ldots, x_n)$ is an R-submodule of X.

Proof. For brevity let $Y = Y(x_1, \ldots, x_n)$. To show that Y is a subgroup of X we first note that $0 \in Y$. Next, for $x \in X$, let $y = (-\alpha_1)x_1 + \ldots + (-\alpha_n)x_n$. Then $y \in Y$ and $x + y = 0$, and hence $y = -x$. Next, let $z = \beta_1 x_1 + \ldots + \beta_n x_n$. Then $x + z = (\alpha_1 + \beta_1)x_1 + \ldots + (\alpha_n + \beta_n)x_n \in Y$. Therefore, by Theorem 2.1.30, Y is a subgroup of X.

Finally, note that for any $a \in R$,

$$ax = a(\alpha_1 x_1 + \ldots + \alpha_n x_n) = a \cdot \alpha_1 x_1 + \ldots + a \cdot \alpha_n x_n \in Y.$$

Thus, by Theorem 2.1.100, Y is an R-submodule of X. ■

We see that $Y(x_1, \ldots, x_n)$ belongs to the family $\mathcal{Y}$ of Definition 2.1.106 if we let $Y = Y(x_1, \ldots, x_n)$, in which case $\bigcap_{Y_i \in \mathcal{Y}} Y_i = Y(x_1, \ldots, x_n)$. This leads to:

2.1.108. Definition. Let X be an R-module, let $x_1, \ldots, x_n \in X$, and let $Y(x_1, \ldots, x_n) = \{x \in X\colon x = \alpha_1 x_1 + \ldots + \alpha_n x_n, \alpha_1, \ldots, \alpha_n \in R\}$. Then $Y(x_1, \ldots, x_n)$ is called the **R-module of X generated** by $x_1, \ldots, x_n$.

Also of interest to us is:

2.1.109. Definition. Let X be an R-module. If there exist elements $x_1, \ldots, x_n \in X$ such that for every $x \in X$ there exist $\alpha_1, \ldots, \alpha_n \in R$ such that $x = \alpha_1 x_1 + \ldots + \alpha_n x_n$, then X is said to be **finitely generated** and $x_1, \ldots, x_n$ are called the **generators** of X.

It can happen that the indexed set $\{\alpha_1, \ldots, \alpha_n\}$ in the above definition is not unique. That is to say, for $x \in X$ we may have $x = \alpha_1 x_1 + \ldots + \alpha_n x_n = \beta_1 x_1 + \ldots + \beta_n x_n$, where $\alpha_i \neq \beta_i$ for some i. However, if it turns out that the above representation of x in terms of $x_1, \ldots, x_n$ is unique, then we have:

2.1.110. Definition. Let X be an R-module which is finitely generated. Let $x_1, \ldots, x_n$ be generators of X. If for every $x \in X$ the relation

$$x = \alpha_1 x_1 + \ldots + \alpha_n x_n = \beta_1 x_1 + \ldots + \beta_n x_n$$

implies that $\alpha_i = \beta_i$ for all $i = 1, \ldots, n$, then the set $\{x_1, \ldots, x_n\}$ is called a **basis for** X.

D. Overview

We conclude this section with the flow chart of Figure B, which attempts to put into perspective most of the algebraic systems considered thus far.

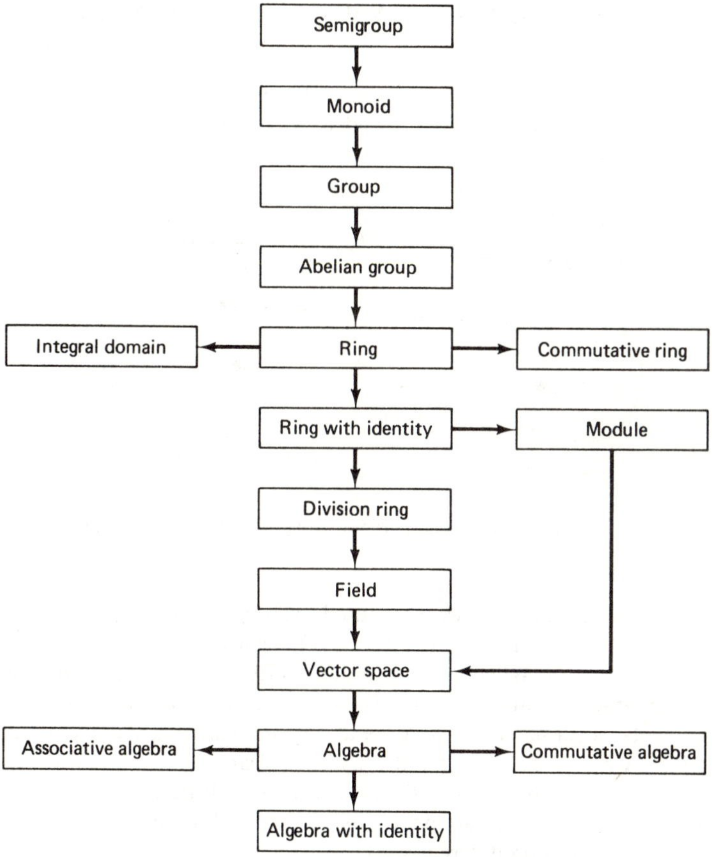

2.1.111. Figure B. Some basic structures of algebra.

2.2. HOMOMORPHISMS

Thus far we have concerned ourselves with various aspects of different mathematical systems (e.g., semigroups, groups, rings, etc.). In the present section we study special types of mappings defined on such algebraic structures. We begin by first considering mappings on semigroups.

2.2.1. Definition. Let $\{X; \alpha\}$ and $\{Y; \beta\}$ be two semigroups (not necessarily distinct). A mapping ρ of set X into set Y is called a **homomorphism** of the semigroup $\{X; \alpha\}$ into the semigroup $\{Y; \beta\}$ if

$$\rho(x \,\alpha\, y) = \rho(x) \beta\, \rho(y) \tag{2.2.2}$$

for every $x, y \in X$. The image of X under ρ, denoted by $\rho(X)$, is called the **homomorphic image of** X. If $x \in X$ then $\rho(x)$ is called the **homomorphic image of** x.

In Figure C, the significance of Eq. (2.2.2) is depicted pictorially. From this figure and from Eq. (2.2.2) it is evident why homomorphisms are said to "preserve the operations α and β."

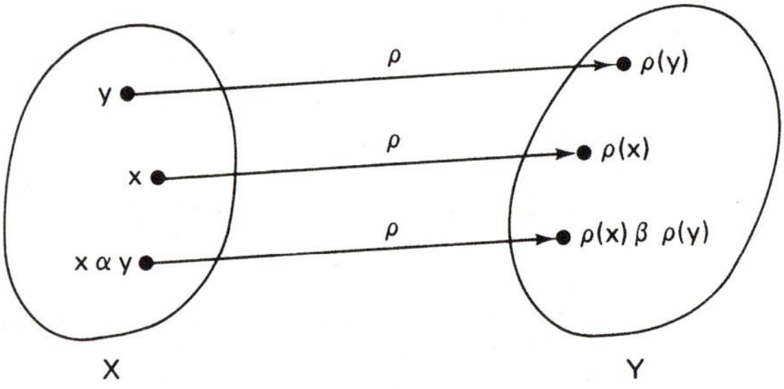

2.2.3. Figure C. Homomorphism of semigroup $\{X; \alpha\}$ into semigroup $\{Y; \beta\}$.

In the above definition we have used arbitrary semigroups $\{X; \alpha\}$ and $\{Y; \beta\}$. As mentioned in Section 2.1, it is often convenient to use the symbol "$+$" for operations. When using the notation $\{X; +\}$ and $\{Y; +\}$ to denote two different semigroups, it should of course be understood that the operation $+$ associated with set X will, in general, be different from the operation $+$ associated with set Y. Since it will usually be clear from context which particular operation is being used, the same symbol will be employed for both semigroups (however, on rare occasions we may wish to distinguish between different operations on different sets).

Using the notation $\{X; +\}$ and $\{Y; +\}$ in Definition 2.2.1, Eq. (2.2.2) now assumes the form

$$\rho(x + y) = \rho(x) + \rho(y) \tag{2.2.4}$$

for every $x, y \in X$. This relation looks very much like the "linearity property" which will be the central topic of a large portion of the remainder of this book, and with which the reader is no doubt familiar. However, we emphasize here that the definition of "linear" will be reserved for a later occasion, and that the term homomorphism is not to be taken as being synonymous with linear. Nevertheless, we will see that many of the subsequent results for homomorphisms will reoccur with appropriate counterparts throughout the this book.

2.2.5. Example. Let R denote the set of real numbers, and let "$+$" and "$\cdot$" denote the usual operations of addition and multiplication on R. Then $\{R; +\}$ and $\{R; \cdot\}$ are semigroups. Let

$$f(x) = e^x$$

for all $x \in R$. Then f is a homomorphism from $\{R; +\}$ to $\{R; \cdot\}$. ■

2.2.6. Exercise. Let $\{X; +\}$ and $\{X; \cdot\}$ denote the semigroups defined in Example 2.1.17. Let $f: X \longrightarrow X$ be defined as follows: $f(0) = 1, f(1) = 3, f(2) = 1$, and $f(3) = 3$. Show that f is a homomorphism from $\{X; +\}$ into $\{X; \cdot\}$.

In order to simplify our notation even further, we will often use the symbol "$\cdot$" in the remainder of the present chapter to denote operations for semigroups (or groups), say $\{X; \cdot\}$, $\{Y; \cdot\}$, and we will often refer to these simply as semigroup (or group) X and Y, respectively. In this case, if ρ denotes a homomorphism of X into Y we write

$$\rho(x \cdot y) = \rho(x) \cdot \rho(y)$$

for all $x, y \in X$.

In Chapter 1 we classified mappings as being *into*, *onto*, *one-to-one and into*, and *one-to-one and onto*. Now if ρ is a homomorphism of a semigroup X into a semigroup Y, we can also classify homomorphisms as being into, onto, one-to-one and into, and one-to-one and onto. This classification gives rise to the following concepts.

2.2.7. Definition. Let ρ be a homomorphism of a semigroup X into a semigroup Y.

(i) If ρ is a mapping of X *onto* Y, we say that X and Y are **homomorphic semigroups**, and we refer to X as being homomorphic to Y.

(ii) If ρ is a *one-to-one* mapping of X into Y, then ρ is called an **isomorphism** of X into Y.

(iii) If ρ is a mapping which is onto and one-to-one, we say that **semigroup X is isomorphic to semigroup Y.**

(iv) If $X = Y$ (i.e., ρ is a homomorphism of semigroup X into itself), then ρ is called an **endomorphism.**

(v) If $X = Y$ and if ρ is an isomorphism (i.e., ρ is an isomorphism of semigroup X into itself), then ρ is called an **automorphism** of X.

We note that since all groups are semigroups, the concepts introduced in the above definition apply necessarily also to groups.

In connection with isomorphic semigroups (or groups) a very *important observation is in order.* We first note that if a semigroup (or group) X is isomorphic to a semigroup Y, then there exists a mapping ρ from X into Y which is one-to-one and onto. Thus, the inverse of ρ, ρ^{-1}, exists and we can associate with each element of X one and only one element of Y, and vice versa. Secondly, we note that ρ is a homomorphism, i.e., ρ preserves the properties of the respective operations associated with semigroup (or group) X and semigroup (or group) Y or, to put it another way, under ρ the (algebraic) properties of semigroups (or groups) X and Y are preserved. Hence, it should be clear that isomorphic semigroups (or groups) are essentially indistinguishable, the homomorphism (which is one-to-one and onto in this case) amounting to a mere relabeling of elements of one set by elements of a second set. We will encounter this type of phenomenon on several other occasions in this book.

We are now ready to prove several results.

2.2.8. Theorem. Let ρ be a homomorphism from a semigroup X into a semigroup Y. Then

(i) $\rho(X)$ is a subsemigroup of Y;

(ii) if X has an identity element, e, $\rho(e)$ is an identity element of $\rho(X)$;

(iii) if X has an identity element, e, and if $x \in X$ has an inverse, x^{-1}, then $\rho(x)$ has an inverse in $\rho(X)$ and, in fact, $[\rho(x)]^{-1} = \rho(x^{-1})$;

(iv) if X_1 is a subsemigroup of X, then $\rho(X_1)$ is a subsemigroup of $\rho(X)$; and

(v) if Y_1 is a subsemigroup of $\rho(X)$, then

$$X_2 = \{x \in X: \rho(x) \in Y_1\}$$

is a subsemigroup of X.

Proof. To prove the first part we must show that the subset $\rho(X)$ of Y is closed relative to the operation "$\cdot$" on Y. Now if $x', y' \in \rho(X)$, then there exists at least one $x \in X$ and at least one $y \in X$ such that $\rho(x) = x'$ and $\rho(y) = y'$. Since ρ is a homomorphism, we have

$$x' \cdot y' = \rho(x) \cdot \rho(y) = \rho(x \cdot y),$$

and since $x \cdot y \in X$ it follows that $x' \cdot y' \in \rho(X)$ because $\rho(x \cdot y) \in \rho(X)$. Thus, $\rho(X)$ is closed and, hence, is a subsemigroup of Y.

To prove the second part, note that since $e \in X$ we have $\rho(e) \in \rho(X)$, and since for any $x' \in \rho(X)$ there exists $x \in X$ such that $\rho(x) = x'$, we have

$$\rho(e) \cdot x' = \rho(e) \cdot \rho(x) = \rho(e \cdot x) = \rho(x) = x'.$$

Since this is true for every $x' \in \rho(X)$, it follows that $\rho(e)$ is a left identity element of $\rho(X)$. Similarly, we can show that $x' \cdot \rho(e) = x'$ for every $x' \in \rho(X)$. Thus, $\rho(e)$ is an identity element of the subsemigroup $\rho(X)$ of Y.

To prove the third part of the theorem, note that since ρ is a homomorphism, we have

$$\rho(x) \cdot \rho(x^{-1}) = \rho(x \cdot x^{-1}) = \rho(e),$$

and

$$\rho(x^{-1}) \cdot \rho(x) = \rho(x^{-1} \cdot x) = \rho(e);$$

i.e., $\rho(e)$ is an identity element of $\rho(X)$. Also, since $\rho(x^{-1}) \in \rho(X)$, $\rho(x)$ has an inverse in $\rho(X)$, and $[\rho(x)]^{-1} = \rho(x^{-1})$.

The proof of parts (iv) and (v) of this theorem are left as an exercise. ■

2.2.9. Exercise. Complete the proof of Theorem 2.2.8.

We emphasize that although $\rho(e)$ in the above theorem is an identity element of the subsemigroup $\rho(X)$ of Y, it is not necessarily true that $\rho(e)$ has to be an identity element of Y.

2.2.10. Definition. Let ρ be a homomorphism of a semigroup X into a semigroup Y. If $\rho(X)$ has identity element, say e', then the subset of X, K_ρ, defined by

$$K_\rho = \{x \in X: \rho(x) = e'\}$$

is called the **kernel** of the homomorphism ρ.

It turns out that K_ρ is a semigroup; i.e., we have:

2.2.11. Theorem. K_ρ is a subsemigroup of X.

2.2.12. Exercise. Prove Theorem 2.2.11.

Now let X and Y be groups (instead of semigroups, as above), and let ρ be a homomorphism of X into Y. We have:

2.2.13. Theorem. Let ρ be a homomorphism from a group X into a group Y. Then

(i) $\rho(X)$ is a subgroup of Y; and
(ii) if e is the identity element of X, then $\rho(e)$ is the identity element of Y.

Proof. To prove the first part, let e denote the identity element of X. By part (i) of Theorem 2.2.8, $\rho(X)$ is a subsemigroup of Y; by part (ii) of Theorem 2.2.8, $\rho(e)$ is an identity element of $\rho(X)$; and by part (iii) of the same theorem, it follows that every element of $\rho(X)$ has an inverse. Thus, $\rho(X)$ is a subgroup of Y.

The second part of this theorem follows from Theorem 2.1.28 and from part (ii) of Theorem 2.2.8. ■

The following result is known as **Cayley's theorem.**

2.2.14. Theorem. Let $\{X; \cdot\}$ be a group, and let $\{P(X); \cdot\}$ denote the permutation group on X. Then X is isomorphic to a subgroup of $P(X)$.

Proof. For each $a \in X$, define the mapping $f_a: X \longrightarrow X$ by $f_a(x) = a \cdot x$ for each $x \in X$. If $x, y \in X$ and $f_a(x) = f_a(y)$, then $a \cdot x = a \cdot y$, and so $x = y$. Hence, f_a is an injective mapping. Now let $y \in X$. Then $a^{-1} \cdot y \in X$ and so $f_a(a^{-1} \cdot y) = y$. This implies that f_a is surjective. Hence, f_a is a (1-1) mapping of X onto X, which implies that f_a is a permutation on X; i.e., $f_a \in P(X)$. Now define the function $\varphi: X \longrightarrow P(X)$ by $\varphi(a) = f_a$ for each $a \in X$. Now let $u, v \in X$. For each $x \in X, f_{u \cdot v}(x) = (u \cdot v) \cdot x = u \cdot (v \cdot x) = f_u(v \cdot x) = f_u(f_v(x)) = f_u \circ f_v(x)$. Thus, $f_{u \cdot v} = f_u \circ f_v$ for all $u, v \in X$. Since $\varphi(u \cdot v) = f_{u \cdot v}$ and $\varphi(u) \circ \varphi(v) = f_u \circ f_v$, it follows that $\varphi(u \cdot v) = \varphi(u) \circ \varphi(v)$, and so φ is a homomorphism. Suppose $u, v \in X$ are such that $\varphi(u) = \varphi(v)$. Then $f_u = f_v$, which implies that $f_u(x) = f_v(x)$ for all $x \in X$. In particular, $f_u(e) = f_v(e)$. Hence, $u \cdot e = v \cdot e$, so that $u = v$. This implies that φ is injective. It follows that φ is a (1-1) mapping of X onto $\varphi(X)$. By Theorem 2.2.13, part (i), $\varphi(X)$ is a subgroup of $P(X)$. This completes the proof. ■

We also have:

2.2.15. Theorem. Let ρ be a homomorphism of a semigroup X into a semigroup Y, and let ρ be an isomorphism of X with $\rho(X)$. Then

(i) ρ^{-1} is an isomorphism of $\rho(X)$ with X; and

(ii) if $\rho(X)$ contains an identity element e', then $\rho^{-1}(e') = e$ is an identity element of X and $K_\rho = \{e\}$ and $K_{\rho^{-1}} = \{e'\}$ (K_ρ denotes the kernel of the homomorphism ρ).

Proof. To prove the first part of the theorem, let $x', y' \in \rho(X)$. Then there exist unique $x, y \in X$ such that $\rho(x) = x'$ and $\rho(y) = y'$, and $\rho^{-1}(x') = x$ and $\rho^{-1}(y') = y$. Since

$$\rho(x \cdot y) = \rho(x) \cdot \rho(y) = x' \cdot y',$$

we have

$$\rho^{-1}(x' \cdot y') = x \cdot y = \rho^{-1}(x') \cdot \rho^{-1}(y').$$

Since this is true for all $x', y' \in \rho(X)$, it follows that ρ^{-1} is an isomorphism of $\rho(X)$ with X.

To prove the second part of the theorem we first note that $\rho(X)$ is a subsemigroup of Y by Theorem 2.2.8. It follows from Theorem 2.2.13 that $e = \rho^{-1}(e')$ is an identity element of X. Now let $\rho(k) = e'$. Since $\rho(e) = e'$, it follows that $k = e$ and that $K_\rho = \{e\}$. We can similarly show that $K_{\rho^{-1}} = \{e'\}$. ■

From the above result we can now conclude that if a semigroup X is isomorphic to a semigroup Y, then the semigroup Y is isomorphic to the semigroup X.

For endomorphisms and automorphisms we have:

2.2.16. Theorem. Let η and ψ be homomorphisms of a semigroup X into itself.

(i) If η and ψ are endomorphisms of X, then the composite mapping $\psi \circ \eta$ is likewise an endomorphism of X.

(ii) If η and ψ are automorphisms of X, then $\psi \circ \eta$ is an automorphism of X.

(iii) If η is an automorphism of X, then η^{-1} is also an automorphism of X.

Proof. To prove the first part, note that η and ψ are both mappings of X into X, and thus $\psi \circ \eta$ is a mapping of X into X. Also, by definition, $(\psi \circ \eta)(x) = \psi(\eta(x))$ for every $x \in X$. Now since $\eta(x \cdot y) = \eta(x) \cdot \eta(y)$ and $\psi(x \cdot y) = \psi(x) \cdot \psi(y)$ for every $x, y \in X$, we have

$$\psi \circ \eta(x \cdot y) = \psi(\eta(x \cdot y)) = \psi(\eta(x) \cdot \eta(y)) = \psi(\eta(x)) \cdot \psi(\eta(x))$$
$$= (\psi \circ \eta(x)) \cdot (\psi \circ \eta(y)).$$

This implies that the mapping $\psi \circ \eta$ is an endomorphism of X.

The proof of the second and third part of this theorem is left as an exercise. ■

2.2.17. Exercise. Complete the proof of the above theorem.

Let us next consider homomorphisms of rings. To this end let, henceforth, X and Y be arbitrary rings, and without loss of generality let the operations of these two rings be denoted by "+" and "·".

2.2.18. Definition. Let X and Y be two rings. A mapping ρ of set X into set Y is called a **homomorphism of the ring X into the ring Y** if

(i) $\rho(x + y) = \rho(x) + \rho(y)$; and

(ii) $\rho(x \cdot y) = \rho(x) \cdot \rho(y)$

for every $x, y \in X$. The image of X into Y, denoted by $\rho(X)$, is called the **homomorphic image of** X.

If a homomorphism ρ is a one-to-one mapping of a ring X into a ring Y, then ρ is called an **isomorphism of** X **into** Y. If the isomorphism ρ is an onto mapping of X into Y, then ρ is called an **isomorphism of** X **with** Y. Furthermore, if ρ is a homomorphism of X into X, then ρ is called an **endomorphism of the ring** X. Finally, an isomorphism of X with itself is called an **automorphism of ring** X.

The properties associated with homomorphisms of groups and semigroups can, of course, be utilized when discussing homomorphisms of rings.

2.2.19. Theorem. Let ρ be a homomorphism of a ring X into a ring Y.

(i) The homomorphic image $\rho(X)$ is a subring of Y.

(ii) If X_1 is a subring of X, then $\rho(X_1)$ is a subring of $\rho(X)$.

(iii) Let Y_1 be a subring of $\rho(X)$. Then the subset $X_1 \subset X$ defined by

$$X_1 = \{x \in X: \rho(x) \in Y_1\}$$

is a subring of X.

(iv) Let Z be a ring and let ψ be a homomorphism of Y into Z. Then the composite mapping $\psi \circ \rho$ is a homomorphism of X into Z.

Proof. To prove the first part of the theorem we note that the homomorphic image $\rho(X)$ is clearly the homomorphic image of the group $\{X; +\}$ and of the semigroup $\{X; \cdot\}$. Since this homomorphic image is a subgroup of $\{Y; +\}$ and subsemigroup of $\{Y; \cdot\}$, it follows from Theorem 2.1.72 that $\rho(X)$ is a subring of Y.

The proofs of the remaining parts of this theorem are left as an exercise. ■

2.2.20. Exercise. Prove parts (ii), (iii), and (iv) of Theorem 2.2.19.

Analogous to 2.2.10, we make the following definition.

2.2.21. Definition. If ρ is a homomorphism of a ring X into a ring Y, then the subset K_ρ of X defined by

$$K_\rho = \{z \in X: \rho(z) = 0\}$$

is called the **kernel** of the homomorphism ρ of the ring X into Y.

We close the present section by introducing one more concept.

2.2.22. Definition. Let $\{R; +, \cdot\}$ be a ring with identity and let X and Y be two R-modules. A mapping $f: X \longrightarrow Y$ is called an **R-homomorphism** if, for all $u, v \in X$ and $\alpha \in R$ the relations

(i) $f(u + v) = f(u) + f(v)$; and

(ii) $f(\alpha u) = \alpha f(u)$

hold.

In the next chapter we will consider in great detail a special class of vector spaces and homomorphisms, and for this reason we will not pursue this subject any further at this time.

2.3. APPLICATION TO POLYNOMIALS

Polynomials play an important role in many branches of mathematics as well as in science and engineering. In the present section we briefly consider applications of some of the concepts of the preceding sections to polynomials.

First, we wish to give an abstract definition for a polynomial function. Basically, we want this function to take the form

$$f(t) = a_0 + a_1 t + \ldots + a_n t^n.$$

However, we are not looking for a way of defining the value of $f(t)$ for each t, but instead we seek a definition of f in terms of the indexed set $\{a_0, \ldots, a_n\}$. To this end we let the a_i belong to some field.

More formally, let F be a field and define a set P as follows. If $a \in P$, then a denotes an infinite sequence of elements from F in which all except a finite number are zero. Thus, if $a \in P$, then

$$a = \{a_0, a_1, \ldots, a_n, 0, 0, \ldots\}.$$

That is to say, there exists some integer $n \geq 0$ such that $a_i = 0$ for all $i > n$. Now let b be another element of P, where

$$b = \{b_0, b_1, \ldots, b_m, 0, 0, \ldots\}.$$

We say that $a = b$ if and only if $a_i = b_i$ for all i. We now define the operation "$+$" on P by

$$a + b = \{a_0 + b_0, a_1 + b_1, \ldots\}.$$

Thus, if $n \geq m$, then $a_i + b_i = 0$ for all $i > n$ and P is clearly closed with respect to "$+$". Next, we define the operation "$\cdot$" on P by

$$a \cdot b = c = \{c_0, c_1, \ldots\},$$

where

$$c_k = \sum_{i=0}^{k} a_i b_{k-i}$$

for all k. In this case $c_k = 0$ for all $k > m + n$, and P is also closed with respect to the operation "$\cdot$". Now let us define

$$0 = \{0, 0, \ldots\}.$$

Then $0 \in P$ and $\{P; +\}$ is clearly an abelian group with identity 0. Next, define

$$e = \{1, 0, 0, \ldots\}.$$

Then $e \in P$ and $\{P; \cdot\}$ is obviously a monoid with e as its identity element. We can now easily prove the following

2.3.1. Theorem. The mathematical system $\{P; +, \cdot\}$ is a commutative ring with identity. It is called the **ring of polynomials** over the field F.

2.3.2. Exercise. Prove Theorem 2.3.1.

Let us next complete the connection between our abstract characterization of polynomials and with the function $f(t)$ we originally introduced. To this end we let

$$t^0 = \{1, 0, 0, \ldots\}$$
$$t^1 = \{0, 1, 0, 0, \ldots\}$$
$$t^2 = \{0, 0, 1, 0, \ldots\}$$
$$t^3 = \{0, 0, 0, 1, 0, \ldots\}$$
$$\ldots\ldots\ldots\ldots\ldots\ldots\ldots$$

At this point we still cannot give meaning to $a_i t^i$, because $a_i \in F$ and $t^i \in P$. However, if we make the obvious identification $\{a_i, 0, 0, \ldots\} \in P$, and if we denote this element simply by $a_i \in P$, then we have

$$f(t) = a_0 \cdot t^0 + a_1 \cdot t^1 + \ldots + a_n \cdot t^n.$$

Thus, we can represent $f(t)$ uniquely by the sequence $\{a_0, a_1, \ldots, a_n, 0, \ldots\}$. By convention, we henceforth omit the symbol "$\cdot$", and write, e.g.,

$$f(t) = a_0 + a_1 t + \ldots + a_n t^n.$$

We assign t appearing in the argument of $f(t)$ a special name.

2.3.3. Definition. Let $\{P; +, \cdot\}$ be the polynomial ring over a field F. The element $t \in P$, $t = \{0, 1, 0, \ldots\}$, is called the **indeterminate** of P.

To simplify notation, we denote by $F[t]$ the ring of polynomials over a field F, and we identify elements of $F[t]$ (i.e., polynomials) by making use of the argument t, e.g., $f(t) \in F[t]$.

2.3.4. Definition. Let $f(t) \in F[t]$, and let $f(t) = \{f_0, f_1, \ldots, f_n, \ldots\} \neq 0$, where $f_i \in F$ for all i. The polynomial $f(t)$ is said to be of **order** n or of **degree** n if $f_n \neq 0$ and if $f_i = 0$ for all $i > n$. In this case we write $\deg f(t) = n$ and we call f_n the **leading coefficient** of f. If $f_n = 1$ and $f_i = 0$ for all $i > n$, then $f(t)$ is said to be **monic**.

If every coefficient of a polynomial f is zero, then $f \triangleq 0$ is called the **zero polynomial**. The order of the zero polynomial is not defined.

2.3.5. Theorem. Let $f(t)$ be a polynomial of order n and let $g(t)$ be a polynomial of order m. Then $f(t)g(t)$ is a polynomial of order $m + n$.

Proof. Let $f(t) = f_0 + f_1 t + \ldots + f_n t^n$, let $g(t) = g_0 + g_1 t + \ldots + g_m t^m$, and let $h(t) = f(t)g(t)$. Then

$$h_k = \sum_{i=0}^{k} f_i g_{k-i}.$$

Since $f_i = 0$ for $i > n$ and $g_j = 0$ for $j > m$, the largest possible value of k such that h_k is non-zero occurs for $k = m + n$; i.e.,

$$h_{m+n} = f_n g_m.$$

Since F is a field, f_n and g_m cannot be zero divisors, and thus $h_{m+n} \neq 0$. Therefore, $h_{m+n} \neq 0$, and $h_k = 0$ for all $k > m + n$. ■

The reader can readily prove the next result.

2.3.6. Theorem. The ring $F(t)$ of polynomials over a field F is an integral domain.

2.3.7. Exercise. Prove Theorem 2.3.6.

Our next result shows that, in general, we cannot go any further than integral domain for $F[t]$.

2.3.8. Theorem. Let $f(t) \in F[t]$. Then $f(t)$ has an inverse relative to "$\cdot$" if and only if $f(t)$ is of order zero.

Proof. Let $f(t) \in F[t]$ be of order n, and assume that $f(t)$ has an inverse relative to "$\cdot$", denoted by $f^{-1}(t)$, which is of order m. Then

$$f(t)f^{-1}(t) = e,$$

where $e = \{1, 0, 0, \ldots\}$ is of order zero. By Theorem 2.3.5 the degree of $f(t)f^{-1}(t)$ is $m + n$. Thus, $m + n = 0$ and since $m \geq 0$ and $n \geq 0$, we must have $m = n = 0$.

Conversely, let $f(t) = f_0 = \{f_0, 0, 0, \ldots\}$, where $f_0 \neq 0$. Then $f^{-1}(t) = f_0^{-1} = \{f_0^{-1}, 0, 0, \ldots\}$. ■

In the case of polynomials of order zero we omit the notation t, and we say $f(t)$ is a **scalar**. Thus, if $c(t)$ is a polynomial of order zero, we have $c(t) = c$, where $c \neq 0$. We see immediately that $cf(t) = cf_0 + cf_1 t + \ldots + cf_n t^n$ for all $f(t) \in F[t]$.

The following result, which we will require in Chapter 4, is sometimes called the **division algorithm**.

2.3.9. Theorem. Let $f(t), g(t) \in F[t]$ and assume that $g(t) \neq 0$. Then there exist unique elements $q(t)$ and $r(t)$ in $F[t]$ such that

$$f(t) = q(t)g(t) + r(t), \tag{2.3.10}$$

where either $r(t) = 0$ or $\deg r(t) < \deg g(t)$.

Proof. If $f(t) = 0$ or if $\deg f(t) < \deg g(t)$, then Eq. (2.3.10) is satisfied with $q(t) = 0$, and $r(t) = f(t)$. If $\deg g(t) = 0$, i.e., $g(t) = c$, then $f(t) = [c^{-1} \cdot f(t)] \cdot c$, and Eq. (2.3.10) holds with $q(t) = c^{-1}f(t)$ and $r(t) = 0$.

Assume now that $\deg f(t) \geq \deg g(t) \geq 1$. The proof is by induction on the degree of the polynomial $f(t)$. Thus, let us assume that Eq. (2.3.10) holds for $\deg f(t) = n$. We first prove our assertion for $n = 1$ and then for $n + 1$.

Assume that $\deg f(t) = 1$, i.e., $f(t) = a_0 + a_1 t$, where $a_1 \neq 0$. We need only consider the case $g(t) = b_0 + b_1 t$, where $b_1 \neq 0$. We readily see that Eq. (2.3.10) is satisfied with $q(t) = a_1 b_1^{-1}$ and $r(t) = a_0 - a_1 b_1^{-1} b_0$.

Now assume that Eq. (2.3.10) holds for $\deg f(t) = k$, where $k = 1, \ldots, n$. We want to show that this implies the validity of Eq. (2.3.10) for $\deg f(t) = n + 1$. Let

$$f(t) = a_0 + a_1 t + \ldots + a_{n+1} t^{n+1},$$

where $a_{n+1} \neq 0$. Let $\deg g(t) = m$. We may assume that $0 < m \leq n + 1$. Let $g(t) = b_0 + b_1 t + \ldots + b_m t^m$, where $b_m \neq 0$. It is now readily verified that

$$f(t) = b_m^{-1} a_n t^{n+1-m} g(t) + [f(t) - b_m^{-1} a_n t^{k+1-m} g(t)]. \tag{2.3.11}$$

Now let $h(t) = f(t) - b_m^{-1} a_n t^{n+1-m} g(t)$. It can readily be verified that the coefficient of t^{n+1} in $h(t)$ is 0. Hence, either $h(t) = 0$ or $\deg h(t) < n + 1$. By our induction hypothesis, this implies there exist polynomials $s(t)$ and $r(t)$ such that $h(t) = s(t)g(t) + r(t)$, where $r(t) = 0$ or $\deg r(t) < \deg g(t)$. Substituting the expression for $h(t)$ into Eq. (2.3.11), we have

$$f(t) = [b_m^{-1} a_n t^{n+1-m} + s(t)]g(t) + r(t).$$

Thus, Eq. (2.3.10) is satisfied and the proof of the existence of $r(t)$ and $q(t)$ is complete.

The proof of the uniqueness of $q(t)$ and $r(t)$ is left as an exercise. ■

2.3.12. Exercise. Prove that $q(t)$ and $r(t)$ in Theorem 2.3.9 are unique.

The preceding result motivates the following definition.

2.3.13. Definition. Let $f(t)$ and $g(t)$ be any non-zero polynomials. Let $q(t)$ and $r(t)$ be the unique polynomials such that $f(t) = q(t)g(t) + r(t)$, where either $r(t) = 0$ or $\deg r(t) < \deg g(t)$. We call $q(t)$ the **quotient** and $r(t)$ the **remainder** in the **division** of $f(t)$ by $g(t)$. If $r(t) = 0$, we say that $g(t)$ **divides** $f(t)$ or is a **factor** of $f(t)$.

Next, we prove:

2.3.14. Theorem. Let $F[t]$ denote the ring of polynomials over a field F. Let $f(t)$ and $g(t)$ be non-zero polynomials in $F[t]$. Then there exists a unique monic polynomial, $d(t)$, such that (i) $d(t)$ divides $f(t)$ and $g(t)$, and (ii) if $d'(t)$ is any polynomial which divides $f(t)$ and $g(t)$, then $d'(t)$ divides $d(t)$.

Proof. Let

$$K[t] = \{x(t) \in F[t]: x(t) = m(t)f(t) + n(t)g(t), \text{ where } m(t), n(t) \in F[t]\}.$$

We note that $f(t), g(t) \in K[t]$. Furthermore, if $a(t)$, $b(t) \in K[t]$, then $a(t) - b(t) \in K[t]$ and $a(t)b(t) \in K[t]$. Also, if c is a scalar, then $ca(t) \in K[t]$ for all $a(t) \in K[t]$. Now let $d(t)$ be a polynomial of lowest degree in $K[t]$. Since all scalar multiples of $d(t)$ belong to $K[t]$, we may assume that $d(t)$ is monic. We now show that for any $h(t) \in K[t]$, there is a $q(t) \in F[t]$ such that $h(t) = d(t)q(t)$. To prove this, we know from Theorem 2.3.9 that there exist unique elements $q(t)$ and $r(t)$ in $F[t]$ such that $h(t) = q(t)d(t) + r(t)$, where either $r(t) = 0$ or $\deg r(t) < \deg d(t)$. Since $d(t) \in K[t]$ and $q(t) \in F[t]$, it follows that $q(t)d(t) \in K(t)$. Also, since $h(t) \in K[t]$, it follows that $r(t) = h(t) - q(t)d(t) \in K[t]$. Since $d(t)$ is a polynomial of smallest degree in $K(t)$, it follows that $r(t) = 0$. Hence, $d(t)$ divides every polynomial in $K[t]$.

To show that $d(t)$ is unique, suppose $d_1(t)$ is another monic polynomial in $K[t]$ which divides every polynomial in $K[t]$. Then $d(t) = a(t)d_1(t)$, and $d_1(t) = b(t)d(t)$ for some $a(t), b(t) \in F[t]$. It can readily be verified that this is true only when $a(t) = b(t) = 1$. Now, since $f(t), g(t) \in K[t]$, part (i) of the theorem has been proven.

To prove part (ii), let $a(t), b(t) \in F[t]$ be such that $f(t) = a(t)d'(t)$ and $g(t) = b(t)d'(t)$. Since $d(t) \in K[t]$, there exist polynomials $m(t), n(t)$ such that $d(t) = m(t)f(t) + n(t)g(t)$. Hence,

$$\begin{aligned} d(t) &= m(t)a(t)d'(t) + n(t)b(t)d'(t) \\ &= [m(t)a(t) + n(t)b(t)]d'(t). \end{aligned}$$

This implies that $d'(t)$ divides $d(t)$ and completes the proof of the theorem. ■

The polynomial $d(t)$ in the preceding theorem is called the **greatest common divisor** of $f(t)$ and $g(t)$. If $d(t) = 1$, then $f(t)$ and $g(t)$ are said to be **relatively prime**.

2.3.15. Exercise. Show that if $d(t)$ is the greatest common divisor of $f(t)$ and $g(t)$, then there exist polynomials $m(t)$ and $n(t)$ such that

$$d(t) = m(t)f(t) + n(t)g(t).$$

If $f(t)$ and $g(t)$ are relatively prime, then

$$1 = m(t)f(t) + n(t)g(t).$$

Now let $f(t) \in F[t]$ be of positive degree. If $f(t) = g(t)h(t)$ implies that either $g(t)$ is a scalar or $h(t)$ is a scalar, then $f(t)$ is said to be **irreducible**.

We close the present section with a statement of the **fundamental theorem of algebra**.

2.3.16. Theorem. Let $f(t) \in F[t]$ be a non-zero polynomial. Let R denote the field of real numbers and let C denote the field of complex numbers.

(i) If $F = C$, then $f(t)$ can be written uniquely, except for order, as a product
$$f(t) = c(t - c_1)(t - c_2) \ldots (t - c_n),$$
where $c, c_1, \ldots, c_n \in C$.

(ii) If $F = R$, then $f(t)$ can be written uniquely, except for order, as a product
$$f(t) = cf_1(t)f_2(t) \ldots f_m(t),$$
where $c \in R$ and the $f_1(t), \ldots, f_m(t)$ are monic irreducible polynomials of degree one or two.

2.4. REFERENCES AND NOTES

There are many excellent texts on abstract algebra. For an introductory exposition of this subject refer, e.g., to Birkhoff and MacLane [2.1], Hanneken [2.2], Hu [2.3], Jacobson [2.4], and McCoy [2.6]. The books by Birkhoff and MacLane and Jacobson are standard references. The texts by Hu and McCoy are very readable. The excellent presentation by Hanneken is concise, somewhat abstract, yet very readable. Polynomials over a field are treated extensively in these references. For a brief summary of the properties of polynomials over a field, refer also to Lipschutz [2.5].

REFERENCES

[2.1] G. Birkhoff and S. MacLane, *A Survey of Modern Algebra*. New York: The Macmillan Company, 1965.

[2.2] C. B. Hanneken, *Introduction to Abstract Algebra*. Belmont, Calif.: Dickenson Publishing Co., Inc., 1968.

[2.3] S. T. Hu, *Elements of Modern Algebra*. San Francisco, Calif.: Holden-Day, Inc., 1965.

[2.4] N. Jacobson, *Lectures in Abstract Algebra*. New York: D. Van Nostrand Company, Inc., 1951.

[2.5] S. Lipschutz, *Linear Algebra*. New York: McGraw-Hill Book Company, 1968.

[2.6] N. H. McCoy, *Fundamentals of Abstract Algebra*. Boston: Allyn & Bacon, Inc., 1972.

3

VECTOR SPACES AND LINEAR TRANSFORMATIONS

In Chapter 1 we considered the set-theoretic structure of mathematical systems, and in Chapter 2 we developed to various degrees of complexity the algebraic structure of mathematical systems. One of the mathematical systems introduced in Chapter 2 was the linear or vector space, a concept of great importance in mathematics and applications.

In the present chapter we further examine properties of linear spaces. Then we consider special types of mappings defined on linear spaces, called linear transformations, and establish several important properties of linear transformations.

In the next chapter we will concern ourselves with finite dimensional vector spaces, and we will consider matrices, which are used to represent linear transformations on finite dimensional vector spaces.

3.1. LINEAR SPACES

We begin by restating the definition of linear space.

3.1.1. Definition. Let X be a non-empty set, let F be a field, let "$+$" denote a mapping of $X \times X$ into X, and let "$\cdot$" denote a mapping of $F \times X$ into X. Let the members $x \in X$ be called **vectors**, let the elements $\alpha \in F$ be called **scalars**, let the operation "$+$" defined on X be called **vector addition,**

and let the mapping "·" be called **scalar multiplication** or **multiplication of vectors by scalars**. Then for each $x, y \in X$ there is a unique element, $x + y \in X$, called the **sum of** x **and** y, and for each $x \in X$ and $\alpha \in F$ there is a unique element, $\alpha \cdot x \triangleq \alpha x \in X$, called the **multiple of** x **by** α. We say that the non-empty set X and the field F, along with the two mappings of vector addition and scalar multiplication constitute a **vector space** or a **linear space** if the following axioms are satisfied:

(i) $x + y = y + x$ for every $x, y \in X$;
(ii) $x + (y + z) = (x + y) + z$ for every $x, y, z \in X$;
(iii) there is a unique vector in X, called the **zero vector** or the **null vector** or the **origin**, which is denoted by 0 and which has the property that $0 + x = x$ for all $x \in X$;
(iv) $\alpha(x + y) = \alpha x + \alpha y$ for all $\alpha \in F$ and for all $x, y \in X$;
(v) $(\alpha + \beta)x = \alpha x + \beta x$ for all $\alpha, \beta \in F$ and for all $x \in X$;
(vi) $(\alpha\beta)x = \alpha(\beta x)$ for all $\alpha, \beta \in F$ and for all $x \in X$;
(vii) $0x = 0$ for all $x \in X$; and
(viii) $1x = x$ for all $x \in X$.

The reader may find it instructive to review the axioms of a field which are summarized in Definition 2.1.63. In (v) the "+" on the left-hand side denotes the operation of addition on F; the "+" on the right-hand side denotes vector addition. Also, in (vi) $\alpha\beta \triangleq \alpha \cdot \beta$, where "·" denotes the operation of mulitplication on F. In (vii) the symbol 0 on the left-hand side is a scalar; the same symbol on the right-hand side denotes a vector. The 1 on the left-hand side of (viii) is the identity element of F relative to "·".

To indicate the relationship between the set of vectors X and the underlying field F, we sometimes refer to a **vector space** X **over field** F. However, usually we speak of a vector space X without making explicit reference to the field F and to the operations of vector addition and scalar multiplication. If F is the field of real numbers we call our vector space a **real vector space**. Similarly, if F is the field of complex numbers, we speak of a **complex vector space**. Throughout this chapter we will usually use lower case Latin letters (e.g., x, y, z) to denote vectors (i.e., elements of X) and lower case Greek letters (e.g., α, β, γ) to denote scalars (i.e., elements of F).

If we agree to denote the element $(-1)x \in X$ simply by $-x$, i.e., $(-1)x \triangleq -x$, then we have $x - x = 1x + (-1)x = (1 - 1)x = 0x = 0$. Thus, if X is a vector space, then for every $x \in X$ there is a unique vector, denoted $-x$, such that $x - x = 0$. There are several other elementary properties of vector spaces which are a direct consequence of the above axioms. Some of these are summarized below. The reader will have no difficulties in verifying these.

3.1.2. Theorem. Let X be a vector space. If x, y, z are elements in X and if α, β are any members of F, then the following hold:

(i) if $\alpha x = \alpha y$ and $\alpha \neq 0$, then $x = y$;
(ii) If $\alpha x = \beta x$ and $x \neq 0$, then $\alpha = \beta$;
(iii) if $x + y = x + z$, then $y = z$;
(iv) $\alpha 0 = 0$;
(v) $\alpha(x - y) = \alpha x - \alpha y$;
(vi) $(\alpha - \beta)x = \alpha x - \beta x$; and
(vii) $x + y = 0$ implies that $x = -y$.

3.1.3. Exercise. Prove Theorem 3.1.2.

We now consider several important examples of vector spaces.

3.1.4. Example. Let X be the set of all "arrows" in the "plane" emanating from a reference point which we call the **origin** or the **zero vector** or the **null vector**, and which we denote by 0. Let F denote the set of real numbers, and let vector addition and scalar multiplication be defined in the usual way, as shown in Figure A.

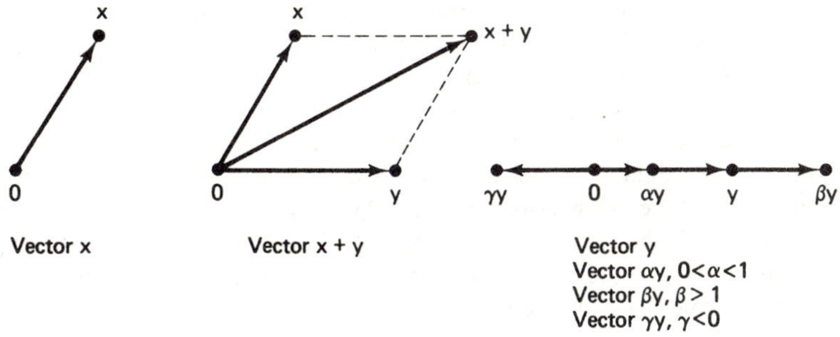

3.1.5. Figure A

The reader can readily verify that, for the space described above, all the axioms of a linear space are satisfied, and hence X is a vector space. ■

The purpose of the above example is to provide an intuitive idea of a linear space. We will utilize this space occasionally for purposes of motivation in our development. We must point out however that the terms "plane" and "arrows" were not formally defined, and thus the space X was not really properly defined. In the examples which follow, we give a more precise formulation of vector spaces.

3.1.6. Example. Let $X = R$ denote the set of real numbers, and let F also denote the set of real numbers. We define vector addition to be the usual addition of real numbers and multiplication of vectors $x \in R$ by scalars $\alpha \in F$ to be multiplication of real numbers. It is a simple matter to show that this space is a linear space. ■

3.1.7. Example. Let $X = F^n$ denote the set of all ordered n-tuples of elements from field F. Thus, if $x \in F^n$, then $x = (\xi_1, \xi_2, \ldots, \xi_n)$, where $\xi_i \in F, i = 1, \ldots, n$. With $x, y \in F^n$ and $\alpha \in F$, let vector addition and scalar multiplication be defined as

$$\begin{aligned} x + y &= (\xi_1, \xi_2, \ldots, \xi_n) + (\eta_1, \eta_2, \ldots, \eta_n) \\ &\triangleq (\xi_1 + \eta_1, \xi_2 + \eta_2, \ldots, \xi_n + \eta_n) \end{aligned} \tag{3.1.8}$$

and

$$\alpha x = \alpha(\xi_1, \xi_2, \ldots, \xi_n) \triangleq (\alpha\xi_1, \alpha\xi_2, \ldots, \alpha\xi_n). \tag{3.1.9}$$

It should be noted that the symbol "+" on the right-hand side of Eq. (3.1.8) denotes addition on the field F, and the symbol "+" on the left-hand side of Eq. (3.1.8) designates vector addition. (See Theorem 2.1.88.)

In the present case the null vector is defined as $0 = (0, 0, \ldots, 0)$ and the vector $-x$ is defined by $-x = -(\xi_1, \xi_2, \ldots, \xi_n) = (-\xi_1, -\xi_2, \ldots, -\xi_n)$. Utilizing the properties of the field F, all axioms of Definition 3.1.1 are readily verified, and F^n is thus a vector space. We call this space the **space F^n of n-tuples of elements of F.** ■

3.1.10. Example. In Example 3.1.7 let $F = R$, the field of real numbers. Then $X = R^n$ denotes the set of all n-tuples of real numbers. We call the vector space R^n the **n-dimensional real coordinate space**. Similarly, in Example 3.1.7 let $F = C$, the field of complex numbers. Then $X = C^n$ designates the set of all n-tuples of complex numbers. The linear space C^n is called the **n-dimensional complex coordinate space.** ■

In the previous example we used the term *dimension*. At a later point in the present chapter the concept of dimension will be defined precisely and some of its properties will be examined in detail.

3.1.11. Example. Let X denote the set of all infinite sequences of real numbers of the form

$$x = (\xi_1, \xi_2, \ldots, \xi_k, \ldots), \tag{3.1.12}$$

let F denote the field of real numbers, let vector addition be defined similarly as in Eq. (3.1.8), and let scalar multiplication be defined similarly as in Eq. (3.1.9). It is again an easy matter to show that this space is a vector space. We point out that this space, which we denote by R^∞, is simply the collection of *all* infinite sequences; i.e., there is no requirement that any type of convergence of the sequence be implied. ■

3.1.13. Example. Let $X = C^\infty$ denote the set of all infinite sequences of complex numbers of the form (3.1.12), let F represent the field of complex numbers, let vector addition be defined similarly as in Eq. (3.1.8), and let scalar multiplication be defined similarly as in Eq. (3.1.9). Then C^∞ is a vector space. ■

3.1.14. Example. Let X denote the set of all sequences of real numbers having only a finite number of non-zero terms. Thus, if $x \in X$, then

$$x = (\xi_1, \xi_2, \dots, \xi_l, 0, \dots 0, \dots) \tag{3.1.15}$$

for some positive integer l. If we define vector addition similarly as in Eq. (3.1.8), if we define scalar multiplication similarly as in Eq. (3.1.9), and if we let F be the field of real numbers, then we can readily show that X is a real vector space. We call this space the **space of finitely non-zero sequences.**

If X denotes the set of all sequences of complex numbers of the form (3.1.15), if vector addition and scalar multiplication are defined similarly as in equations (3.1.8) and (3.1.9), respectively, then X is again a vector space (a complex vector space). ■

3.1.16. Example. Let X be the set of infinite sequences of real numbers of the form (3.1.12), with the property that $\lim_{n \to \infty} \xi_n = 0$. If F is the field of real numbers, if vector addition is defined similarly as in Eq. (3.1.8), and if scalar multiplication is defined similarly as in Eq. (3.1.9), then X is a vector space. This is so because the sum of two sequences which converge to zero also converges to zero, and because the scalar multiple of a sequence converging to zero also converges to zero. ■

3.1.17. Example. Let X be the set of infinite sequences of real numbers of the form (3.1.12) which are bounded. If vector addition and scalar multiplication are again defined similarly as in (3.1.8) and (3.1.9), respectively, and if F denotes the field of real numbers, then X is a vector space. This space is called the **space of bounded real sequences.**

There also exists, of course, a complex counterpart to this space, the **space of bounded complex sequences.** ■

3.1.18. Example. Let X denote the set of infinite sequences of real numbers of the form (3.1.12), with the property that $\sum_{i=1}^{\infty} |\xi_i| < \infty$. Let F be the field of real numbers, let vector addition be defined similarly as in (3.1.8), and let scalar multiplication be defined similarly as in Eq. (3.1.9). Then X is a vector space. ■

3.1.19. Example. Let X be the set of all real-valued continuous functions defined on the interval $[a, b]$. Thus, if $x \in X$, then $x: [a, b] \to R$ is a real, continuous function defined for all $a \leq t \leq b$. We note that $x = y$ if and only if $x(t) = y(t)$ for all $t \in [a, b]$, and that the null vector is the function which is zero for all $t \in [a, b]$. Let F denote the field of real numbers, let $\alpha \in F$, and let vector addition and scalar multiplication be defined pointwise by

$$(x + y)(t) = x(t) + y(t) \text{ for all } t \in [a, b] \tag{3.1.20}$$

and

$$(\alpha x)(t) = \alpha x(t) \text{ for all } t \in [a, b]. \tag{3.1.21}$$

Then clearly $x + y \in X$ whenever $x, y \in X$, $\alpha x \in X$ whenever $\alpha \in F$ and $x \in X$, and all the axioms of a vector space are satisfied. We call this vector space the **space of real-valued continuous functions on** $[a, b]$ and we denote it by $\mathfrak{C}[a, b]$. ■

3.1.22. Example. Let X be the set of all real-valued functions defined on the interval $[a, b]$ such that

$$\int_a^b |x(t)|\, dt < \infty,$$

where integration is taken in the Riemann sense. Let F denote the field of real numbers, and let vector addition and scalar multiplication be defined as in equations (3.1.20) and (3.1.21), respectively. We can readily verify that X is a vector space. ■

3.1.23. Example. Let X denote the set of all real-valued polynomials defined on the interval $[a, b]$, let F be the field of real numbers, and let vector addition and scalar multiplication be defined as in equations (3.1.20) and (3.1.21), respectively. We note that the null vector is the function which is zero for all $t \in [a, b]$, and also, if $x(t)$ is a polynomial, then so is $-x(t)$. Furthermore, we observe that the sum of two polynomials is again a polynomial, and that a scalar multiple of a polynomial is also a polynomial. We can now readily verify that X is a linear space. ■

3.1.24. Example. Let X denote the set of real numbers between $-a < 0$ and $+a > 0$; i.e., if $x \in X$ then $x \in [-a, a]$. Let F be the field of real numbers. Let vector addition and scalar multiplication be as defined in Example 3.1.6. Now, if $\alpha \in F$ is such that $\alpha > 1$, then $\alpha a > a$ and $\alpha a \notin X$. From this it follows that X is *not* a vector space. ■

Vector spaces such as those encountered in Examples 3.1.19, 3.1.22, and 3.1.23 are called **function spaces.** In Chapter 6 we will consider some additional linear spaces.

3.1.25. Exercise. Verify the assertions made in Examples 3.1.6, 3.1.7, 3.1.10, 3.1.11, 3.1.13, 3.1.14, 3.1.16, 3.1.17, 3.1.18, 3.1.19, 3.1.22, and 3.1.23.

3.2. LINEAR SUBSPACES AND DIRECT SUMS

We first introduce the notion of linear subspace. (See also Definition 2.1.102.)

3.2.1. Definition. A non-empty subset Y of a vector space X is called a **linear manifold** or a **linear subspace** in X if (i) $x + y$ is in Y whenever x and y are in Y, and (ii) αx is in Y whenever $\alpha \in F$ and $x \in Y$.

It is an easy matter to verify that a linear manifold Y satisfies all the axioms of a vector space and may as such be regarded as a linear space itself.

3.2.2. Example. The set consisting of the null vector 0 is a linear subspace; i.e., the set $Y = \{0\}$ is a linear subspace. Also, the vector space X is a linear subspace of itself. If a linear subspace Y is not all of X, then we say that Y is a **proper subspace** of X. ∎

3.2.3. Example. The set of all real-valued polynomials defined on the interval $[a, b]$ (see Example 3.1.23) is a linear subspace of the vector space consisting of all real-valued continuous functions defined on the interval $[a, b]$ (see Example 3.1.19). ∎

Concerning linear subspaces we now state and prove the following result.

3.2.4. Theorem. Let Y and Z be linear subspaces of a vector space X. The intersection of Y and Z, $Y \cap Z$, is also a linear subspace of X.

Proof. Since Y and Z are linear subspaces, it follows that $0 \in Y$ and $0 \in Z$, and thus $0 \in Y \cap Z$. Hence, $Y \cap Z$ is non-empty. Now let $\alpha, \beta \in F$, let $x, y \in Y$, and let $x, y \in Z$. Then $\alpha x + \beta y \in Y$ and also $\alpha x + \beta y \in Z$, because Y and Z are both linear subspaces. Hence, $\alpha x + \beta y \in Y \cap Z$ and $Y \cap Z$ is a linear subspace of X. ∎

We can extend the above theorem to a more general result.

3.2.5. Theorem. Let X be a vector space and let X_i be a linear subspace of X for every $i \in I$, where I denotes some index set. Then $\bigcap_{i \in I} X_i$ is a linear subspace of X.

3.2.6. Exercise. Prove Theorem 3.2.5.

Now consider in the vector space of Example 3.1.4 the subsets Y and Z consisting of two lines intersecting at the origin 0, as shown in Figure B. Clearly, Y and Z are linear subspaces of the vector space X. On the other hand, the union of Y and Z, $Y \cup Z$, obviously does not contain arbitrary sums $\alpha y + \beta z$, where $\alpha, \beta \in F$ and $y \in Y$ and $z \in Z$. From this it follows that if Y and Z are linear subspaces then, in general, the union $Y \cup Z$ is *not* a linear subspace of X.

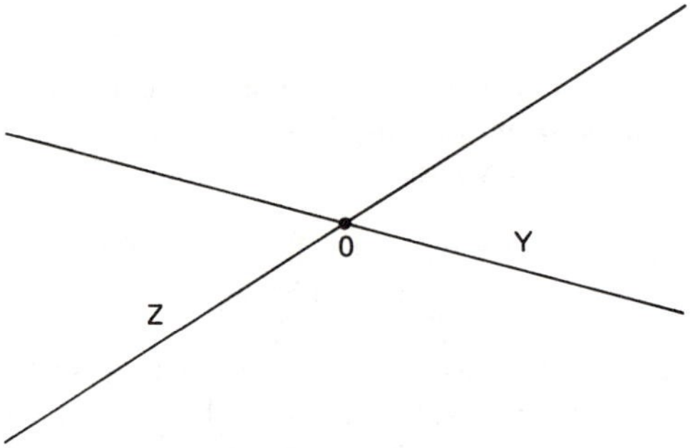

3.2.7. Figure B

3.2.8. Definition. Let X be a linear space, and let Y and Z be arbitrary subsets of X. The **sum** of sets Y and Z, denoted by $Y + Z$, is the set of all vectors in X which are of the form $y + z$, where $y \in Y$ and $z \in Z$.

The above concept is depicted pictorially in Figure C by utilizing the vector space of Example 3.1.4. With the aid of our next result we can generate various linear subspaces.

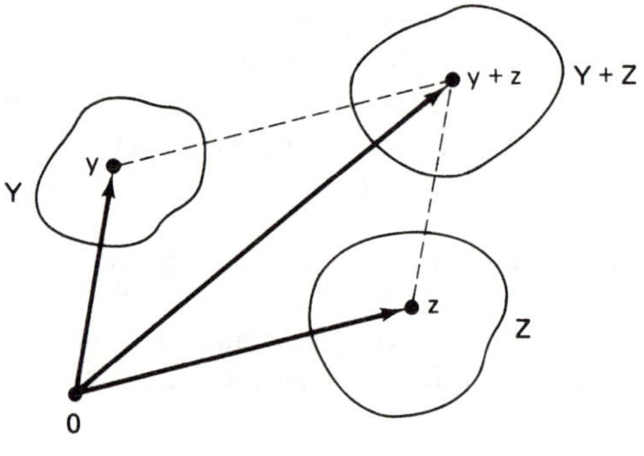

3.2.9. Figure C. Sum of two Subsets.

3.2.10. Theorem. Let Y and Z be linear subspaces of a vector space X. Then their sum, $Y + Z$, is also a linear subspace of X.

3.2.11. Exercise. Prove Theorem 3.2.10.

Now let Y and Z be linear subspaces of a vector space X. If $Y \cap Z = \{0\}$, we say that the spaces Y and Z are **disjoint**. We emphasize that this terminology is not consistent with that used in connection with sets. We now have:

3.2.12. Theorem. Let Y and Z be linear subspaces of a vector space X. Then for every $x \in Y + Z$ there exist unique elements $y \in Y$ and $z \in Z$ such that $x = y + z$ if and only if $Y \cap Z = \{0\}$.

Proof. Let $x \in Y + Z$ be such that $x = y_1 + z_1 = y_2 + z_2$, where $y_1, y_2 \in Y$ and where $z_1, z_2 \in Z$. Then clearly $y_1 - y_2 = z_2 - z_1$. Now $y_1 - y_2 \in Y$ and $z_2 - z_1 \in Z$, and since by assumption $Y \cap Z = \{0\}$, it follows that $y_1 - y_2 = 0$ and $z_2 - z_1 = 0$, $y_1 = y_2$ and $z_1 = z_2$. Thus, every $x \in Y + Z$ has a unique representation $x = y + z$, where $y \in Y$ and $z \in Z$, provided that $Y \cap Z = \{0\}$.

Conversely, let us assume that for each $x = y + z \in Y + Z$ the $y \in Y$ and the $z \in Z$ are uniquely determined. Let us further assume that the linear subspaces Y and Z are not disjoint. Then there exists a non-zero vector $v \in Y \cap Z$. In this case we can write $x = y + z = y + z + \alpha v - \alpha v = (y + \alpha v) + (z - \alpha v)$ for all $\alpha \in F$. But this implies that y and z are not unique, which is a contradiction to our hypothesis. Hence, the spaces Y and Z must be disjoint. ■

Theorem 3.2.10 is readily extended to any number of linear subspaces of X. Specifically, if $X_1, \ldots, X_r$ are linear subspaces of X, then $X_1 + \ldots + X_r$ is also a linear subspace of X. This enables us to introduce the following:

3.2.13. Definition. Let $X_1, \ldots, X_r$ be linear subspaces of the vector space X. The sum $X_1 + \ldots + X_r$ is said to be a **direct sum** if for each $x \in X_1 + \ldots + X_r$ there is a unique set of $x_i \in X_i$, $i = 1, \ldots, r$ such that $x = x_1 + \ldots + x_r$. We denote the direct sum of $X_1, \ldots, X_r$ by $X_1 \oplus \ldots \oplus X_r$.

There is a connection between the Cartesian product of two vector spaces and their direct sum. Let Y and Z be two arbitrary linear spaces over the same field F and let $V = Y \times Z$. Thus, if $v \in V$, then v is the ordered pair

$$v = (y, z),$$

where $y \in Y$ and $z \in Z$. Now let us define vector addition as

$$(y_1, z_1) + (y_2, z_2) = (y_1 + y_2, z_1 + z_2) \tag{3.2.14}$$

and scalar multiplication as

$$\alpha(y, z) = (\alpha y, \alpha z), \tag{3.2.15}$$

where $(y_1, z_1), (y_2, z_2) \in V = Y \times Z$ and where $\alpha \in F$. Noting that for each vector $(y, z) \in V$ there is a vector $-(y, z) = (-y, -z) \in V$, and observing that $(0, 0) = (y, z) - (y, z)$ for all elements in V, it is an easy matter to show that the space $V = Y \times Z$ is a linear space. We note that Y is not a linear subspace of V, because, in fact, it is not even a subset of V. However, if we let

$$Y' = \{(y, 0) : y \in Y\},$$

and

$$Z' = \{(0, z) : z \in Z\},$$

Then Y' and Z' are linear subspaces of V and $V = Y' \oplus Z'$. By abuse of notation, we frequently express this simply as $V = Y \oplus Z$.

Once more, making use of Example 3.1.4, let Y and Z denote two lines intersecting at the origin 0, as shown in Figure D. The direct sum of linear subspaces Y and Z is in this case the "entire plane."

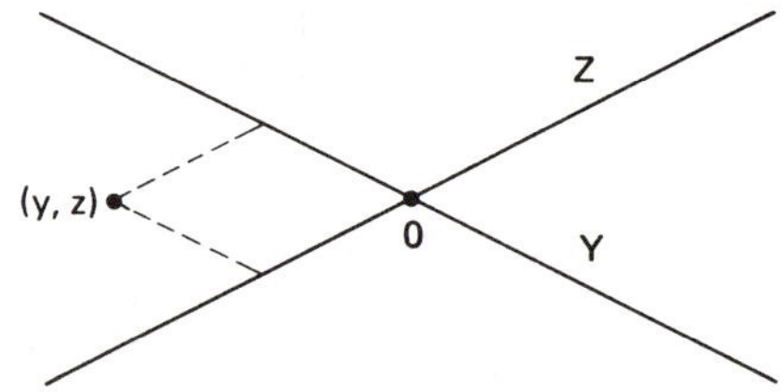

3.2.16. Figure D

In order that a subset be a linear subspace of a vector space, it is necessary that this subset contain the null vector. Thus, in Figure D, the lines Y and Z passing through the origin 0 are linear subspaces of the plane (see Example 3.1.4). In many applications this requirement is too restrictive and a generalization is called for. We have:

3.2.17. Definition. Let Y be a linear subspace of a vector space X, and let x be a fixed vector in X. We call the translation

$$Z = x + Y \triangleq \{z \in X: z = x + y, y \in Y\}$$

a **linear variety** or a **flat** or an **affine linear subspace** of X.

In Figure E, an example of a linear variety is given for the vector space of Example 3.1.4.

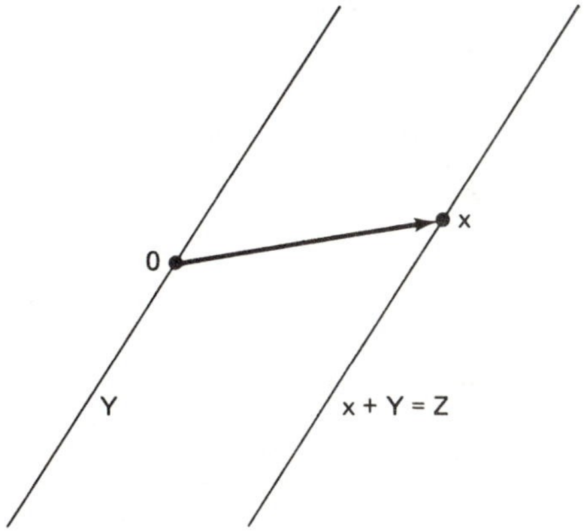

3.2.18. Figure E

3.3. LINEAR INDEPENDENCE, BASES, AND DIMENSION

Throughout the remainder of this and in the following chapter we use the following notation: $\{\alpha_1, \ldots, \alpha_n\}$, $\alpha_i \in F$, denotes an indexed set of scalars, and $\{x_1, \ldots, x_n\}$, $x_i \in X$, denotes an indexed set of vectors.

Before introducing the notions of linear dependence and independence of a set of vectors in a linear space X, we first consider the following.

3.3.1. Definition. Let Y be a set in a linear space X (Y may be a finite set or an infinite set). We say that a vector $x \in X$ is a **finite linear combination of vectors** in Y if there is a finite set of elements $\{y_1, y_2, \ldots, y_n\}$ in Y and a finite set of scalars $\{\alpha_1, \alpha_2, \ldots, \alpha_n\}$ in F such that

$$x = \alpha_1 y_1 + \alpha_2 y_2 + \ldots + \alpha_n y_n. \tag{3.3.2}$$

In Eq. (3.3.2) vector addition has been extended in an obvious way from the case of two vectors to the case of n vectors. In later chapters we will consider linear combinations which are not necessarily finite. The representation of x in Eq. (3.3.2) is, of course, not necessarily unique. Thus, in the case of Example 3.1.10, if $X = R^2$ and if $x = (1, 1)$, then x can be represented as

$$x = \alpha_1 y_1 + \alpha_2 y_2 = 1(1, 0) + 1(0, 1)$$

or as

$$x = \beta_1 z_1 + \beta_2 z_2 = 2(\tfrac{1}{2}, 0) + 3(0, \tfrac{1}{3}),$$

etc. This situation is depicted in Figure F.

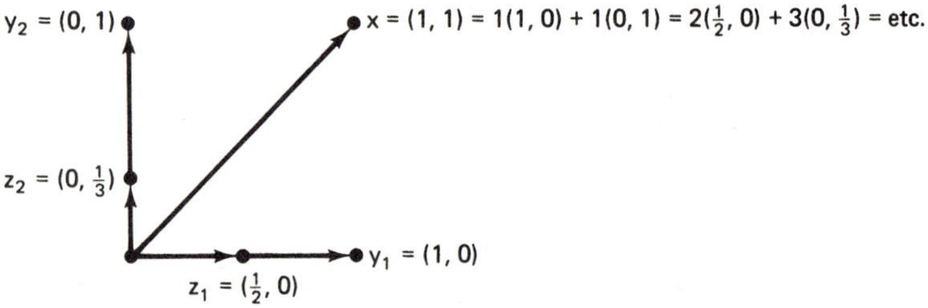

3.3.3. Figure F

3.3.4. Theorem. Let Y be a non-empty subset of a linear space X. Let $V(Y)$ be the set of all finite linear combinations of the vectors from Y; i.e., $y \in V(Y)$ if and only if there is some set of scalars $\{\alpha_1, \ldots, \alpha_m\}$ and some finite subset $\{y_1, \ldots, y_m\}$ of Y such that

$$y = \alpha_1 y_1 + \alpha_2 y_2 + \ldots + \alpha_m y_m,$$

where m may be any positive integer. Then $V(Y)$ is a linear subspace of X.

3.3.5. Exercise. Prove Theorem 3.3.4.

Our previous result motivates the following concepts.

3.3.6. Definition. We say the linear space $V(Y)$ in Theorem 3.3.4 is the **linear subspace generated** by the set Y.

3.3.7. Definition. Let Z be a linear subspace of a vector space X. If there exists a set of vectors $Y \subset X$ such that the linear space $V(Y)$ generated by Y is Z, then we say Y **spans** Z.

If, in particular, the space of Example 3.1.4 is considered and if V and W are linear subspaces of X as depicted in Figure G, then the set $Y = \{e_1\}$ spans W, the set $Z = \{e_2\}$ spans V, and the set $M = \{e_1, e_2\}$ spans the vector space X. The set $N = \{e_1, e_2, e_3\}$ also spans the vector space X.

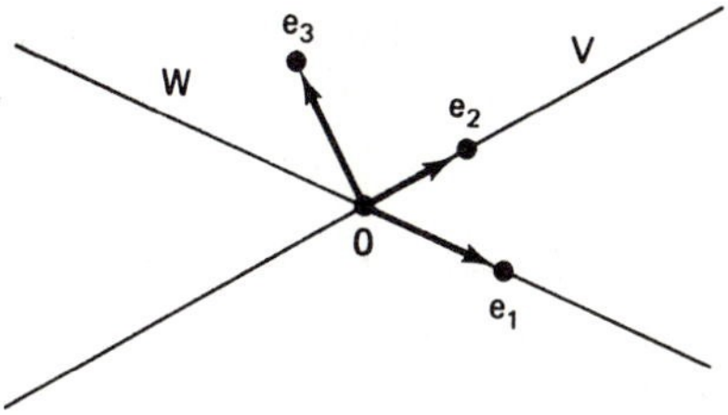

3.3.8. **Figure G.** V and W are Lines Intersecting at Origin O.

3.3.9. Exercise. Show that $V(Y)$ is the smallest linear subspace of a vector space X containing the subset Y of X. Specifically, show that if Z is a linear subspace of X and if Z contains Y, then Z also contains $V(Y)$.

And now the important notion of linear dependence.

3.3.10. Definition. Let $\{x_1, x_2, \ldots, x_m\}$ be a finite non-empty set in a linear space X. If there exist scalars $\alpha_1, \ldots, \alpha_m \in F$, not all zero, such that

$$\alpha_1 x_1 + \ldots + \alpha_m x_m = 0 \tag{3.3.11}$$

then the set $\{x_1, x_2, \ldots, x_m\}$ is said to be **linearly dependent**. If a set is not linearly dependent, then it is said to be **linearly independent**. In this case the relation (3.3.11) implies that $\alpha_1 = \alpha_2 = \ldots = \alpha_m = 0$. An infinite set of vectors Y in X is said to be linearly independent if every finite subset of Y is linearly independent.

Note that the null vector cannot be contained in a set which is linearly independent. Also, if a set of vectors contains a linearly dependent subset, then the whole set is linearly dependent.

If X denotes the space of Example 3.1.4, the set of vectors $\{y, z\}$ in Figure H is linearly independent, while the set of vectors $\{u, v\}$ is linearly dependent.

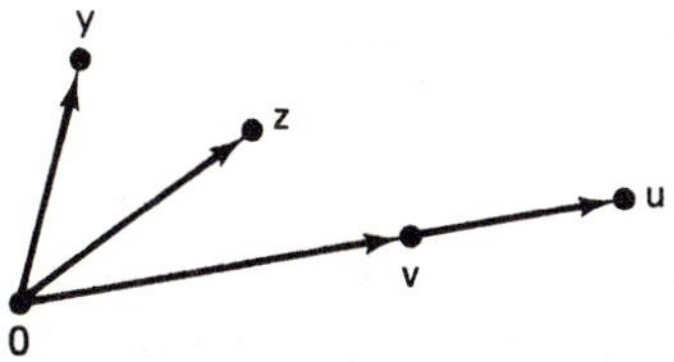

3.3.12. **Figure H.** Linearly Independent and Linearly Dependent Vectors.

3.3.13. Exercise. Let $X = \mathcal{C}[a, b]$, the set of all real-valued continuous functions on $[a, b]$, where $b > a$. As we saw in Example 3.1.19, this set forms a vector space. Let n be a fixed positive integer, and let us define $x_i \in X$ for $i = 0, 1, 2, \ldots, n$, as follows. For all $t \in [a, b]$, let

$$x_0(t) = 1$$

and

$$x_i(t) = t^i, \quad i = 1, \ldots, n.$$

Let $Y = \{x_0, x_1, \ldots, x_n\}$. Then $V(Y)$ is the set of all polynomials on $[a, b]$ of degree less than or equal to n.

(a) Show that Y is a linearly independent set in X.

(b) Let $X_i = \{x_i\}$, $i = 0, 1, \ldots, n$; i.e., each X_i is a singleton subset of X. Show that

$$V(Y) = V(X_0) \oplus V(X_1) \oplus \ldots \oplus V(X_n).$$

(c) Let $z_0(t) = 1$ for all $t \in [a, b]$ and let

$$z_k(t) = 1 + t + \ldots + t^k$$

for all $t \in [a, b]$ and $k = 1, \ldots, n$. Show that $Z = \{z_0, z_1, \ldots, z_n\}$ is a linearly independent set in $V(Y)$.

3.3.14. Theorem. Let $\{x_1, x_2, \ldots, x_m\}$ be a linearly independent set in a vector space X. If $\sum_{i=1}^{m} \alpha_i x_i = \sum_{i=1}^{m} \beta_i x_i$, then $\alpha_i = \beta_i$ for all $i = 1, 2, \ldots, m$.

Proof. If $\sum_{i=1}^{m} \alpha_i x_i = \sum_{i=1}^{m} \beta_i x_i$ then $\sum_{i=1}^{m} (\alpha_i - \beta_i) x_i = 0$. Since the set $\{x_1, \ldots, x_m\}$ is linearly independent, we have $(\alpha_i - \beta_i) = 0$ for all $i = 1, \ldots, m$. Therefore $\alpha_i = \beta_i$ for all i. ■

The next result provides us with an alternate way of defining linear dependence.

3.3.15. Theorem. A set of vectors $\{x_1, x_2, \ldots, x_m\}$ in a linear space X is linearly dependent if and only if for some index i, $1 \leq i \leq m$, we can find scalars $\alpha_1, \ldots, \alpha_{i-1}, \alpha_{i+1}, \ldots, \alpha_m$ such that

$$x_i = \alpha_1 x_1 + \ldots + \alpha_{i-1} x_{i-1} + \alpha_{i+1} x_{i+1} + \ldots + \alpha_m x_m. \tag{3.3.16}$$

Proof. Assume that Eq. (3.3.16) is satisfied. Then

$$\alpha_1 x_1 + \ldots + \alpha_{i-1} x_{i-1} + (-1) x_i + \alpha_{i+1} x_{i+1} + \ldots + \alpha_m x_m = 0.$$

Thus, $\alpha_i = -1 \neq 0$ is a non-trivial choice of coefficient for which Eq. (3.3.11) holds, and therefore the set $\{x_1, x_2, \ldots, x_m\}$ is linearly dependent.

Conversely, assume that the set $\{x_1, x_2, \ldots, x_m\}$ is linearly dependent. Then there exist coefficients $\alpha_1, \ldots, \alpha_m$ which are not all zero, such that

$$\alpha_1 x_1 + \alpha_2 x_2 + \ldots + \alpha_m x_m = 0. \tag{3.3.17}$$

Suppose that index i is chosen such that $\alpha_i \neq 0$. Rearranging Eq. (3.3.17) to

$$-\alpha_i x_i = \alpha_1 x_1 + \ldots + \alpha_{i-1} x_{i-1} + \alpha_{i+1} x_{i+1} + \ldots + \alpha_m x_m \quad (3.3.18)$$

and multiplying both sides of Eq. (3.3.18) by $-1/\alpha_i$, we obtain

$$x_i = \beta_1 x_1 + \beta_2 x_2 + \ldots + \beta_{i-1} x_{i-1} + \beta_{i+1} x_{i+1} + \ldots + \beta_m x_m,$$

where $\beta_k = -\alpha_k/\alpha_i$, $k = 1, \ldots, i-1, i+1, \ldots, m$. This concludes our proof. ■

The proof of the next result is left as an exercise.

3.3.19. Theorem. A finite non-empty set Y in a linear space X is linearly indenpendent if and only if for each $y \in V(Y)$, $y \neq 0$, there is a unique finite subset of Y, say $\{x_1, x_2, \ldots, x_m\}$ and a unique set of scalars $\{\alpha_1, \alpha_2, \ldots, \alpha_m\}$, such that

$$y = \alpha_1 x_1 + \ldots + \alpha_m x_m.$$

3.3.20. Exercise. Prove Theorem 3.3.19.

3.3.21. Exercise. Let Y be a finite set in a linear space X. Show that Y is linearly independent if and only if there is no proper subset Z of Y such that $V(Z) = V(Y)$.

A concept which is of utmost importance in the study of vector spaces is that of basis of a linear space.

3.3.22. Definition. A set Y in a linear space X is called a **Hamel basis**, or simply a **basis**, for X if

(i) Y is linearly independent; and
(ii) the span of Y is the linear space X itself; i.e., $V(Y) = X$.

As an immediate consequence of this definition we have:

3.3.23. Theorem. Let X be a linear space, and let Y be a linearly independent set in X. Then Y is a basis for $V(Y)$.

3.3.24. Exercise. Prove Theorem 3.3.23.

In order to introduce the notion of dimension of a vector space we show that if a linear space X is generated by a finite number of linearly independent elements, then this number of elements must be unique. We first prove the following result.

3.3.25. Theorem. Let $\{x_1, x_2, \ldots, x_n\}$ be a basis for a linear space X. Then for each vector $x \in X$ there exist *unique* scalars $\alpha_1, \ldots, \alpha_n$ such that

$$x = \alpha_1 x_1 + \ldots + \alpha_n x_n.$$

Proof. Since $x_1, \ldots, x_n$ span X, every vector $x \in X$ can be expressed as a linear combination of them; i.e.,

$$x = \alpha_1 x_1 + \alpha_2 x_2 + \ldots + \alpha_n x_n$$

for some choice of scalars $\alpha_1, \ldots, \alpha_n$. We now must show that these scalars are unique. To this end, suppose that

$$x = \alpha_1 x_1 + \alpha_2 x_2 + \ldots + \alpha_n x_n$$

and

$$x = \beta_1 x_1 + \beta_2 x_2 + \ldots + \beta_n x_n.$$

Then

$$\begin{aligned} x + (-x) &= (\alpha_1 x_1 + \alpha_2 x_2 + \ldots + \alpha_n x_n) + (-\beta_1 x_1 - \beta_2 x_2 \\ &\quad - \ldots - \beta_n x_n) \\ &= (\alpha_1 - \beta_1)x_1 + (\alpha_2 - \beta_2)x_2 + \ldots + (\alpha_n - \beta_n)x_n = 0. \end{aligned}$$

Since the vectors $x_1, x_2, \ldots, x_n$ form a basis for X, it follows that they are linearly independent, and therefore we must have $(\alpha_i - \beta_i) = 0$ for $i = 1, \ldots, n$. From this it follows that $\alpha_1 = \beta_1, \alpha_2 = \beta_2, \ldots, \alpha_n = \beta_n$. ■

We also have:

3.3.26. Theorem. Let $\{x_1, x_2, \ldots, x_n\}$ be a basis for vector space X, and let $\{y_1, \ldots, y_m\}$ be any linearly independent set of vectors. Then $m \leq n$.

Proof. We need to consider only the case $m \geq n$ and prove that then we actually have $m = n$. Consider the set of vectors $\{y_1, x_1, \ldots, x_n\}$. Since the vectors $x_1, \ldots, x_n$ span X, y_1 can be expressed as a linear combination of them. Thus, the set $\{y_1, x_1, \ldots, x_n\}$ is not linearly independent. Therefore, there exist scalars $\beta_1, \alpha_1, \ldots, \alpha_n$, not all zero, such that

$$\beta_1 y_1 + \alpha_1 x_1 + \ldots + \alpha_n x_n = 0. \tag{3.3.27}$$

If all the α_i are zero, then $\beta_1 \neq 0$ and $\beta_1 y_1 = 0$. Thus, we can write

$$\beta_1 y_1 + 0 \cdot y_2 + \ldots + 0 \cdot y_m = 0.$$

But this contradicts the hypothesis of the theorem and can't happen because the $y_1, \ldots, y_m$ are linearly independent. Therefore, at least one of the $\alpha_i \neq 0$. Renumbering all the x_i, if necessary, we can assume that $\alpha_n \neq 0$. Solving for x_n we now obtain

$$x_n = \left(\frac{-\beta_1}{\alpha_n}\right)y_1 + \left(\frac{-\alpha_1}{\alpha_n}\right)x_1 + \ldots + \left(\frac{-\alpha_{n-1}}{\alpha_n}\right)x_{n-1}. \tag{3.3.28}$$

Now we show that the set $\{y_1, x_1, \ldots, x_{n-1}\}$ is also a basis for X. Since $\{x_1, \ldots, x_n\}$ is a basis for X, we have $\xi_1, \xi_2, \ldots, \xi_n \in F$ such that

$$x = \xi_1 x_1 + \ldots + \xi_n x_n.$$

Substituting (3.3.28) into the above expression we note that

$$x = \xi_1 x_1 + \xi_2 x_2 + \ldots + \xi_n\left[\left(\frac{-\beta_1}{\alpha_n}\right)y_1 + \ldots + \left(\frac{-\alpha_{n-1}}{\alpha_n}\right)x_{n-1}\right]$$
$$= \gamma y_1 + \gamma_1 x_1 + \ldots + \gamma_{n-1}x_{n-1},$$

where γ and γ_i are defined in an obvious way. In any case, every $x \in X$ can be expressed as a linear combination of the set of vectors $\{y_1, x_1, \ldots, x_{n-1}\}$, and thus this set must span X. To show that this set is also linearly independent, let us assume that there are scalars $\lambda, \lambda_1, \ldots, \lambda_{n-1}$ such that

$$\lambda y_1 + \lambda_1 x_1 + \ldots + \lambda_{n-1}x_{n-1} = 0,$$

and assume that $\lambda \neq 0$. Then

$$y_1 = \left(\frac{-\lambda_1}{\lambda}\right)x_1 + \ldots + \left(\frac{-\lambda_{n-1}}{\lambda}\right)x_{n-1} + 0 \cdot x_n. \tag{3.3.29}$$

In view of Eq. (3.3.27) we have, since $\beta_1 \neq 0$, the relation

$$y_1 = \left(\frac{-\alpha_1}{\beta_1}\right)x_1 + \ldots + \left(\frac{-\alpha_{n-1}}{\beta_1}\right)x_{n-1} + \left(\frac{-\alpha_n}{\beta_1}\right)x_n. \tag{3.3.30}$$

Now the term $(-\alpha_n/\beta_1)x_n$ in Eq. (3.3.30) is not zero, because we solved for x_n in Eq. (3.3.28); yet the coefficient multiplying x_n in Eq. (3.3.29) is zero. Since $\{x_1, \ldots, x_n\}$ is a basis, we have arrived at a contradiction, in view of Theorem 3.3.25. Therefore, we must have $\lambda = 0$. Thus, we have

$$\lambda_1 x_1 + \ldots + \lambda_{n-1}x_{n-1} + 0 \cdot x_n = 0$$

and since $\{x_1, \ldots, x_n\}$ is a linearly independent set it follows that $\lambda_1 = 0, \ldots, \lambda_{n-1} = 0$. Therefore, the set $\{y_1, x_1, \ldots, x_{n-1}\}$ is indeed a basis for X.

By a similar argument as the preceding one we can show that the set $\{y_2, y_1, x_1, \ldots, x_{n-2}\}$ is a basis for X, that the set $\{y_3, y_2, y_1, x_1, \ldots, x_{n-3}\}$ is a basis for X, etc. Now if $m > n$, then we would not utilize y_{n+1} in our process. Since $\{y_n, \ldots, y_1\}$ is a basis by the preceding argument, there exist coefficients $\eta_n, \ldots, \eta_1$ such that

$$y_{n+1} = \eta_n y_n + \ldots + \eta_1 y_1.$$

But by Theorem 3.3.15 this means the $y_i, i = 1, \ldots, n + 1$ are linearly dependent, a contradiction to the hypothesis of our theorem. From this it now follows that if $m \geq n$, then we must have $m = n$. This concludes the proof of the theorem. ■

As a direct consequence of Theorem 3.3.26 we have:

3.3.31. Theorem. If a linear space X has a basis containing a finite number of vectors n, then any other basis for X consists of exactly n elements.

Proof. Let $\{x_1, \ldots, x_n\}$ be a basis for X, and let also $\{y_1, \ldots, y_m\}$ be a basis for X. Then in view of Theorem 3.3.26 we have $m \leq n$. Interchanging the role of the x_i and y_i we also have $n \leq m$. Hence, $m = n$. ■

Our preceding result enables us to make the following definition.

3.3.32. Definition. If a linear space X has a basis consisting of a finite number of vectors, say $\{x_1, \ldots, x_n\}$, then X is said to be a **finite-dimensional vector space** and the **dimension of** X is n, abbreviated dim $X = n$. In this case we speak of an n-**dimensional vector space**. If X is not a finite-dimensional vector space, it is said to be an **infinite-dimensional vector space**.

We will agree that the linear space consisting of the null vector is finite dimensional, and we will say that the dimension of this space is zero.

Our next result provides us with an alternate characterization of (finite) dimension of a linear space.

3.3.33. Theorem. Let X be a vector space which contains n linearly independent vectors. If every set of $n + 1$ vectors in X is linearly dependent, then X is finite dimensional and dim $X = n$.

Proof. Let $\{x_1, \ldots, x_n\}$ be a linearly independent set in X, and let $x \in X$. Then there exists a set of scalars $\{\alpha_1, \ldots, \alpha_{n+1}\}$ not all zero, such that

$$\alpha_1 x_1 + \ldots + \alpha_n x_n + \alpha_{n+1} x = 0.$$

Now $\alpha_{n+1} \neq 0$, otherwise we would contradict the fact that $x_1, \ldots, x_n$ are linearly independent. Hence,

$$x = -\left(\frac{\alpha_1}{\alpha_{n+1}}\right)x_1 - \ldots - \left(\frac{\alpha_n}{\alpha_{n+1}}\right)x_n$$

and $x \in V(\{x_1, \ldots, x_n\})$; i.e., $\{x_1, \ldots, x_n\}$ is a basis for X. Therefore, X is n-dimensional. ■

From our preceding result follows:

3.3.34. Corollary. Let X be a vector space. If for given n every set of $n + 1$ vectors in X is linearly dependent, then X is finite dimensional and dim $X \leq n$.

3.3.35. Exercise. Prove Corollary 3.3.34.

We are now in a position to speak of coordinates of a vector. We have:

3.3.36. Definition. Let X be a finite-dimensional vector space, and let $\{x_1, \ldots, x_n\}$ be a basis for X. Let $x \in X$ be represented by

$$x = \xi_1 x_1 + \ldots + \xi_n x_n.$$

The unique scalars $\xi_1, \xi_2, \ldots, \xi_n$ are called the **coordinates of** x **with respect to the basis** $\{x_1, x_2, \ldots, x_n\}$.

It is possible to prove results similar to Theorems 3.3.26 and 3.3.31 for infinite-dimensional linear spaces. Since we will not make further use of these results in this book, their proofs will be omitted. In the following theorems, X is an arbitrary vector space (i.e., finite dimensional or infinite dimensional).

3.3.37. Theorem. If Y is a linearly independent set in a linear space X, then there exists a Hamel basis Z for X such that $Y \subset Z$.

3.3.38. Theorem. If Y and Z are Hamel bases for a linear space X, then Y and Z have the same cardinal number.

The notion of Hamel basis is not the only concept of basis with which we will deal. Such other concepts (to be specified later) reduce to Hamel basis on finite-dimensional vector spaces but differ significantly on infinite-dimensional spaces. We will find that on infinite-dimensional spaces the concept of Hamel basis is not very useful. However, in the case of finite-dimensional spaces the concept of Hamel basis is most crucial.

In view of the results presented thus far, the reader can readily prove the following facts.

3.3.39. Theorem. Let X be a finite-dimensional linear space with $\dim X = n$.

(i) No linearly independent set in X contains more than n vectors.
(ii) A linearly independent set in X is a basis if and only if it contains exactly n vectors.
(iii) Every spanning or generating set for X contains a basis for X.
(iv) Every set of vectors which spans X contains at least n vectors.
(v) Every linearly independent set of vectors in X is contained in a basis for X.
(vi) If Y is a linear subspace of X, then Y is finite dimensional and $\dim Y \leq n$.
(vii) If Y is a linear subspace of X and if $\dim X = \dim Y$, then $Y = X$.

3.3.40. Exercise. Prove Theorem 3.3.39.

From Theorem 3.3.39 follows directly our next result.

3.3.41. Theorem. Let X be a finite-dimensional linear space of dimension n, and let Y be a collection of vectors in X. Then any two of the three conditions listed below imply the third condition:

(i) the vectors in Y are linearly independent;
(ii) the vectors in Y span X; and
(iii) the number of vectors in Y is n.

3.3.42. Exercise. Prove Theorem 3.3.41.

Another way of restating Theorem 3.3.41 is as follows:

(a) the dimension of a finite-dimensional linear space X is equal to the smallest number of vectors that can be used to span X; and
(b) the dimension of a finite-dimensional linear space X is the largest number of vectors that can be linearly independent in X.

For the direct sum of two linear subspaces we have the following result.

3.3.43. Theorem. Let X be a finite-dimensional vector space. If there exist linear subspaces Y and Z of X such that $X = Y \oplus Z$, then $\dim(X) = \dim(Y) + \dim(Z)$.

Proof. Since X is finite dimensional it follows from part (vi) of Theorem 3.3.39 that Y and Z are finite-dimensional linear spaces. Thus, there exists a basis, say $\{y_1, \ldots, y_n\}$ for Y, and a basis, say $\{z_1, \ldots, z_m\}$, for Z. Let $W = \{y_1, \ldots, y_n, z_1, \ldots, z_m\}$. We must show that W is a linearly independent set in X and that $V(W) = X$. Now suppose that

$$0 = \sum_{i=1}^{n} \alpha_i y_i + \sum_{i=1}^{m} \beta_i z_i.$$

Since the representation for $0 \in X$ must be unique in terms of its components in Y and Z, we must have

$$\sum_{i=1}^{n} \alpha_i y_i = 0$$

and

$$\sum_{i=1}^{m} \beta_i z_i = 0.$$

But this implies that $\alpha_1 = \alpha_2 = \ldots = \alpha_n = \beta_1 = \beta_2 = \ldots = \beta_m = 0$. Thus, W is a linearly independent set in X. Since X is the direct sum of Y and Z, it is clear that W generates X. Thus, $\dim X = m + n$. This completes the proof of the theorem. ■

We conclude the present section with the following results.

3.3.44. Theorem. Let X be an n-dimensional vector space, and let $\{y_1, \ldots, y_m\}$ be a linearly independent set of vectors in X, where $m < n$. Then it is possible to form a basis for X consisting of n vectors $x_1, \ldots, x_n$, where $x_i = y_i$ for $i = 1, \ldots, m$.

Proof. Let $\{e_1, \ldots, e_n\}$ be a basis for X. Let S_1 be the set of vectors $\{y_1, \ldots, y_m, e_1, \ldots, e_n\}$, where $\{y_1, \ldots, y_m\}$ is a linearly independent set of vectors in X and where $m < n$. We note that S_1 spans X and is linearly dependent, since it contains more than n vectors. Now let

$$\sum_{i=1}^{m} \alpha_i y_i + \sum_{i=1}^{n} \beta_i e_i = 0.$$

Then there must be some $\beta_j \neq 0$, otherwise the linear independence of $\{y_1, \ldots, y_m\}$ would be contradicted. But this means that e_j is a linear combination of the set of vectors $S_2 = \{y_1, \ldots, y_m, e_1, \ldots, e_{j-1}, e_{j+1}, \ldots, e_n\}$; i.e., S_2 is the set S_1 with e_j eliminated. Clearly, S_2 still spans X. Now either S_2 contains n vectors or else it is a linearly dependent set. If it contains n vectors, then by Theorem 3.3.41 these vectors must be linearly independent in which case S_2 is a basis for X. We then let $x_n = e_j$, and the theorem is proved. On the other hand, if S_2 contains more than n vectors, then we continue the above procedure to eliminate vectors from the remaining e_i's until exactly $n - m$ of them are left. Letting $e_{j_1}, \ldots, e_{j_{n-m}}$ be the remaining vectors and letting $x_{m+1} = e_{j_1}, \ldots, x_n = e_{j_{n-m}}$, we have completed the proof of the theorem. ■

3.3.45. Corollary. Let X be an n-dimensional vector space, and let Y be an m-dimensional subspace of X. Then there exists a subspace Z of X of dimension $(n - m)$ such that $X = Y \oplus Z$.

3.3.46. Exercise. Prove Corollary 3.3.45.

Referring to Figure 3.3.8, it is easy to see that the subspace Z in Corollary 3.3.45 need not be unique.

3.4. LINEAR TRANSFORMATIONS

Among the most important notions which we will encounter are special types of mappings on vector spaces, called **linear transformations**.

3.4.1. Definition. A mapping T of a linear space X into a linear space Y, where X and Y are vector spaces over the same field F, is called a **linear transformation** or **linear operator** provided that

(i) $T(x + y) = T(x) + T(y)$ for all $x, y \in X$; and
(ii) $T(\alpha x) = \alpha T(x)$ for all $x \in X$ and for all $\alpha \in F$.

A transformation which is not linear is called a **non-linear transformation**.

We will find it convenient to write $T \in L(X, Y)$ to indicate that T is a linear transformation from a linear space X into a linear space Y (i.e.,

$L(X, Y)$ denotes the set of all linear transformations from linear space X into linear space Y).

It follows immediately from the above definition that T is a linear transformation from a linear space X into a linear space Y if and only if $T\left(\sum_{i=1}^{n} \alpha_i x_i\right) = \sum_{i=1}^{n} \alpha_i T(x_i)$ for all $x_i \in X$ and for all $\alpha_i \in F$, $i = 1, \ldots, n$. In engineering and science this is called the **principle of superposition** and is among the most important concepts in those disciplines.

3.4.2. Example. Let $X = Y$ denote the space of real-valued continuous functions on the interval $[a, b]$ as described in Example 3.1.19. Let $T: X \longrightarrow Y$ be defined by

$$[Tx](t) = \int_a^t x(s)ds, \quad a \leq t \leq b,$$

where integration is in the Riemann sense. By the properties of integrals it follows readily that T is a linear transformation. ■

3.4.3. Example. Let $X = \mathcal{C}^n(a, b)$ denote the set of functions $x(t)$ with n continuous derivatives on the interval (a, b), and let vector addition and scalar multiplication be defined by equations (3.1.20) and (3.1.21), respectively. It is readily verified that $\mathcal{C}^n(a, b)$ is a linear space. Now let $T: \mathcal{C}^n(a, b) \longrightarrow \mathcal{C}^{n-1}(a, b)$ be defined by

$$[Tx](t) = \frac{dx(t)}{dt}.$$

From the properties of derivatives it follows that T is a linear transformation from $\mathcal{C}^n(a, b)$ to $\mathcal{C}^{n-1}(a, b)$. ■

3.4.4. Example. Let X denote the space of all complex-valued functions $x(t)$ defined on the half-open interval $[0, \infty)$ such that $x(t)$ is Riemann integrable and such that

$$\lim_{t \to \infty} |x(t)| < ke^{at},$$

where k is some positive constant and a is any real number. Defining vector addition and scalar multiplication as in Eqs. (3.1.20) and (3.1.21), respectively, it is easily shown that X is a linear space. Now let Y denote the linear space of complex functions of a complex variable s ($s = \sigma + i\omega$, $i = \sqrt{-1}$). The reader can readily verify that the mapping $T: X \longrightarrow Y$ defined by

$$[Tx](s) = \int_0^\infty e^{-st}x(t)\, dt \tag{3.4.5}$$

is a linear transformation (called the **Laplace transform** of $x(t)$). ■

3.4.6. Example. Let X be the space of real-valued continuous functions on $[a, b]$ as described in Example 3.1.19. Let $k(s, t)$ be a real-valued function defined for $a \leq s \leq b, a \leq t \leq b$, such that for each $x \in X$ the Riemann integral

$$\int_a^b k(s, t)x(t)\,dt \tag{3.4.7}$$

exists and defines a continuous function of s on $[a, b]$. Let $T_1: X \longrightarrow X$ be defined by

$$[T_1 x](s) = y(s) = \int_a^b k(s, t)x(t)\,dt. \tag{3.4.8}$$

It is readily shown that $T_1 \in L(X, X)$. The equation (3.4.8) is called the **Fredholm integral equation of the first type.** ■

3.4.9. Example. If in place of (3.4.8) we define $T_2: X \longrightarrow X$ by

$$[T_2 x](s) = y(s) = x(s) - \int_a^b k(s, t)x(t)\,dt, \tag{3.4.10}$$

then it is again readily shown that $T_2 \in L(X, X)$. Equation (3.4.10) is known as the **Fredholm integral equation of the second type.** ■

3.4.11. Example. In Examples 3.4.6 and 3.4.9, assume that $k(s, t) = 0$ when $t > s$. In place of (3.4.7) we now have

$$\int_a^s k(s, t)x(t)dt. \tag{3.4.12}$$

Equations (3.4.8) and (3.4.10) now become

$$[T_3 x](s) = y(s) = \int_a^s k(s, t)x(t)\,dt \tag{3.4.13}$$

and

$$[T_4 x](s) = y(s) = x(s) - \int_a^s k(s, t)x(t)\,dt, \tag{3.4.14}$$

respectively. Equations (3.4.13) and (3.4.14) are called **Volterra integral equations** (of the **first type** and the **second type**, respectively). Again, the mappings T_3 and T_4 are linear transformations from X into X. ■

3.4.15. Example. Let $X = C$, the set of complex numbers. If $x \in C$, let $\bar{x}$ denote the complex conjugate of x. Define $T: X \longrightarrow X$ as

$$T(x) = \bar{x}.$$

Then, clearly, $T(x + y) = \overline{x + y} = \bar{x} + \bar{y} = T(x) + T(y)$. Now if $F = C$, the field of complex numbers, and if $\alpha \in F$, then

$$T(\alpha x) = \overline{\alpha x} = \overline{\alpha x} = \bar{\alpha}T(x) \neq \alpha T(x).$$

Therefore, *T is not a linear transformation.* ■

Example 3.4.15 demonstrates the important fact that condition (i) of Definition 3.4.1 does *not* imply condition (ii) of this definition.

Henceforth, where dealing with linear transformations $T: X \longrightarrow Y$, we will write Tx in place of $T(x)$.

3.4.16. Definition. Let $T \in L(X, Y)$. We call the set

$$\mathfrak{N}(T) = \{x \in X: Tx = 0\} \tag{3.4.17}$$

the **null space of** T. The set

$$\mathfrak{R}(T) = \{y \in Y: y = Tx, x \in X\} \tag{3.4.18}$$

is called the **range space of** T.

Since $T0 = 0$ it follows that $\mathfrak{N}(T)$ and $\mathfrak{R}(T)$ are never empty. The next two important assertions are readily proved.

3.4.19. Theorem. Let $T \in L(X, Y)$. Then

(i) the null space $\mathfrak{N}(T)$ is a linear subspace of X; and
(ii) the range space $\mathfrak{R}(T)$ is a linear subspace of Y.

3.4.20. Exercise. Prove Theorem 3.4.19.

For the dimension of the range space $\mathfrak{R}(T)$ we have

3.4.21. Theorem. Let $T \in L(X, Y)$. If X is finite dimensional with dimension n, then $\mathfrak{R}(T)$ is finite dimensional and $\dim [\mathfrak{R}(T)] \leq n$.

Proof. We assume that $\mathfrak{R}(T) \neq \{0\}$ and $X \neq \{0\}$, for if $\mathfrak{R}(T) = \{0\}$ or $X = \{0\}$, then $\dim \{\mathfrak{R}(T)\} = 0$, and the theorem is proved. Thus, assume that $n > 0$ and let $y_1, \ldots, y_{n+1} \in \mathfrak{R}(T)$. Then there exist $x_1, \ldots, x_{n+1} \in X$ such that $Tx_i = y_i$ for $i = 1, \ldots, n + 1$. Since X is of dimension n, there exist $\alpha_1, \ldots, \alpha_{n+1} \in F$ such that not all $\alpha_i = 0$ and

$$\alpha_1 x_1 + \ldots + \alpha_{n+1} x_{n+1} = 0.$$

This implies that

$$T(\alpha_1 x_1 + \ldots + \alpha_{n+1} x_{n+1}) = 0.$$

Thus,

$$\alpha_1 T x_1 + \ldots + \alpha_{n+1} T x_{n+1} = 0$$

or

$$\alpha_1 y_1 + \ldots + \alpha_{n+1} y_{n+1} = 0.$$

Therefore, by Corollary 3.3.34, $\mathfrak{R}(T)$ is finite dimensional and $\dim [\mathfrak{R}(T)] \leq n$. ■

3.4.22. Example. Let $T: R^2 \longrightarrow R^\infty$, where R^2 and R^∞ are defined in Examples 3.1.10 and 3.1.11, respectively. For $x \in R^2$ we write $x = (\xi_1, \xi_2)$. Define

T by

$$T(\xi_1, \xi_2) = (0, \xi_1, 0, \xi_2, 0, 0, \ldots).$$

The mapping T is clearly a linear transformation. The vectors $(0, 1, 0, \ldots)$ and $(0, 0, 0, 1, 0, 0, \ldots)$ span $\mathcal{R}(T)$ and $\dim [\mathcal{R}(T)] = 2 = \dim [R^2]$. ■

We also have:

3.4.23. Theorem. Let $T \in L(X, Y)$, and let X be finite dimensional. Let $\{y_1, \ldots, y_n\}$ be a basis for $\mathcal{R}(T)$ and let x_i be such that $Tx_i = y_i$ for $i = 1, \ldots, n$. Then $x_1, \ldots, x_n$ are linearly independent in X.

3.4.24. Exercise. Prove Theorem 3.4.23.

Our next result, which as we will see is of utmost importance, is sometimes called the **fundamental theorem of linear equations.**

3.4.25. Theorem. Let $T \in L(X, Y)$. If X is finite dimensional, then

$$\dim \mathfrak{N}(T) + \dim \mathcal{R}(T) = \dim X. \tag{3.4.26}$$

Proof. Let $\dim X = n$, let $\dim \mathfrak{N}(T) = s$, and let $r = n - s$. We must show that $\dim \mathcal{R}(T) = r$.

First, let us assume that $0 < s < n$, and let $\{e_1, e_2, \ldots, e_n\}$ be a basis for X chosen in such a way that the last s vectors, $e_{r+1}, e_{r+2}, \ldots, e_n$, form a basis for the linear subspace $\mathfrak{N}(T)$ (see Theorem 3.3.44). Then the vectors $Te_1, Te_2, \ldots, Te_r, Te_{r+1}, \ldots, Te_n$ generate the linear subspace $\mathcal{R}(T)$. But $e_{r+1}, e_{r+2}, \ldots, e_n$ are vectors in $\mathfrak{N}(T)$, and thus $Te_{r+1} = 0, \ldots, Te_n = 0$. From this it now follows that the vectors $Te_1, Te_2, \ldots, Te_r$ must generate $\mathcal{R}(T)$. Now let $f_1 = Te_1, f_2 = Te_2, \ldots, f_r = Te_r$. We must show that the vectors $\{f_1, f_2, \ldots, f_r\}$ are linearly independent and as such form a basis for $\mathcal{R}(T)$.

Next, we observe that $\gamma_1 f_1 + \gamma_2 f_2 + \ldots + \gamma_r f_r \in \mathcal{R}(T)$. If the $\gamma_1, \gamma_2, \ldots, \gamma_r$ are chosen in such a fashion that $\gamma_1 f_1 + \gamma_2 f_2 + \ldots + \gamma_r f_r = 0$, then

$$\begin{aligned} 0 &= \gamma_1 f_1 + \gamma_2 f_2 + \ldots + \gamma_r f_r = \gamma_1 Te_1 + \gamma_2 Te_2 + \ldots + \gamma_r Te_r \\ &= T(\gamma_1 e_1 + \gamma_2 e_2 + \ldots + \gamma_r e_r), \end{aligned}$$

and from this it follows that $x = \gamma_1 e_1 + \gamma_2 e_2 + \ldots + \gamma_r e_r \in \mathfrak{N}(T)$. Now, by assumption, the set $\{e_{r+1}, \ldots, e_n\}$ is a basis for $\mathfrak{N}(T)$. Thus there must exist scalars $\gamma_{r+1}, \gamma_{r+2}, \ldots, \gamma_n$ such that

$$\gamma_1 e_1 + \gamma_2 e_2 + \ldots + \gamma_r e_r = \gamma_{r+1} e_{r+1} + \ldots + \gamma_n e_n.$$

This can be rewritten as

$$\gamma_1 e_1 + \gamma_2 e_2 + \ldots + \gamma_r e_r + (-\gamma_{r+1}) e_{r+1} + \ldots + (-\gamma_n) e_n = 0.$$

But $\{e_1, e_2, \ldots, e_n\}$ is a basis for X. From this it follows that $\gamma_1 = \gamma_2 = \ldots = \gamma_r = \gamma_{r+1} = \ldots = \gamma_n = 0$. Hence, $f_1, f_2, \ldots, f_r$ are linearly independent and therefore dim $\mathcal{R}(T) = r$. If $s = 0$, the preceding proof remains valid if we let $\{e_1, \ldots, e_n\}$ be any basis for X and ignore the remarks about the vectors $\{e_{r+1}, \ldots, e_n\}$. If $s = n$, then $\mathfrak{N}(T) = X$. Hence, $\mathcal{R}(T) = \{0\}$ and so dim $\mathcal{R}(T) = 0$. This concludes the proof of the theorem. ■

Our preceding result gives rise to the next definition.

3.4.27. Definition. The **rank** $\rho(T)$ of a linear transformation T of a finite-dimensional vector space X into a vector space Y is the dimension of the range space $\mathcal{R}(T)$. The **nullity** $\nu(T)$ of the linear transformation T is the dimension of the nullspace $\mathfrak{N}(T)$.

The reader is now in a position to prove the next result.

3.4.28. Theorem. Let $T \in L(X, Y)$. Let X be finite dimensional, and let $s = \dim \mathfrak{N}(T)$. Let $\{x_1, \ldots, x_s\}$ be a basis for $\mathfrak{N}(T)$. Then

(i) a vector $x \in X$ satisfies the equation

$$Tx = 0$$

if and only if $x = \alpha_1 x_1 + \ldots + \alpha_s x_s$ for some set of scalars $\{\alpha_1, \ldots, \alpha_s\}$. Furthermore, for each $x \in X$ such that $Tx = 0$ is satisfied, the set of scalars $\{\alpha_1, \ldots, \alpha_s\}$ is unique;

(ii) if y_0 is a fixed vector in Y, then $Tx = y_0$ holds for at least one $x \in X$ (called a **solution** of the equation $Tx = y_0$) if and only if $y_0 \in \mathcal{R}(T)$; and

(iii) if y_0 is any fixed vector in Y and if x_0 is some vector in X such that $Tx_0 = y_0$ (i.e., x_0 is a solution of the equation $Tx_0 = y_0$), then a vector $x \in X$ satisfies $Tx = y_0$ if and only if $x = x_0 + \beta_1 x_1 + \ldots + \beta_s x_s$ for some set of scalars $\{\beta_1, \beta_2, \ldots, \beta_s\}$. Furthermore, for each $x \in X$ such that $Tx = y_0$, the set of scalars $\{\beta_1, \beta_2, \ldots, \beta_s\}$ is unique.

3.4.29. Exercise. Prove Theorem 3.4.28.

Since a linear transformation T of a linear space X into a linear space Y is a mapping, we can distinguish, as in Chapter 1, between linear transformations that are **surjective** (i.e., onto), **injective** (i.e., one-to-one), and **bijective** (i.e., onto and one-to-one). We will often be particularly interested in knowing when a linear transformation T has an inverse, which we denote by T^{-1}. In this connection, the following terms are used interchangeably: **T^{-1} exists**, **T has an inverse**, T is **invertible**, and T is **non-singular**. Also, a linear transformation which is not non-singular is said to be **singular**. We recall, if T has an inverse, then

$$T^{-1}(Tx) = x \text{ for all } x \in X \tag{3.4.30}$$

and

$$T(T^{-1}y) = y \text{ for all } y \in \mathcal{R}(T). \tag{3.4.31}$$

The following theorem is a fundamental result concerning inverses of linear transformations.

3.4.32. Theorem. Let $T \in L(X, Y)$.

(i) The inverse of T exists if and only if $Tx = 0$ implies $x = 0$.

(ii) If T^{-1} exists, then T^{-1} is a linear transformation from $\mathcal{R}(T)$ onto X.

Proof. To prove part (i), assume first that $Tx = 0$ implies $x = 0$. Let $x_1, x_2 \in X$ with $Tx_1 = Tx_2$. Then $T(x_1 - x_2) = 0$ and therefore $x_1 - x_2 = 0$. Thus, $x_1 = x_2$ and T has an inverse.

Conversely, assume that T has an inverse. Let $Tx = 0$. Since $T0 = 0$, we have $T0 = Tx$. Since T has an inverse, $x = 0$.

To prove part (ii), assume that T^{-1} exists. To establish the linearity of T^{-1}, let $y_1 = Tx_1$ and $y_2 = Tx_2$, where $y_1, y_2 \in \mathcal{R}(T)$ and $x_1, x_2 \in X$ are such that $y_1 = Tx_1$ and $y_2 = Tx_2$. Then

$$\begin{aligned} T^{-1}(y_1 + y_2) &= T^{-1}(Tx_1 + Tx_2) = T^{-1}T(x_1 + x_2) = x_1 + x_2 \\ &= T^{-1}(y_1) + T^{-1}(y_2). \end{aligned}$$

Also, for $\alpha \in F$ we have

$$T^{-1}(\alpha y_1) = T^{-1}(\alpha T x_1) = T^{-1}(T(\alpha x_1)) = \alpha x_1 = \alpha T^{-1}(y_1).$$

Thus, T^{-1} is linear. It is also a mapping onto X, since every $y \in \mathcal{R}(T)$ is the image of some $x \in X$. For, if $x \in X$, then there is a $y \in \mathcal{R}(T)$ such that $Tx = y$. Hence, $x = T^{-1}y$ and $x \in \mathcal{R}(T^{-1})$. ■

3.4.33. Example. Consider the linear transformation $T: R^2 \longrightarrow R^\infty$ of Example 3.4.22. Since $Tx = 0$ implies $x = 0$, T has an inverse. We see that T is not a mapping of R^2 onto R^∞; however, T is clearly a one-to-one mapping of R^2 onto $\mathcal{R}(T)$. ■

For finite-dimensional vector spaces we have:

3.4.34. Theorem. Let $T \in L(X, Y)$. If X is finite dimensional, T has an inverse if and only if $\mathcal{R}(T)$ has the same dimension as X; i.e., $\rho(T) = \dim X$.

Proof. By Theorem 3.4.25 we have

$$\dim \mathfrak{N}(T) + \dim \mathcal{R}(T) = \dim X.$$

Since T has an inverse if and only if $\mathfrak{N}(T) = \{0\}$, it follows that $\rho(T) = \dim X$ if and only if T has an inverse. ■

For finite-dimensional linear spaces we also have:

3.4.35. Theorem. Let X and Y be finite-dimensional vector spaces of the same dimension, say $\dim X = \dim Y = n$. Let $T \in L(X, Y)$. Then $\mathfrak{R}(T) = Y$ if and only if T has an inverse.

Proof. Assume that T has an inverse. By Theorem 3.4.34 we know that $\dim \mathfrak{R}(T) = n$. Thus, $\dim \mathfrak{R}(T) = \dim Y$ and if follows from Theorem 3.3.39, part (vii), that $\mathfrak{R}(T) = Y$.

Conversely, assume that $\mathfrak{R}(T) = Y$. Let $\{y_1, y_2, \ldots, y_n\}$ be a basis for $\mathfrak{R}(T)$. Let x_i be such that $Tx_i = y_i$ for $i = 1, \ldots, n$. Then, by Theorem 3.4.23, the vectors $x_1, \ldots, x_n$ are linearly independent. Since the dimension of X is n, it follows that the vectors $x_1, \ldots, x_n$ span X. Now let $Tx = 0$ for some $x \in X$. We can represent x as $x = \alpha_1 x_1 + \ldots + \alpha_n x_n$. Hence, $0 = Tx = \alpha_1 y_1 + \ldots + \alpha_n y_n$. Since the vectors $y_1, \ldots, y_n$ are linearly independent, we must have $\alpha_1 = \ldots = \alpha_n = 0$, and thus $x = 0$. This implies that T has an inverse. ■

At this point we find it instructive to summarize the preceding results which characterize injective, surjective, and bijective linear transformations. In so doing, it is useful to keep Figure J in mind.

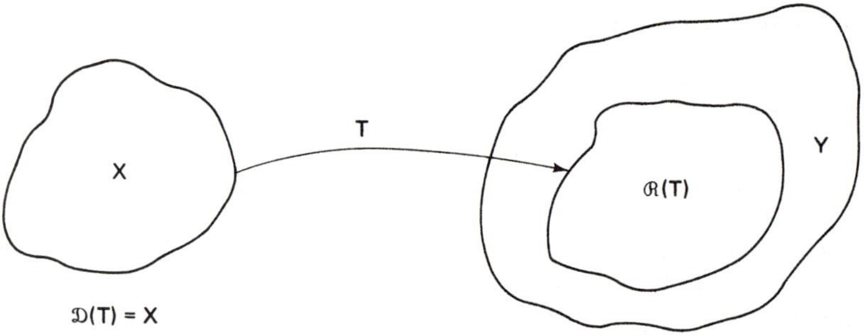

3.4.36. **Figure J.** Linear transformation T from vector space X into vector space Y.

3.4.37. Summary (Injective Linear Transformations). Let X and Y be vector spaces over the same field F, and let $T \in L(X, Y)$. The following are equivalent:

(i) T is injective;

(ii) T has an inverse;

(iii) $Tx = 0$ implies $x = 0$;
(iv) for each $y \in \mathcal{R}(T)$, there is a unique $x \in X$ such that $Tx = y$;
(v) if $Tx_1 = Tx_2$, then $x_1 = x_2$; and
(vi) if $x_1 \neq x_2$, then $Tx_1 \neq Tx_2$.

If X is finite dimensional, then the following are equivalent:

(i) T is injective; and
(ii) $\rho(T) = \dim X$.

3.4.38. Summary (Surjective Linear Transformations). Let X and Y be vector spaces over the same field F, and let $T \in L(X, Y)$. The following are equivalent:

(i) T is surjective; and
(ii) for each $y \in Y$, there is an $x \in X$ such that $Tx = y$.

If X and Y are finite dimensional, then the following are equivalent:

(i) T is surjective; and
(ii) $\dim Y = \rho(T)$.

3.4.39. Summary (Bijective Linear Transformations). Let X and Y be vector spaces over the same field F, and let $T \in L(X, Y)$. The following are equivalent:

(i) T is bijective; and
(ii) for every $y \in Y$ there is a unique $x \in X$ such that $Tx = y$.

If X and Y are finite dimensional, then the following are equivalent:

(i) T is bijective; and
(ii) $\dim X = \dim Y = \rho(T)$.

3.4.40. Summary (Injective, Surjective, and Bijective Linear Transformations). Let X and Y be finite-dimensional vector spaces, over the same field F, and let $\dim X = \dim Y$. (Note: this is true if, e.g., $X = Y$.) The following are equivalent:

(i) T is injective;
(ii) T is surjective:
(iii) T is bijective; and
(iv) T has an inverse.

3.4.41. Exercise. Verify the assertions made in summaries (3.4.37)–(3.4.40).

Let us next examine some of the properties of the set $L(X, Y)$, the set of all linear transformations from a vector space X into a vector space Y. As before, we assume that X and Y are linear spaces over the same field F.

Let $S, T \in L(X, Y)$, and define the **sum of S and T** by

$$(S + T)x \triangleq Sx + Tx \tag{3.4.42}$$

for all $x \in X$. Also, with $\alpha \in F$ and $T \in L(X, Y)$, define **multiplication of T by a scalar α** as

$$(\alpha T)x \triangleq \alpha Tx \tag{3.4.43}$$

for all $x \in X$. It is an easy matter to show that $(S + T) \in L(X, Y)$ and also that $\alpha T \in L(X, Y)$. Let us further note that there exists a zero element in $L(X, Y)$, called the **zero transformation** and denoted by 0, which is defined by

$$0x = 0 \tag{3.4.44}$$

for all $x \in X$. Moreover, to each $T \in L(X, Y)$ there corresponds a unique linear transformation $-T \in L(X, Y)$ defined by

$$(-T)x = -Tx \tag{3.4.45}$$

for all $x \in X$. In this case it follows trivially that $-T + T = 0$.

3.4.46. Exercise. Let X be a finite-dimensional space, and let $T \in L(X, Y)$. Let $\{e_1, \ldots, e_n\}$ be a basis for X. Then $Te_i = 0$ for $i = 1, \ldots, n$ if and only if $T = 0$ (i.e., T is the zero transformation).

With the above definitions it is now easy to establish the following result.

3.4.47. Theorem. Let X and Y be two linear spaces over the same field of scalars F, and let $L(X, Y)$ denote the set of all linear transformations from X into Y. Then $L(X, Y)$ is itself a linear space over F, called the **space of linear transformations** (here, vector addition is defined by Eq. (3.4.42) and multiplication of vectors by scalars is defined by Eq. (3.4.43)).

3.4.48. Exercise. Prove Theorem 3.4.47.

Next, let us recall the definition of an algebra, considered in Chapter 2.

3.4.49. Definition. A set X is called an **algebra** if it is a linear space and if in addition to each $x, y \in X$ there corresponds an element in X, denoted by $x \cdot y$ and called the **product** of x times y, satisfying the following axioms:

(i) $x \cdot (y + z) = x \cdot y + x \cdot z$ for all $x, y, z \in X$;

(ii) $(x + y) \cdot z = x \cdot z + y \cdot z$ for all $x, y, z \in X$; and

(iii) $(\alpha x) \cdot (\beta y) = (\alpha\beta)(x \cdot y)$ for all $x, y \in X$ and for all $\alpha, \beta \in F$.

If in addition to the above,

(iv) $(x \cdot y) \cdot z = x \cdot (y \cdot z)$ for all $x, y, z \in X$,

then X is called an **associative algebra**.

If there exists an element $i \in X$ such that $i \cdot x = x \cdot i = x$ for every $x \in X$, then i is called the **identity** of the algebra. It can be readily shown that if i exists, then it is unique. Furthermore, if $x \cdot y = y \cdot x$ for all $x, y \in X$, then X is said to be a **commutative algebra**. Finally, if Y is a subset of X (X is an algebra) and (a) if $x + y \in Y$ whenever $x, y \in Y$, and (b) if $\alpha x \in Y$ whenever $\alpha \in F$ and $x \in Y$, and (c) if $x \cdot y \in Y$ whenever $x, y \in Y$, then Y is called a **subalgebra** of X.

Now let us return to the subject on hand. Let X, Y, and Z be linear spaces over F, and consider the vector spaces $L(X, Y)$ and $L(Y, Z)$. If $S \in L(Y, Z)$ and if $T \in L(X, Y)$, then we define the **product** ST as the mapping of X into Z characterized by

$$(ST)x = S(Tx) \tag{3.4.50}$$

for all $x \in X$. The reader can readily verify that $ST \in L(X, Z)$.

Next, let $X = Y = Z$. If $S, T, U \in L(X, X)$ and if $\alpha, \beta \in F$, then it is easily shown that

$$S(TU) = (ST)U, \tag{3.4.51}$$

$$S(T + U) = ST + SU, \tag{3.4.52}$$

$$(S + T)U = SU + TU, \tag{3.4.53}$$

and

$$(\alpha S)(\beta T) = (\alpha\beta)ST. \tag{3.4.54}$$

For example, to verify (3.4.52), we observe that

$$\begin{aligned}[S(T + U)]x &= S[(T + U)x] = S[Tx + Ux] \\ &= (ST)x + (SU)x = (ST + SU)x\end{aligned}$$

for all $x \in X$, and hence Eq. (3.4.52) follows.

We emphasize at this point that, in general, commutativity of linear transformations does not hold; i.e., in general,

$$ST \neq TS. \tag{3.4.55}$$

There is a special mapping from a linear space X into X, called the **identity transformation**, defined by

$$Ix = x \tag{3.4.56}$$

for all $x \in X$. We note that I is linear, i.e., $I \in L(X, X)$, that $I \neq 0$ if and only if $X \neq \{0\}$, that I is unique, and that

$$TI = IT = T \tag{3.4.57}$$

for all $T \in L(X, X)$. Also, we can readily verify that the transformation

αI, $\alpha \in F$, defined by

$$(\alpha I)x = \alpha Ix = \alpha x \tag{3.4.58}$$

is also a linear transformation.

The above discussion gives rise to the following result.

3.4.59. Theorem. The set of linear transformations of a linear space X into X, denoted by $L(X, X)$, is an associative algebra with identity I. This algebra is, in general, not commutative.

We further have:

3.4.60. Theorem. Let $T \in L(X, X)$. If T is bijective, then $T^{-1} \in L(X, X)$ and

$$T^{-1}T = TT^{-1} = I, \tag{3.4.61}$$

where I denotes the identity transformation defined in Eq. (3.4.56).

3.4.62. Exercise. Prove Theorem 3.4.60.

For invertible linear transformations defined on finite-dimensional linear spaces we have the following result.

3.4.63. Theorem. Let X be a finite-dimensional vector space, and let $T \in L(X, X)$. Then the following are equivalent:

(i) T is invertible;
(ii) rank $T = \dim X$;
(iii) T is one-to-one;
(iv) T is onto; and
(v) $Tx = 0$ implies $x = 0$.

3.4.64. Exercise. Prove Theorem 3.4.63.

Bijective linear transformations are further characterized by our next result.

3.4.65. Theorem. Let X be a linear space, and let $S, T, U \in L(X, X)$. Let $I \in L(X, X)$ denote the identity transformation.

(i) If $ST = US = I$, then S is bijective and $S^{-1} = T = U$.
(ii) If S and T are bijective, then ST is bijective, and $(ST)^{-1} = T^{-1}S^{-1}$.
(iii) If S is bijective, then $(S^{-1})^{-1} = S$.
(iv) If S is bijective, then αS is bijective and $(\alpha S)^{-1} = \dfrac{1}{\alpha}S^{-1}$ for all $\alpha \in F$ and $\alpha \neq 0$.

3.4.66. Exercise. Prove Theorem 3.4.65.

With the aid of the above concepts and results we can now construct certain classes of functions of linear transformations. Since relation (3.4.51) allows us to write the product of three or more linear transformations without the use of parentheses, we can define T^n, where $T \in L(X, X)$ and n is a positive integer, as

$$T^n \triangleq \underbrace{T \cdot T \cdot \ldots \cdot T}_{n \text{ times}}. \tag{3.4.67}$$

Similarly, if T^{-1} is the inverse of T, then we can define T^{-m}, where m is a positive integer, as

$$T^{-m} \triangleq (T^{-1})^m = \underbrace{T^{-1} \cdot T^{-1} \ldots \cdot T^{-1}}_{m \text{ times}}. \tag{3.4.68}$$

Using these definitions, the usual laws of exponents can be verified. Thus,

$$\begin{aligned} T^m \cdot T^n &= \underbrace{(T \cdot T \cdot \ldots \cdot T)}_{m \text{ times}} \cdot \underbrace{(T \cdot T \cdot \ldots \cdot T)}_{n \text{ times}} \\ &= \underbrace{(T \cdot T \cdot \ldots \cdot T)}_{m+n \text{ times}} \\ &= T^{m+n} = \underbrace{(T \cdot T \cdot \ldots \cdot T)}_{n \text{ times}} \cdot \underbrace{(T \cdot T \cdot \ldots \cdot T)}_{m \text{ times}} \\ &= T^n \cdot T^m. \end{aligned} \tag{3.4.69}$$

In a similar fashion we have

$$(T^m)^n = T^{mn} = T^{nm} = (T^n)^m \tag{3.4.70}$$

and

$$T^m \cdot T^{-n} = T^{m-n}, \tag{3.4.71}$$

where m and n are positive integers. Consistent with this notation we also have

$$T^1 = T \tag{3.4.72}$$

and

$$T^0 = I. \tag{3.4.73}$$

We are now in a position to consider polynomials of linear transformations. Thus, if $f(\lambda)$ is a polynomial, i.e.,

$$f(\lambda) = \alpha_0 + \alpha_1 \lambda + \ldots + \alpha_n \lambda^n, \tag{3.4.74}$$

where $\alpha_0, \ldots, \alpha_n \in F$, then by $f(T)$ we mean

$$f(T) = \alpha_0 I + \alpha_1 T + \ldots + \alpha_n T^n. \tag{3.4.75}$$

The reader is cautioned that the above concept can, in general, not be extended to functions of two or more linear transformations, because linear transformations in general do not commute.

Next, we consider the important concept of isomorphic linear spaces. In Chapter 2 we encountered the notion of isomorphisms of groups and rings. We saw that such mappings, if they exist, preserve the algebraic properties of groups and rings. Thus, in many cases two algebraic systems (such as groups or rings) may differ only in the nature of the elements of the underlying set and may thus be considered as being the same in all other respects. We now extend this concept to linear spaces.

3.4.76. Definition. Let X and Y be vector spaces over the same field F. If there exists $T \in L(X, Y)$ such that T is a one-to-one mapping of X into Y, then T is said to be an **isomorphism** of X into Y. If in addition, T maps X onto Y then X and Y are said to be **isomorphic**.

Note that if X and Y are isomorphic, then clearly Y and X are isomorphic. Our next result shows that all n-dimensional linear spaces over the same field are isomorphic.

3.4.77. Theorem. Every n-dimensional vector space X over a field F is isomorphic to F^n.

Proof. Let $\{e_1, \ldots, e_n\}$ be a basis for X. Then every $x \in X$ has the unique representation

$$x = \xi_1 e_1 + \ldots + \xi_n e_n,$$

where $\{\xi_1, \xi_2, \ldots, \xi_n\}$ is a unique set of scalars (belonging to F). Now let us define a linear transformation T from X into F^n by

$$Tx = (\xi_1, \xi_2, \ldots, \xi_n).$$

It is an easy matter to verify that T is a linear transformation of X onto F^n, and that it is one-to-one (the reader is invited to do so). Thus, X is isomorphic to F^n. ■

It is not difficult to establish the next result.

3.4.78. Theorem. Two finite-dimensional vector spaces X and Y over the same field F are isomorphic if and only if $\dim X = \dim Y$.

3.4.79. Exercise. Prove Theorem 3.4.78.

Theorem 3.4.77 points out the importance of the spaces R^n and C^n. Namely, every n-dimensional vector space over the field of real numbers is isomorphic to R^n and every n-dimensional vector space over the field of complex numbers is isomorphic to C^n (see Example 3.1.10).

3.5. LINEAR FUNCTIONALS

There is a special type of linear transformation which is so important that we give it a special name: **linear functional.**

We showed in Example 3.1.7 that if F is a field, then F^n is a vector space over F. If, in particular, $n = 1$, then we may view F as being a vector space over itself. This enables us to consider linear transformations of a vector space X over F into F.

3.5.1. Definition. Let X be a vector space over a field F. A mapping f of X into F is called a **functional** on X. If f is a linear transformation of X into F, then we call f a **linear functional** on X.

We cite some specific examples of linear functionals.

3.5.2. Example. Consider the space $\mathcal{C}[a, b]$. Then the mapping

$$f_1(x) = \int_a^b x(s)\, ds, \quad x \in \mathcal{C}[a, b] \tag{3.5.3}$$

is a linear functional on $\mathcal{C}[a, b]$. Also, the function defined by

$$f_2(x) = x(s_0), \quad x \in \mathcal{C}[a, b], \quad s_0 \in [a, b] \tag{3.5.4}$$

is also a linear functional on $\mathcal{C}[a, b]$. Furthermore, the mapping

$$f_3(x) = \int_a^b x(s)x_0(s)\, ds, \tag{3.5.5}$$

where x_0 is a fixed element of $\mathcal{C}[a, b]$ and where x is any element in $\mathcal{C}[a, b]$, is also a linear functional on $\mathcal{C}[a, b]$. ■

3.5.6. Example. Let $X = F^n$, and denote $x \in X$ by $x = (\xi_1, \ldots, \xi_n)$. The mapping f_4 defined by

$$f_4(x) = \xi_1 \tag{3.5.7}$$

is a linear functional on X. A more general form of f_4 is as follows. Let $a = (\alpha_1, \ldots, \alpha_n) \in X$ be fixed and let $x = (\xi_1, \ldots, \xi_n)$ be an arbitrary element of X. It is readily shown that the function

$$f_5(x) = \sum_{i=1}^{n} \alpha_i \xi_i \tag{3.5.8}$$

is a linear functional on X. ■

3.5.9. Exercise. Show that the mappings (3.5.3), (3.5.4), (3.5.5), (3.5.7), and (3.5.8) are linear functionals.

Now let X be a linear space and let X^f denote the set of all linear functionals on X. If $f \in X^f$ is evaluated at a point $x \in X$, we write $f(x)$. Frequently we will also find the notation

$$f(x) \triangleq \langle x, f \rangle \tag{3.5.10}$$

useful. In addition to Eq. (3.5.10), the notation $x'(x)$ or $x'x$ is sometimes used. In this case Eq. (3.5.10) becomes

$$f(x) = \langle x, f \rangle = \langle x, x' \rangle, \tag{3.5.11}$$

where x' is used in place of f. Now let $f_1 = x'_1, f_2 = x'_2$ belong to X^f, and let $\alpha \in F$. Let us define $f_1 + f_2 = x'_1 + x'_2$ and $\alpha f = \alpha x'$ by

$$\begin{aligned}(f_1 + f_2)(x) &= \langle x, x'_1 + x'_2 \rangle \triangleq \langle x, x'_1 \rangle + \langle x, x'_2 \rangle \\ &= f_1(x) + f_2(x),\end{aligned} \tag{3.5.12}$$

and

$$(\alpha f)(x) = \langle x, \alpha x' \rangle \triangleq \alpha \langle x, x' \rangle = \alpha f(x), \tag{3.5.13}$$

respectively. We denote the functional $f = x'$ such that $f(x) = x'(x) = 0$ for all $x \in X$ by 0. If f is a linear functional then we note that

$$\begin{aligned}f(x_1 + x_2) &= \langle x_1 + x_2, x' \rangle \\ &= \langle x_1, x' \rangle + \langle x_2, x' \rangle = f(x_1) + f(x_2),\end{aligned} \tag{3.5.14}$$

and also,

$$f(\alpha x) = \langle \alpha x, x' \rangle = \alpha \langle x, x' \rangle = \alpha f(x). \tag{3.5.15}$$

It is now a simple matter to prove the following:

3.5.16. Theorem. The space X^f with vector addition and multiplication of vectors by scalars defined by equations (3.5.12) and (3.5.13), respectively, is a vector space over F.

3.5.17. Exercise. Prove Theorem 3.5.16.

3.5.18. Definition. The linear space X^f is called the **algebraic conjugate** of X.

Let us now examine some of the propeties of X^f for the case of finite-dimensional linear spaces. We have:

3.5.19. Theorem. Let X be a finite-dimensional vector space, and let $\{e_1, \ldots, e_n\}$ be a basis for X. If $\{\alpha_1, \ldots, \alpha_n\}$ is an arbitrary set of scalars, then there is a unique linear functional $x' \in X^f$ such that $\langle e_i, x' \rangle = \alpha_i$ for $i = 1, \ldots, n$.

Proof. For every $x \in X$, we have

$$x = \xi_1 e_1 + \xi_2 e_2 + \ldots + \xi_n e_n.$$

Now let $x' \in X^f$ be given by

$$\langle x, x' \rangle = \sum_{i=1}^{n} \alpha_i \xi_i.$$

If $x = e_i$ for some i, we have $\xi_i = 1$ and $\xi_j = 0$ if $i \neq j$. Thus, $\langle e_i, x' \rangle = \alpha_i$ for $i = 1, \ldots, n$. To show that x' is unique, suppose there is an $x'_1 \in X^f$ such that $\langle e_i, x'_1 \rangle = \alpha_i$ for $i = 1, \ldots, n$. It then follows that $\langle e_i, x'_1 \rangle - \langle e_i, x' \rangle = 0$ for $i = 1, \ldots, n$, and so $\langle e_i, x'_1 - x \rangle = 0$ for $i = 1, \ldots, n$. This implies $x'_1 - x = 0$; i.e., $x'_1 = x$. ■

In our next result and on several other occasions throughout this book, we make use of the Kronecker delta.

3.5.20. Definition. Let

$$\delta_{ij} = \begin{cases} 1 & \text{if } i = j \\ 0 & \text{if } i \neq j \end{cases} \tag{3.5.21}$$

for $i, j = 1, \ldots, n$. Then δ_{ij} is called the **Kronecker delta**.

We now have:

3.5.22. Theorem. Let X be a finite-dimensional vector space. If $\{e_1, e_2, \ldots, e_n\}$ is a basis for X, then there is a unique basis $\{e'_1, e'_2, \ldots, e'_n\}$ in X^f with the property that $\langle e_i, e'_j \rangle = \delta_{ij}$. From this it follows that if X is n-dimensional, then so is X^f.

Proof. From Theorem 3.5.19 it follows that for each $j = 1, \ldots, n$, a unique $e'_j \in X^f$ can be found such that $\langle e_i, e'_j \rangle = \delta_{ij}$. Thus, we only have to show that the set $\{e'_1, e'_2, \ldots, e'_n\}$ is a linearly independent set which spans X^f.

To show that $\{e'_1, e'_2, \ldots, e'_n\}$ is linearly independent, let

$$\beta_1 e'_1 + \beta_2 e'_2 + \ldots + \beta_n e'_n = 0.$$

Then

$$0 = \left\langle e_j, \sum_{i=1}^{n} \beta_i e'_i \right\rangle = \sum_{i=1}^{n} \beta_i \langle e_j, e'_i \rangle = \sum_{i=1}^{n} \beta_i \delta_{ij} = \beta_j,$$

and therefore we have $\beta_1 = \beta_2 = \ldots = \beta_n = 0$. This proves that $\{e'_1, e'_2, \ldots, e'_n\}$ is a linearly independent set.

To show that the set $\{e'_1, e'_2, \ldots, e'_n\}$ spans X^f, let $x' \in X^f$ and define $\alpha_i = \langle e_i, x' \rangle$. Let $x = \sum_{i=1}^{n} \xi_i e_i$. We then have

$$\begin{aligned} \langle x, x' \rangle &= \langle \xi_1 e_1 + \ldots + \xi_n e_n, x' \rangle = \langle \xi_1 e_1, x' \rangle + \ldots + \langle \xi_n e_n, x' \rangle \\ &= \xi_1 \langle e_1, x' \rangle + \ldots + \xi_n \langle e_n, x' \rangle = \xi_1 \alpha_1 + \ldots + \xi_n \alpha_n. \end{aligned}$$

Also,

$$\langle x, e'_j \rangle = \sum_{i=1}^{n} \xi_i \langle e_i, e'_j \rangle = \xi_j.$$

Combining the above relations we now have

$$\langle x, x' \rangle = \alpha_1 \langle x, e_1' \rangle + \ldots + \alpha_n \langle x, e_n' \rangle$$
$$= \langle x, \alpha_1 e_1' + \ldots + \alpha_n e_n' \rangle.$$

From this it now follows that for any $x' \in X^f$ we have

$$x' = \alpha_1 e_1' + \ldots + \alpha_n e_n',$$

which proves our theorem. ■

The previous result motivates the following definition.

3.5.23. Definition. The basis $\{e_1', e_2', \ldots, e_n'\}$ of X^f in Theorem 3.5.22 is called the **dual basis** of $\{e_1, e_2, \ldots, e_n\}$.

We are now in a position to consider the algebraic transpose of a linear transformation. Let S be a linear transformation of a linear space X into a linear space Y and let X^f and Y^f denote the algebraic conjugates of X and Y, respectively (the spaces X and Y need not be finite dimensional). For each $y' \in Y^f$ let us establish a correspondence with an element $x' \in X^f$ according to the rule

$$x'(x) = \langle x, x' \rangle = \langle Sx, y' \rangle = y'(Sx), \tag{3.5.24}$$

where $x \in X$. Let us denote the mapping defined in this way by S^T: $S^T y' = x'$ and let us rewrite Eq. (3.5.24) as

$$\langle x, S^T y' \rangle = \langle Sx, y' \rangle, \quad x \in X, y' \in Y^f, \tag{3.5.25}$$

to define S^T. It should be noted that if S is a mapping of X into Y, then S^T is a mapping of Y^f into X^f, as depicted in Figure K. We now state the following formal definition.

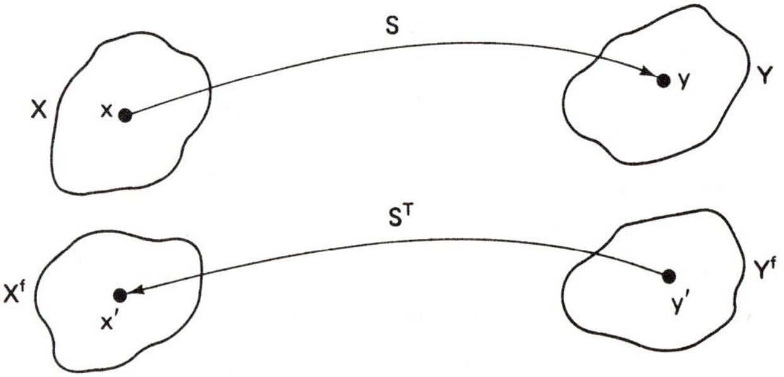

3.5.26. Figure K. Transpose of a linear transformation.

3.5.27. Definition. Let S be a linear transformation of a linear space X into a linear space Y over the same field F and let X^f and Y^f denote the algebraic conjugates of X and Y, respectively. A transformation S^T from Y^f into X^f such that

$$\langle x, S^T y' \rangle = \langle Sx, y' \rangle$$

for all $x \in X$ and all $y' \in Y^f$ is called the (algebraic) **transpose of** S.

We now show that S^T is a linear transformation.

3.5.28. Theorem. Let $S \in L(X, Y)$, and let S^T be the transpose of S. Then S^T is a linear transformation from Y^f into X^f.

Proof. Let $\alpha \in F$, and let $y'_1, y'_2 \in Y^f$. Then for all $x \in X$,

$$\langle x, S^T(y'_1 + y'_2) \rangle = \langle Sx, (y'_1 + y'_2) \rangle = \langle Sx, y'_1 \rangle + \langle Sx, y'_2 \rangle$$
$$= \langle x, S^T y'_1 \rangle + \langle x, S^T y'_2 \rangle.$$

Thus, $S^T(y'_1 + y'_2) = S^T(y'_1) + S^T(y'_2)$. Also,

$$\langle x, S^T(\alpha y'_1) \rangle = \langle Sx, \alpha y'_1 \rangle = \alpha \langle Sx, y'_1 \rangle$$
$$= \alpha \langle x, S^T y'_1 \rangle = \langle x, \alpha S^T y'_1 \rangle.$$

Hence, $S^T(\alpha y'_1) = \alpha S^T(y'_1)$. Therefore, $S^T \in L(Y^f, X^f)$. ■

The reader should have now no difficulties in proving the following results.

3.5.29. Theorem. Let $R, S \in L(X, Y)$, and let $T \in L(Y, Z)$. Let R^T, S^T, and T^T be the transpose transformations of R, S, and T, respectively. Then,

(i) $(R + S)^T = R^T + S^T$; and
(ii) $(TS)^T = S^T T^T$.

3.5.30. Theorem. Let I denote the identity element of $L(X, X)$. Then I^T is the identity element of $L(X^f, X^f)$.

3.5.31. Theorem. Let 0 be the null transformation in $L(X, Y)$. Then 0^T is the null transformation in $L(Y^f, X^f)$.

3.5.32. Exercise. Prove Theorems 3.5.29–3.5.31.

We will consider an important class of transpose linear transformations in Chapter 4 (transpose of a matrix).

3.6. BILINEAR FUNCTIONALS

In the present section we introduce the notion of bilinear functional and examine some of the properties of this concept. Throughout the present section we concern ourselves only with real vector spaces or complex vector spaces. Thus, if X is a linear space over a field F, it will be assumed that F is either the field of real numbers, R, or the field of complex numbers, C.

3.6.1. Definition. Let X be a vector space over C. A mapping g from X into C is said to be a **conjugate functional** if

$$g(\alpha x + \beta y) = \bar{\alpha} g(x) + \bar{\beta} g(y) \tag{3.6.2}$$

for all $x, y \in X$ and for all $\alpha, \beta \in C$, where $\bar{\alpha}$ denotes the complex conjugate of α and $\bar{\beta}$ denotes the complex conjugate of β.

If in Definition 3.6.1 the complex vector space is replaced by a real linear space, then the concept of conjugate functional reduces to that of linear functional, for in this case Eq. (3.6.2) assumes the form

$$g(\alpha x + \beta y) = \alpha g(x) + \beta g(y) \tag{3.6.3}$$

for all $x, y \in X$ and for all $\alpha, \beta \in R$.

3.6.4. Definition. Let X be a vector space over C. A mapping g of $X \times X$ into C is called a **bilinear functional** or a **bilinear form** if

(i) for each fixed y, $g(x, y)$ is a linear functional in x; and
(ii) for each fixed x, $g(x, y)$ is a conjugate functional in y.

Thus, if g is a bilinear functional, then

(a) $g(\alpha x + \beta y, z) = \alpha g(x, z) + \beta g(y, z)$; and
(b) $g(x, \alpha y + \beta z) = \bar{\alpha} g(x, y) + \bar{\beta} g(x, z)$

for all $x, y, z \in X$ and for all $\alpha, \beta \in C$.

For the case of real linear spaces the definition of bilinear functional is modified in an obvious way by deleting in Definition 3.6.4 the symbol for complex conjugates.

We leave it as an exercise to verify that the examples cited below are bilinear functionals.

3.6.5. Example. Let $x, y \in C^2$, where C^2 denotes the linear space of ordered pairs of complex numbers (if $x, y \in C^2$, then $x = (\xi_1, \xi_2)$ and $y = (\eta_1, \eta_2)$). The function

$$g(x, y) = \xi_1 \bar{\eta}_1 + \xi_2 \bar{\eta}_2$$

is a bilinear functional. ■

3.6.6. Example. Let $x, y \in R^2$, where R^2 denotes the linear space of ordered pairs of real numbers (if $x, y \in R^2$, then $x = (\xi_1, \xi_2)$ and $y = (\eta_1, \eta_2)$). Let θ denote the angle between $x, y \in R^2$. The **dot product** of two vectors, defined by

$$g(x, y) = \xi_1\eta_1 + \xi_2\eta_2 = (\xi_1^2 + \xi_2^2)^{1/2}(\eta_1^2 + \eta_2^2)^{1/2} \cos\theta$$

is a bilinear functional. ■

3.6.7. Example. Let X be an arbitrary linear space over C, and let $L(x)$ and $P(y)$ denote two linear functionals on X. The transformation

$$g(x, y) = L(x)\overline{P(y)}$$

is a bilinear functional. ■

3.6.8. Example. Let X be any linear space over C, and let g be a bilinear functional. The transformation h defined by

$$h(x, y) = \overline{g(x, y)}$$

is a bilinear functional. ■

3.6.9. Exercise. Verify that the transformations given in Examples 3.6.5 through 3.6.8 are bilinear functionals.

We note that for any bilinear functional, g, we have $g(0, y) = g(0 \cdot 0, y) = 0 \cdot g(0, y) = 0$ for all $y \in X$. Also, $g(x, 0) = 0$ for all $x \in X$.

Frequently, we find it convenient to impose certain restrictions on bilinear functionals.

3.6.10. Definition. Let X be a complex linear space. A bilinear functional g is said to be **symmetric** if $g(x, y) = \overline{g(y, x)}$ for all $x, y \in X$. If $g(x, x) \geq 0$ for all $x \in X$, then g is said to be **positive**. If $g(x, x) > 0$ for all $x \neq 0$, then g is said to be **strictly positive**.

3.6.11. Definition. Let X be a complex vector space, and let g be a bilinear functional. We call the function $\hat{g}: X \longrightarrow C$ defined by

$$\hat{g}(x) = g(x, x)$$

for all $x \in X$, the **quadratic form induced by** g (we frequently omit the phrase "induced by g").

For example, if $g(x, y) = \xi_1\bar{\eta}_1 + \xi_2\bar{\eta}_2$, as in Example 3.6.5, then $\hat{g}(x) = \xi_1\bar{\xi}_1 + \xi_2\bar{\xi}_2 = |\xi_1|^2 + |\xi_2|^2$. This is a quadratic form as studied in analytic geometry.

For real linear spaces, Definitions 3.6.10 and 3.6.11 are again modified in an obvious way by ignoring complex conjugates.

3.6.12. Theorem. If $\hat{g}$ is the quadratic form induced by a bilinear functional

g, then

$$\frac{1}{2}[g(x, y) + g(y, x)] = \hat{g}\left(\frac{x + y}{2}\right) - \hat{g}\left(\frac{x - y}{2}\right).$$

Proof. By direct expansion we have,

$$\hat{g}\left(\frac{x + y}{2}\right) = g\left(\frac{x + y}{2}, \frac{x + y}{2}\right) = \frac{1}{4}g(x + y, x + y)$$

$$= \frac{1}{4}[g(x, x + y) + g(y, x + y)]$$

$$= \frac{1}{4}[g(x, x) + g(x, y) + g(y, x) + g(y, y)],$$

and also,

$$\hat{g}\left(\frac{x - y}{2}\right) = \frac{1}{4}[g(x, x) - g(x, y) - g(y, x) + g(y, y)].$$

Thus,

$$\frac{1}{2}[g(x, y) + g(y, x)] = \hat{g}\left(\frac{x + y}{2}\right) - \hat{g}\left(\frac{x - y}{2}\right). \quad \blacksquare$$

Our next result is commonly referred to as **polarization**.

3.6.13. Theorem. If $\hat{g}$ is the quadratic form induced by a bilinear form g on a complex vector space X, then

$$g(x, y) = \hat{g}[\tfrac{1}{2}(x + y)] - \hat{g}[\tfrac{1}{2}(x - y)] + i\hat{g}[\tfrac{1}{2}(x + iy)] - i\hat{g}[\tfrac{1}{2}(x - iy)] \qquad (3.6.14)$$

for every $x, y \in X$ (here $i = \sqrt{-1}$).

Proof. From the proof of the last theorem we have

$$\hat{g}\left(\frac{x + y}{2}\right) = \frac{1}{4}[g(x, x) + g(x, y) + g(y, x) + g(y, y)]$$

and

$$\hat{g}\left(\frac{x - y}{2}\right) = \frac{1}{4}[g(x, x) - g(x, y) - g(y, x) + g(y, y)].$$

Also,

$$i\hat{g}\left(\frac{x + iy}{2}\right) = \frac{i}{4}[g(x, x) - ig(x, y) + ig(y, x) + g(y, y)]$$

and

$$i\hat{g}\left(\frac{x - iy}{2}\right) = \frac{i}{4}[g(x, x) + ig(x, y) - ig(y, x) + g(y, y)].$$

After combining the above four expressions, Eq. (3.6.14) results. $\blacksquare$

The reader can prove the next result readily.

3.6.15. Theorem. Let X be a complex vector space. If two bilinear functionals g and h are such that $\hat{g} = \hat{h}$, then $g = h$.

3.6.16. Exercise. Prove Theorem 3.6.15.

For symmetric bilinear functionals we have:

3.6.17. Theorem. A bilinear functional g on a complex vector space X is symmetric if and only if $\hat{g}$ is real (i.e., $\hat{g}(x)$ is real for all $x \in X$).

Proof. Suppose that g is symmetric; i.e., suppose that

$$g(x, y) = \overline{g(y, x)}$$

for all $x, y \in X$. Setting $x = y$, we obtain

$$\hat{g}(x) = g(x, x) = \overline{g(x, x)} = \overline{\hat{g}(x)}$$

for all $x \in X$. But this implies that $\hat{g}$ is real.

Conversely, if $\hat{g}(x)$ is real for all $x \in X$, then for $h(x, y) = \overline{g(y, x)}$ we have $\hat{h}(x) = \overline{g(x, x)} = g(x, x) = \hat{g}(x)$. Since $\hat{h} = \hat{g}$, it now follows from Theorem 3.6.15 that $h = g$, and thus

$$g(x, y) = \overline{g(y, x)}. \quad \blacksquare$$

Note that Theorems 3.6.13, 3.6.15, and 3.6.17 hold only for complex vector spaces. Theorem 3.6.15 implies that a bilinear form is uniquely determined by its induced quadratic form, and Theorem 3.6.13 gives an explicit connection between g and $\hat{g}$. In the case of real spaces, these conclusions do not follow.

3.6.18. Example. Let $X = R^2$ with $x = (\xi_1, \xi_2) \in R^2$ and $y = (\eta_1, \eta_2) \in R^2$. Define the bilinear functionals g and h by

$$g(x, y) = \xi_1\eta_1 + 2\xi_2\eta_1 + 4\xi_1\eta_2 + \xi_2\eta_2$$

and

$$h(x, y) = \xi_1\eta_1 + 3\xi_2\eta_1 + 3\xi_1\eta_2 + \xi_2\eta_2.$$

Then $\hat{g}(x) = \hat{h}(x)$, but $g \neq h$. Note that h is symmetric whereas g is not. $\blacksquare$

Using bilinear functionals, we now introduce the very important concept of inner product.

3.6.19. Definition. A strictly positive, symmetric bilinear functional g on a complex linear space X is called an **inner product**.

For the case of real linear spaces, the definition of inner product is identical to the above definition.

Since in a given discussion the particular bilinear functional g is always specified, we will write (x, y) in place of $g(x, y)$ to denote an inner product. Utilizing this notation, the inner product can alternatively be defined as a rule which assigns a scalar (x, y) to every $x, y \in X$ (X is a complex vector space), having the following properties:

(i) $(x, x) > 0$ for all $x \neq 0$ and $(x, x) = 0$ if $x = 0$;
(ii) $(x, y) = \overline{(y, x)}$ for all $x, y \in X$;
(iii) $(\alpha x + \beta y, z) = \alpha(x, z) + \beta(y, z)$ for all $x, y, z \in X$ and for all $\alpha, \beta \in C$; and
(iv) $(x, \alpha y + \beta z) = \bar{\alpha}(x, y) + \bar{\beta}(x, z)$ for all $x, y, z \in X$ and for all $\alpha, \beta \in C$.

In the case of real linear spaces, the preceding characterization of inner product is identical, except, of course, that we omit conjugates in (i)–(iv).

We are now in a position to introduce the concept of inner product space.

3.6.20. Definition. A complex (real) linear space X on which a complex (real) inner product, $(\cdot, \cdot)$, is defined is called a complex (real) **inner product space**. In general, we denote this space by $\{X; (\cdot, \cdot)\}$. If the particular inner product is understood, we simply write X to denote such a space (and we usually speak of an inner product space rather than a complex or real inner product space).

It should be noted that if two different inner products are defined on the same linear space X, say $(\cdot, \cdot)_1$ and $(\cdot, \cdot)_2$, then we have two different inner product spaces, namely, $\{X; (\cdot, \cdot)_1\}$ and $\{X; (\cdot, \cdot)_2\}$.

Now let $\{X; (\cdot, \cdot)'\}$ be an inner product space, let Y be a linear subspace of X, and let $(\cdot, \cdot)''$ denote the inner product on Y induced by the inner product on X; i.e.,

$$(x, y)' = (x, y)'' \tag{3.6.21}$$

for all $x, y \in Y \subset X$. Then $\{Y; (\cdot, \cdot)''\}$ is an inner product space in its own right, and we say that Y is an **inner product subspace** of X.

Using the concept of inner product, we are in a position to introduce the notion of orthogonality. We have:

3.6.22. Definition. Let X be an inner product space. The vectors $x, y \in X$ are said to be **orthogonal** if $(x, y) = 0$. In this case we write $x \perp y$. If a vector $x \in X$ is orthogonal to every vector of a set $Y \subset X$, then x is said to be orthogonal to set Y, and we write $x \perp Y$. If every vector of set $Y \subset X$ is orthogonal to every vector of set $Z \subset X$, then set Y is said to be orthogonal to set Z, and we write $Y \perp Z$.

Clearly, if x is orthogonal to y, then y is orthogonal to x. Note that if $x \neq 0$, then it is not possible that $x \perp x$, because $(x, x) > 0$ for all $x \neq 0$. Also note that $0 \perp x$ for all $x \in X$.

Before closing the present section, let us consider a few specific examples.

3.6.23. Example. Let $X = R^n$. For $x = (\xi_1, \ldots, \xi_n) \in R^n$ and $y = (\eta_1, \ldots, \eta_n) \in R^n$, we can readily verify that

$$(x, y) = \sum_{i=1}^{n} \xi_i \eta_i$$

is an inner product, and $\{X; (\cdot, \cdot)\}$ is a real inner product space. ■

3.6.24. Example. Let $X = C^n$. For $x = (\xi_1, \ldots, \xi_n) \in C^n$ and $y = (\eta_1, \ldots, \eta_n) \in C^n$, let

$$(x, y) = \sum_{i=1}^{n} \xi_i \bar{\eta}_i.$$

Then (x, y) is an inner product and $\{X; (\cdot, \cdot)\}$ is a complex inner product space. ■

3.6.25. Example. Let X denote the space of continuous complex valued functions on the interval [0, 1]. The reader can readily show that for $f, g \in X$,

$$(f, g) = \int_0^1 f(t)\overline{g(t)}\, dt$$

is an inner product. Now consider the family of functions $\{f_n\}$ defined by

$$f_n(t) = e^{2\pi n t i}, \quad t \in [0, 1],$$

$n = 0, \pm 1, \pm 2, \ldots$. Clearly, $f_n \in X$ for all n. It is easily shown that $(f_m, f_n) = 0$ if $m \neq n$. Thus, $f_m \perp f_n$ if $m \neq n$. ■

3.7. PROJECTIONS

In the present section we consider another special class of linear transformations, called **projections**. Such transformations which utilize direct sums (introduced in Section 3.2) as their natural setting will find wide applications in later parts of this book.

3.7.1. Definition. Let X be the direct sum of linear spaces X_1 and X_2; i.e., let $X = X_1 \oplus X_2$. Let $x = x_1 + x_2$ be the unique representation of $x \in X$, where $x_1 \in X_1$ and $x_2 \in X_2$. We say that the **projection on X_1 along X_2** is the transformation defined by

$$P(x) = x_1.$$

Referring to Figure L, we note that elements in the plane X can uniquely be represented as $x = x_1 + x_2$, where $x_1 \in X_1$ and $x_2 \in X_2$ (X_1 and X_2 are one-dimensional linear spaces represented by the indicated lines intersecting at the origin 0). In this case, a projection P can be defined as that

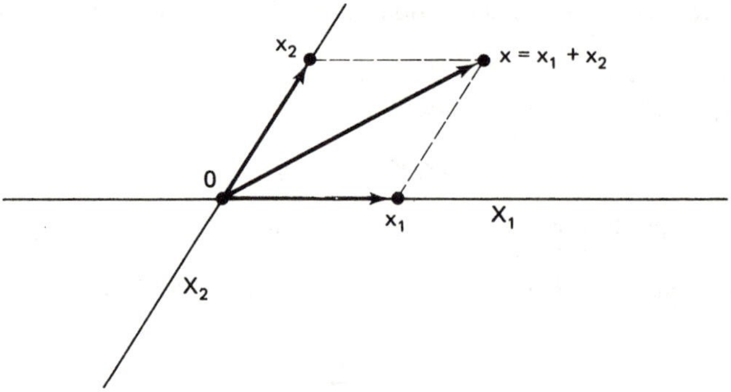

3.7.2. **Figure L.** Projection on X_1 along X_2.

transformation which maps every point x in the plane X onto the subspace X_1 along the subspace X_2.

3.7.3. Theorem. Let X be the direct sum of two linear subspaces X_1 and X_2, and let P be the projection on X_1 along X_2. Then

(i) $P \in L(X, X)$;
(ii) $\mathcal{R}(P) = X_1$; and
(iii) $\mathfrak{N}(P) = X_2$.

Proof. To prove the first part, note that if $x = x_1 + x_2$ and $y = y_1 + y_2$, where $x_1, y_1 \in X_1$ and $x_2, y_2 \in X_2$, then clearly

$$\begin{aligned} P(\alpha x + \beta y) &= P(\alpha x_1 + \alpha x_2 + \beta y_1 + \beta y_2) = \alpha x_1 + \beta y_1 \\ &= \alpha P(x_1) + \beta P(y_1) = \alpha P(x_1 + x_2) + \beta P(y_1 + y_2) \\ &= \alpha P(x) + \beta P(y), \end{aligned}$$

and therefore P is a linear transformation.

To prove the second part of the theorem, we note that from the definition of P it follows that $\mathcal{R}(P) \subset X_1$. Now assume that $x_1 \in X_1$. Then $Px_1 = x_1$, and thus $x_1 \in \mathcal{R}(P)$. This implies that $X_1 \subset \mathcal{R}(P)$ and proves that $\mathcal{R}(P) = X_1$.

To prove the last part of the theorem, let $x_2 \in X_2$. Then $Px_2 = 0$ so that $X_2 \subset \mathfrak{N}(P)$. On the other hand, if $x \in \mathfrak{N}(P)$, then $Px = 0$. Since $x = x_1 + x_2$, where $x_1 \in X_1$ and $x_2 \in X_2$, it follows that $x_1 = 0$ and $x \in X_2$. Thus, $X_2 \supset \mathfrak{N}(P)$. Therefore, $X_2 = \mathfrak{N}(P)$. ■

Our next result enables us to characterize projections in an alternative way.

3.7.4. Theorem. Let $P \in L(X, X)$. Then P is a projection on $\mathcal{R}(P)$ along $\mathfrak{N}(P)$ if and only if $PP = P^2 = P$.

Proof. Assume that P is the projection on the linear subspace X_1 of X along the linear subspace X_2, where $X = X_1 \oplus X_2$. By the preceding theorem, $X_1 = \mathcal{R}(P)$ and $X_2 = \mathfrak{N}(P)$. For $x \in X$, we have $x = x_1 + x_2$, where $x_1 \in X_1$ and $x_2 \in X_2$. Then

$$P^2x = P(Px) = Px_1 = x_1 = Px,$$

and thus $P^2 = P$.

Conversely, let us assume that $P^2 = P$. Let $X_2 = \mathfrak{N}(P)$ and let $X_1 = \mathcal{R}(P)$. Clearly, $\mathfrak{N}(P)$ and $\mathcal{R}(P)$ are linear subspaces of X. We must show that $X = \mathcal{R}(P) \oplus \mathfrak{N}(P) = X_1 \oplus X_2$. In particular, we must show that $\mathcal{R}(P) \cap \mathfrak{N}(P) = \{0\}$ and that $\mathcal{R}(P)$ and $\mathfrak{N}(P)$ span X.

Now if $y \in \mathcal{R}(P)$ there exists an $x \in X$ such that $Px = y$. Thus, $P^2x = Py = Px = y$. If $y \in \mathfrak{N}(P)$ then $Py = 0$. Thus, if y is in both $\mathcal{R}(P)$ and $\mathfrak{N}(P)$, then we must have $y = 0$; i.e., $\mathcal{R}(P) \cap \mathfrak{N}(P) = \{0\}$.

Next, let x be an arbitrary element in X. Then we have

$$x = Px + (I - P)x.$$

Letting $Px = x_1$ and $(I - P)x = x_2$, we have $Px_1 = P^2x = Px = x_1$ and also $Px_2 = P(I - P)x = Px - P^2x = Px - Px = 0$; i.e., $x_1 \in X_1$ and $x_2 \in X_2$. From this it follows that $X = X_1 \oplus X_2$ and that the projection on X_1 along X_2 is P. ■

The preceding result gives rise to the following:

3.7.5. Definition. Let $P \in L(X, X)$. Then P is said to be **idempotent** if $P^2 = P$.

Now let P be the projection on a linear subspace X_1 along a linear subspace X_2. Then the projection on X_2 along X_1 is characterized in the following way.

3.7.6. Theorem. A linear transformation P is a projection on a linear subspace if and only if $(I - P)$ is a projection. If P is the projection on X_1 along X_2, then $(I - P)$ is the projection on X_2 along X_1.

3.7.7. Exercise. Prove Theorem 3.7.6.

In view of the preceding results there is no ambiguity in simply saying a transformation P is a projection (rather than P is a projection on X_1 along X_2).

We emphasize here that if P is a projection, then

$$X = \mathcal{R}(P) \oplus \mathfrak{N}(P). \tag{3.7.8}$$

This is not necessarily the case for arbitrary linear transformations $T \in L(X, X)$ for, in general, $\mathcal{R}(T)$ and $\mathfrak{N}(T)$ need not be disjoint. For example, if there exists a vector $x \in X$ such that $Tx \neq 0$ and such that $T^2x = 0$, then $Tx \in \mathcal{R}(T)$ and $Tx \in \mathfrak{N}(T)$.

Let us now consider:

3.7.9. Definition. Let $T \in L(X, X)$. A linear subspace Y of a vector space X is said to be **invariant under the linear transformation** T if $y \in Y$ implies that $Ty \in Y$.

Note that this definition does *not* imply that every element in Y can be written in the form $z = Ty$, with $y \in Y$. It is not even assumed that $Ty \in Y$ implies $y \in Y$.

For invariant subspaces under a transformation $T \in L(X, X)$ we can readily prove the following result.

3.7.10. Theorem. Let $T \in L(X, X)$. Then

(i) X is an invariant subspace under T;
(ii) $\{0\}$ is an invariant subspace under T;
(iii) $\mathcal{R}(T)$ is an invariant subspace under T; and
(iv) $\mathfrak{N}(T)$ is an invariant subspace under T.

3.7.11. Exercise. Prove Theorem 3.7.10.

Next we consider:

3.7.12. Definition. Let X be a linear space which is the direct sum of two linear subspaces Y and Z; i.e., $X = Y \oplus Z$. If Y and Z are both invariant under a linear transformation T, then T is said to be **reduced** by Y and Z.

We are now in a position to prove the following result.

3.7.13. Theorem. Let Y and Z be two linear subspaces of a vector space X such that $X = Y \oplus Z$. Let $T \in L(X, X)$. Then T is reduced by Y and Z if and only if $PT = TP$, where P is the projection on Y along Z.

Proof. Assume that $PT = TP$. If $y \in Y$, then $Ty = TPy = PTy$ so that $Ty \in Y$ and Y is invariant under T. Now let $y \in Z$. Then $Py = 0$ and $PTy = TPy = T0 = 0$. Thus, $Ty \in Z$ and Z is also invariant under T. Hence, T is reduced by Y and Z.

Conversely, let us assume that T is reduced by Y and Z. If $x \in X$, then $x = y + z$, where $y \in Y$ and $z \in Z$. Then $Px = y$ and $TPx = Ty \in Y$. Hence, $PTPx = Ty = TPx$; i.e.,

$$PTPx = TPx \tag{3.7.14}$$

for all $x \in X$. On the other hand, since Y and Z are invariant under T, we have $Tx = Ty + Tz$ with $Ty \in Y$ and $Tz \in Z$. Hence, $PTx = Ty = PTy = PTPx$; i.e.,

$$PTPx = PTx \tag{3.7.15}$$

for all $x \in X$. Equations (3.7.14) and (3.7.15) imply that $PT = TP$. ■

We close the present section by considering the following special type of projection.

3.7.16. Definition. A projection P on an inner product space X is said to be an **orthogonal projection** if the range of P and the null space of P are orthogonal; i.e., if $\mathcal{R}(P) \perp \mathfrak{N}(P)$.

We will consider examples and additional properties of projections in much greater detail in Chapters 4 and 7.

3.8. NOTES AND REFERENCES

The material of the present chapter as well as that of the next chapter is usually referred to as linear algebra. Thus, these two chapters should be viewed as one package. For this reason, applications (dealing with ordinary differential equations) are presented at the end of the next chapter.

There are many textbooks and reference works dealing with vector spaces and linear transformations. Some of these which we have found to be very useful are cited in the references for this chapter. The reader should consult these for further study.

REFERENCES

[3.1] P. R. Halmos, *Finite Dimensional Vector Spaces*. Princeton, N.J.: D. Van Nostrand Company, Inc., 1958.

[3.2] K. Hoffman and R. Kunze, *Linear Algebra*. Englewood Cliffs, N.J.: Prentice-Hall, Inc., 1971.

[3.3] A. W. Naylor and G. R. Sell, *Linear Operator Theory in Engineering and Science*. New York: Holt, Rinehart and Winston, 1971.

[3.4] A. E. Taylor, *Introduction to Functional Analysis*. New York: John Wiley & Sons, Inc., 1966.

4

FINITE-DIMENSIONAL VECTOR SPACES AND MATRICES

In the present chapter we examine some of the properties of finite-dimensional linear spaces. We will show how elements of such spaces are represented by coordinate vectors and how linear transformations on such spaces are represented by means of matrices. We then will study some of the important properties of matrices. Also, we will investigate in some detail a special type of vector space, called the Euclidean space. This space is one of the most important spaces encountered in applied mathematics.

Throughout this chapter $\{\alpha_1, \ldots, \alpha_n\}$, $\alpha_i \in F$, and $\{x_1, \ldots, x_n\}$, $x_i \in X$, denote an indexed set of scalars and an indexed set of vectors, respectively.

4.1. COORDINATE REPRESENTATION OF VECTORS

Let X be a finite-dimensional linear space over a field F, and let $\{x_1, \ldots, x_n\}$ be a basis for X. Now if $x \in X$, then according to Theorem 3.3.25 and Definition 3.3.36, there exist *unique* scalars $\xi_1, \ldots, \xi_n$, called **coordinates** of x with respect to this basis such that

$$x = \xi_1 x_1 + \ldots + \xi_n x_n. \tag{4.1.1}$$

This enables us to represent x unambiguously in terms of its coordinates as

$$\mathbf{x} = \begin{bmatrix} \xi_1 \\ \cdot \\ \cdot \\ \cdot \\ \xi_n \end{bmatrix} \tag{4.1.2}$$

or as

$$\mathbf{x}^T = (\xi_1, \ldots, \xi_n). \tag{4.1.3}$$

We call $\mathbf{x}$ (or $\mathbf{x}^T$) the **coordinate representation** of the **underlying object** (vector) x with respect to the basis $\{x_1, \ldots, x_n\}$. We call $\mathbf{x}$ a **column vector** and $\mathbf{x}^T$ a **row vector**. Also, we say that $\mathbf{x}^T$ is the **transpose vector**, or simply the **transpose** of the vector $\mathbf{x}$. Furthermore, we define $(\mathbf{x}^T)^T$ to be $\mathbf{x}$.

It is important to note that in the coordinate representation (4.1.2) or (4.1.3) of the vector (4.1.1), an "ordering" of the basis $\{x_1, \ldots, x_n\}$ is employed (i.e., the coefficient ξ_i of x_i is the ith entry in Eqs. (4.1.2) and (4.1.3)). If the members of this basis were to be relabeled, thus specifying a different "ordering," then the corresponding coordinate representation of the vector x would have to be altered, to reflect this change. However, this does not pose any difficulties, because in a given discussion we will always agree on a particular "ordering" of the basis vectors.

Now let $\alpha \in F$. Then

$$\alpha x = \alpha(\xi_1 x_1 + \ldots + \xi_n x_n) = (\alpha\xi_1)x_1 + \ldots + (\alpha\xi_n)x_n. \tag{4.1.4}$$

In view of Eqs. (4.1.1)–(4.1.4) it now follows that the coordinate representation of αx with respect to the basis $\{x_1, \ldots, x_n\}$ is given by

$$\alpha\mathbf{x} = \alpha\begin{bmatrix} \xi_1 \\ \xi_2 \\ \cdot \\ \cdot \\ \cdot \\ \xi_n \end{bmatrix} = \begin{bmatrix} \alpha\xi_1 \\ \alpha\xi_2 \\ \cdot \\ \cdot \\ \cdot \\ \alpha\xi_n \end{bmatrix} \tag{4.1.5}$$

or

$$\alpha\mathbf{x}^T = \alpha(\xi_1, \xi_2, \ldots, \xi_n) = (\alpha\xi_1, \alpha\xi_2, \ldots, \alpha\xi_n). \tag{4.1.6}$$

Next, let $y \in X$, where

$$y = \eta_1 x_1 + \eta_2 x_2 + \ldots + \eta_n x_n. \tag{4.1.7}$$

The coordinate representation of y with respect to the basis $\{x_1, \ldots, x_n\}$ is, of course,

$$\mathbf{y} = \begin{bmatrix} \eta_1 \\ \eta_2 \\ \cdot \\ \cdot \\ \cdot \\ \eta_n \end{bmatrix} \tag{4.1.8}$$

or

$$\mathbf{y}^T = (\eta_1, \ldots, \eta_n). \tag{4.1.9}$$

Now

$$\begin{aligned} x + y &= (\xi_1 x_1 + \ldots + \xi_n x_n) + (\eta_1 x_1 + \ldots + \eta_n x_n) \\ &= (\xi_1 + \eta_1)x_1 + \ldots + (\xi_n + \eta_n)x_n. \end{aligned} \tag{4.1.10}$$

From Eq. (4.1.10) it now follows that the coordinate representation of the vector $x + y \in X$ with respect to the basis $\{x_1, \ldots, x_n\}$ is given by

$$\mathbf{x} + \mathbf{y} = \begin{bmatrix} \xi_1 \\ \cdot \\ \cdot \\ \cdot \\ \xi_n \end{bmatrix} + \begin{bmatrix} \eta_1 \\ \cdot \\ \cdot \\ \cdot \\ \eta_n \end{bmatrix} = \begin{bmatrix} \xi_1 + \eta_1 \\ \cdot \\ \cdot \\ \cdot \\ \xi_n + \eta_n \end{bmatrix} \tag{4.1.11}$$

or

$$\begin{aligned} \mathbf{x}^T + \mathbf{y}^T &= (\xi_1, \ldots, \xi_n) + (\eta_1, \ldots, \eta_n) \\ &= (\xi_1 + \eta_1, \ldots, \xi_n + \eta_n). \end{aligned} \tag{4.1.12}$$

Next, let $\{u_1, \ldots, u_n\}$ and $\{v_1, \ldots, v_n\}$ be two different bases for the linear space X. Then clearly there exist two different but unique sets of scalars (i.e., coordinates) $\{\alpha_1, \ldots, \alpha_n\}$ and $\{\beta_1, \ldots, \beta_n\}$ such that

$$x = \alpha_1 u_1 + \ldots + \alpha_n u_n = \beta_1 v_1 + \ldots + \beta_n v_n. \tag{4.1.13}$$

This enables us to represent the same vector $x \in X$ with respect to two different bases in terms of two different but unique sets of coordinates, namely,

$$\begin{bmatrix} \alpha_1 \\ \cdot \\ \cdot \\ \cdot \\ \alpha_n \end{bmatrix} \quad \text{and} \quad \begin{bmatrix} \beta_1 \\ \cdot \\ \cdot \\ \cdot \\ \beta_n \end{bmatrix}. \tag{4.1.14}$$

The next two examples are intended to throw additional light on the above discussion.

4.1.15. Example. Let $x = (\xi_1, \ldots, \xi_n) \in R^n$. Let $u_1 = (1, 0, \ldots, 0)$, $u_2 = (0, 1, 0, \ldots, 0), \ldots, u_n = (0, \ldots, 0, 1)$. It is readily shown that the set $\{u_1, \ldots, u_n\}$ is a basis for R^n. We call this basis the **natural basis** for R^n. Noting that

$$x = \xi_1 u_1 + \ldots + \xi_n u_n, \tag{4.1.16}$$

the unambiguous coordinate representation of $x \in R^n$ with respect to the natural basis of R^n is

$$\mathbf{x} = \begin{bmatrix} \xi_1 \\ \cdot \\ \cdot \\ \cdot \\ \xi_n \end{bmatrix} \tag{4.1.17}$$

or

$$\mathbf{x}^T = (\xi_1, \ldots, \xi_n).$$

Moreover, the coordinate representations of the basis vectors $u_1, \ldots, u_n$ are

$$\mathbf{u}_1 = \begin{bmatrix} 1 \\ 0 \\ \cdot \\ \cdot \\ \cdot \\ 0 \end{bmatrix}, \quad \mathbf{u}_2 = \begin{bmatrix} 0 \\ 1 \\ 0 \\ \cdot \\ \cdot \\ 0 \end{bmatrix}, \quad \ldots, \quad \mathbf{u}_n = \begin{bmatrix} 0 \\ \cdot \\ \cdot \\ \cdot \\ 0 \\ 1 \end{bmatrix}, \tag{4.1.18}$$

respectively. We call the coordinates in Eq. (4.1.17) the **natural coordinates** of $x \in R^n$. (The natural basis for F^n and the natural coordinates of $x \in F^n$ are similarly defined.)

Next, consider the set of vectors $\{v_1, \ldots, v_n\}$, given by $v_1 = (1, 0, \ldots, 0)$, $v_2 = (1, 1, 0, \ldots, 0), \ldots, v_n = (1, \ldots, 1)$. We see that the vectors $\{v_1, \ldots, v_n\}$ form a basis for R^n. We can express the vector x given in Eq. (4.1.16) in terms of this basis by

$$x = \alpha_1 v_1 + \ldots + \alpha_n v_n, \tag{4.1.19}$$

where $\alpha_n = \xi_n$ and $\alpha_i = \xi_i - \xi_{i+1}$ for $i = 1, 2, \ldots, n-1$. Thus, the coordinate representation of x relative to $\{v_1, \ldots, v_n\}$ is given by

$$\begin{bmatrix} \alpha_1 \\ \alpha_2 \\ \cdot \\ \cdot \\ \cdot \\ \alpha_{n-1} \\ \alpha_n \end{bmatrix} = \begin{bmatrix} \xi_1 - \xi_2 \\ \xi_2 - \xi_3 \\ \cdot \\ \cdot \\ \cdot \\ \xi_{n-1} - \xi_n \\ \xi_n \end{bmatrix}. \tag{4.1.20}$$

Hence, we have represented the same vector $x \in R^n$ by two different coordinate vectors with respect to two different bases for R^n. ■

4.1.21. Example. Let $X = \mathcal{C}[a, b]$, the set of all real-valued continuous functions on the interval $[a, b]$. Let $Y = \{x_0, x_1, \ldots, x_n\} \subset X$, where $x_0(t) = 1$ and $x_i(t) = t^i$ for all $t \in [a, b]$, $i = 1, \ldots, n$. As we saw in Exercise 3.3.13, Y is a linearly independent set in X and as such it is a basis for $V(Y)$.

Hence, for any $y \in V(Y)$ there exists a unique set of scalars $\{\eta_0, \eta_1, \ldots, \eta_n\}$ such that

$$y = \eta_0 x_0 + \ldots + \eta_n x_n. \tag{4.1.22}$$

Since y is a polynomial in t we can write, more explicitly,

$$y(t) = \eta_0 + \eta_1 t + \ldots + \eta_n t^n, \quad t \in [a, b]. \tag{4.1.23}$$

In the present example there is also a coordinate representation; i.e., we can represent $y \in V(Y)$ by

$$\begin{bmatrix} \eta_0 \\ \eta_1 \\ \cdot \\ \cdot \\ \cdot \\ \eta_n \end{bmatrix}. \tag{4.1.24}$$

This representation is with respect to the basis $\{x_0, x_1, \ldots, x_n\}$ in $V(Y)$. We could, of course, also have used another basis fo- $V(Y)$. For example, let us choose the basis $\{z_0, z_1, \ldots, z_n\}$ for $V(Y)$ given in Exercise 3.3.13. Then we have

$$y = \alpha_0 z_0 + \alpha_1 z_1 + \ldots + \alpha_n z_n, \tag{4.1.25}$$

where $\alpha_n = \eta_n$ and $\alpha_i = \eta_i - \eta_{i+1}$, $i = 0, 1, \ldots, n-1$. Thus, $y \in V(Y)$ may also be represented with respect to the basis $\{z_0, z_1, \ldots, z_n\}$ by

$$\begin{bmatrix} \alpha_0 \\ \alpha_1 \\ \cdot \\ \cdot \\ \cdot \\ \alpha_{n-1} \\ \alpha_n \end{bmatrix} = \begin{bmatrix} \eta_0 - \eta_1 \\ \eta_1 - \eta_2 \\ \cdot \\ \cdot \\ \cdot \\ \eta_{n-1} - \eta_n \\ \eta_n \end{bmatrix}. \tag{4.1.26}$$

Thus, two different coordinate vectors were used above in representing the same vector $y \in V(Y)$ with respect to two different bases for $V(Y)$. ■

Summarizing, we observe:

1. Every vector x belonging to an n-dimensional linear space X over a field F can be represented in terms of a coordinate vector $\mathbf{x}$, or its transpose $\mathbf{x}^T$, with respect to a given basis $\{e_1, \ldots, e_n\} \subset X$. We note that $\mathbf{x}^T \in F^n$ (the space F^n is defined in Example 3.1.7). By *convention* we will henceforth also write $\mathbf{x} \in F^n$. To indicate the coordinate representation of $x \in X$ by $\mathbf{x} \in F^n$, we write $x \leftrightarrow \mathbf{x}$.
2. In representing x by $\mathbf{x}$, an "ordering" of the basis $\{e_1, \ldots, e_n\} \subset X$ is implied.

3. Usage of different bases for X results in different coordinate representations of $x \in X$.

4.2. MATRICES

In this section we will first concern ourselves with the representation of linear transformations on finite-dimensional vector spaces. Such representations of linear transformations are called **matrices**. We will then examine the properties of matrices in great detail. *Throughout the present section X will denote an n-dimensional vector space and Y an m-dimensional vector space over the same field F.*

A. Representation of Linear Transformations by Matrices

We first prove the following result.

4.2.1. Theorem. Let $\{e_1, e_2, \ldots, e_n\}$ be a basis for a linear space X.

(i) Let A be a linear transformation from X into vector space Y and set $e'_1 = Ae_1, e'_2 = Ae_2, \ldots, e'_n = Ae_n$. If x is any vector in X and if $(\xi_1, \xi_2, \ldots, \xi_n)$ are the coordinates of x with respect to $\{e_1, e_2, \ldots, e_n\}$, then $Ax = \xi_1 e'_1 + \xi_2 e'_2 + \ldots + \xi_n e'_n$.

(ii) Let $\{e'_1, e'_2, \ldots, e'_n\}$ be any set of vectors in Y. Then there exists a unique linear transformation A from X into Y such that $Ae_1 = e'_1$, $Ae_2 = e'_2, \ldots, Ae_n = e'_n$.

Proof. To prove (i) we note that

$$\begin{aligned} Ax &= A(\xi_1 e_1 + \xi_2 e_2 + \ldots + \xi_n e_n) = \xi_1 Ae_1 + \xi_2 Ae_2 + \ldots + \xi_n Ae_n \\ &= \xi_1 e'_1 + \xi_2 e'_2 + \ldots + \xi_n e'_n. \end{aligned}$$

To prove (ii), we first observe that for each $x \in X$ we have unique scalars $\xi_1, \xi_2, \ldots, \xi_n$ such that

$$x = \xi_1 e_1 + \xi_2 e_2 + \ldots + \xi_n e_n.$$

Now define a mapping A from X into Y as

$$A(x) = \xi_1 e'_1 + \xi_2 e'_2 + \ldots + \xi_n e'_n.$$

Clearly, $A(e_i) = e'_i$ for $i = 1, \ldots, n$. We first must show that A is linear. Given $x = \xi_1 e_1 + \xi_2 e_2 + \ldots + \xi_n e_n$ and $y = \eta_1 e_1 + \eta_2 e_2 + \ldots + \eta_n e_n$, we have

$$\begin{aligned} A(x + y) &= A[(\xi_1 + \eta_1)e_1 + \ldots + (\xi_n + \eta_n)e_n] \\ &= (\xi_1 + \eta_1)e'_1 + \ldots + (\xi_n + \eta_n)e'_n. \end{aligned}$$

On the other hand,

$$A(x) = \xi_1 e'_1 + \xi_2 e'_2 + \ldots + \xi_n e'_n$$

and

$$A(y) = \eta_1 e'_1 + \eta_2 e'_2 + \ldots + \eta_n e'_n.$$

Thus,

$$\begin{aligned} A(x) + A(y) &= \xi_1 e'_1 + \xi_2 e'_2 + \ldots + \xi_n e'_n + \eta_1 e'_1 + \eta_2 e'_2 + \ldots + \eta_n e'_n \\ &= (\xi_1 + \eta_1)e'_1 + (\xi_2 + \eta_2)e'_2 + \ldots + (\xi_n + \eta_n)e'_n \\ &= A(x + y). \end{aligned}$$

In an identical way we establish that

$$\alpha A(x) = A(\alpha x)$$

for all $x \in X$ and all $\alpha \in F$. It thus follows that $A \in L(X, Y)$.

To show that A is unique, suppose there exists a $B \in L(X, Y)$ such that $Be_i = e'_i$ for $i = 1, \ldots, n$. It follows that $(A - B)e_i = 0$ for all $i = 1, \ldots, n$, and thus it follows from Exercise 3.4.46 that $A = B$. ■

We point out that part (i) of Theorem 4.2.1 implies that *a linear transformation is completely determined by knowing how it transforms the basis vectors in its domain*, and part (ii) of Theorem 4.2.1 states that this linear transformation is uniquely determined in this way. We will utilize these facts in the following.

Now let X be an n-dimensional vector space, and let $\{e_1, e_2, \ldots, e_n\}$ be a basis for X. Let Y be an m-dimensional vector space, and let $\{f_1, f_2, \ldots, f_m\}$ be a basis for Y. Let $A \in L(X, Y)$, and let $e'_i = Ae_i$ for $i = 1, \ldots, n$. Since $\{f_1, f_2, \ldots, f_m\}$ is a basis for Y, there are unique scalars $\{a_{ij}\}$, $i = 1, \ldots, m$, $j = 1, \ldots, n$, such that

$$\begin{aligned} Ae_1 &= e'_1 = a_{11} f_1 + a_{21} f_2 + \ldots + a_{m1} f_m \\ Ae_2 &= e'_2 = a_{12} f_1 + a_{22} f_2 + \ldots + a_{m2} f_m \\ &\cdots\cdots\cdots\cdots\cdots\cdots\cdots\cdots \\ Ae_n &= e'_n = a_{1n} f_1 + a_{2n} f_2 + \ldots + a_{mn} f_m. \end{aligned} \tag{4.2.2}$$

Now let $x \in X$. Then x has the unique representation

$$x = \xi_1 e_1 + \xi_2 e_2 + \ldots + \xi_n e_n$$

with respect to the basis $\{e_1, \ldots, e_n\}$. In view of part (i) of Theorem 4.2.1 we have

$$Ax = \xi_1 e'_1 + \ldots + \xi_n e'_n. \tag{4.2.3}$$

Since $Ax \in Y$, Ax has a unique representation with respect to the basis $\{f_1, f_2, \ldots, f_m\}$, say,

$$Ax = \eta_1 f_1 + \eta_2 f_2 + \ldots + \eta_m f_m. \tag{4.2.4}$$

Combining Equations (4.2.2) and (4.2.3), we have

$$\begin{aligned} Ax = {} & \xi_1(a_{11}f_1 + \ldots + a_{m1}f_m) \\ & + \xi_2(a_{12}f_1 + \ldots + a_{m2}f_m) \\ & + \cdots\cdots\cdots\cdots\cdots\cdots \\ & + \xi_n(a_{1n}f_1 + \ldots + a_{mn}f_m). \end{aligned}$$

Rearranging the last expression we have

$$\begin{aligned} Ax = {} & (a_{11}\xi_1 + a_{12}\xi_2 + \ldots + a_{1n}\xi_n)f_1 \\ & + (a_{21}\xi_1 + a_{22}\xi_2 + \ldots + a_{2n}\xi_n)f_2 \\ & \cdots\cdots\cdots\cdots\cdots\cdots\cdots\cdots \\ & + (a_{m1}\xi_1 + a_{m2}\xi_2 + \ldots + a_{mn}\xi_n)f_m. \end{aligned}$$

However, in view of the uniqueness of the representation in Eq. (4.2.4) we have

$$\begin{aligned} \eta_1 &= a_{11}\xi_1 + a_{12}\xi_2 + \ldots + a_{1n}\xi_n, \\ \eta_2 &= a_{21}\xi_1 + a_{22}\xi_2 + \ldots + a_{2n}\xi_n, \\ & \cdots\cdots\cdots\cdots\cdots\cdots\cdots\cdots \\ \eta_m &= a_{m1}\xi_1 + a_{m2}\xi_2 + \ldots + a_{mn}\xi_n. \end{aligned} \tag{4.2.5}$$

This set of equations enables us to represent the linear transformation A from linear space X into linear space Y by the unique scalars $\{a_{ij}\}$, $i = 1, \ldots, m, j = 1, \ldots, n$. For convenience we let

$$\mathbf{A} = [a_{ij}] = \begin{bmatrix} a_{11} & a_{12} & \cdots & a_{1n} \\ a_{21} & a_{22} & \cdots & a_{2n} \\ \cdots & \cdots & \cdots & \cdots \\ a_{m1} & a_{m2} & \cdots & a_{mn} \end{bmatrix}. \tag{4.2.6}$$

We see that once the bases $\{e_1, e_2, \ldots, e_n\}$, $\{f_1, f_2, \ldots, f_m\}$ are fixed, we can represent the linear transformation A by the array of scalars in Eq. (4.2.6) which are uniquely determined by Eq. (4.2.2).

In view of part (ii) of Theorem 4.2.1, the converse to the preceding also holds. Specifically, with the bases for X and Y still fixed, the array given in Eq. (4.2.6) is uniquely associated with the linear transformation A of X into Y.

The above discussion justifies the following important definition.

4.2.7. Definition. The array given in Eq. (4.2.6) is called the **matrix A of the linear transformation** A from linear space X into linear space Y with respect to the basis $\{e_1, \ldots, e_n\}$ of X and the basis $\{f_1, \ldots, f_m\}$ of Y.

If, in Definition 4.2.7, $X = Y$, and if for both X and Y the same basis $\{e_1, \ldots, e_n\}$ is used, then we simply speak of the matrix **A** of the linear transformation A with respect to the basis $\{e_1, \ldots, e_n\}$.

In Eq. (4.2.6), the scalars $(a_{i1}, a_{i2}, \ldots, a_{in})$ form the ith **row** of **A** and the scalars $(a_{1j}, a_{2j}, \ldots, a_{mj})$ form the jth **column** of $\mathbf{A}$. The scalar a_{ij} refers to that element of matrix $\mathbf{A}$ which can be found in the ith row and jth column of $\mathbf{A}$. The array in Eq. (4.2.6) is said to be an $(m \times n)$ **matrix**. If $m = n$, we speak of a **square matrix** (i.e., an $(n \times n)$ matrix).

In accordance with our discussion of Section 4.1, an $(n \times 1)$ matrix is called a **column vector, column matrix**, or **n-vector**, and a $(1 \times n)$ matrix is called a **row vector**.

We say that two $(m \times n)$ matrices $\mathbf{A} = [a_{ij}]$ and $\mathbf{B} = [b_{ij}]$ are **equal** if and only if $a_{ij} = b_{ij}$ for all $i = 1, \ldots, m$ and for all $j = 1, \ldots, n$.

From the preceding discussion it should be clear that the same linear transformation A from linear space X into linear space Y may be represented by different matrices, depending on the particular choice of bases in X and Y. Since it is always clear from context which particular bases are being used, we usually don't refer to them explicitly, thus avoiding cumbersome notation.

Now let A^T denote the transpose of $A \in L(X, Y)$ (refer to Definition 3.5.27). Our next result provides the matrix representation of A^T.

4.2.8. Theorem. Let $A \in L(X, Y)$ and let $\mathbf{A}$ denote the matrix of A with respect to the bases $\{e_1, \ldots, e_n\}$ in X and $\{f_1, \ldots, f_m\}$ in Y. Let X^f and Y^f be the algebraic conjugates of X and Y, respectively. Let $A^T \in L(Y^f, X^f)$ be the transpose of A. Let $\{f'_1, \ldots, f'_m\}$ and $\{e'_1, \ldots, e'_n\}$, denote the dual bases of $\{f_1, \ldots, f_m\}$ and $\{e_1, \ldots, e_n\}$, respectively. If the matrix $\mathbf{A}$ is given by Eq. (4.2.6), then the matrix of A^T with respect to $\{f'_1, \ldots, f'_m\}$ of Y^f and $\{e'_1, \ldots, e'_n\}$ of X^f is given by

$$\mathbf{A}^T = \begin{bmatrix} a_{11} & a_{21} & \cdots & a_{m1} \\ a_{12} & a_{22} & \cdots & a_{m2} \\ \cdots & \cdots & \cdots & \cdots \\ a_{1n} & a_{2n} & \cdots & a_{mn} \end{bmatrix}. \tag{4.2.9}$$

Proof. Let $\mathbf{B} = [b_{ij}]$ denote the $(n \times m)$ matrix of the linear transformation A^T with respect to the bases $\{f'_1, \ldots, f'_m\}$ and $\{e'_1, \ldots, e'_n\}$. We want to show that $\mathbf{B}$ is the matrix in Eq. (4.2.9). By Eq. (4.2.2) we have

$$Ae_i = \sum_{j=1}^{m} a_{ji} f_j$$

for $i = 1, \ldots, n$, and

$$A^T f'_j = \sum_{k=1}^{n} b_{kj} e'_k$$

for $j = 1, \ldots, m$. By Theorem 3.5.22, $\langle e_i, e'_k \rangle = \delta_{ik}$ and $\langle f_k, f'_j \rangle = \delta_{kj}$. Therefore,

$$\langle Ae_i, f'_j \rangle = \left\langle \sum_{k=1}^{m} a_{ki} f_k, f'_j \right\rangle = \sum_{k=1}^{m} a_{ki} \langle f_k, f'_j \rangle = a_{ji}.$$

Also,

$$\langle Ae_i, f'_j \rangle = \langle e_i, A^T f'_j \rangle = \left\langle e_i, \sum_{k=1}^{n} b_{kj} e'_k \right\rangle$$
$$= \sum_{k=1}^{n} b_{kj} \langle e_i, e'_k \rangle = b_{ij}.$$

Therefore, $b_{ij} = a_{ji}$, which proves the theorem. ■

The preceding result gives rise to the following concept.

4.2.10. Definition. The matrix $\mathbf{A}^T$ in Eq. (4.2.9) is called the **transpose** of matrix $\mathbf{A}$.

Our next result follows trivially from the discussion leading up to Definition 4.2.7.

4.2.11. Theorem. Let A be a linear transformation of an n-dimensional vector space X into an m-dimensional vector space Y, and let $y = Ax$. Let the coordinates of x with respect to the basis $\{e_1, e_2, \ldots, e_n\}$ be $(\xi_1, \xi_2, \ldots, \xi_n)$, and let the coordinates of y with respect to the basis $\{f_1, f_2, \ldots, f_m\}$ be $(\eta_1, \eta_2, \ldots, \eta_m)$. Let

$$\mathbf{A} = [a_{ij}] = \begin{bmatrix} a_{11} & a_{12} & \cdots & a_{1n} \\ a_{21} & a_{22} & \cdots & a_{2n} \\ \vdots & \vdots & & \vdots \\ a_{m1} & a_{m2} & \cdots & a_{mn} \end{bmatrix} \tag{4.2.12}$$

be the matrix of A with respect to the bases $\{e_1, e_2, \ldots, e_n\}$ and $\{f_1, f_2, \ldots, f_m\}$. Then

$$\begin{aligned} a_{11}\xi_1 + a_{12}\xi_2 + \ldots + a_{1n}\xi_n &= \eta_1, \\ a_{21}\xi_1 + a_{22}\xi_2 + \ldots + a_{2n}\xi_n &= \eta_2, \\ \cdots\cdots\cdots\cdots\cdots\cdots &\cdots, \\ a_{m1}\xi_1 + a_{m2}\xi_2 + \ldots + a_{mn}\xi_n &= \eta_m, \end{aligned} \tag{4.2.13}$$

or, equivalently,

$$\eta_i = \sum_{j=1}^{n} a_{ij}\xi_j, \quad i = 1, \ldots, m. \tag{4.2.14}$$

4.2.15. Exercise. Prove Theorem 4.2.11.

Using matrix and vector notation, let us agree to express the system of linear equations given by Eq. (4.2.13) equivalently as

$$\begin{bmatrix} a_{11} & a_{12} & \cdots & a_{1n} \\ a_{21} & a_{22} & \cdots & a_{2n} \\ \vdots & \vdots & & \vdots \\ a_{m1} & a_{m2} & \cdots & a_{mn} \end{bmatrix} \begin{bmatrix} \xi_1 \\ \xi_2 \\ \vdots \\ \xi_n \end{bmatrix} = \begin{bmatrix} \eta_1 \\ \eta_2 \\ \vdots \\ \eta_m \end{bmatrix}, \tag{4.2.16}$$

or, more succinctly, as

$$\mathbf{A}\mathbf{x} = \mathbf{y}, \tag{4.2.17}$$

where $\mathbf{x}^T = (\xi_1, \xi_2, \ldots, \xi_n)$ and $\mathbf{y}^T = (\eta_1, \eta_2, \ldots, \eta_m)$.

In terms of $\mathbf{x}^T$, $\mathbf{y}^T$, and $\mathbf{A}^T$, let us agree to express Eq. (4.2.13) equivalently as

$$(\xi_1, \xi_2, \ldots, \xi_n) \begin{bmatrix} a_{11} & a_{21} & \cdots & a_{m1} \\ a_{12} & a_{22} & \cdots & a_{m2} \\ \vdots & \vdots & & \vdots \\ a_{1n} & a_{2n} & \cdots & a_{mn} \end{bmatrix} = (\eta_1, \eta_2, \ldots, \eta_m) \tag{4.2.18}$$

or, in short, as

$$\mathbf{x}^T\mathbf{A}^T = \mathbf{y}^T. \tag{4.2.19}$$

We note that in Eq. (4.2.17), $\mathbf{x} \in F^n$, $\mathbf{y} \in F^m$, and $\mathbf{A}$ is an $m \times n$ matrix.

From our discussion thus far it should be clear that we can utilize matrices to study systems of linear equations which are of the form of Eq. (4.2.13). It should also be clear that an $m \times n$ matrix $\mathbf{A}$ is nothing more than a unique representation of a linear transformation A of an n-dimensional vector space X into an m-dimensional vector space Y over the same field F. As such, $\mathbf{A}$ possesses all the properties of such transformations. We could, in fact, utilize matrices in place of general linear transformations to establish many facts concerning linear transformations defined on finite-dimensional linear spaces. However, since a given matrix is dependent upon the selection of two particular sets of bases (not necessarily distinct), such practice will, in general, be avoided whenever possible.

We emphasize that a matrix and a linear transformation are not one and the same thing. In many texts no distinction in symbols is made between linear transformations and their matrix representation. We will not follow this custom.

B. Rank of a Matrix

We begin by proving the following result.

4.2.20. Theorem. Let A be a linear transformation from X into Y. Then A has rank r if and only if it is possible to choose a basis $\{e_1, e_2, \ldots, e_n\}$

for X and a basis $\{f_1, \ldots, f_m\}$ for Y such that the matrix $\mathbf{A}$ of A with respect to these bases is of the form

$$\mathbf{A} = \underbrace{\begin{bmatrix} \overbrace{1 & 0 & 0 & \cdots & 0}^{r} & 0 & 0 & \cdots & 0 \\ 0 & 1 & 0 & \cdots & 0 & 0 & 0 & \cdots & 0 \\ \cdots & \cdots & \cdots & \cdots & \cdots & \cdots & \cdots & \cdots & \cdots \\ \cdots & \cdots & \cdots & \cdots & \cdots & \cdots & \cdots & \cdots & \cdots \\ 0 & 0 & 0 & \cdots & 1 & 0 & 0 & \cdots & 0 \\ 0 & 0 & 0 & \cdots & 0 & 0 & 0 & \cdots & 0 \\ \cdots & \cdots & \cdots & \cdots & \cdots & \cdots & \cdots & \cdots & \cdots \\ 0 & 0 & 0 & \cdots & 0 & 0 & 0 & \cdots & 0 \end{bmatrix}}_{n = \dim X} \Bigg\} m = \dim Y. \tag{4.2.21}$$

Proof. We choose a basis for X of the form $\{e_1, e_2, \ldots, e_r, e_{r+1}, \ldots, e_n\}$, where $\{e_{r+1}, \ldots, e_n\}$ is a basis for $\mathfrak{N}(A)$. If $f_1 = Ae_1, f_2 = Ae_2, \ldots, f_r = Ae_r$, then $\{f_1, f_2, \ldots, f_r\}$ is a basis for $\mathfrak{R}(A)$, as we saw in the proof of Theorem 3.4.25. Now choose vectors $f_{r+1}, \ldots, f_m$ in Y such that the set of vectors $\{f_1, f_2, \ldots, f_m\}$ forms a basis for Y (see Theorem 3.3.44). Then

$$\begin{aligned} f_1 &= Ae_1 = (1)f_1 + (0)f_2 + \ldots + (0)f_r + (0)f_{r+1} + \ldots + (0)f_m, \\ f_2 &= Ae_2 = (0)f_1 + (1)f_2 + \ldots + (0)f_r + (0)f_{r+1} + \ldots + (0)f_m, \\ &\cdots\cdots\cdots\cdots\cdots\cdots\cdots\cdots\cdots\cdots, \\ f_r &= Ae_r = (0)f_1 + (0)f_2 + \ldots + (1)f_r + (0)f_{r+1} + \ldots + (0)f_m, \\ 0 &= Ae_{r+1} = (0)f_1 + (0)f_2 + \ldots + (0)f_r + (0)f_{r+1} + \ldots + (0)f_m, \\ &\cdots\cdots\cdots\cdots\cdots\cdots\cdots\cdots\cdots\cdots, \\ 0 &= Ae_n = (0)f_1 + (0)f_2 + \ldots + (0)f_r + (0)f_{r+1} + \ldots + (0)f_m. \end{aligned} \tag{4.2.22}$$

The necessity is proven by applying Definition 4.2.7 (and also Eq. (4.2.2)) to the set of equations (4.2.22); the desired result given by Eq. (4.2.21) follows.

Sufficiency follows from the fact that the basis for $\mathfrak{R}(A)$ contains r linearly independent vectors. ■

A question of practical significance is the following: if $\mathbf{A}$ is the matrix of a linear transformation A from linear space X into linear space Y with respect to arbitrary bases $\{e_1, \ldots, e_n\}$ for X and $\{f_1, \ldots, f_m\}$ for Y, what is the rank of A in terms of matrix $\mathbf{A}$? Let $\mathfrak{R}(A)$ be the subspace of Y generated by $Ae_1, Ae_2, \ldots, Ae_n$. Then, in view of Eq. (4.2.2), the coordinate representation of Ae_i, $i = 1, \ldots, n$, in Y with respect to $\{f_1, \ldots, f_m\}$ is given by

$$Ae_1 \longleftrightarrow \begin{bmatrix} a_{11} \\ a_{21} \\ \cdot \\ \cdot \\ \cdot \\ a_{m1} \end{bmatrix}, \quad Ae_2 \longleftrightarrow \begin{bmatrix} a_{12} \\ a_{22} \\ \cdot \\ \cdot \\ \cdot \\ a_{m2} \end{bmatrix}, \quad \ldots, \quad Ae_n \longleftrightarrow \begin{bmatrix} a_{1n} \\ a_{2n} \\ \cdot \\ \cdot \\ \cdot \\ a_{mn} \end{bmatrix}.$$

From this it follows that $\mathcal{R}(A)$ consists of vectors y whose coordinate representation is

$$\mathbf{y} = \begin{bmatrix} \eta_1 \\ \eta_2 \\ \cdot \\ \cdot \\ \cdot \\ \eta_m \end{bmatrix} = \gamma_1 \begin{bmatrix} a_{11} \\ a_{21} \\ \cdot \\ \cdot \\ \cdot \\ a_{m1} \end{bmatrix} + \gamma_2 \begin{bmatrix} a_{12} \\ a_{22} \\ \cdot \\ \cdot \\ \cdot \\ a_{m2} \end{bmatrix} + \ldots + \gamma_n \begin{bmatrix} a_{1n} \\ a_{2n} \\ \cdot \\ \cdot \\ \cdot \\ a_{mn} \end{bmatrix}, \tag{4.2.23}$$

where $\gamma_1, \ldots, \gamma_n$ are scalars. Since every spanning or generating set of a linear space contains a basis, we are able to select from among the vectors Ae_1, $Ae_2, \ldots, Ae_n$ a basis for $\mathcal{R}(A)$. Suppose that the set $\{Ae_1, Ae_2, \ldots, Ae_k\}$ is this basis. Then the vectors $Ae_1, Ae_2, \ldots, Ae_k$ are linearly independent, and the vectors $Ae_{k+1}, \ldots, Ae_n$ are linear combinations of the vectors Ae_1, $Ae_2, \ldots, Ae_k$. From this there now follows:

4.2.24. Theorem. Let $A \in L(X, Y)$, and let $\mathbf{A}$ be the matrix of A with respect to the (arbitrary) basis $\{e_1, e_2, \ldots, e_n\}$ for X and with respect to the (arbitrary) basis $\{f_1, f_2, \ldots, f_m\}$ for Y. Let the coordinate representation of $y = Ax$ be $\mathbf{y} = \mathbf{A}\mathbf{x}$. Then

(i) the rank of A is the number of vectors in the largest possible linearly independent set of columns of $\mathbf{A}$; and

(ii) the rank of A is the number of vectors in the smallest possible set of columns of $\mathbf{A}$ which has the property that all columns not in it can be expressed as linear combinations of the columns in it.

In view of this result we make the following definition.

4.2.25. Definition. The **rank** of an $m \times n$ matrix $\mathbf{A}$ is the largest number of linearly independent columns of $\mathbf{A}$.

C. Properties of Matrices

Now let X be an n-dimensional linear space, let Y be an m-dimensional linear space, let F be the field for X and Y, and let A and B be linear transformations of X into Y. Let $\mathbf{A} = [a_{ij}]$ be the matrix of A, and let $\mathbf{B} = [b_{ij}]$ be the matrix of B with respect to the bases $\{e_1, e_2, \ldots, e_n\}$ in X and $\{f_1, f_2,$

$\ldots, f_m\}$ in Y. Using Eq. (3.4.42) as well as Definition 4.2.7, the reader can readily verify that the matrix of $A + B$, denoted by $\mathbf{C} \triangleq \mathbf{A} + \mathbf{B}$, is given by

$$\mathbf{A} + \mathbf{B} = [a_{ij}] + [b_{ij}] = [a_{ij} + b_{ij}] = [c_{ij}] = \mathbf{C}. \tag{4.2.26}$$

Using Eq. (3.4.43) and Definition 4.2.7, the reader can also easily show that the matrix of αA, denoted by $\mathbf{D} \triangleq \alpha\mathbf{A}$, is given by

$$\alpha\mathbf{A} = \alpha[a_{ij}] = [\alpha a_{ij}] = [d_{ij}] = \mathbf{D}. \tag{4.2.27}$$

From Eq. (4.2.26) we note that, in order to be able to add two matrices $\mathbf{A}$ and $\mathbf{B}$, they must have the same number of rows and columns. In this case we say that $\mathbf{A}$ and $\mathbf{B}$ are **comparable matrices**. Also, from Eq. (4.2.27) it is clear that if $\mathbf{A}$ is an $m \times n$ matrix, then so is $\alpha\mathbf{A}$.

Next, let Z be an r-dimensional vector space, let $A \in L(X, Y)$, and let $B \in L(Y, Z)$. Let $\mathbf{A}$ be the matrix of A with respect to the basis $\{e_1, e_2, \ldots, e_n\}$ in X and with respect to the basis $\{f_1, f_2, \ldots, f_m\}$ in Y. Let $\mathbf{B}$ be the matrix of B with respect to the basis $\{f_1, f_2, \ldots, f_m\}$ in Y and with respect to the basis $\{g_1, g_2, \ldots, g_r\}$ in Z. The product mapping BA as defined by Eq. (3.4.50) is a linear transformation of X into Z. We now ask: what is the matrix $\mathbf{C}$ of BA with respect to the bases $\{e_1, e_2, \ldots, e_n\}$ of X and $\{g_1, g_2, \ldots, g_r\}$ of Z? By definition of matrices $\mathbf{A}$ and $\mathbf{B}$ (see Eq. (4.2.2)), we have

$$Ae_k = \sum_{j=1}^{m} a_{jk} f_j, \quad k = 1, \ldots, n$$

and

$$Bf_j = \sum_{i=1}^{r} b_{ij} g_i, \quad j = 1, \ldots, m.$$

Now

$$\begin{aligned} BAe_k = B(Ae_k) &= B\left(\sum_{j=1}^{m} a_{jk} f_j\right) \\ &= \sum_{j=1}^{m} a_{jk} Bf_j = \sum_{j=1}^{m} a_{jk}\left(\sum_{i=1}^{r} b_{ij} g_i\right) \\ &= \sum_{i=1}^{r}\sum_{j=1}^{m} b_{ij}\, a_{jk} g_i, \end{aligned}$$

for $k = 1, \ldots, n$. Thus, the matrix $\mathbf{C}$ of BA with respect to basis $\{e_1, \ldots, e_n\}$ in X and $\{g_1, \ldots, g_r\}$ in Z is $[c_{ij}]$, where

$$c_{ij} = \sum_{k=1}^{m} b_{ik} a_{kj}, \tag{4.2.28}$$

for $i = 1, \ldots, r$ and $j = 1, \ldots, n$. We write this as

$$\mathbf{C} = \mathbf{BA}. \tag{4.2.29}$$

From the preceding discussion it is clear that two matrices $\mathbf{A}$ and $\mathbf{B}$ can be multiplied to form the product $\mathbf{BA}$ if and only if the number of columns of $\mathbf{B}$ is equal to the number of rows of $\mathbf{A}$. In this case we say that the matrices $\mathbf{B}$ and $\mathbf{A}$ are **conformal matrices**.

In arriving at Equations (4.2.28) and (4.2.29) we established the result given below.

4.2.30. Theorem. Let $\mathbf{A}$ be the matrix of $A \in L(X, Y)$ with respect to the basis $\{e_1, e_2, \ldots, e_n\}$ in X and basis $\{f_1, f_2, \ldots, f_m\}$ in Y. Let $\mathbf{B}$ be the matrix of $B \in L(Y, Z)$ with respect to basis $\{f_1, f_2, \ldots, f_m\}$ in Y and basis $\{g_1, g_2, \ldots, g_r\}$ in Z. Then $\mathbf{BA}$ is the matrix of BA.

We now summarize the above discussion in the following definition.

4.2.31. Definition. Let $\mathbf{A} = [a_{ij}]$ and $\mathbf{B} = [b_{ij}]$ be two $m \times n$ matrices, let $\mathbf{C} = [c_{ij}]$ be an $n \times r$ matrix, and let $\alpha \in F$. Then

(i) the **sum** of $\mathbf{A}$ and $\mathbf{B}$ is the $m \times n$ matrix

$$\mathbf{D} = \mathbf{A} + \mathbf{B}$$

where

$$d_{ij} = a_{ij} + b_{ij}$$

for all $i = 1, \ldots, m$ and for all $j = 1, \ldots, n$;

(ii) the product of matrix $\mathbf{A}$ by scalar α is the $m \times n$ matrix

$$\mathbf{E} = \alpha\mathbf{A}$$

where

$$e_{ij} = \alpha a_{ij}$$

for all $i = 1, \ldots, m$ and for all $j = 1, \ldots, n$; and

(iii) the product of matrix $\mathbf{A}$ and matrix $\mathbf{C}$ is the $m \times r$ matrix

$$\mathbf{G} = \mathbf{AC},$$

where

$$g_{ij} = \sum_{k=1}^{n} a_{ik} c_{kj}$$

for each $i = 1, \ldots, m$ and for each $j = 1, \ldots, r$.

The properties of general linear transformations established in Section 3.4 hold, of course, in the case of their matrix representation. We summarize some of these in the remainder of the present section.

4.2.32. Theorem.

(i) Let $\mathbf{A}$ and $\mathbf{B}$ be $(m \times n)$ matrices, and let $\mathbf{C}$ be an $(n \times r)$ matrix. Then

$$(\mathbf{A} + \mathbf{B})\mathbf{C} = \mathbf{AC} + \mathbf{BC}. \tag{4.2.33}$$

(ii) Let $\mathbf{A}$ be an $(m \times n)$ matrix, and let $\mathbf{B}$ and $\mathbf{C}$ be $(n \times r)$ matrices. Then

$$\mathbf{A}(\mathbf{B} + \mathbf{C}) = \mathbf{AB} + \mathbf{AC}. \tag{4.2.34}$$

(iii) Let $\mathbf{A}$ be an $(m \times n)$ matrix, let $\mathbf{B}$ be an $(n \times r)$ matrix, and let $\mathbf{C}$ be an $(r \times s)$ matrix. Then

$$\mathbf{A}(\mathbf{BC}) = (\mathbf{AB})\mathbf{C}. \tag{4.2.35}$$

(iv) Let $\alpha, \beta \in F$, and let $\mathbf{A}$ be an $(m \times n)$ matrix. Then

$$(\alpha + \beta)\mathbf{A} = \alpha\mathbf{A} + \beta\mathbf{A}. \tag{4.2.36}$$

(v) Let $\alpha \in F$, and let $\mathbf{A}$ and $\mathbf{B}$ be $(m \times n)$ matrices. Then

$$\alpha(\mathbf{A} + \mathbf{B}) = \alpha\mathbf{A} + \alpha\mathbf{B}. \tag{4.2.37}$$

(vi) Let $\alpha, \beta \in F$, let $\mathbf{A}$ be an $(m \times n)$ matrix, and let $\mathbf{B}$ be an $(n \times r)$ matrix. Then

$$(\alpha\mathbf{A})(\beta\mathbf{B}) = (\alpha\beta)(\mathbf{AB}). \tag{4.2.38}$$

(vii) Let $\mathbf{A}$ and $\mathbf{B}$ be $(m \times n)$ matrices. Then

$$\mathbf{A} + \mathbf{B} = \mathbf{B} + \mathbf{A}. \tag{4.2.39}$$

(viii) Let $\mathbf{A}$, $\mathbf{B}$, and $\mathbf{C}$ be $(m \times n)$ matrices. Then

$$(\mathbf{A} + \mathbf{B}) + \mathbf{C} = \mathbf{A} + (\mathbf{B} + \mathbf{C}). \tag{4.2.40}$$

The proofs of the next two results are left as an exercise.

4.2.41. Theorem. Let $0 \in L(X, Y)$ be the zero transformation defined by Eq. (3.4.44). Then for *any* bases $\{e_1, \ldots, e_n\}$ and $\{f_1, \ldots, f_m\}$ for X and Y, respectively, the linear transformation 0 is represented by the $(m \times n)$ matrix

$$\mathbf{0} = \begin{bmatrix} 0 & 0 & \cdots & 0 \\ 0 & 0 & \cdots & 0 \\ \cdots & \cdots & \cdots & \cdots \\ 0 & 0 & \cdots & 0 \end{bmatrix}. \tag{4.2.42}$$

The matrix $\mathbf{0}$ is called the **null matrix.**

4.2.43. Theorem. Let $I \in L(X, X)$ be the identity transformation defined by Eq. (3.4.56). Let $\{e_1, \ldots, e_n\}$ be an arbitrary basis for X. Then the matrix representation of the linear transformation I from X into X with respect to the basis $\{e_1, \ldots, e_n\}$ is given by

$$\mathbf{I} = \begin{bmatrix} 1 & 0 & 0 & \cdots & 0 \\ 0 & 1 & 0 & \cdots & 0 \\ \cdots & \cdots & \cdots & \cdots & \cdots \\ 0 & 0 & 0 & \cdots & 1 \end{bmatrix}. \tag{4.2.44}$$

$\mathbf{I}$ is called the $n \times n$ **identity matrix.**

4.2.45. Exercise. Prove Theorems 4.2.32, 4.2.41, and 4.2.43.

For any $(m \times n)$ matrix $\mathbf{A}$ we have

$$\mathbf{A} + \mathbf{0} = \mathbf{0} + \mathbf{A} = \mathbf{A} \tag{4.2.46}$$

and for any $(n \times n)$ matrix $\mathbf{B}$ we have

$$\mathbf{BI} = \mathbf{IB} = \mathbf{B} \tag{4.2.47}$$

where $\mathbf{I}$ is the $(n \times n)$ identity matrix.

If $\mathbf{A} = [a_{ij}]$ is a matrix of the linear transformation A, then correspondingly, $-\mathbf{A}$ is a matrix of the linear transformation $-A$, where

$$-\mathbf{A} = (-1)\mathbf{A} = (-1)\begin{bmatrix} a_{11} & a_{12} & \cdots & a_{1n} \\ a_{21} & a_{22} & \cdots & a_{2n} \\ \cdot & \cdot & & \cdot \\ \cdot & \cdot & & \cdot \\ \cdot & \cdot & & \cdot \\ a_{m1} & a_{m2} & \cdots & a_{mn} \end{bmatrix} \tag{4.2.48}$$

$$= \begin{bmatrix} -a_{11} & -a_{12} & \cdots & -a_{1n} \\ -a_{21} & -a_{22} & \cdots & -a_{2n} \\ \cdot & \cdot & & \cdot \\ \cdot & \cdot & & \cdot \\ \cdot & \cdot & & \cdot \\ -a_{m1} & -a_{m2} & \cdots & -a_{mn} \end{bmatrix}.$$

It follows immediately that $\mathbf{A} + (-\mathbf{A}) = \mathbf{0}$, where $\mathbf{0}$ denotes the null matrix. By convention we usually write $\mathbf{A} + (-\mathbf{A}) = \mathbf{A} - \mathbf{A}$.

Let $\mathbf{A}$ and $\mathbf{B}$ be $(n \times n)$ matrices. Then we have, in general,

$$\mathbf{AB} \neq \mathbf{BA}, \tag{4.2.49}$$

as was the case in Eq. (3.4.55).

Next, let $A \in L(X, X)$ and assume that A is non-singular. Let A^{-1} denote the inverse of A. Then, by Theorem 3.4.60, $AA^{-1} = A^{-1}A = I$. Now if $\mathbf{A}$ is the $(n \times n)$ matrix of A with respect to the basis $\{e_1, \ldots, e_n\}$ in X, then there is an $(n \times n)$ matrix $\mathbf{B}$ of A^{-1} with respect to the basis $\{e_1, \ldots, e_n\}$ in X, such that

$$\mathbf{BA} = \mathbf{AB} = \mathbf{I}. \tag{4.2.50}$$

We call $\mathbf{B}$ the **inverse** of $\mathbf{A}$ and we denote it by $\mathbf{A}^{-1}$. In this connection we use the following terms interchangeably: **$\mathbf{A}^{-1}$ exists, $\mathbf{A}$ has an inverse, $\mathbf{A}$ is invertible**, or **$\mathbf{A}$ is non-singular**. If $\mathbf{A}$ is not non-singular, we say **$\mathbf{A}$ is singular**.

With the aid of Theorem 3.4.63 the reader can readily establish the following result for matrices.

4.2.51. Theorem. Let $\mathbf{A}$ be an $(n \times n)$ matrix. The following are equivalent:

(i) rank $\mathbf{A} = n$;

(ii) $\mathbf{Ax} = \mathbf{0}$ implies $\mathbf{x} = \mathbf{0}$;

(iii) for every $\mathbf{y}_0 \in F^n$, there is a unique $\mathbf{x}_0 \in F^n$ such that $\mathbf{y}_0 = \mathbf{A}\mathbf{x}_0$;
(iv) the columns of $\mathbf{A}$ are linearly independent; and
(v) $\mathbf{A}^{-1}$ exists.

4.2.52. Exercise. Prove Theorem 4.2.51.

We have shown that we can represent n linear equations by the matrix equation (4.2.17). Now let $\mathbf{A}$ be a non-singular $(n \times n)$ matrix and consider the equation

$$\mathbf{y} = \mathbf{A}\mathbf{x}. \tag{4.2.53}$$

If we premultiply both sides of this equation by $\mathbf{A}^{-1}$ we obtain

$$\mathbf{x} = \mathbf{A}^{-1}\mathbf{y}, \tag{4.2.54}$$

the solution to Eq. (4.2.53). Thus, knowledge of the inverse of $\mathbf{A}$ enables us to solve the system of linear equations (4.2.53).

In our next result, which is readily verified, some of the important properties of non-singular matrices are given.

4.2.55. Theorem.

(i) An $(n \times n)$ non-singular matrix has one and only one inverse.
(ii) If $\mathbf{A}$ and $\mathbf{B}$ are non-singular $(n \times n)$ matrices, then $(\mathbf{AB})^{-1} = \mathbf{B}^{-1}\mathbf{A}^{-1}$.
(iii) If $\mathbf{A}$ and $\mathbf{B}$ are $(n \times n)$ matrices and if $\mathbf{AB}$ is non-singular, then so are $\mathbf{A}$ and $\mathbf{B}$.

4.2.56. Exercise. Prove Theorem 4.2.55.

Our next theorem summarizes some of the important properties of the transpose of matrices. The proof of this theorem is a direct consequence of the definition of the transpose of a matrix (see Eq. (4.2.9)).

4.2.57. Theorem.

(i) For any matrix $\mathbf{A}$, $(\mathbf{A}^T)^T = \mathbf{A}$.
(ii) Let $\mathbf{A}$ and $\mathbf{B}$ be conformal matrices. Then $(\mathbf{AB})^T = \mathbf{B}^T\mathbf{A}^T$.
(iii) Let $\mathbf{A}$ be a non-singular matrix. Then $(\mathbf{A}^T)^{-1} = (\mathbf{A}^{-1})^T$.
(iv) Let $\mathbf{A}$ be an $(n \times n)$ matrix. Then $\mathbf{A}^T$ is non-singular if and only if $\mathbf{A}$ is non-singular.
(v) Let $\mathbf{A}$ and $\mathbf{B}$ be comparable matrices. Then $(\mathbf{A} + \mathbf{B})^T = \mathbf{A}^T + \mathbf{B}^T$.
(vi) Let $\alpha \in F$ and $\mathbf{A}$ be a matrix. Then $(\alpha\mathbf{A})^T = \alpha\mathbf{A}^T$.

4.2.58. Exercise. Prove Theorem 4.2.57.

Now let $\mathbf{A}$ be an $(n \times n)$ matrix, and let m be a positive integer. Similarly as in Eq. (3.4.67) we define the $(n \times n)$ matrix $\mathbf{A}^m$ by

$$\mathbf{A}^m = \underbrace{\mathbf{A} \cdot \mathbf{A} \cdot \ldots \cdot \mathbf{A}}_{m \text{ times}}, \tag{4.2.59}$$

and if $\mathbf{A}^{-1}$ exists, then similarly as in Eq. (3.4.68), we define the $(n \times n)$ matrix $\mathbf{A}^{-m}$ as

$$\mathbf{A}^{-m} = (\mathbf{A}^{-1})^m = \underbrace{\mathbf{A}^{-1} \cdot \mathbf{A}^{-1} \cdot \ldots \cdot \mathbf{A}^{-1}}_{m \text{ times}}. \tag{4.2.60}$$

As in the case of Eqs. (3.4.69) through (3.4.71), the usual laws of exponents follow from the above definitions. Specifically, if $\mathbf{A}$ is an $(n \times n)$ matrix and if r and s are positive integers, then

$$\mathbf{A}^r \cdot \mathbf{A}^s = \mathbf{A}^{r+s} = \mathbf{A}^{s+r} = \mathbf{A}^s \cdot \mathbf{A}^r, \tag{4.2.61}$$

$$(\mathbf{A}^r)^s = \mathbf{A}^{rs} = \mathbf{A}^{sr} = (\mathbf{A}^s)^r, \tag{4.2.62}$$

and if $\mathbf{A}^{-1}$ exists, then

$$\mathbf{A}^r \cdot \mathbf{A}^{-s} = \mathbf{A}^{r-s}. \tag{4.2.63}$$

Consistent with the above notation we have

$$\mathbf{A}^1 = \mathbf{A} \tag{4.2.64}$$

and

$$\mathbf{A}^0 = \mathbf{I}. \tag{4.2.65}$$

We are now once more in a position to consider functions of linear transformations, where in the present case the linear transformations are represented by matrices For example, if $f(\lambda)$ is the polynomial in λ given in Eq. (3.4.74), and if $\mathbf{A}$ is any $(n \times n)$ matrix, then by $f(\mathbf{A})$ we mean

$$f(\mathbf{A}) = \alpha_0 \mathbf{I} + \alpha_1 \mathbf{A} + \ldots + \alpha_n \mathbf{A}^n. \tag{4.2.66}$$

4.2.67. Exercise. Let $A \in L(X, X)$, and let $\mathbf{A}$ be the matrix of A with respect to the basis $\{e_1, \ldots, e_n\}$ in X. Let $f(\lambda)$ be given by Eq. (3.4.74). Show that $f(\mathbf{A})$ is the matrix of $f(A)$ with respect to the basis $\{e_1, \ldots, e_n\}$.

We noted earlier that in general linear transformations and matrices do not commute (see (3.4.55) and (4.2.49)). However, in the case of square matrices, the reader can verify the following result easily.

4.2.68. Theorem. Let $\mathbf{A}, \mathbf{B}, \mathbf{C}$ denote $(n \times n)$ matrices, let $\mathbf{0}$ denote the $(n \times n)$ null matrix, and let $\mathbf{I}$ denote the $(n \times n)$ identity matrix. Then,

(i) $\mathbf{0}$ commutes with any $\mathbf{A}$;

(ii) $\mathbf{A}^p$ commutes with $\mathbf{A}^q$, where p and q are positive integers;

(iii) $\alpha\mathbf{I}$ commutes with any $\mathbf{A}$, where $\alpha \in F$; and

(iv) if $\mathbf{A}$ commutes with $\mathbf{B}$ and if $\mathbf{A}$ commutes with $\mathbf{C}$, then $\mathbf{A}$ commutes with $\alpha\mathbf{B} + \beta\mathbf{C}$, where $\alpha, \beta \in F$.

4.2.69. Exercise. Prove Theorem 4.2.68.

Let us now consider some specific examples.

4.2.70. Example. Let F denote the field of real numbers, and let

$$\mathbf{A} = \begin{bmatrix} 2 & 4 & 1 \\ 4 & 5 & 6 \\ 1 & 6 & 3 \\ 2 & 0 & 1 \end{bmatrix} \quad \text{and} \quad \mathbf{B} = \begin{bmatrix} 3 & 1 & 2 \\ 1 & 2 & 3 \\ 2 & 1 & 3 \\ 1 & 0 & 2 \end{bmatrix}.$$

Then

$$\mathbf{A} + \mathbf{B} = \begin{bmatrix} 5 & 5 & 3 \\ 5 & 7 & 9 \\ 3 & 7 & 6 \\ 3 & 0 & 3 \end{bmatrix} \quad \text{and} \quad \mathbf{A} - \mathbf{B} = \begin{bmatrix} -1 & 3 & -1 \\ 3 & 3 & 3 \\ -1 & 5 & 0 \\ 1 & 0 & -1 \end{bmatrix}.$$

If $\alpha = 3$, then

$$\alpha\mathbf{A} = \begin{bmatrix} 6 & 12 & 3 \\ 12 & 15 & 18 \\ 3 & 18 & 9 \\ 6 & 0 & 3 \end{bmatrix}. \quad \blacksquare$$

4.2.71. Example. Let F denote the field of complex numbers, let $i = \sqrt{-1}$, let

$$\mathbf{C} = \begin{bmatrix} 1 & -i & 3 + 2i \\ 8 & 7i & 6 \\ 1 + i & 3 & -5 \end{bmatrix} \quad \text{and} \quad \mathbf{D} = \begin{bmatrix} 1 & 3 & 2 \\ 4 & 7 & 5 \\ 6 & 2 & 1 \end{bmatrix}.$$

Then

$$\mathbf{C} + \mathbf{D} = \begin{bmatrix} 2 & 3 - i & 5 + 2i \\ 12 & 7 + 7i & 11 \\ 7 + i & 5 & -4 \end{bmatrix}.$$

If $\alpha = -i$, then

$$\alpha\mathbf{C} = \begin{bmatrix} -i & -1 & 2 - 3i \\ -8i & 7 & -6i \\ 1 - i & -3i & +5i \end{bmatrix}. \quad \blacksquare$$

4.2.72. Example. Let F denote the field of real numbers, let

$$\mathbf{G} = \begin{bmatrix} 1 & 2 \\ 3 & 4 \\ 5 & 0 \end{bmatrix} \quad \text{and} \quad \mathbf{H} = \begin{bmatrix} 2 & 3 \\ 4 & 1 \end{bmatrix}.$$

Then

$$\mathbf{GH} = \begin{bmatrix} 10 & 5 \\ 22 & 13 \\ 10 & 15 \end{bmatrix}.$$

Notice that in this case **HG** is *not* defined. ■

4.2.73. Example. Let F be the field of real numbers, let

$$\mathbf{K} = \begin{bmatrix} 1 & 2 \\ 3 & 4 \end{bmatrix} \quad \text{and} \quad \mathbf{L} = \begin{bmatrix} 2 & 3 \\ 4 & 1 \end{bmatrix}.$$

Then

$$\mathbf{KL} = \begin{bmatrix} 10 & 5 \\ 22 & 13 \end{bmatrix} \quad \text{and} \quad \mathbf{LK} = \begin{bmatrix} 11 & 16 \\ 7 & 12 \end{bmatrix}.$$

Clearly, $\mathbf{KL} \neq \mathbf{LK}$. ■

4.2.74. Example. Let

$$\mathbf{M} = \begin{bmatrix} 0 & 1 \\ 0 & 0 \end{bmatrix} \quad \text{and} \quad \mathbf{N} = \begin{bmatrix} 1 & 0 \\ 0 & 0 \end{bmatrix}.$$

Then

$$\mathbf{MN} = \begin{bmatrix} 0 & 0 \\ 0 & 0 \end{bmatrix} = \mathbf{0},$$

i.e., $\mathbf{MN} = \mathbf{0}$, even though $\mathbf{M} \neq \mathbf{0}$ and $\mathbf{N} \neq \mathbf{0}$. ■

4.2.75. Example. If **A** is as defined in Example 4.2.70, then

$$\mathbf{A}^T = \begin{bmatrix} 2 & 4 & 1 & 2 \\ 4 & 5 & 6 & 0 \\ 1 & 6 & 3 & 1 \end{bmatrix}. \quad ■$$

4.2.76. Example. Let

$$\mathbf{P} = \begin{bmatrix} 2 & 5 & 1 \\ 3 & -6 & 2 \\ 4 & 7 & 3 \end{bmatrix} \quad \text{and} \quad \mathbf{Q} = \begin{bmatrix} \frac{32}{24} & \frac{8}{24} & \frac{-16}{24} \\ \frac{1}{24} & \frac{-2}{24} & \frac{1}{24} \\ \frac{-45}{24} & \frac{-6}{24} & \frac{27}{24} \end{bmatrix}.$$

Then

$$\mathbf{P} \cdot \mathbf{Q} = \mathbf{Q} \cdot \mathbf{P} = \begin{bmatrix} 1 & 0 & 0 \\ 0 & 1 & 0 \\ 0 & 0 & 1 \end{bmatrix} = \mathbf{I},$$

i.e., $\mathbf{Q} = \mathbf{P}^{-1}$ or, equivalently, $\mathbf{P} = \mathbf{Q}^{-1}$. ■

4.2.77. Example. Consider the set of simultaneous linear equations

$$\begin{aligned} 4\xi_1 + 2\xi_2 + \xi_3 + 3\xi_4 &= 0, \\ 6\xi_1 + 3\xi_2 + \xi_3 + 4\xi_4 &= 0, \\ 2\xi_1 + \xi_2 + 0 \cdot \xi_3 + \xi_4 &= 0. \end{aligned} \tag{4.2.78}$$

Equation (4.2.78) can be rewritten as

$$\begin{bmatrix} 4 & 2 & 1 & 3 \\ 6 & 3 & 1 & 4 \\ 2 & 1 & 0 & 1 \end{bmatrix} \begin{bmatrix} \xi_1 \\ \xi_2 \\ \xi_3 \\ \xi_4 \end{bmatrix} = \begin{bmatrix} 0 \\ 0 \\ 0 \end{bmatrix}. \tag{4.2.79}$$

Let

$$\mathbf{A} = \begin{bmatrix} 4 & 2 & 1 & 3 \\ 6 & 3 & 1 & 4 \\ 2 & 1 & 0 & 1 \end{bmatrix}. \tag{4.2.80}$$

Matrix $\mathbf{A}$ is the coordinate representation of a linear transformation $A \in L(X, Y)$. In this case $\dim X = 4$ and $\dim Y = 3$. Observe now that the first column of $\mathbf{A}$ is a linear combination of the second column of $\mathbf{A}$. Also, by adding the third column of $\mathbf{A}$ to the second column we obtain the fourth column of $\mathbf{A}$. It follows that $\mathbf{A}$ has only two linearly independent columns. Hence, the rank of A is 2. Now since $\dim X = \dim \mathfrak{N}(A) + \dim \mathfrak{R}(A)$, the nullity of A is also 2. ■

Next, we discuss briefly partitioned vectors and matrices. Such vectors and matrices arise in a natural way when linear transformations acting on the direct sum of linear spaces are considered.

Let X be an n-dimensional vector space, and let Y be an m-dimensional vector space. Suppose that $X = U \oplus W$, where U is an r-dimensional linear subspace of X, and suppose that $Y = R \oplus Q$, where R is a p-dimensional linear subspace of Y. Let $A \in L(X, Y)$, let $\{e_1, \ldots, e_n\}$ be a basis for X such that $\{e_1, \ldots, e_r\}$ is a basis for U, and let $\{f_1, \ldots, f_m\}$ be a basis for Y such that $\{f_1, \ldots, f_p\}$ is a basis for R. Let $\mathbf{A}$ be the matrix of A with respect to these bases. Now if $\mathbf{x} \in F^n$ is the coordinate representation of $x \in X$ with respect to the basis $\{e_1, \ldots, e_n\}$, we can partition $\mathbf{x}$ into two components,

$$\mathbf{x} = \begin{bmatrix} \xi_1 \\ \cdot \\ \cdot \\ \cdot \\ \xi_r \\ \hdashline \xi_{r+1} \\ \cdot \\ \cdot \\ \cdot \\ \xi_n \end{bmatrix} = \begin{bmatrix} \mathbf{u} \\ \mathbf{w} \end{bmatrix}, \tag{4.2.81}$$

where $\mathbf{u} \in F^r$, $\mathbf{w} \in F^{(n-r)}$ and where

$$\mathbf{u} = \begin{bmatrix} \xi_1 \\ \cdot \\ \cdot \\ \cdot \\ \xi_r \end{bmatrix} \text{ and } \mathbf{w} = \begin{bmatrix} \xi_{r+1} \\ \cdot \\ \cdot \\ \cdot \\ \xi_n \end{bmatrix}.$$

Similarly, we can express $\mathbf{y} \in F^m$ as

$$\mathbf{y} = \begin{bmatrix} \eta_1 \\ \cdot \\ \cdot \\ \cdot \\ \eta_p \\ \hdashline \eta_{p+1} \\ \cdot \\ \cdot \\ \cdot \\ \eta_m \end{bmatrix} = \begin{bmatrix} \mathbf{r} \\ \mathbf{q} \end{bmatrix}, \tag{4.2.82}$$

where $\mathbf{y}$ is the coordinate representation of y with respect to $\{f_1, \ldots, f_m\}$ and where $\mathbf{r} \in F^p$ and $\mathbf{q} \in F^{m-p}$. We say the vector $\mathbf{x}$ in Eq. (4.2.81) is **partitioned** into components $\mathbf{u}$ and $\mathbf{w}$. Clearly, the vector $\mathbf{u}$ is determined by the coordinates of $\mathbf{x}$ corresponding to the basis vectors $\{e_1, \ldots, e_r\}$ in U.

We can similarly divide the matrix $\mathbf{A}$ into the partition

$$\mathbf{A} = \left[\begin{array}{c:c} \mathbf{A}_{11} & \mathbf{A}_{12} \\ \hdashline \mathbf{A}_{21} & \mathbf{A}_{22} \end{array}\right], \tag{4.2.83}$$

where $\mathbf{A}_{11}$ is a $(p \times r)$ matrix, $\mathbf{A}_{12}$ is a $(p \times (n-r))$ matrix, $\mathbf{A}_{21}$ is an $((m-p) \times r)$ matrix, and $\mathbf{A}_{22}$ is an $((m-p) \times (n-r))$ matrix. In this case, the equation

$$\mathbf{y} = \mathbf{A}\mathbf{x} \tag{4.2.84}$$

is equivalent to the pair of equations

$$\left.\begin{aligned} \mathbf{r} &= \mathbf{A}_{11}\mathbf{u} + \mathbf{A}_{12}\mathbf{w} \\ \mathbf{q} &= \mathbf{A}_{21}\mathbf{u} + \mathbf{A}_{22}\mathbf{w} \end{aligned}\right\}. \tag{4.2.85}$$

A matrix in the form of Eq. (4.2.83) is called a **partitioned matrix**. The matrices $\mathbf{A}_{11}$, $\mathbf{A}_{12}$, $\mathbf{A}_{21}$, and $\mathbf{A}_{22}$ are called **submatrices of A.**

The generalization of partitioning the matrix **A** into more than four submatrices is accomplished in an obvious way, when the linear space X and/or the linear space Y are the direct sum of more than two linear subspaces.

Now let the linear spaces X and Y and the linear transformation A and the matrix **A** of A still be defined as in the preceding discussion. Let Z be a k-dimensional vector space (the spaces X, Y, and Z are vector spaces over the same field F). Let $Z = M \oplus N$, where M is a j-dimensional linear subspace of Z. Let $B \in L(Y, Z)$. In a manner analogous to our preceding discussion, we represent B by the partitioned matrix

$$\mathbf{B} = \left[\begin{array}{c|c} \mathbf{B}_{11} & \mathbf{B}_{12} \\ \hline \mathbf{B}_{21} & \mathbf{B}_{22} \end{array}\right]. \tag{4.2.86}$$

It is now a simple matter to show that the linear transformation $BA \in L(X, Z)$ is represented by the partitioned matrix

$$\mathbf{BA} = \left[\begin{array}{c|c} \mathbf{B}_{11}\mathbf{A}_{11} + \mathbf{B}_{12}\mathbf{A}_{21} & \mathbf{B}_{11}\mathbf{A}_{12} + \mathbf{B}_{12}\mathbf{A}_{22} \\ \hline \mathbf{B}_{21}\mathbf{A}_{11} + \mathbf{B}_{22}\mathbf{A}_{21} & \mathbf{B}_{21}\mathbf{A}_{12} + \mathbf{B}_{22}\mathbf{A}_{22} \end{array}\right]. \tag{4.2.87}$$

We now prove:

4.2.88. Theorem. Let X be an n-dimensional vector space, and let $P \in L(X, X)$. If P is a projection, then there exists a basis $\{e_1, \ldots, e_n\}$ for X such that the matrix **P** of P with respect to this basis is of the form

$$\mathbf{P} = \left[\begin{array}{cccc|ccc} 1 & 0 & \cdots & 0 & & & \\ 0 & 1 & \cdots & 0 & & & \\ \vdots & \vdots & & \vdots & & \mathbf{0} & \\ 0 & 0 & \cdots & 1 & & & \\ \hline & & & & 0 & \cdots & 0 \\ & \mathbf{0} & & & \vdots & & \vdots \\ & & & & 0 & \cdots & 0 \end{array}\right] \begin{array}{l} \left.\vphantom{\begin{array}{c}1\\0\\\vdots\\0\end{array}}\right\} r \\ \left.\vphantom{\begin{array}{c}0\\\vdots\\0\end{array}}\right\} n-r \end{array} \tag{4.2.89}$$

where $r = \dim \mathcal{R}(P)$.

Proof. Since P is a projection we have, from Eq. (3.7.8),

$$X = \mathcal{R}(P) \oplus \mathfrak{N}(P).$$

Now let $r = \dim \mathcal{R}(P)$, and let $\{e_1, \ldots, e_n\}$ be a basis for X such that $\{e_1, \ldots, e_r\}$ is a basis for $\mathcal{R}(P)$. Let **P** be the matrix of P with respect to this basis, and the theorem follows. ■

We leave the next result as an exercise.

4.2.90. Theorem. Let X be a finite-dimensional vector space, and let $A \in L(X, X)$. If W is a p-dimensional invariant subspace of X and if $X = W \oplus Z$, then there exists a basis for X such that the matrix $\mathbf{A}$ of A with respect to this basis has the form

$$\mathbf{A} = \left[\begin{array}{c|c} \mathbf{A}_{11} & \mathbf{A}_{12} \\ \hline \mathbf{0} & \mathbf{A}_{22} \end{array}\right],$$

where $\mathbf{A}_{11}$ is a $(p \times p)$ matrix and the remaining submatrices are of appropriate dimension.

4.2.91. Exercise. Prove Theorem 4.2.90.

4.3. EQUIVALENCE AND SIMILARITY

From the previous section it is clear that a linear transformation A of a finite-dimensional vector space X into a finite-dimensional vector space Y can be represented by means of different matrices, depending on the particular choice of bases in X and Y. The choice of bases may in different cases result in matrices that are "easy" or "hard" to utilize. Many of the resulting "standard" forms of matrices, called **canonical forms**, arise because of practical considerations. Such canonical forms often exhibit inherent characteristics of the underlying transformation A. Before we can consider some of the more important canonical forms of matrices, we need to introduce several new concepts which are of great importance in their own right.

Throughout the present section X and Y are finite-dimensional vector spaces over the same field F, dim $X = n$ *and* dim $Y = m$. We begin our discussion with the following result.

4.3.1. Theorem. Let $\{e_1, \ldots, e_n\}$ be a basis for a linear space X, and let $\{e'_1, \ldots, e'_n\}$ be a set of vectors in X given by

$$e'_i = \sum_{j=1}^{n} p_{ji} e_j, \quad i = 1, \ldots, n, \tag{4.3.2}$$

where $p_{ij} \in F$ for all $i, j = 1, \ldots, n$. The set $\{e'_1, \ldots, e'_n\}$ forms a basis for X if and only if $\mathbf{P} = [p_{ij}]$ is non-singular.

Proof. Let $\{e'_1, \ldots, e'_n\}$ be linearly independent, and let $\mathbf{p}_j$ denote the jth column vector of $\mathbf{P}$. Let

$$\sum_{j=1}^{n} \alpha_j \mathbf{p}_j = \mathbf{0}$$

for some scalars $\alpha_1, \ldots, \alpha_n \in F$. This implies that

$$\sum_{j=1}^{n} \alpha_j p_{ij} = 0, \quad i = 1, \ldots, n.$$

It follows that

$$\sum_{i=1}^{n} \sum_{j=1}^{n} \alpha_j p_{ij} e_i = 0.$$

Rearranging, we have

$$\sum_{j=1}^{n} \alpha_j \left[\sum_{i=1}^{n} p_{ij} e_i \right] = 0$$

or

$$\sum_{j=1}^{n} \alpha_j e'_j = 0.$$

Since $e'_1, \ldots, e'_n$ are linearly independent, it follows that $\alpha_1 = \ldots = \alpha_n = 0$. Thus, the columns of $\mathbf{P}$ are linearly independent. Therefore, $\mathbf{P}$ is non-singular.

Conversely, let $\mathbf{P}$ be non-singular, i.e., let $\{\mathbf{p}_1, \ldots, \mathbf{p}_n\}$ be a linearly independent set of vectors in X. Let $\sum_{i=1}^{n} \alpha_i e'_i = 0$ for some scalars $\alpha_1, \ldots, \alpha_n \in F$. Then

$$0 = \sum_{i=1}^{n} \alpha_i \left[\sum_{j=1}^{n} p_{ji} e_j \right] = \sum_{j=1}^{n} \left(\sum_{i=1}^{n} \alpha_i p_{ji} \right) e_j.$$

Since $\{e_1, \ldots, e_n\}$ is a linearly independent set, it follows that $\sum_{i=1}^{n} \alpha_i p_{ji} = 0$ for $j = 1, \ldots, n$, and thus, $\sum_{i=1}^{n} \alpha_i \mathbf{p}_i = \mathbf{0}$. Since $\{\mathbf{p}_1, \ldots, \mathbf{p}_n\}$ is a linearly independent set, it now follows that $\alpha_1 = \ldots = \alpha_n = 0$, and therefore $\{e'_1, \ldots, e'_n\}$ is a linearly independent set. ■

The preceding result gives rise to:

4.3.3. Definition. The matrix $\mathbf{P}$ of Theorem 4.3.1 is called the **matrix of basis** $\{e'_1, \ldots, e'_n\}$ **with respect to basis** $\{e_1, \ldots, e_n\}$.

We note that since $\mathbf{P}$ is non-singular, $\mathbf{P}^{-1}$ exists. Thus, we can readily prove the next result.

4.3.4. Theorem. Let $\{e_1, \ldots, e_n\}$ and $\{e'_1, \ldots, e'_n\}$ be two bases for X, and let $\mathbf{P}$ be the matrix of basis $\{e'_1, \ldots, e'_n\}$ with respect to basis $\{e_1, \ldots, e_n\}$. Then $\mathbf{P}^{-1}$ is the matrix of basis $\{e_1, \ldots, e_n\}$ with respect to the basis $\{e'_1, \ldots, e'_n\}$.

4.3.5. Exercise. Prove Theorem 4.3.4.

The next result is also easily verified.

4.3.6. Theorem. Let X be a linear space, and let the sets of vectors $\{e_1, \ldots, e_n\}$, $\{e'_1, \ldots, e'_n\}$, and $\{e''_1, \ldots, e''_n\}$ be bases for X. If $\mathbf{P}$ is the matrix of basis $\{e'_1, \ldots, e'_n\}$ with respect to basis $\{e_1, \ldots, e_n\}$ and if $\mathbf{Q}$ is the matrix of basis $\{e''_1, \ldots, e''_n\}$ with respect to basis $\{e'_1, \ldots, e'_n\}$, then $\mathbf{PQ}$ is the matrix of basis $\{e''_1, \ldots, e''_n\}$ with respect to basis $\{e_1, \ldots, e_n\}$.

4.3.7. Exercise. Prove Theorem 4.3.6.

We now prove:

4.3.8. Theorem. Let $\{e_1, \ldots, e_n\}$ and $\{e'_1, \ldots, e'_n\}$ be two bases for a linear space X, and let $\mathbf{P}$ be the matrix of basis $\{e'_1, \ldots, e'_n\}$ with respect to basis $\{e_1, \ldots, e_n\}$. Let $x \in X$ and let $\mathbf{x}$ denote the coordinate representation of x with respect to the basis $\{e_1, \ldots, e_n\}$. Let $\mathbf{x}'$ denote the coordinate representation of x with respect to the basis $\{e'_1, \ldots, e'_n\}$. Then $\mathbf{Px}' = \mathbf{x}$.

Proof. Let $\mathbf{x}^T = (\xi_1, \ldots, \xi_n)$, and let $(\mathbf{x}')^T = (\xi'_1, \ldots, \xi'_n)$. Then

$$x = \sum_{i=1}^{n} \xi_i e_i$$

and

$$x = \sum_{j=1}^{n} \xi'_j e'_j.$$

Thus,

$$\sum_{j=1}^{n} \xi'_j e'_j = \sum_{j=1}^{n} \xi'_j \left[\sum_{i=1}^{n} p_{ij} e_i\right] = \sum_{i=1}^{n} \left(\sum_{j=1}^{n} p_{ij} \xi'_j\right) e_i$$

which implies that

$$\xi_i = \sum_{j=1}^{n} p_{ij} \xi'_j, \quad i = 1, \ldots, n.$$

Therefore,

$$\mathbf{x} = \mathbf{Px}'. \quad \blacksquare$$

4.3.9. Exercise. Let $X = R^n$ and let $\{u_1, \ldots, u_n\}$ be the natural basis for R^n (see Example 4.1.15). Let $\{e_1, \ldots, e_n\}$ be another basis for R^n, and let $\mathbf{e}_1, \ldots, \mathbf{e}_n$ be the coordinate representations of $e_1, \ldots, e_n$, respectively, with respect to the natural basis. Show that the matrix of basis $\{e_1, \ldots, e_n\}$ with respect to basis $\{u_1, \ldots, u_n\}$ is given by $\mathbf{P} = [\mathbf{e}_1, \mathbf{e}_2, \ldots, \mathbf{e}_n]$, i.e., the matrix whose columns are the column vectors $\mathbf{e}_1, \ldots, \mathbf{e}_n$.

4.3.10. Theorem. Let $A \in L(X, Y)$, and let $\{e_1, \ldots, e_n\}$ and $\{f_1, \ldots, f_m\}$ be bases for X and Y, respectively. Let $\mathbf{A}$ be the matrix of A with respect to the bases $\{e_1, \ldots, e_n\}$ in X and $\{f_1, \ldots, f_m\}$ in Y. Let $\{e'_1, \ldots, e'_n\}$ be another basis for X, and let the matrix of $\{e'_1, \ldots, e'_n\}$ with respect to $\{e_1, \ldots, e_n\}$ be $\mathbf{P}$. Let $\{f'_1, \ldots, f'_m\}$ be another basis for Y, and let $\mathbf{Q}$ be the matrix of $\{f_1, \ldots, f_m\}$ with respect to $\{f'_1, \ldots, f'_m\}$. Let $\mathbf{A}'$ be the matrix of A with respect to the bases $\{e'_1, \ldots, e'_n\}$ in X and $\{f'_1, \ldots, f'_m\}$ in Y. Then.

$$\mathbf{A}' = \mathbf{QAP}.$$

Proof. We have

$$Ae'_i = A\left(\sum_{k=1}^{n} p_{ki}e_k\right) = \sum_{k=1}^{n} p_{ki}Ae_k = \sum_{k=1}^{n} p_{ki}\left(\sum_{\ell=1}^{m} a_{\ell k}f_\ell\right)$$

$$= \sum_{k=1}^{n} p_{ki}\left[\sum_{\ell=1}^{m} a_{\ell k}\left(\sum_{j=1}^{m} q_{j\ell}f'_j\right)\right] = \sum_{j=1}^{m}\left(\sum_{\ell=1}^{m}\sum_{k=1}^{n} q_{j\ell}a_{\ell k}p_{ki}\right)f'_j.$$

Now, by definition, $Ae'_i = \sum_{j=1}^{m} a'_{ji}f'_j$. Since a matrix of a linear transformation is uniquely determined once the bases are specified, we conclude that

$$a'_{ij} = \sum_{\ell=1}^{m}\sum_{k=1}^{n} q_{i\ell}a_{\ell k}p_{kj}$$

for $i = 1, \ldots, m$ and $j = 1, \ldots, n$. Therefore, $\mathbf{A}' = \mathbf{QAP}$. ■

In Figure A, Theorem 4.3.10 is depicted schematically.

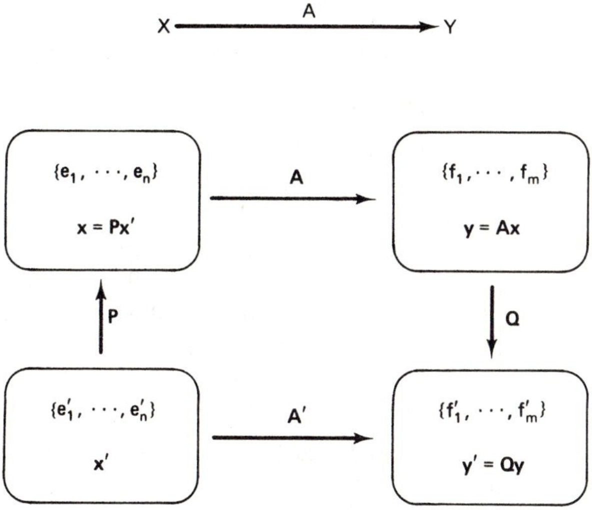

4.3.11. **Figure A.** Schematic diagram of Theorem 4.3.10.

The preceding result motivates the following definition.

4.3.12. Definition. An $(m \times n)$ matrix $\mathbf{A}'$ is said to be **equivalent** to an $(m \times n)$ matrix $\mathbf{A}$ if there exists an $(m \times m)$ non-singular matrix $\mathbf{Q}$ and an

$(n \times n)$ non-singular matrix **P** such that

$$\mathbf{A}' = \mathbf{QAP}. \tag{4.3.13}$$

If **A**′ is equivalent to **A**, we write $\mathbf{A}' \sim \mathbf{A}$.

Thus, an $(m \times n)$ matrix **A**′ is equivalent to an $(m \times n)$ matrix **A** if and only if **A** and **A**′ can be interpreted as both being matrices of the same linear transformation A of a linear space X into a linear space Y, but with respect to possibly different choices of bases.

Our next result shows that $\sim$ is reflexive, symmetric, and transitive, and as such is an equivalence relation.

4.3.14. Theorem. Let **A**, **B**, and **C** be $(m \times n)$ matrices. Then

(i) **A** is always equivalent to **A**;
(ii) if **A** is equivalent to **B**, then **B** is equivalent to **A**; and
(iii) if **A** is equivalent to **B** and **B** is equivalent to **C**, then **A** is equivalent to **C**.

4.3.15. Exercise. Prove Theorem 4.3.14.

The reader can prove the next result readily.

4.3.16. Theorem. Let **A** and **B** be $m \times n$ matrices. Then

(i) every matrix **A** is equivalent to a matrix of the form

$$\left.\begin{bmatrix} 1 & 0 & 0 & \cdots & & \cdots & & \cdots & 0 \\ 0 & 1 & 0 & \cdots & & \cdots & & \cdots & 0 \\ \cdots & \cdots & \cdots & \cdots & \cdots & \cdots & \cdots & \cdots & \cdots \\ \cdots & \cdots & \cdots & \cdots & \cdots & \cdots & \cdots & \cdots & \cdots \\ 0 & 0 & 0 & \cdots & 1 & 0 & 0 & \cdots & 0 \\ 0 & 0 & 0 & \cdots & 0 & 0 & 0 & \cdots & 0 \\ \cdots & \cdots & \cdots & \cdots & \cdots & \cdots & \cdots & \cdots & \cdots \\ 0 & 0 & 0 & \cdots & 0 & 0 & 0 & \cdots & 0 \end{bmatrix}\right\} r = \text{rank } \mathbf{A} \tag{4.3.17}$$

(ii) two $(m \times n)$ matrices **A** and **B** are equivalent if and only if they have the same rank; and
(iii) **A** and $\mathbf{A}^T$ have the same rank.

4.3.18. Exercise. Prove Theorem 4.3.16.

Our definition of rank of a matrix given in the last section (Definition 4.2.25) is sometimes called the **column rank of a matrix**. Sometimes, an analogous definition for **row rank of a matrix** is also considered. *The above theorem shows that the row rank of a matrix is equal to its column rank.*

Next, let us consider the special case when $X = Y$. We have:

4.3.19. Theorem. Let $A \in L(X, X)$, let $\{e_1, \ldots, e_n\}$ be a basis for X, and let $\mathbf{A}$ be the matrix of A with respect to $\{e_1, \ldots, e_n\}$. Let $\{e'_1, \ldots, e'_n\}$ be another basis for X whose matrix with respect to $\{e_1, \ldots, e_n\}$ is $\mathbf{P}$. Let $\mathbf{A}'$ be the matrix of A with respect to $\{e'_1, \ldots, e'_n\}$. Then

$$\mathbf{A}' = \mathbf{P}^{-1}\mathbf{A}\mathbf{P}. \tag{4.3.20}$$

The meaning of the above theorem is depicted schematically in Figure B. The proof of this theorem is just a special application of Theorem 4.3.10.

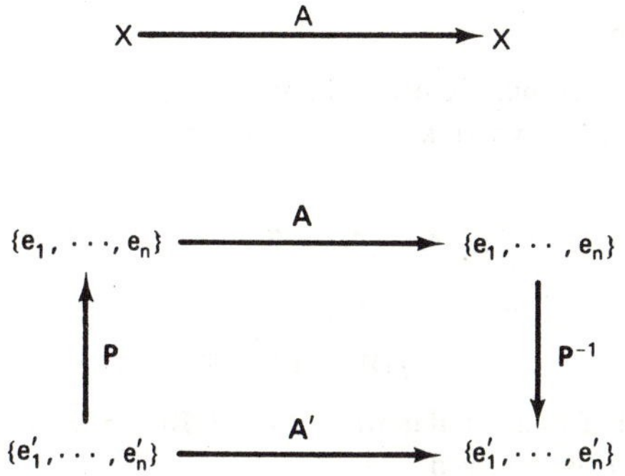

4.3.21. **Figure B.** Schematic diagram of Theorem 4.3.19.

Theorem 4.3.19 gives rise to the following concept.

4.3.22. Definition. An $(n \times n)$ matrix $\mathbf{A}'$ is said to be **similar** to an $(n \times n)$ matrix $\mathbf{A}$ if there exists an $(n \times n)$ non-singular matrix $\mathbf{P}$ such that

$$\mathbf{A}' = \mathbf{P}^{-1}\mathbf{A}\mathbf{P}. \tag{4.3.23}$$

If $\mathbf{A}'$ is similar to $\mathbf{A}$, we write $\mathbf{A}' \sim \mathbf{A}$. We call $\mathbf{P}$ a **similarity transformation.**

It is a simple matter to prove the following:

4.3.24. Theorem. Let $\mathbf{A}'$ be similar to $\mathbf{A}$; i.e., $\mathbf{A}' = \mathbf{P}^{-1}\mathbf{A}\mathbf{P}$, where $\mathbf{P}$ is non-singular. Then $\mathbf{A}$ is similar to $\mathbf{A}'$ and $\mathbf{A} = \mathbf{P}\mathbf{A}'\mathbf{P}^{-1}$.

In view of this result, there is no ambiguity in saying two matrices are similar.

To sum up, if two matrices $\mathbf{A}$ and $\mathbf{A}'$ represent the same linear transformation $A \in L(X, X)$, possibly with respect to two different bases for X, then $\mathbf{A}$ and $\mathbf{A}'$ are similar matrices.

Our next result shows that $\sim$ given in Definition 4.3.22 is an equivalence relation.

4.3.25. Theorem. Let $\mathbf{A}$, $\mathbf{B}$, and $\mathbf{C}$ be $(n \times n)$ matrices. Then

(i) $\mathbf{A}$ is similar to $\mathbf{A}$;
(ii) if $\mathbf{A}$ is similar to $\mathbf{B}$, then $\mathbf{B}$ is similar to $\mathbf{A}$; and
(iii) if $\mathbf{A}$ is similar to $\mathbf{B}$ and if $\mathbf{B}$ is similar to $\mathbf{C}$, then $\mathbf{A}$ is similar to $\mathbf{C}$.

4.3.26. Exercise. Prove Theorem 4.3.25.

For similar matrices we also have the following result.

4.3.27. Theorem.

(i) If an $(n \times n)$ matrix $\mathbf{A}$ is similar to an $(n \times n)$ matrix $\mathbf{B}$, then $\mathbf{A}^k$ is similar to $\mathbf{B}^k$, where k is a positive integer.
(ii) Let

$$f(\lambda) = \sum_{i=0}^{m} \alpha_i \lambda^i = \alpha_0 + \alpha_1 \lambda + \ldots + \alpha_m \lambda^m, \tag{4.3.28}$$

where $\alpha_0, \ldots, \alpha_m \in F$. Then

$$f(\mathbf{P}^{-1}\mathbf{A}\mathbf{P}) = \mathbf{P}^{-1} f(\mathbf{A}) \mathbf{P}. \tag{4.3.29}$$

This implies that if $\mathbf{B}$ is similar to $\mathbf{A}$, then $f(\mathbf{B})$ is similar to $f(\mathbf{A})$. In fact, the same matrix $\mathbf{P}$ is involved.

(iii) Let $\mathbf{A}'$ be similar to $\mathbf{A}$, and let $f(\lambda)$ denote the polynomial of Eq. (4.3.28). Then $f(\mathbf{A}) = \mathbf{0}$ if and only if $f(\mathbf{A}') = \mathbf{0}$.
(iv) Let $A \in L(X, X)$, and let $\mathbf{A}$ be the matrix of A with respect to a basis $\{e_1, \ldots, e_n\}$ in X. Let $f(\lambda)$ denote the polynomial of Eq. (4.3.28). Then $f(\mathbf{A})$ is the matrix of $f(A)$ with respect to the basis $\{e_1, \ldots, e_n\}$.
(v) Let $A \in L(X, X)$, and let $f(\lambda)$ denote the polynomial of Eq. (4.3.28). Let $\mathbf{A}$ be any matrix of A. Then $f(A) = 0$ if and only if $f(\mathbf{A}) = \mathbf{0}$.

4.3.30. Exercise. Prove Theorem 4.3.27.

We can use results such as the preceding ones to good advantage. For example, let $\mathbf{A}'$ denote the matrix

$$\mathbf{A}' = \begin{bmatrix} \lambda_1 & 0 & 0 & \cdots\cdots\cdots & 0 \\ 0 & \lambda_2 & 0 & \cdots\cdots\cdots & 0 \\ \cdot & \cdot & & & \cdot \\ \cdot & \cdot & & & \cdot \\ \cdot & \cdot & & & \cdot \\ 0 & 0 & 0 & \cdots \quad \lambda_{n-1} & 0 \\ 0 & 0 & 0 & \cdots \quad 0 & \lambda_n \end{bmatrix}. \tag{4.3.31}$$

Then

$$(\mathbf{A}')^k = \begin{bmatrix} \lambda_1^k & 0 & 0 & \cdots\cdots\cdots & 0 \\ 0 & \lambda_2^k & 0 & \cdots\cdots\cdots & 0 \\ \cdot & \cdot & & & \cdot \\ \cdot & \cdot & & & \cdot \\ \cdot & \cdot & & & \cdot \\ 0 & 0 & 0 & \cdots \quad \lambda_{n-1}^k & 0 \\ 0 & 0 & 0 & \cdots \quad 0 & \lambda_n^k \end{bmatrix}.$$

Now let $f(\lambda)$ be given by Eq. (4.3.28). Then

$$f(\mathbf{A}') = \alpha_0 \begin{bmatrix} 1 & 0 & \cdots & & 0 \\ 0 & 1 & \cdots & & 0 \\ \cdot & \cdot & & & \cdot \\ \cdot & \cdot & & & \cdot \\ \cdot & \cdot & & & \cdot \\ 0 & 0 & \cdots & 1 & 0 \\ 0 & 0 & \cdots & & 1 \end{bmatrix} + \alpha_1 \begin{bmatrix} \lambda_1 & 0 & \cdots\cdots\cdots & & 0 \\ 0 & \lambda_2 & \cdots\cdots\cdots & & 0 \\ \cdot & \cdot & & & \cdot \\ \cdot & \cdot & & & \cdot \\ \cdot & \cdot & & & \cdot \\ 0 & 0 & \cdots & \lambda_{n-1} & 0 \\ 0 & 0 & \cdots & 0 & \lambda_n \end{bmatrix} + \cdots$$

$$+ \alpha_m \begin{bmatrix} \lambda_1^m & 0 & \cdots\cdots\cdots & & 0 \\ 0 & \lambda_2^m & \cdots\cdots\cdots & & 0 \\ \cdot & \cdot & & & \cdot \\ \cdot & \cdot & & & \cdot \\ \cdot & \cdot & & & \cdot \\ 0 & 0 & \cdots & \lambda_{n-1}^m & 0 \\ 0 & 0 & \cdots & 0 & \lambda_n^m \end{bmatrix} = \begin{bmatrix} f(\lambda_1) & 0 & \cdots\cdots\cdots & & 0 \\ 0 & f(\lambda_2) & \cdots\cdots\cdots & & 0 \\ \cdot & & & & \cdot \\ \cdot & & & & \cdot \\ \cdot & & & & \cdot \\ 0 & 0 & \cdots & f(\lambda_{n-1}) & 0 \\ 0 & 0 & \cdots & 0 & f(\lambda_n) \end{bmatrix}.$$

We conclude the present section with the following definition.

4.3.32. Definition. We call a matrix of the form (4.3.31) a **diagonal matrix**. Specifically, a square $(n \times n)$ matrix $\mathbf{A} = [a_{ij}]$ is said to be a diagonal matrix if $a_{ij} = 0$ for all $i \neq j$. In this case we write $\mathbf{A} = \text{diag}\,(a_{11}, a_{22}, \ldots, a_{nn})$.

4.4. DETERMINANTS OF MATRICES

At this point of our development we need to consider the important topic of determinants. After stating the definition of the determinant of a matrix, we explore some of the commonly used properties of determinants. We then characterize singular and non-singular linear transformations on finite-dimensional vector spaces in terms of determinants. Finally, we give a method of determining the inverse of non-singular matrices.

Let $N = \{1, 2, \ldots, n\}$. We recall (see Definition 1.2.28) that a permutation on N is a one-to-one mapping of N onto itself. For example, if σ denotes a permutation on N, then we can represent it as

$$\sigma = \begin{pmatrix} 1 & 2 & \cdots & n \\ j_1 & j_2 & \cdots & j_n \end{pmatrix},$$

where $j_i \in N$ for $i = 1, \ldots, n$ and $j_i \neq j_k$ for $i \neq k$. Henceforth, we represent σ given above, more compactly, as

$$\sigma = j_1 j_2 \ldots j_n.$$

Clearly, there are $n!$ possible permutations on N. We let $P(N)$ denote the set of all permutations on N, and we distinguish between **odd** and **even permutations**. Specifically, if there is an even number of pairs (i, k) such that $i > k$ but i precedes k in σ, then we say that σ is even. Otherwise σ is said to be odd. Finally, we define the function sgn from $P(N)$ into F by

$$\operatorname{sgn}(\sigma) = \begin{cases} +1 & \sigma \text{ is even} \\ -1 & \sigma \text{ is odd} \end{cases}$$

for all $\sigma \in P(N)$.

Before giving the definition of the determinant of a matrix, let us consider a specific example.

4.4.1. Example. As indicated in the accompanying table, there are six permutations on $N = \{1, 2, 3\}$. In this table the odd and even permutations are identified and the function sgn is given.

σ	(j_1, j_2)	(j_1, j_3)	(j_2, j_3)	σ is odd or even	sgn σ
123	(1, 2)	(1, 3)	(2, 3)	even	+1
132	(1, 3)	(1, 2)	(3, 2)	odd	−1
213	(2, 1)	(2, 3)	(1, 3)	odd	−1
231	(2, 3)	(2, 1)	(3, 1)	even	+1
312	(3, 1)	(3, 2)	(1, 2)	even	+1
321	(3, 2)	(3, 1)	(2, 1)	odd	−1

Now let $\mathbf{A}$ denote the $(n \times n)$ matrix

$$\mathbf{A} = \begin{bmatrix} a_{11} & a_{12} & \cdots & a_{1n} \\ a_{21} & a_{22} & \cdots & a_{2n} \\ \cdots & \cdots & \cdots & \cdots \\ a_{n1} & a_{n2} & \cdots & a_{nn} \end{bmatrix}.$$

We form the product of n elements from $\mathbf{A}$ by taking one and only one element from each row and one and only one element from each column. We represent this product as

$$a_{1j_1} \cdot a_{2j_2} \cdot \ldots \cdot a_{nj_n},$$

where $(j_1 j_2 \ldots j_n) \in P(N)$. It is possible to find $n!$ such products, one for each $\sigma \in P(N)$. We now define the **determinant of A**, denoted by det (**A**), by the sum

$$\det(\mathbf{A}) = \sum_{\sigma \in P(N)} \operatorname{sgn}(\sigma) \cdot a_{1j_1} \cdot a_{2j_2} \cdot \ldots \cdot a_{nj_n}, \tag{4.4.2}$$

where $\sigma = j_1 \ldots j_n$. We also denote the determinant of **A** by writing

$$\det(\mathbf{A}) = \begin{vmatrix} a_{11} & a_{12} & \cdots & a_{1n} \\ a_{21} & a_{22} & \cdots & a_{2n} \\ \cdots & \cdots & \cdots & \cdots \\ a_{n1} & a_{n2} & \cdots & a_{nn} \end{vmatrix}. \tag{4.4.3}$$

We now present some of the fundamental properties of determinants.

4.4.4. Theorem. Let **A** and **B** be $(n \times n)$ matrices.

(i) $\det(\mathbf{A}^T) = \det(\mathbf{A})$.
(ii) If all elements of a column (or row) of **A** are zero, then $\det(\mathbf{A}) = 0$.
(iii) If **B** is the matrix obtained by multiplying every element in a column (or row) of **A** by a constant α, while all other columns of **B** are the same as those in **A**, then $\det(\mathbf{B}) = \alpha \det(\mathbf{A})$.
(iv) If **B** is the same as **A**, except that two columns (or rows) are interchanged, then $\det(\mathbf{B}) = -\det(\mathbf{A})$.
(v) If two columns (or rows) of **A** are identical, then $\det(\mathbf{A}) = 0$.
(vi) If the columns (or rows) of **A** are linearly dependent, then $\det(\mathbf{A}) = 0$.

Proof. To prove the first part, we note first that each product in the sum given in Eq. (4.4.2) has as a factor one and only one element from each column and each row of **A**. Thus, transposing matrix **A** will not affect the $n!$ products appearing in the summation. We now must check to see that the sign of each term is the same.

For $\sigma \in P(N)$, the term in det (**A**) corresponding to σ is $\operatorname{sgn}(\sigma)a_{1j_1}a_{2j_2} \ldots a_{nj_n}$. There is a product term in $\det(\mathbf{A}^T)$ of the form $a_{j_1'1}a_{j_2'2} \ldots a_{j_n'n}$ such that $a_{1j_1}a_{2j_2} \ldots a_{nj_n} = a_{j_1'1}a_{j_2'2} \ldots a_{j_n'n}$. The right-hand side of this equation is just a rearrangement of the left-hand side. The number of $j_i' > j_{i+1}'$ for $i = 1, \ldots, n-1$ is the same as the number of $j_i > j_{i+1}$ for $i = 1, \ldots, n-1$. Thus, if $\sigma' = (j_1' j_2' \ldots j_n')$ then $\operatorname{sgn}(\sigma') = \operatorname{sgn}(\sigma)$, which means $\det(\mathbf{A}^T) = \det(\mathbf{A})$. Note that this result implies that any property below which is proved for columns holds equally as well for rows.

To prove the second part, we note from Eq. (4.4.2) that if for some i, $a_{ik} = 0$ for all k, then $\det(\mathbf{A}) = 0$. This proves that if every element in a row of **A** is zero, then $\det(\mathbf{A}) = 0$. By part (i) it follows that this result holds also for columns. ■

4.4.5. Exercise. Prove parts (iii)–(vi) of Theorem 4.4.4.

We now introduce some additional concepts for determinants.

4.4.6. Definition. Let $\mathbf{A} = [a_{ij}]$ be an $n \times n$ matrix. If the ith row and jth column of $\mathbf{A}$ are deleted, the remaining $(n - 1)$ rows and $(n - 1)$ columns can be used to form another matrix $\mathbf{M}_{ij}$ whose determinant is $\det(\mathbf{M}_{ij})$. We call $\det(\mathbf{M}_{ij})$ the **minor** of a_{ij}. If the diagonal elements of $\mathbf{M}_{ij}$ are diagonal elements of $\mathbf{A}$, i.e., $i = j$, then we speak of a **principal minor of A**. The **cofactor** of a_{ij} is defined as $(-1)^{i+j} \det(\mathbf{M}_{ij})$.

For example, if $\mathbf{A}$ is a (3×3) matrix, then

$$\det(\mathbf{A}) = \begin{vmatrix} a_{11} & a_{12} & a_{13} \\ a_{21} & a_{22} & a_{23} \\ a_{31} & a_{32} & a_{33} \end{vmatrix},$$

the minor of element a_{23} is

$$\det(\mathbf{M}_{23}) = \begin{vmatrix} a_{11} & a_{12} \\ a_{31} & a_{32} \end{vmatrix},$$

and the cofactor of a_{23} is

$$c_{23} = (-1)\begin{vmatrix} a_{11} & a_{12} \\ a_{31} & a_{32} \end{vmatrix}.$$

The next result provides us with a convenient method of evaluating determinants.

4.4.7. Theorem. Let $\mathbf{A}$ be an $n \times n$ matrix. Let c_{ij} denote the cofactor of a_{ij}, $i, j = 1, \ldots, n$. Then the determinant of $\mathbf{A}$ is equal to the sum of the products of the elements of any column (or row) of $\mathbf{A}$, each by its own cofactor. Specifically,

$$\det(\mathbf{A}) = \sum_{i=1}^{n} a_{ij}c_{ij}, \tag{4.4.8}$$

for $j = 1, \ldots, n$, and,

$$\det(\mathbf{A}) = \sum_{j=1}^{n} a_{ij}c_{ij}, \tag{4.4.9}$$

for $i = 1, \ldots, n$.

For example, if $\mathbf{A}$ is a (2×2) matrix, then we have

$$\det(\mathbf{A}) = \begin{vmatrix} a_{11} & a_{12} \\ a_{21} & a_{22} \end{vmatrix} = a_{11}a_{22} - a_{12}a_{21}.$$

If **A** is a (3×3) matrix, then we have

$$\det(\mathbf{A}) = \begin{vmatrix} a_{11} & a_{12} & a_{13} \\ a_{21} & a_{22} & a_{23} \\ a_{31} & a_{32} & a_{33} \end{vmatrix}$$

$$= a_{11}\begin{vmatrix} a_{22} & a_{23} \\ a_{32} & a_{33} \end{vmatrix} - a_{12}\begin{vmatrix} a_{21} & a_{23} \\ a_{31} & a_{33} \end{vmatrix} + a_{13}\begin{vmatrix} a_{21} & a_{22} \\ a_{31} & a_{32} \end{vmatrix}$$

$$= a_{11}c_{11} + a_{12}c_{12} + a_{13}c_{13}.$$

In this case five other possibilities exist. For example, we also have

$$\det(\mathbf{A}) = a_{11}c_{11} + a_{21}c_{21} + a_{31}c_{31}.$$

4.4.10. Exercise. Prove Theorem 4.4.7.

We also have:

4.4.11. Theorem. If the ith row of an $(n \times n)$ matrix **A** consists of elements of the form $a_{i1} + a'_{i1}, a_{i2} + a'_{i2}, \ldots, a_{in} + a'_{in}$; i.e., if

$$A = \begin{bmatrix} a_{11} & a_{12} & \cdots & a_{1n} \\ a_{21} & a_{22} & \cdots & a_{2n} \\ \cdots & \cdots & \cdots & \cdots \\ (a_{i1} + a'_{i1}) & (a_{i2} + a'_{i2}) & \cdots & (a_{in} + a'_{in}) \\ \cdots & \cdots & \cdots & \cdots \\ a_{n1} & a_{n2} & \cdots & a_{nn} \end{bmatrix},$$

then

$$\det(\mathbf{A}) = \begin{vmatrix} a_{11} & a_{12} & \cdots & a_{1n} \\ a_{21} & a_{22} & \cdots & a_{2n} \\ \cdots & \cdots & \cdots & \cdots \\ a_{i1} & a_{i2} & \cdots & a_{in} \\ \cdots & \cdots & \cdots & \cdots \\ a_{n1} & a_{n2} & \cdots & a_{nn} \end{vmatrix} + \begin{vmatrix} a_{11} & a_{12} & \cdots & a_{1n} \\ a_{21} & a_{22} & \cdots & a_{2n} \\ \cdots & \cdots & \cdots & \cdots \\ a'_{i1} & a'_{i2} & \cdots & a'_{in} \\ \cdots & \cdots & \cdots & \cdots \\ a_{n1} & a_{n2} & \cdots & a_{nn} \end{vmatrix}.$$

4.4.12. Exercise. Prove Theorem 4.4.11.

Furthermore, we have:

4.4.13. Theorem. Let **A** and **B** be $(n \times n)$ matrices. If **B** is obtained from the matrix **A** by adding a constant α times any column (or row) to any other column (or row) of **A**, then $\det(\mathbf{B}) = \det(\mathbf{A})$.

4.4.14. Exercise. Prove Theorem 4.4.13.

In addition, we can prove:

4.4.15. Theorem. Let $\mathbf{A}$ be an $(n \times n)$ matrix, and let c_{ij} denote the cofactor of a_{ij}, $i, j = 1, \ldots, n$. Then the sum of products of the elements of any column (or row) by the corresponding cofactors of the elements of any other column (or row) is zero. That is,

$$\sum_{i=1}^{n} a_{ij} c_{ik} = 0 \text{ for } j \neq k \tag{4.4.16a}$$

and

$$\sum_{j=1}^{n} a_{ij} c_{kj} = 0 \text{ for } i \neq k. \tag{4.4.16b}$$

4.4.17. Exercise. Prove Theorem 4.4.15.

We can combine Eqs. (4.4.8) and (4.4.16a) to obtain

$$\sum_{i=1}^{n} a_{ij} c_{ik} = \det(\mathbf{A}) \delta_{jk}, \tag{4.4.18}$$

$j, k = 1, \ldots, n$, where δ_{jk} denotes the Kronecker delta. Similarly, we can combine Eqs. (4.4.9) and (4.4.16b) to obtain

$$\sum_{j=1}^{n} a_{ij} c_{kj} = \det(\mathbf{A}) \delta_{ik}, \tag{4.4.19}$$

$i, k = 1, \ldots, n$.

We are now in a position to prove the following important result.

4.4.20. Theorem. Let $\mathbf{A}$ and $\mathbf{B}$ be $(n \times n)$ matrices. Then

$$\det(\mathbf{AB}) = \det(\mathbf{A}) \det(\mathbf{B}). \tag{4.4.21}$$

Proof. We have

$$\det(\mathbf{AB}) = \begin{vmatrix} \sum_{i_1=1}^{n} a_{1i_1} b_{i_1 1} & \cdots & \sum_{i_n=1}^{n} a_{1i_n} b_{i_n n} \\ \cdots & \cdots & \cdots \\ \sum_{i_1=1}^{n} a_{ni_1} b_{i_1 1} & \cdots & \sum_{i_n=1}^{n} a_{ni_n} b_{i_n n} \end{vmatrix}.$$

By Theorem 4.4.11 and Theorem 4.4.4, part (iii), we have

$$\det(\mathbf{AB}) = \sum_{i_1=1}^{n} \cdots \sum_{i_n=1}^{n} b_{i_1 1} b_{i_2 2} \ldots b_{i_n n} \begin{vmatrix} a_{1i_1} & \cdots & a_{1i_n} \\ \cdot & & \cdot \\ \cdot & & \cdot \\ \cdot & & \cdot \\ a_{ni_1} & \cdots & a_{ni_n} \end{vmatrix}.$$

This determinant will vanish whenever two or more of the indices i_j, $j = 1, \ldots, n$, are identical. Thus, we need to sum only over $\sigma \in P(N)$. We have

$$\det(\mathbf{AB}) = \sum_{\sigma \in P(N)} b_{i_1 1} b_{i_2 2} \ldots b_{i_n n} \begin{vmatrix} a_{1i_1} & \cdots & a_{1i_n} \\ \cdot & & \cdot \\ \cdot & & \cdot \\ \cdot & & \cdot \\ a_{ni_1} & \cdots & a_{ni_n} \end{vmatrix},$$

where $\sigma = i_1 i_2 \ldots i_n$ and $P(N)$ is the set of all permutations of $N = \{1, \ldots, n\}$. It is now straightforward to show that

$$\begin{vmatrix} a_{1i_1} & \cdots & a_{1i_n} \\ \cdot & & \cdot \\ \cdot & & \cdot \\ \cdot & & \cdot \\ a_{ni_1} & \cdots & a_{ni_n} \end{vmatrix} = \operatorname{sgn}(\sigma) \det(\mathbf{A}),$$

and hence it follows that

$$\det(\mathbf{AB}) = \det(\mathbf{A}) \det(\mathbf{B}). \quad \blacksquare$$

Our next result is readily verified.

4.4.22. Theorem. Let $\mathbf{I}$ be the $(n \times n)$ identity matrix, and let $\mathbf{0}$ be the $(n \times n)$ zero matrix. Then $\det(\mathbf{I}) = 1$ and $\det(\mathbf{0}) = 0$.

4.4.23. Exercise. Prove Theorem 4.4.22.

The next theorem allows us to characterize non-singular matrices in terms of their determinants.

4.4.24. Theorem. An $(n \times n)$ matrix $\mathbf{A}$ is non-singular if and only if $\det(\mathbf{A}) \neq 0$.

Proof. Suppose that $\mathbf{A}$ is non-singular. Then $\mathbf{A}^{-1}$ exists and $\mathbf{A}^{-1}\mathbf{A} = \mathbf{A}\mathbf{A}^{-1} = \mathbf{I}$. From this it follows that $\det(\mathbf{A}^{-1}\mathbf{A}) = 1 \neq 0$, and thus, in view of Eq. (4.4.21), $\det(\mathbf{A}^{-1}) \neq 0$ and $\det(\mathbf{A}) \neq 0$.

Next, assume that $\mathbf{A}$ is singular. By Theorem 4.3.16, there exist nonsingular matrices $\mathbf{Q}$ and $\mathbf{P}$ such that

$$\mathbf{A}' = \mathbf{QAP} = \begin{bmatrix} 1 & & & & & & & \\ & \cdot & & & & & \mathbf{0} & \\ & & \cdot & & & & & \\ & & & \cdot & & & & \\ & & & & 1 & & & \\ & & & & & 0 & & \\ & & & & & & \cdot & \\ & \mathbf{0} & & & & & & \cdot \\ & & & & & & & \cdot \\ & & & & & & & & 0 \end{bmatrix}.$$

This shows that rank $\mathbf{A} < n$ and $\det(\mathbf{A}') = 0$. But

$$\det(\mathbf{QAP}) = [\det(\mathbf{Q})] \cdot [\det(\mathbf{A})] \cdot [\det(\mathbf{P})] = 0,$$

and $\det(\mathbf{P}) \neq 0$ and $\det(\mathbf{Q}) \neq 0$. Therefore, if $\mathbf{A}$ is singular, then $\det(\mathbf{A}) = 0$. $\blacksquare$

Let us now turn to the problem of finding the inverse $\mathbf{A}^{-1}$ of a non-singular matrix $\mathbf{A}$. In doing so, we need to introduce the classical adjoint of $\mathbf{A}$.

4.4.25. Definition. Let $\mathbf{A}$ be an $(n \times n)$ matrix, and let c_{ij} be the cofactor of a_{ij} for $i, j = 1, \ldots, n$. Let $\mathbf{C}$ be the matrix formed by the cofactors of $\mathbf{A}$; i.e., $\mathbf{C} = [c_{ij}]$. The matrix $\mathbf{C}^T$ is called the **classical adjoint of A**. We write adj $(\mathbf{A})$ to denote the classical adjoint of $\mathbf{A}$.

We now have:

4.4.26. Theorem. Let $\mathbf{A}$ be an $(n \times n)$ matrix. Then

$$\mathbf{A}[\text{adj}\,(\mathbf{A})] = [\text{adj}\,(\mathbf{A})]\mathbf{A} = [\det\,(\mathbf{A})] \cdot \mathbf{I}.$$

Proof. The proof follows by direct computation, using Eqs. (4.4.18) and (4.4.19). ■

As an immediate consequence of Theorem 4.4.26 we now have the following practical result.

4.4.27. Corollary. Let $\mathbf{A}$ be a non-singular $(n \times n)$ matrix. Then

$$\mathbf{A}^{-1} = \frac{1}{\det\,(\mathbf{A})}\,\text{adj}\,(\mathbf{A}). \tag{4.4.28}$$

4.4.29. Example. Consider the matrix

$$\mathbf{A} = \begin{bmatrix} 0 & 1 & 1 \\ 1 & 2 & 2 \\ 1 & -1 & 0 \end{bmatrix}.$$

We have $\det\,(\mathbf{A}) = -1$,

$$\text{adj}\,(\mathbf{A}) = \begin{bmatrix} 2 & -1 & 0 \\ 2 & -1 & 1 \\ -3 & 1 & -1 \end{bmatrix},$$

and

$$\mathbf{A}^{-1} = \begin{bmatrix} -2 & 1 & 0 \\ -2 & 1 & -1 \\ 3 & -1 & 1 \end{bmatrix}. \quad ■$$

The proofs of the next two theorems are left as an exercise.

4.4.30. Theorem. If $\mathbf{A}$ and $\mathbf{B}$ are similar matrices, then $\det\,(\mathbf{A}) = \det\,(\mathbf{B})$.

4.4.31. Theorem. Let $A \in L(X, X)$. Let $\mathbf{A}$ be the matrix of A with respect to a basis $\{e_1, \ldots, e_n\}$ in X, and let $\mathbf{A}'$ be the matrix of A with respect to another basis $\{e'_1, \ldots, e'_n\}$ in X. Then $\det\,(\mathbf{A}) = \det\,(\mathbf{A}')$.

4.4.32. Exercise. Prove Theorems 4.4.30 and 4.4.31.

In view of the preceding results, there is no ambiguity in the following definition.

4.4.33. Definition. The **determinant of a linear transformation** A of a finite-dimensional vector space X into X is the determinant of any matrix $\mathbf{A}$ representing it; i.e., $\det(A) \triangleq \det(\mathbf{A})$.

The last result of the present section is a consequence of Theorems 4.4.20 and 4.4.24.

4.4.34. Theorem. Let X be a finite-dimensional vector space, and let $A, B \in L(X, X)$. Then A is non-singular if and only if $\det(A) \neq 0$. Also, $\det(AB) = [\det(A)] \cdot [\det(B)]$.

4.5. EIGENVALUES AND EIGENVECTORS

In the present section we consider eigenvalues and eigenvectors of linear transformations defined on finite-dimensional vector spaces. Later, in Chapter 7, we will reconsider these concepts in a more general setting. Eigenvalues and eigenvectors play, of course, a crucial role in the study of linear transformations.

Throughout the present section, X denotes an n-dimensional vector space over a field F.

Let $A \in L(X, X)$, and let us assume that there exist sets of vectors $\{e_1, \ldots, e_n\}$ and $\{e'_1, \ldots, e'_n\}$, which are bases for X such that

$$\begin{aligned} e'_1 &= Ae_1 = \lambda_1 e_1, \\ e'_2 &= Ae_2 = \lambda_2 e_2, \\ &\cdots\cdots\cdots\cdots, \\ e'_n &= Ae_n = \lambda_n e_n, \end{aligned} \tag{4.5.1}$$

where $\lambda_i \in F$, $i = 1, \ldots, n$. If this is the case, then the matrix $\mathbf{A}'$ of A with respect to the given basis is

$$\mathbf{A}' = \begin{bmatrix} \lambda_1 & & & & \\ & \lambda_2 & & 0 & \\ & & \cdot & & \\ & 0 & & \cdot & \\ & & & & \lambda_n \end{bmatrix}.$$

This motivates the following result.

4.5.2. Theorem. Let $A \in L(X, X)$, and let $\lambda \in F$. Then the set of all $x \in X$ such that

$$Ax = \lambda x \tag{4.5.3}$$

is a linear subspace of X. In fact, it is the null space of the linear transformation $(A - \lambda I)$, where I is the identity element of $L(X, X)$.

Proof. Since the zero vector satisfies Eq. (4.5.3) for any $\lambda \in F$, the set is non-void. If the zero vector is the only such vector, then we are done, for $\{0\}$ is a linear subspace of X (of dimension zero). In any case, Eq. (4.5.3) holds if and only if $(A - \lambda I)x = 0$. Thus, x belongs to the null space of $A - \lambda I$, and it follows from Theorem 3.4.19 that the set of all $x \in X$ satisfying Eq. (4.5.3) is a linear subspace of X. ■

Henceforth we let

$$\mathfrak{N}_\lambda = \{x \in X: (A - \lambda I)x = 0\}. \tag{4.5.4}$$

The preceding result gives rise to several important concepts which we introduce in the following definition.

4.5.5. Definition. Let X, $A \in L(X, X)$, and $\mathfrak{N}_\lambda$ be defined as in Theorem 4.5.2 and Eq. (4.5.4). A scalar λ such that $\mathfrak{N}_\lambda$ contains more than just the zero vector is called an **eigenvalue** of A (i.e., if there is an $x \neq 0$ such that $Ax = \lambda x$, then λ is called an eigenvalue of A). When λ is an eigenvalue of A, then each $x \neq 0$ in $\mathfrak{N}_\lambda$ is called an **eigenvector** of A corresponding to the eigenvalue λ. The dimension of the linear subspace $\mathfrak{N}_\lambda$ is called the **multiplicity of the eigenvalue** λ. If $\mathfrak{N}_\lambda$ is of dimension one, then λ is called a **simple eigenvalue**. The set of all eigenvalues of A is called the **spectrum** of A.

Some authors call an eigenvalue a **proper value** or a **characteristic value** or a **latent value** or a **secular value**. Similarly, other names for eigenvector are **proper vector** or **characteristic vector**. The space $\mathfrak{N}_\lambda$ is called the λth **proper subspace** of X.

For matrices we give the following corresponding definition.

4.5.6. Definition. Let $\mathbf{A}$ be an $(n \times n)$ matrix whose elements belong to the field F. If there exists $\lambda \in F$ and a non-zero vector $\mathbf{x} \in F^n$ such that

$$\mathbf{A}\mathbf{x} = \lambda \mathbf{x} \tag{4.5.7}$$

then λ is called an **eigenvalue** of $\mathbf{A}$ and $\mathbf{x}$ is called an **eigenvector** of $\mathbf{A}$ corresponding to the eigenvalue λ.

Our next result provides the connection between Definitions 4.5.5 and 4.5.6.

4.5.8. Theorem. Let $A \in L(X, X)$, and let $\mathbf{A}$ be the matrix of A with respect to the basis $\{e_1, \ldots, e_n\}$. Then λ is an eigenvalue of A if and only if λ is an eigenvalue of $\mathbf{A}$. Also, $x \in X$ is an eigenvector of A corresponding to λ if and only if the coordinate representation of x with respect to the basis $\{e_1, \ldots, e_n\}$, $\mathbf{x}$, is an eigenvector of $\mathbf{A}$ corresponding to λ.

4.5.9. Exercise. Prove Theorem 4.5.8.

Note that if x (or $\mathbf{x}$) is an eigenvector of A (of $\mathbf{A}$), then any non-zero multiple of x (of $\mathbf{x}$) is also an eigenvector of A (of $\mathbf{A}$).

In the next result, the proof of which is left as an exercise, we use determinants to characterize eigenvalues. We have:

4.5.10. Theorem. Let $A \in L(X, X)$. Then $\lambda \in F$ is an eigenvalue of A if and only if $\det(A - \lambda I) = 0$.

4.5.11. Exercise. Prove Theorem 4.5.10.

Let us next examine the equation

$$\det(A - \lambda I) = 0 \tag{4.5.12}$$

in terms of the parameter λ. We ask: Can we determine which values of λ, if any, satisfy Eq. (4.5.12)? Let $\{e_1, \ldots, e_n\}$ be an arbitrary basis for X and let $\mathbf{A}$ be the matrix of A with respect to this basis. We then have

$$\det(A - \lambda I) = \det(\mathbf{A} - \lambda \mathbf{I}). \tag{4.5.13}$$

The right-hand side of Eq. (4.5.13) may be rewritten as

$$\det(\mathbf{A} - \lambda \mathbf{I}) = \begin{vmatrix} (a_{11} - \lambda) & a_{12} & \cdots & a_{1n} \\ a_{21} & (a_{22} - \lambda) & \cdots & a_{2n} \\ \vdots & \vdots & \cdots & \vdots \\ a_{n1} & a_{n2} & \cdots & (a_{nn} - \lambda) \end{vmatrix}. \tag{4.5.14}$$

It is clear from Eq. (4.4.2) that expansion of the determinant (4.5.14) yields a polynomial in λ of degree n. In order for λ to be an eigenvalue of A it must (a) satisfy Eq. (4.5.12), and (b) it must belong to F. Requirement (b) warrants further comment: note that there is no guarantee that there exists $\lambda \in F$ such that Eq. (4.5.12) is satisfied, or equivalently we have no assurance that the nth-order polynomial equation

$$\det(\mathbf{A} - \lambda \mathbf{I}) = 0$$

has any roots in F. There is, however, a special class of fields for which requirement (b) is automatically satisfied. We have:

4.5.15. Definition. A field F is said to be **algebraically closed** if for every polynomial $p(\lambda)$ there is at least one $\lambda \in F$ such that

$$p(\lambda) = 0. \tag{4.5.16}$$

Any λ which satisfies Eq. (4.5.16) is said to be a **root** of the polynomial equation (4.5.16).

In particular, the field of complex numbers is algebraically closed, whereas the field of real numbers is not (e.g., consider the equation $\lambda^2 + 1 = 0$).

There are other fields besides the field of complex numbers which are algebraically closed. However, since we will not develop these, we will restrict ourselves to the field of complex numbers, C, whenever the algebraic closure property of Definition 4.5.15 is required. When considering results that are valid for a vector space over an arbitrary field, we will (as before) make usage of the symbol F or frequently (as before) make no reference to F at all.

We summarize the above discussion in the following theorem.

4.5.17. Theorem. Let $A \in L(X, X)$. Then

(i) $\det(A - \lambda I)$ is a polynomial of degree n in the parameter λ; i.e., there exist scalars $\alpha_0, \alpha_1, \ldots, \alpha_n$, depending only on A, such that

$$\det(A - \lambda I) = \alpha_0 + \alpha_1\lambda + \alpha_2\lambda^2 + \ldots + \alpha_n\lambda^n \quad (4.5.18)$$

(note that $\alpha_0 = \det(A)$ and $\alpha_n = (-1)^n$);

(ii) the eigenvalues of A are precisely the roots of the equation $\det(A - \lambda I) = 0$; i.e., they are the roots of

$$\alpha_0 + \alpha_1\lambda + \alpha_2\lambda^2 + \ldots + \alpha_n\lambda^n = 0; \text{ and} \quad (4.5.19)$$

(iii) A has, at most, n distinct eigenvalues.

The above result motivates the following definition.

4.5.20. Definition. Let $A \in L(X, X)$, and let $\mathbf{A}$ be a matrix of A. We call

$$\det(A - \lambda I) = \det(\mathbf{A} - \lambda\mathbf{I}) = \alpha_0 + \alpha_1\lambda + \ldots + \alpha_n\lambda^n \quad (4.5.21)$$

the **characteristic polynomial** of A (or of $\mathbf{A}$) and

$$\det(A - \lambda I) = \det(\mathbf{A} - \lambda\mathbf{I}) = 0 \quad (4.5.22)$$

the **characteristic equation** of A (or of $\mathbf{A}$).

From the fundamental properties of polynomials over the field of complex numbers there now follows:

4.5.23. Theorem. If X is an n-dimensional vector space over C and if $A \in L(X, X)$, then it is possible to write the characteristic polynomial of A in the form

$$\det(A - \lambda I) = (\lambda_1 - \lambda)^{m_1}(\lambda_2 - \lambda)^{m_2} \ldots (\lambda_p - \lambda)^{m_p}, \quad (4.5.24)$$

where λ_i, $i = 1, \ldots, p$, are the distinct roots of Eq. (4.5.19) (i.e., $\lambda_i \neq \lambda_j$ for $i \neq j$). In Eq. (4.5.24), m_i is called the **algebraic multiplicity** of the root λ_i. The m_i are positive integers, and $\sum_{i=1}^{p} m_i = n$.

Note the distinction between the concept of algebraic multiplicity of λ_i given in Theorem 4.5.23 and the multiplicity of λ_i as given in Definition 4.5.5. In general, these need not be the same, as will be seen later.

We now state and prove one of the most important results of linear algebra, the **Cayley-Hamilton theorem**.

4.5.25. Theorem. Let $\mathbf{A}$ be an $n \times n$ matrix, and let $p(\lambda) = \det(\mathbf{A} - \lambda\mathbf{I})$ be the characteristic polynomial of $\mathbf{A}$. Then

$$p(\mathbf{A}) = \mathbf{0}.$$

Proof. Let the characteristic polynomial for $\mathbf{A}$ be

$$p(\lambda) = \alpha_0 + \alpha_1\lambda + \ldots + \alpha_n\lambda^n.$$

Now let $\mathbf{B}(\lambda)$ be the classical adjoint of $(\mathbf{A} - \lambda\mathbf{I})$. Since the elements $b_{ij}(\lambda)$ of $\mathbf{B}(\lambda)$ are cofactors of the matrix $\mathbf{A} - \lambda\mathbf{I}$, they are polynomials in λ of degree not more than $n - 1$. Thus,

$$b_{ij}(\lambda) = \beta_{ij0} + \beta_{ij1}\lambda + \ldots + \beta_{ij(n-1)}\lambda^{n-1}.$$

Letting $\mathbf{B}_k = [\beta_{ijk}]$ for $k = 0, 1, \ldots, n - 1$, we have

$$\mathbf{B}(\lambda) = \mathbf{B}_0 + \lambda\mathbf{B}_1 + \ldots + \lambda^{n-1}\mathbf{B}_{n-1}.$$

By Theorem 4.4.26,

$$(\mathbf{A} - \lambda\mathbf{I})\mathbf{B}(\lambda) = [\det(\mathbf{A} - \lambda\mathbf{I})]\mathbf{I}.$$

Thus,

$$(\mathbf{A} - \lambda\mathbf{I})[\mathbf{B}_0 + \lambda\mathbf{B}_1 + \ldots + \lambda^{n-1}\mathbf{B}_{n-1}] = (\alpha_0 + \alpha_1\lambda + \ldots + \alpha_n\lambda^n)\mathbf{I}.$$

Expanding the left-hand side of this equation and equating like powers of λ, we have

$$-\mathbf{B}_{n-1} = \alpha_n\mathbf{I}, \quad \mathbf{A}\mathbf{B}_{n-1} - \mathbf{B}_{n-2} = \alpha_{n-1}\mathbf{I}, \quad \ldots, \quad \mathbf{A}\mathbf{B}_1 - \mathbf{B}_0 = \alpha_1\mathbf{I},$$
$$\mathbf{A}\mathbf{B}_0 = \alpha_0\mathbf{I}.$$

Premultiplying the above matrix equations by $\mathbf{A}^n, \mathbf{A}^{n-1}, \ldots, \mathbf{A}, \mathbf{I}$, respectively, we have

$$-\mathbf{A}^n\mathbf{B}_{n-1} = \alpha_n\mathbf{A}^n, \quad \mathbf{A}^n\mathbf{B}_{n-1} - \mathbf{A}^{n-1}\mathbf{B}_{n-2} = \alpha_{n-1}\mathbf{A}^{n-1}, \quad \ldots,$$
$$\mathbf{A}^2\mathbf{B}_1 - \mathbf{A}\mathbf{B}_0 = \alpha_1\mathbf{A}, \quad \mathbf{A}\mathbf{B}_0 = \alpha_0\mathbf{I}.$$

Adding these matrix equations, we obtain

$$\mathbf{0} = \alpha_0\mathbf{I} + \alpha_1\mathbf{A} + \ldots + \alpha_n\mathbf{A}^n = p(\mathbf{A}),$$

which was to be shown. ∎

As an immediate consequence of the Cayley-Hamilton theorem, we have:

4.5.26. Theorem. Let $\mathbf{A}$ be an $(n \times n)$ matrix with characteristic polynomial given by Eq. (4.5.21). Then

(i) $\mathbf{A}^n = (-1)^{n+1}[\alpha_0\mathbf{I} + \alpha_1\mathbf{A} + \ldots + \alpha_{n-1}\mathbf{A}^{n-1}]$; and

(ii) if $f(\lambda)$ is any polynomial in λ, then there exist $\beta_0, \beta_1, \ldots, \beta_{n-1} \in F$ such that

$$f(\mathbf{A}) = \beta_0\mathbf{I} + \beta_1\mathbf{A} + \ldots + \beta_{n-1}\mathbf{A}^{n-1}.$$

Proof. Part (i) follows from Theorem 4.5.25 and from the fact that $\alpha_n = (-1)^n$.

To prove part (ii), let $f(\lambda)$ be any polynomial in λ and let $p(\lambda)$ denote the characteristic polynomial of $\mathbf{A}$. Then there exist two polynomials $g(\lambda)$ and $r(\lambda)$ (see Theorem 2.3.9) such that

$$f(\lambda) = p(\lambda)g(\lambda) + r(\lambda), \tag{4.5.27}$$

where $\deg[r(\lambda)] \leq n - 1$. Using the fact that $p(\mathbf{A}) = \mathbf{0}$, we have $f(\mathbf{A}) = r(\mathbf{A})$, and the theorem follows. ■

The Cayley-Hamilton theorem holds also in the case of linear transformations. Specifically, we have the following result.

4.5.28. Theorem. Let $A \in L(X, X)$, and let $p(\lambda)$ denote the characteristic polynomial of A. Then $p(A) = 0$.

4.5.29. Exercise. Prove Theorem 4.5.28.

Let us now consider a specific example.

4.5.30. Example. Consider the matrix

$$\mathbf{A} = \begin{bmatrix} 1 & 0 \\ 1 & 2 \end{bmatrix}.$$

Let us use Theorem 4.5.26 to evaluate $\mathbf{A}^{37}$. Since $n = 2$, we assume that $\mathbf{A}^{37}$ is of the form

$$\mathbf{A}^{37} = \beta_0\mathbf{I} + \beta_1\mathbf{A}.$$

The characteristic polynomial of $\mathbf{A}$ is

$$p(\lambda) = (1 - \lambda)(2 - \lambda)$$

and the eigenvalues of $\mathbf{A}$ are $\lambda_1 = 1$ and $\lambda_2 = 2$. In the present case $f(\lambda) = \lambda^{37}$ and $r(\lambda)$ in Eq. (4.5.27) is

$$r(\lambda) = \beta_0 + \beta_1\lambda.$$

We must determine β_0 and β_1. Using the fact that $p(\lambda_1) = p(\lambda_2) = 0$, it follows that $f(\lambda_1) = r(\lambda_1)$ and $f(\lambda_2) = r(\lambda_2)$. Thus, we have

$$\beta_0 + \beta_1 = 1^{37} = 1, \quad \beta_0 + 2\beta_1 = 2^{37}.$$

Hence, $\beta_1 = 2^{37} - 1$ and $\beta_0 = 2 - 2^{37}$. Therefore,

$$\mathbf{A}^{37} = (2 - 2^{37})\mathbf{I} + (2^{37} - 1)\mathbf{A},$$

or,

$$\mathbf{A}^{37} = \begin{bmatrix} 1 & 0 \\ 2^{37} - 1 & 2^{37} \end{bmatrix}. \quad \blacksquare$$

Before closing the present section, let us introduce another important concept for matrices.

4.5.31. Definition. If **A** is an $(n \times n)$ matrix, then the **trace** of **A**, denoted by trace **A** or by tr **A**, is defined as

$$\text{trace } \mathbf{A} = \text{tr } \mathbf{A} = a_{11} + a_{22} + \ldots + a_{nn} \tag{4.5.32}$$

(i.e., the trace of a square matrix is the sum of its diagonal elements).

It turns out that if $F = C$, the field of complex numbers, then there is a relationship between the trace, determinant, and eigenvalues of an $(n \times n)$ matrix **A**. We have:

4.5.33. Theorem. Let X be a vector space over C. Let **A** be a matrix of $A \in L(X, X)$ and let $\det(A - \lambda I)$ be given by Eq. (4.5.24). Then

(i) $\det(\mathbf{A}) = \prod_{j=1}^{p} \lambda_j^{m_j}$;

(ii) $\text{trace}(\mathbf{A}) = \sum_{j=1}^{p} m_j \lambda_j$;

(iii) if **B** is any matrix similar to **A**, then trace (**B**) = trace (**A**); and

(iv) let $f(\lambda)$ denote the polynomial

$$f(\lambda) = \gamma_0 + \gamma_1 \lambda + \ldots + \gamma_m \lambda^m;$$

then the roots of the characteristic polynomial of $f(A)$ are $f(\lambda_1), \ldots, f(\lambda_p)$ and

$$\det[f(A) - \lambda I] = [f(\lambda_1) - \lambda]^{m_1} \ldots [f(\lambda_p) - \lambda]^{m_p}.$$

4.5.34. Exercise. Prove Theorem 4.5.33.

4.6. SOME CANONICAL FORMS OF MATRICES

In the present section we investigate under which conditions a linear transformation of a vector space into itself can be represented by special types of matrices, namely, by (a) a diagonal matrix, (b) a so-called triangular matrix, and (c) a so-called "block diagonal matrix." We will also investigate when a linear transformation cannot be represented by a diagonal matrix.

Throughout the present section X denotes an n-dimensional vector space over a field F.

4.6.1. Theorem. Let $\lambda_1, \ldots, \lambda_p$ be distinct eigenvalues of a linear transformation $A \in L(X, X)$. Let $e'_1 \neq 0, \ldots, e'_p \neq 0$ be eigenvectors of A corresponding to $\lambda_1, \ldots, \lambda_p$, respectively. Then the set $\{e'_1, \ldots, e'_p\}$ is linearly independent.

Proof. The proof is by contradiction. Assume that the set $\{e'_1, \ldots, e'_p\}$ is linearly dependent so that there exist scalars $\alpha_1, \ldots, \alpha_p$, not all zero, such that

$$\alpha_1 e'_1 + \ldots + \alpha_p e'_p = 0.$$

We assume that these scalars have been chosen in such a fashion that as few of them as possible are non-zero. Relabeling, if necessary, we thus have

$$\alpha_1 e'_1 + \ldots + \alpha_r e'_r = 0, \tag{4.6.2}$$

where $\alpha_1 \neq 0, \ldots, \alpha_r \neq 0$ and where $r \leq p$ is the smallest number for which we can get such an expression.

Since $\lambda_1, \ldots, \lambda_r$ are eigenvalues and since $e'_1, \ldots, e'_r$ are eigenvectors, we have

$$\begin{aligned} 0 = A(0) = A(\alpha_1 e'_1 + \ldots + \alpha_r e'_r) &= \alpha_1 A e'_1 + \ldots + \alpha_r A e'_r \\ &= (\alpha_1 \lambda_1) e'_1 + \ldots + (\alpha_r \lambda_r) e'_r. \end{aligned} \tag{4.6.3}$$

Also,

$$\begin{aligned} 0 = \lambda_r \cdot 0 &= \lambda_r(\alpha_r e'_1 + \ldots + \alpha_r e'_r) \\ &= (\alpha_1 \lambda_r) e'_1 + \ldots + (\alpha_r \lambda_r) e'_r. \end{aligned} \tag{4.6.4}$$

Subtracting Eq. (4.6.4) from Eq. (4.6.3) we obtain

$$0 = \alpha_1(\lambda_1 - \lambda_r) e'_1 + \ldots + \alpha_r(\lambda_r - \lambda_r) e'_r.$$

Since by assumption the λ_i's are distinct, we have found an expression involving only $(r - 1)$ vectors satisfying Eq. (4.6.2). But r was chosen to be the smallest number for which Eq. (4.6.2) holds. We have thus arrived at a contradiction, and our theorem is proved. ■

We note that if, in the above theorem, A has n distinct eigenvalues, then the corresponding n eigenvectors span the linear space X (recall that dim $X = n$).

Our next result enables us to represent a linear transformation with n distinct eigenvalues in a very convenient form.

4.6.5. Theorem. Let $A \in L(X, X)$. Assume that the characteristic polynomial of A has n distinct roots, so that

$$\det(A - \lambda I) = (\lambda_1 - \lambda)(\lambda_2 - \lambda) \ldots (\lambda_n - \lambda),$$

where $\lambda_1, \lambda_2, \ldots, \lambda_n$ are distinct eigenvalues. Then there exists a basis $\{e'_1, e'_2, \ldots, e'_n\}$ of X such that e'_i is an eigenvector corresponding to λ_i for $i = 1, 2, \ldots, n$. The matrix $\mathbf{A}'$ of A with respect to the basis $\{e'_1, e'_2, \ldots, e'_n\}$ is

$$\mathbf{A}' = \begin{bmatrix} \lambda_1 & & & \\ & \lambda_2 & & 0 \\ & & \ddots & \\ & 0 & & \\ & & & \lambda_n \end{bmatrix} = \operatorname{diag}(\lambda_1, \lambda_2, \ldots, \lambda_n). \tag{4.6.6}$$

Proof. Let e'_i denote the eigenvector corresponding to the eigenvalue λ_i. In view of Theorem 4.6.1, the set $\{e'_1, e'_2, \ldots, e'_n\}$ is linearly independent because $\lambda_1, \lambda_2, \ldots, \lambda_n$ are all different. Moreover, since there are n of the e'_i, the set $\{e'_1, e'_2, \ldots, e'_n\}$ forms a basis for the n-dimensional space X. Also, from the definition of eigenvalue and eigenvector, we have

$$\begin{aligned} Ae'_1 &= \lambda_1 e'_1 \\ Ae'_2 &= \lambda_2 e'_2 \\ &\cdots\cdots\cdots \\ Ae'_n &= \lambda_n e'_n. \end{aligned} \tag{4.6.7}$$

From Eq. (4.6.7) we obtain the desired matrix given in Eq. (4.6.6). ■

The reader can prove the following useful result readily.

4.6.8. Theorem. Let $A \in L(X, X)$, and let $\mathbf{A}$ be the matrix of A with respect to a basis $\{e_1, e_2, \ldots, e_n\}$. If the characteristic polynomial

$$\det(A - \lambda I) = \alpha_0 + \alpha_1\lambda + \alpha_2\lambda^2 + \ldots + \alpha_n\lambda^n$$

has n distinct roots, $\lambda_1, \ldots, \lambda_n$, then $\mathbf{A}$ is similar to the matrix $\mathbf{A}'$ of A with respect to a basis $\{e'_1, \ldots, e'_n\}$, where

$$\mathbf{A}' = \begin{bmatrix} \lambda_1 & & & \\ & \lambda_2 & & 0 \\ & & \ddots & \\ & 0 & & \\ & & & \lambda_n \end{bmatrix}. \tag{4.6.9}$$

In this case there eixsts a non-singular matrix $\mathbf{P}$ such that

$$\mathbf{A}' = \mathbf{P}^{-1}\mathbf{A}\mathbf{P}. \tag{4.6.10}$$

The matrix $\mathbf{P}$ is the matrix of basis $\{e'_1, e'_2, \ldots, e'_n\}$ with respect to basis $\{e_1, e_2, \ldots, e_n\}$, and $\mathbf{P}^{-1}$ is the matrix of basis $\{e_1, \ldots, e_n\}$ with respect to basis $\{e'_1, \ldots, e'_n\}$. The matrix $\mathbf{P}$ can be constructed by letting its columns be eigenvectors of $\mathbf{A}$ corresponding to $\lambda_1, \ldots, \lambda_n$, respectively. That is,

$$\mathbf{P} = [\mathbf{x}_1, \mathbf{x}_2, \ldots, \mathbf{x}_n], \tag{4.6.11}$$

where $\mathbf{x}_1, \ldots, \mathbf{x}_n$ are eigenvectors of $\mathbf{A}$ corresponding to the eigenvalues $\lambda_1, \ldots, \lambda_n$, respectively.

The similarity transformation $\mathbf{P}$ given in Eq. (4.6.11) is called a **modal matrix**. If the conditions of Theorem 4.6.8 are satisfied and if, in particular, Eq. (4.6.9) holds, then we say that **matrix A has been diagonalized.**

4.6.12. Exercise. Prove Theorem 4.6.8.

Let us now consider some specific examples.

4.6.13. Example. Let X be a two-dimensional vector space over the field of real numbers. Let $A \in L(X, X)$, and let $\{e_1, e_2\}$ be a basis for X. Suppose the matrix $\mathbf{A}$ of A with respect to this basis is given by

$$\mathbf{A} = \begin{bmatrix} -2 & 4 \\ 1 & 1 \end{bmatrix}.$$

The characteristic polynomial of A is

$$p(\lambda) = \det(A - \lambda I) = \det(\mathbf{A} - \lambda\mathbf{I}) = \lambda^2 + \lambda - 6.$$

Now $\det(\mathbf{A} - \lambda\mathbf{I}) = 0$ if and only if $\lambda^2 + \lambda - 6 = 0$, or $(\lambda - 2)(\lambda + 3) = 0$. Thus, the eigenvalues of A are $\lambda_1 = 2$ and $\lambda_2 = -3$. To find an eigenvector corresponding to λ_1, we solve the equation $(\mathbf{A} - \lambda_1\mathbf{I})\mathbf{x} = \mathbf{0}$, or

$$\begin{bmatrix} -4 & 4 \\ 1 & -1 \end{bmatrix}\begin{bmatrix} \xi_1 \\ \xi_2 \end{bmatrix} = \begin{bmatrix} 0 \\ 0 \end{bmatrix}.$$

The last equation yields the equations

$$-4\xi_1 + 4\xi_2 = 0, \quad \xi_1 - \xi_2 = 0.$$

These are satisfied whenever $\xi_1 = \xi_2$. Thus, any vector of the form

$$\begin{bmatrix} \xi \\ \xi \end{bmatrix}, \quad \xi \neq 0,$$

is an eigenvector of $\mathbf{A}$ corresponding to the eigenvalue λ_1. For convenience, let us choose $\xi = 1$. Then

$$\mathbf{x}_1 = \begin{bmatrix} 1 \\ 1 \end{bmatrix}$$

is an eigenvector. In a similar fashion we obtain an eigenvector $\mathbf{x}_2$ corresponding to λ_2, given by

$$\mathbf{x}_2 = \begin{bmatrix} 4 \\ -1 \end{bmatrix}.$$

The diagonal matrix $\mathbf{A}'$ given in Eq. (4.6.9) is, in the present case,

$$\mathbf{A}' = \begin{bmatrix} \lambda_1 & 0 \\ 0 & \lambda_2 \end{bmatrix} = \begin{bmatrix} 2 & 0 \\ 0 & -3 \end{bmatrix}.$$

We can arrive at $\mathbf{A}'$, using Eq. (4.6.10). Specifically, let

$$\mathbf{P} = [\mathbf{x}_1, \mathbf{x}_2] = \begin{bmatrix} 1 & 4 \\ 1 & -1 \end{bmatrix}.$$

Then

$$\mathbf{P}^{-1} = \begin{bmatrix} .2 & .8 \\ .2 & -.2 \end{bmatrix}$$

and

$$\mathbf{P}^{-1}\mathbf{A}\mathbf{P} = \begin{bmatrix} 2 & 0 \\ 0 & -3 \end{bmatrix} = \begin{bmatrix} \lambda_1 & 0 \\ 0 & \lambda_2 \end{bmatrix}.$$

By Eq. (4.3.2), the basis $\{e_1', e_2'\} \subset X$ with respect to which $\mathbf{A}'$ represents A is given by

$$e_1' = \sum_{j=1}^{2} p_{j1}e_j = e_1 + e_2, \quad e_2' = \sum_{j=1}^{2} p_{j2}e_j = 4e_1 - e_2.$$

In view of Theorem 4.3.8, if $\mathbf{x}$ is the coordinate representation of x with respect to $\{e_1, e_2\}$, then $\mathbf{x}' = \mathbf{P}^{-1}\mathbf{x}$ is the coordinate representation of x with respect to $\{e_1', e_2'\}$. The vectors e_1', e_2' are, of course, eigenvectors of A corresponding to λ_1 and λ_2, respectively. ■

When the algebraic multiplicity of one or more of the eigenvalues of a linear transformation is greater than one, then the linear transformation is said to have **repeated eigenvalues**. Unfortunately, in this case it is not always possible to represent the linear transformation by a diagonal matrix. To put it another way, if a square matrix has repeated eigenvalues, then it is not always possible to diagonalize it. However, from the preceding results of the present section it should be clear that a linear transformation with repeated eigenvalues can be represented by a diagonal matrix if the number of linearly independent eigenvectors corresponding to any eigenvalue is the same as the algebraic multiplicity of the eigenvalue. The following examples throw additional light on these comments.

4.6.14. Example. The characteristic equation of the matrix

$$\mathbf{A} = \begin{bmatrix} 1 & 3 & -2 \\ 0 & 4 & -2 \\ 0 & 3 & -1 \end{bmatrix}$$

is

$$\det(\mathbf{A} - \lambda\mathbf{I}) = (1 - \lambda)^2(2 - \lambda) = 0,$$

and the eigenvalues of $\mathbf{A}$ are $\lambda_1 = 1$ and $\lambda_2 = 2$. The algebraic multiplicity of λ_1 is two. Corresponding to λ_1 we can find two linearly independent eigenvectors

$$\begin{bmatrix} 1 \\ 2 \\ 3 \end{bmatrix} \quad \text{and} \quad \begin{bmatrix} 1 \\ 0 \\ 0 \end{bmatrix}.$$

Corresponding to λ_2 we have an eigenvector

$$\begin{bmatrix} 1 \\ 1 \\ 1 \end{bmatrix}.$$

Letting $\mathbf{P}$ denote a modal matrix, we have

$$\mathbf{P} = \begin{bmatrix} 1 & 1 & 1 \\ 2 & 0 & 1 \\ 3 & 0 & 1 \end{bmatrix} \quad \text{and} \quad \mathbf{P}^{-1} = \begin{bmatrix} 0 & -1 & 1 \\ 1 & -2 & 1 \\ 0 & 3 & -2 \end{bmatrix}$$

and

$$\mathbf{A}' = \mathbf{P}^{-1}\mathbf{A}\mathbf{P} = \begin{bmatrix} 1 & 0 & 0 \\ 0 & 1 & 0 \\ 0 & 0 & 2 \end{bmatrix}.$$

In this example, $\dim \mathfrak{N}_{\lambda_1} = 2$, which happens to be the same as the algebraic multiplicity of λ_1. For this reason we were able to diagonalize matrix $\mathbf{A}$. ■

The next example shows that the multiplicity of an eigenvalue need not be the same as its algebraic multiplicity. In this case we are not able to diagonalize the matrix.

4.6.15. Example. The characteristic equation of the matrix

$$\mathbf{A} = \begin{bmatrix} 2 & 1 & -2 \\ 0 & 2 & -1 \\ 0 & 0 & 1 \end{bmatrix}$$

is

$$\det(\mathbf{A} - \lambda\mathbf{I}) = (1 - \lambda)(2 - \lambda)^2 = 0$$

and the eigenvalues of $\mathbf{A}$ are $\lambda_1 = 1$ and $\lambda_2 = 2$. The algebraic multiplicity of λ_2 is two. An eigenvector corresponding to λ_1 is $\mathbf{x}_1^T = (1, 1, 1)$. An eigenvector corresponding to λ_2 must be of the form

$$\begin{bmatrix} \xi \\ 0 \\ 0 \end{bmatrix}, \quad \xi \neq 0.$$

Setting $\mathbf{x}_2^T = (1, 0, 0)$, we see that $\dim \mathfrak{N}_{\lambda_2} = 1$, and thus we have not been able to determine a basis for R^3, consisting of eigenvectors. Consequently, we have not been able to diagonalize $\mathbf{A}$. ■

When a matrix cannot be diagonalized we seek, for practical reasons, to represent a linear transformation by a matrix which is as nearly diagonal as possible. Our next result provides the basis of representing linear transformations by such matrices, which we call **block diagonal** matrices. In the next section we will consider the "simplest" type of block diagonal matrix, called the **Jordan canonical form**.

4.6.16. Theorem. Let X be an n-dimensional vector space, and let $A \in L(X, X)$. Let Y and Z be linear subspaces of X such that $X = Y \oplus Z$ and such that A is reduced by Y and Z. Then there exists a basis for X such that the matrix $\mathbf{A}$ of A with respect to this basis has the form

$$\mathbf{A} = \left[\begin{array}{c|c} \mathbf{A}_1 & \mathbf{0} \\ \hline \mathbf{0} & \mathbf{A}_2 \end{array}\right]$$

where $\dim Y = r$, $\mathbf{A}_1$ is an $(r \times r)$ matrix and $\mathbf{A}_2$ is an $(n - r) \times (n - r)$ matrix.

4.6.17. Exercise. Prove Theorem 4.6.16.

We can generalize the preceding result. Suppose that X is the direct sum of linear subspaces $X_1, \ldots, X_p$ that are invariant under $A \in L(X, X)$. We can define linear transformations $A_i \in L(X_i, X_i)$, $i = 1, \ldots, p$, by $A_i x = Ax$ for $x \in X_i$. That is to say, A_i is the restriction of A to X_i. We now can find for each A_i a matrix representation $\mathbf{A}_i$, which will lead us to the following result.

4.6.18. Theorem. Let X be a finite-dimensional vector space, and let $A \in L(X, X)$. If X is the direct sum of p linear subspaces, $X_1, \ldots, X_p$, which are invariant under A, then there exists a basis for X such that the matrix representation for A is in the **block diagonal form** given by

$$\mathbf{A} = \begin{bmatrix} \mathbf{A}_1 & & & & \\ & \mathbf{A}_2 & & & \mathbf{0} \\ & & \cdot & & \\ & & & \cdot & \\ & & & & \cdot \\ & \mathbf{0} & & & \mathbf{A}_p \end{bmatrix}.$$

Moreover, $\mathbf{A}_i$ is a matrix representation of A_i, the restriction of A to X_i,

$i = 1, \ldots, p$. Also,

$$\det(\mathbf{A}) = \prod_{i=1}^{p} \det(\mathbf{A}_i).$$

4.6.19. Exercise. Prove Theorem 4.6.18.

From the preceding it is clear that, in order to carry out the block diagonalization of a matrix **A**, we need to find an appropriate set of invariant subspaces of X and, furthermore, to find a simple matrix representation on each of these subspaces.

4.6.20. Example. Let X be an n-dimensional vector space. If $A \in L(X, X)$ has n distinct eigenvalues, $\lambda_1, \ldots, \lambda_n$, and if we let

$$\mathfrak{N}_j = \{x: (A - \lambda_j I)x = 0\}, \quad j = 1, \ldots, n,$$

then $\mathfrak{N}_j$ is an invariant linear subspace under A and

$$X = \mathfrak{N}_1 \oplus \ldots \oplus \mathfrak{N}_n.$$

For any $x \in \mathfrak{N}_j$, we have $Ax = \lambda_j x$, and hence $A_j x = \lambda_j x$ for $x \in \mathfrak{N}_j$. A basis for $\mathfrak{N}_j$ is any non-zero $x_j \in \mathfrak{N}_j$. Thus, with respect to this basis, A_j is represented by the matrix λ_j (in this case, simply a scalar). With respect to a basis of n linearly independent eigenvectors, $\{x_1, \ldots, x_n\}$, A is represented by Eq. (4.6.6). ■

In addition to the diagonal form and the block diagonal form, there are many other useful forms for matrices to represent linear transformations on finite-dimensional vector spaces. One of these canonical forms involves triangular matrices, which we consider in the last result of the present section.

We say that an $(n \times n)$ matrix is a **triangular matrix** if it either has the form

$$\begin{bmatrix} a_{11} & a_{12} & a_{13} & \cdots & a_{1n} \\ 0 & a_{22} & a_{23} & \cdots & a_{2n} \\ \cdots & \cdots & \cdots & \cdots & \cdots \\ 0 & 0 & 0 & \cdots & a_{n-1,n} \\ 0 & 0 & 0 & \cdots & a_{nn} \end{bmatrix} \tag{4.6.21}$$

or the form

$$\begin{bmatrix} a_{11} & 0 & 0 & \cdots & 0 \\ a_{21} & a_{22} & 0 & \cdots & 0 \\ \cdots & \cdots & \cdots & \cdots & \cdots \\ a_{n-1,1} & a_{n-1,2} & a_{n-1,3} & \cdots & 0 \\ a_{n1} & a_{n2} & a_{n3} & \cdots & a_{nn} \end{bmatrix}. \tag{4.6.22}$$

In case of Eq. (4.6.21) we speak of an **upper triangular matrix**, whereas in case of Eq. (4.6.22) we say the matrix is in the **lower triangular form.**

4.6.23. Theorem. Let X be an n-dimensional vector space over C, and let $A \in L(X, X)$. Then there exists a basis for X such that A is represented by an upper triangular matrix.

Proof. We will show that if $\mathbf{A}$ is a matrix of A, then $\mathbf{A}$ is similar to an upper triangular matrix $\mathbf{A}'$. Our proof is by induction on n.

If $n = 1$, then the assertion is clearly true. Now assume that for $n = k$, and $\mathbf{C}$ any $k \times k$ matrix, there exists a non-singular matrix $\mathbf{Q}$ such that $\mathbf{C}' = \mathbf{Q}^{-1}\mathbf{C}\mathbf{Q}$ is an upper triangular matrix. We now must show the validity of the assertion for $n = k + 1$. Let X be a $(k + 1)$-dimensional vector space over C. Let λ_1 be an eigenvalue of A, and let f_1 be a corresponding eigenvector. Let $\{f_2, \ldots, f_{k+1}\}$ be any set of vectors in X such that $\{f_1, \ldots, f_{k+1}\}$ is a basis for X. Let $\mathbf{B}$ be the matrix of A with respect to the basis $\{f_1, \ldots, f_{k+1}\}$. Since $Af_1 = \lambda_1 f_1$, $\mathbf{B}$ must be of the form

$$\mathbf{B} = \begin{bmatrix} \lambda_1 & b_{12} & \cdots & b_{1,k+1} \\ 0 & b_{22} & \cdots & b_{2,k+1} \\ \cdots & \cdots & \cdots & \cdots \\ 0 & b_{k+1,2} & \cdots & b_{k+1,k+1} \end{bmatrix}.$$

Now let $\mathbf{C}$ be the $k \times k$ matrix

$$\mathbf{C} = \begin{bmatrix} b_{22} & \cdots & b_{2,k+1} \\ \vdots & & \vdots \\ b_{k+1,2} & \cdots & b_{k+1,k+1} \end{bmatrix}.$$

By our induction hypothesis, there exists a non-singular matrix $\mathbf{Q}$ such that

$$\mathbf{C}' = \mathbf{Q}^{-1}\mathbf{C}\mathbf{Q},$$

where $\mathbf{C}'$ is an upper triangular matrix. Now let

$$\mathbf{P} = \left[\begin{array}{c|ccc} 1 & 0 & \cdots & 0 \\ \hline 0 & & & \\ \vdots & & \mathbf{Q} & \\ 0 & & & \end{array}\right].$$

By direct computation we have

$$\mathbf{P}^{-1} = \left[\begin{array}{c|ccc} 1 & 0 & \cdots & 0 \\ \hline 0 & & & \\ \vdots & & \mathbf{Q}^{-1} & \\ 0 & & & \end{array}\right]$$

and

$$\mathbf{P}^{-1}\mathbf{B}\mathbf{P} = \left[\begin{array}{c|cccc} \lambda_1 & * & * & \cdots & * \\ \hline 0 & & & & \\ \cdot & & & & \\ \cdot & & \mathbf{Q}^{-1}\mathbf{C}\mathbf{Q} & & \\ \cdot & & & & \\ 0 & & & & \end{array}\right],$$

where the $*$'s denote elements which may be non-zero. Letting $\mathbf{A} = \mathbf{P}^{-1}\mathbf{B}\mathbf{P}$, it follows that $\mathbf{A}$ is upper triangular and is similar to $\mathbf{B}$. Hence, any $(k + 1) \times (k + 1)$ matrix which represents $A \in L(X, X)$ is similar to the upper triangular matrix $\mathbf{A}$, by Theorem 4.3.19. This completes the proof of the theorem. ■

Note that if $\mathbf{A}$ is in the triangular form of either Eq. (4.6.21) or (4.6.22), then

$$\det(\mathbf{A} - \lambda\mathbf{I}) = (a_{11} - \lambda)(a_{22} - \lambda) \ldots (a_{nn} - \lambda).$$

In this case the diagonal elements of $\mathbf{A}$ are the eigenvalues of $\mathbf{A}$.

4.7. MINIMAL POLYNOMIALS, NILPOTENT OPERATORS, AND THE JORDAN CANONICAL FORM

In the present section we develop the Jordan canonical form of a matrix. To do so, we need to introduce the concepts of minimal polynomial and nilpotent operator and to study some of the properties of such polynomials and operators. *Unless otherwise specified, X denotes an n-dimensional vector space over a field F throughout the present section.*

A. *Minimal Polynomials*

For purposes of motivation, consider the matrix

$$\mathbf{A} = \begin{bmatrix} 1 & 3 & -2 \\ 0 & 4 & -2 \\ 0 & 3 & -1 \end{bmatrix}.$$

The characteristic polynomial of $\mathbf{A}$ is

$$p(\lambda) = (1 - \lambda)^2(2 - \lambda),$$

and we know from the Cayley-Hamilton theorem that

$$p(\mathbf{A}) = \mathbf{0}. \tag{4.7.1}$$

Now let us consider the polynomial

$$m(\lambda) = (1 - \lambda)(2 - \lambda) = 2 - 3\lambda + \lambda^2.$$

Then

$$m(\mathbf{A}) = 2\mathbf{I} - 3\mathbf{A} + \mathbf{A}^2 = \mathbf{0}. \tag{4.7.2}$$

Thus, matrix **A** satisfies Eq. (4.7.2), which is of lower degree than Eq. (4.7.1), the characteristic equation of **A**.

Before stating our first result, we recall that an nth-order polynomial in λ is said to be monic if the coefficient of λ^n is unity (see Definition 2.3.4).

4.7.3. Theorem. Let **A** be an $(n \times n)$ matrix. Then there exists a unique polynomial $m(\lambda)$ such that

(i) $m(\mathbf{A}) = \mathbf{0}$;

(ii) $m(\lambda)$ is monic; and,

(iii) if $m'(\lambda)$ is any other polynomial such that $m'(\mathbf{A}) = \mathbf{0}$, then the degree of $m(\lambda)$ is less or equal to the degree of $m'(\lambda)$ (i.e., $m(\lambda)$ is of the lowest degree such that $m(\mathbf{A}) = \mathbf{0}$).

Proof. We know that a polynomial, $p(\lambda)$, exists such that $p(\mathbf{A}) = \mathbf{0}$, namely, the characteristic polynomial. Furthermore, the degree of $p(\lambda)$ is n. Thus, there exists a polynomial, say $f(\lambda)$, of degree $m \leq n$ such that $f(\mathbf{A}) = \mathbf{0}$. Let us choose m to be the lowest degree for which $f(\mathbf{A}) = \mathbf{0}$. Since $f(\lambda)$ is of degree m, we may divide $f(\lambda)$ by the coefficient of λ^m, thus obtaining a monic polynomial, $m(\lambda)$, such that $m(\mathbf{A}) = \mathbf{0}$. To show that $m(\lambda)$ is unique, suppose there is another monic polynomial $m'(\lambda)$ of degree m such that $m'(\mathbf{A}) = \mathbf{0}$. Then $m(\lambda) - m'(\lambda)$ is a polynomial of degree less than m. Furthermore, $m(\mathbf{A}) - m'(\mathbf{A}) = \mathbf{0}$, which contradicts our assumption that $m(\lambda)$ is the polynomial of lowest degree such that $m(\mathbf{A}) = \mathbf{0}$. This completes the proof. ■

The preceding result gives rise to the notion of minimal polynomial.

4.7.4. Definition. The polynomial $m(\lambda)$ defined in Theorem 4.7.3 is called the **minimal polynomial of A**.

Other names for minimal polynomial are **minimum polynomial** and **reduced characteristic function**. In the following we will develop an explicit form for the minimal polynomial of **A**, which makes it possible to determine it systematically, rather than by trial and error.

In the remainder of this section we let **A** *denote an* $(n \times n)$ *matrix, we let* $p(\lambda)$ *denote the characteristic polynomial of* **A**, *and we let* $m(\lambda)$ *denote the minimal polynomial of* **A**.

4.7.5. Theorem. Let $f(\lambda)$ be any polynomial such that $f(\mathbf{A}) = \mathbf{0}$. Then $m(\lambda)$ divides $f(\lambda)$.

Proof. Let ν denote the degree of $m(\lambda)$. Then there exist polynomials $q(\lambda)$ and $r(\lambda)$ such that (see Theorem 2.3.9)

$$f(\lambda) = q(\lambda)m(\lambda) + r(\lambda),$$

where $\deg [r(\lambda)] < \nu$ or $r(\lambda) = 0$. Since $f(\mathbf{A}) = \mathbf{0}$, we have

$$\mathbf{0} = q(\mathbf{A})m(\mathbf{A}) + r(\mathbf{A}),$$

and hence $r(\mathbf{A}) = \mathbf{0}$. This means $r(\lambda) = 0$, for otherwise we would have a contradiction to the fact that $m(\lambda)$ is the minimal polynomial of $\mathbf{A}$. Hence, $f(\lambda) = q(\lambda)m(\lambda)$ and $m(\lambda)$ divides $f(\lambda)$. ■

4.7.6. Corollary. The minimal polynomial of $\mathbf{A}$, $m(\lambda)$, divides the characteristic polynomial of $\mathbf{A}$, $p(\lambda)$.

4.7.7. Exercise. Prove Corollary 4.7.6.

We now prove:

4.7.8. Theorem. The polynomial $p(\lambda)$ divides $[m(\lambda)]^n$.

Proof. We want to show that $[m(\lambda)]^n = p(\lambda)q(\lambda)$ for some polynomial $q(\lambda)$. Let $m(\lambda)$ be of degree ν and be given by

$$m(\lambda) = \lambda^\nu + \beta_1\lambda^{\nu-1} + \ldots + \beta_\nu.$$

Let us now define the matrices $\mathbf{B}_0, \mathbf{B}_1, \ldots, \mathbf{B}_{\nu-1}$ as

$$\mathbf{B}_0 = \mathbf{I}, \quad \mathbf{B}_1 = \mathbf{A} + \beta_1\mathbf{I}, \quad \mathbf{B}_2 = \mathbf{A}^2 + \beta_1\mathbf{A} + \beta_2\mathbf{I}, \quad \ldots,$$
$$\mathbf{B}_{\nu-1} = \mathbf{A}^{\nu-1} + \beta_1\mathbf{A}^{\nu-2} + \ldots + \beta_{\nu-1}\mathbf{I}.$$

Then

$$\mathbf{B}_0 = \mathbf{I}, \quad \mathbf{B}_1 - \mathbf{A}\mathbf{B}_0 = \beta_1\mathbf{I}, \quad \mathbf{B}_2 - \mathbf{A}\mathbf{B}_1 = \beta_2\mathbf{I}, \quad \ldots,$$
$$\mathbf{B}_{\nu-1} - \mathbf{A}\mathbf{B}_{\nu-2} = \beta_{\nu-1}\mathbf{I},$$

and

$$-\mathbf{A}\mathbf{B}_{\nu-1} = \beta_\nu\mathbf{I} - [\mathbf{A}^\nu + \beta_1\mathbf{A}^{\nu-1} + \ldots + \beta_\nu\mathbf{I}]$$
$$= \beta_\nu\mathbf{I} - m(\mathbf{A}) = \beta_\nu\mathbf{I}.$$

Now let

$$\mathbf{B}(\lambda) = -\lambda^{\nu-1}\mathbf{B}_0 - \lambda^{\nu-2}\mathbf{B}_1 - \ldots - \mathbf{B}_{\nu-1}.$$

Then

$$(\mathbf{A} - \lambda\mathbf{I})\mathbf{B}(\lambda) = \lambda^\nu\mathbf{B}_0 + \lambda^{\nu-1}\mathbf{B}_1 + \ldots + \lambda\mathbf{B}_{\nu-1} - [\lambda^{\nu-1}\mathbf{A}\mathbf{B}_0 + \lambda^{\nu-2}\mathbf{A}\mathbf{B}_1 + \ldots + \mathbf{A}\mathbf{B}_{\nu-1}]$$
$$= \lambda^\nu\mathbf{B}_0 + \lambda^{\nu-1}[\mathbf{B}_1 - \mathbf{A}\mathbf{B}_0] + \lambda^{\nu-2}[\mathbf{B}_2 - \mathbf{A}\mathbf{B}_1] + \ldots + \lambda[\mathbf{B}_{\nu-1} - \mathbf{A}\mathbf{B}_{\nu-2}] - \mathbf{A}\mathbf{B}_{\nu-1}$$
$$= \lambda^\nu\mathbf{I} + \beta_1\lambda^{\nu-1}\mathbf{I} + \ldots + \beta_{\nu-1}\lambda\mathbf{I} + \beta_\nu\mathbf{I} = m(\lambda)\mathbf{I}.$$

Taking the determinant of both sides of this equation we have

$$[\det(\mathbf{A} - \lambda\mathbf{I})] \cdot [\det \mathbf{B}(\lambda)] = [m(\lambda)]^n.$$

But $\det \mathbf{B}(\lambda)$ is a polynomial in λ, say $q(\lambda)$. Thus, we have proved that $p(\lambda)q(\lambda) = [m(\lambda)]^n$. ■

The next result establishes the form of the minimal polynomial.

4.7.9. Theorem. Let $p(\lambda)$ be given by Eq. (4.5.24); i.e.,

$$p(\lambda) = (\lambda_1 - \lambda)^{m_1}(\lambda_2 - \lambda)^{m_2} \ldots (\lambda_p - \lambda)^{m_p},$$

where $m_1, \ldots, m_p$ are the algebraic multiplicities of the distinct eigenvalues $\lambda_1, \ldots, \lambda_p$ of $\mathbf{A}$, respectively. Then

$$m(\lambda) = (\lambda - \lambda_1)^{\nu_1}(\lambda - \lambda_2)^{\nu_2} \ldots (\lambda - \lambda_p)^{\nu_p}, \tag{4.7.10}$$

where $1 \leq \nu_i \leq m_i$ for $i = 1, \ldots, p$.

4.7.11. Exercise. Prove Theorem 4.7.9. (Hint: Assume that $m(\lambda) = (\lambda - \rho_1)^{\nu_1} \ldots (\lambda - \rho_q)^{\nu_q}$, and use Corollary 4.7.6 and Theorem 4.7.8).

The only unknowns left to determine the minimal polynomial of $\mathbf{A}$ are $\nu_1, \ldots, \nu_p$ in Eq. (4.7.10). These can be determined in several ways.

Our next result is an immediate consequence of Theorem 4.3.27.

4.7.12. Theorem. Let $\mathbf{A}'$ be similar to $\mathbf{A}$, and let $m'(\lambda)$ be the minimal polynomial of $\mathbf{A}'$. Then $m'(\lambda) = m(\lambda)$.

This result justifies the following definition.

4.7.13. Definition. Let $A \in L(X, X)$. The **minimal polynomial of** A is the minimal polynomial of any matrix $\mathbf{A}$ which represents A.

In order to develop the Jordan canonical form (for linear transformations with repeated eigenvalues), we need to establish several additional preliminary results which are important in their own right.

4.7.14. Theorem. Let $A \in L(X, X)$, and let $f(\lambda)$ be any polynomial in λ. Let $\mathfrak{N}_f = \{x : f(A)x = 0\}$. Then $\mathfrak{N}_f$ is an invariant linear subspace of X under A.

Proof. The proof that $\mathfrak{N}_f$ is a linear subspace of X is straightforward and is left as an exercise. To show that $\mathfrak{N}_f$ is invariant under A, let $x \in \mathfrak{N}_f$, so that $f(A)x = 0$. We want to show that $Ax \in \mathfrak{N}_f$. Let

$$f(\lambda) = \sum_{k=0}^{m} \beta_k \lambda^k.$$

Then

$$f(A)x = \sum_{k=0}^{m} \beta_k A^k x = 0$$

and

$$f(A)Ax = \sum_{k=0}^{m} \beta_k A^k Ax = A \sum_{k=0}^{m} \beta_k A^k x = Af(A)x = 0,$$

which completes the proof. ■

Before proceeding further, we establish some additional notation. Let $\lambda_1, \ldots, \lambda_p$ be distinct eigenvalues of $A \in L(X, X)$. For $j = 1, \ldots, p$ and for any positive integer q, let

$$\mathfrak{N}_j^q = \{x : (A - \lambda_j I)^q x = 0\}. \tag{4.7.15}$$

Note that this notation is consistent with that used in Example 4.6.20 if we define

$$\mathfrak{N}_j^1 = \mathfrak{N}_j.$$

Note also that, in view of Theorem 4.7.14, $\mathfrak{N}_j^q$ is an invariant linear subspace of X under A.

We will need the following result concerning the restriction of a linear transformation.

4.7.16. Theorem. Let $A \in L(X, X)$. Let X_1 and X_2 be linear subspaces of X such that $X = X_1 \oplus X_2$ and let A_1 be the restriction of A to X_1. Let $f(\lambda)$ be any polynomial in λ. If A is reduced by X_1 and X_2 then, for all $x_1 \in X_1$, $f(A_1)x_1 = f(A)x_1$.

4.7.17. Exercise. Prove Theorem 4.7.16.

Next we prove:

4.7.18. Theorem. Let X be a vector space over C, and let $A \in L(X, X)$. Let $m(\lambda)$ be the minimal polynomial of A as given in Eq. (4.7.10). Let $g(\lambda) = (\lambda - \lambda_1)^{\nu_1}$, let $h(\lambda) = (\lambda - \lambda_2)^{\nu_2} \ldots (\lambda - \lambda_p)^{\nu_p}$ if $p \geq 2$, let $h(\lambda) = 1$ if $p = 1$. Let A_1 be the restriction of A to $\mathfrak{N}_1^{\nu_1}$, i.e., $A_1 x = Ax$ for all $x \in \mathfrak{N}_1^{\nu_1}$. Let $\mathfrak{M} = \{x \in X : h(A)x = 0\}$. Then

(i) $X = \mathfrak{N}_1^{\nu_1} \oplus \mathfrak{M}$; and
(ii) $(\lambda - \lambda_1)^{\nu_1}$ is the minimal polynomial for A_1.

Proof. By Theorem 4.7.14, $\mathfrak{M}$ and $\mathfrak{N}_1^{\nu_1}$ are invariant linear subspaces under A. Since $g(\lambda)$ and $h(\lambda)$ are relatively prime, there exist polynomials $q(\lambda)$ and $r(\lambda)$ such that (see Exercise 2.3.15)

$$q(\lambda)g(\lambda) + r(\lambda)h(\lambda) = 1.$$

Hence, for the linear transformation A we have

$$q(A)g(A) + r(A)h(A) = I. \tag{4.7.19}$$

Thus, for $x \in X$, we have

$$x = q(A)g(A)x + r(A)h(A)x.$$

Now since

$$h(A)q(A)g(A)x = q(A)g(A)h(A)x = q(A)m(A)x = q(A)0x = 0,$$

it follows that $q(A)g(A)x \in \mathfrak{M}$. We can similarly show that $r(A)h(A)x \in \mathfrak{N}_1^{\nu_1}$. Thus, for every $x \in X$ we have $x = x_1 + x_2$, where $x_1 \in \mathfrak{N}_1^{\nu_1}$ and $x_2 \in \mathfrak{M}$. Let us now show that this representation of x is unique. Let $x = x_1 + x_2 = x_1' + x_2'$, where $x_1, x_1' \in \mathfrak{N}_1^{\nu_1}$ and $x_2, x_2' \in \mathfrak{M}$. Then

$$r(A)h(A)x = r(A)h(A)x_1 = r(A)h(A)x_1'.$$

Applying Eq. (4.7.19) to x_1 and x_1' we get

$$x_1 = r(A)h(A)x_1$$

and

$$x_1' = r(A)h(A)x_1'.$$

From this we conclude that $x_1 = x_1'$. Similarly, we can show that $x_2 = x_2'$. Therefore, $X = \mathfrak{N}_1^{\nu_1} \oplus \mathfrak{M}$.

To prove the second part of the theorem, let A_1 be the restriction of A to $\mathfrak{N}_1^{\nu_1}$ and let A_2 be the restriction of A to $\mathfrak{M}$. Let $m_1(\lambda)$ and $m_2(\lambda)$ be the minimal polynomials for A_1 and A_2, respectively. Since $g(A_1) = 0$ and $h(A_2) = 0$, it follows that $m_1(\lambda)$ divides $g(\lambda)$ and $m_2(\lambda)$ divides $h(\lambda)$, by Theorem 4.7.5. Hence, we can write

$$m_1(\lambda) = (\lambda - \lambda_1)^{k_1}$$

and

$$m_2(\lambda) = (\lambda - \lambda_2)^{k_2} \cdots (\lambda - \lambda_p)^{k_p},$$

where $0 \leq k_i \leq \nu_i$ for $i = 1, \ldots, p$. Now let $f(\lambda) = m_1(\lambda)m_2(\lambda)$. Then $f(A) = m_1(A)m_2(A)$. Let $x \in X$ with $x = x_1 + x_2$, where $x_1 \in \mathfrak{N}_1^{\nu_1}$ and $x_2 \in \mathfrak{M}$. Then

$$f(A)x = m_1(A)m_2(A)x_1 + m_1(A)m_2(A)x_2 = m_2(A)m_1(A)x_1 = 0.$$

Therefore, $f(A) = 0$. But this implies that $m(\lambda)$ divides $f(\lambda)$ and $0 \leq \nu_i \leq k_i$, $i = 1, \ldots, p$.

We thus conclude that $k_i = \nu_i$ for $i = 1, \ldots, p$, which completes the proof of the theorem. ■

We are now in a position to prove the following important result, called the **primary decomposition theorem**.

4.7.20. Theorem. Let X be an n-dimensional vector space over C, let $\lambda_1, \ldots, \lambda_p$ be the distinct eigenvalues of $A \in L(X, X)$, let the characteristic polynomial of A be

$$p(\lambda) = (\lambda_1 - \lambda)^{m_1} \dots (\lambda_p - \lambda)^{m_p}, \tag{4.7.21}$$

and let the minimal polynomial of A be

$$m(\lambda) = (\lambda - \lambda_1)^{\nu_1} \dots (\lambda - \lambda_p)^{\nu_p}. \tag{4.7.22}$$

Let

$$X_i = \{x: (A - \lambda_i I)^{\nu_i} x = 0\}, \quad i = 1, \dots, p.$$

Then

(i) $X_i, i = 1, \dots, p$ are invariant linear subspaces of X under A;
(ii) $X = X_1 \oplus \dots \oplus X_p$;
(iii) $(\lambda - \lambda_i)^{\nu_i}$ is the minimal polynomial of A_i, where A_i is the restriction of A to X_i; and,
(iv) $\dim X_i = m_i, i = 1, \dots, p$.

Proof. The proofs of parts (i), (ii), and (iii) follow from the preceding theorem by a simple induction argument and are left as an exercise.

To prove the last part of the theorem, we first show that the only eigenvalue of $A_i \in L(X_i, X_i)$ is $\lambda_i, i = 1, \dots, p$. Let $v \in X_i$, $v \neq 0$, and consider $(A_i - \lambda I)v = 0$. From part (iii) it follows that

$$\begin{aligned} 0 &= (A_i - \lambda_i I)^{\nu_i} v = (A_i - \lambda_i I)^{\nu_i - 1}(A_i - \lambda_i I)v \\ &= (A_i - \lambda_i I)^{\nu_i - 1}(\lambda - \lambda_i)v = (\lambda - \lambda_i)(A_i - \lambda_i I)^{\nu_i - 2}(A_i - \lambda_i I)v \\ &= (\lambda - \lambda_i)^2(A_i - \lambda_i I)^{\nu_i - 2} v = \dots = (\lambda - \lambda_i)^{\nu_i} v. \end{aligned}$$

From this we conclude that $\lambda = \lambda_i$.

We can now find a matrix representation of A in the form given in Theorem 4.6.18. Furthermore, from this theorem it follows that

$$p(\lambda) = \det(A - \lambda I) = \prod_{i=1}^{p} \det(A_i - \lambda I).$$

Now since the only eigenvalue of A_i is λ_i, the determinant of $A_i - \lambda I$ must be of the form

$$\det(A_i - \lambda I) = (\lambda_i - \lambda)^{q_i},$$

where $q_i = \dim X_i$. Since $p(\lambda)$ is given by Eq. (4.7.21), we must have

$$(\lambda_1 - \lambda)^{m_1} \dots (\lambda_p - \lambda)^{m_p} = (\lambda_1 - \lambda)^{q_1} \dots (\lambda_p - \lambda)^{q_p},$$

from which we conclude that $m_i = q_i$. Thus, $\dim X_i = m_i, i = 1, \dots, p$.

This concludes the proof of the theorem. ■

4.7.23. Exercise. Prove parts (i)–(iii) of Theorem 4.7.20.

The preceding result shows that we can always represent $A \in L(X, X)$ by a matrix in block diagonal form, where the number of diagonal blocks (in the matrix **A** of Theorem 4.6.18) is equal to the number of distinct eigenvalues of A. We will next find a convenient representation for each of the diagonal submatrices $\mathbf{A}_i$. It may turn out that one or more of the submatrices $\mathbf{A}_i$ will be diagonal. Our next result tells us specifically when $A \in L(X, X)$ is representable by a diagonal matrix.

4.7.24. Theorem. Let X be an n-dimensional vector space over C, and let $A \in L(X, X)$. Let $\lambda_1, \ldots, \lambda_p$, $p \leq n$, be the distinct eigenvalues of A. Then there exists a basis for X such that the matrix **A** of A with respect to this basis is diagonal if and only if the minimal polynomial for A is of the form

$$m(\lambda) = (\lambda - \lambda_1)(\lambda - \lambda_2)\ldots(\lambda - \lambda_p).$$

4.7.25. Exercise. Prove Theorem 4.7.24.

4.7.26. Exercise. Apply the above theorem to the matrices in Examples 4.6.14 and 4.6.15.

B. Nilpotent Operators

Let us now proceed to find a representation for each of the $A_i \in L(X_i, X_i)$ in Theorem 4.7.20 so that the block diagonal matrix representation of $A \in L(X, X)$ (see Theorem 4.6.18) is as simple as possible. To accomplish this, we first need to define and examine so-called nilpotent operators.

4.7.27. Definition. Let $N \in L(X, X)$. Then N is said to be **nilpotent** if there exists an integer $q > 0$ such that $N^q = 0$. A nilpotent operator is said to be of **index** q if $N^q = 0$ but $N^{q-1} \neq 0$.

Recall now that Theorem 4.7.20 enables us to write $X = X_1 \oplus X_2 \oplus \ldots \oplus X_p$. Furthermore, the linear transformation $(A_i - \lambda_i I)$ is nilpotent on X_i. If we let $N_i = A_i - \lambda_i I$, then $A_i = \lambda_i I + N_i$. Now $\lambda_i I$ is clearly represented by a diagonal matrix. However, the transformation N_i forces the matrix representation of A_i to be in general non-diagonal. So our next task is to seek a simple representation of the nilpotent operator N_i.

In the next few results, which are concerned with properties of nilpotent operators, we drop for convenience the subscript i.

4.7.28. Theorem. Let $N \in L(V, V)$, where V is an m-dimensional vector space. If N is a nilpotent linear transformation of index q and if $x \in V$ is such that $N^{q-1}x \neq 0$, then the vectors $x, Nx, \ldots, N^{q-1}x$ in V are linearly independent.

Proof. We first note that if $N^{q-1}x \neq 0$, then $N^j x \neq 0$ for $j = 0, 1, \ldots, q - 1$. Our proof is now by contradiction. Suppose that

$$\sum_{i=0}^{q-1} \alpha_i N^i x = 0.$$

Let j be the smallest integer such that $\alpha_j \neq 0$. Then we can write

$$N^j x = -\sum_{i=j+1}^{q-1} \frac{\alpha_i}{\alpha_j} N^i x \neq 0.$$

Thus,

$$N^j x = N^{j+1}\left[\sum_{i=j+1}^{q-1} \left(-\frac{\alpha_i}{\alpha_j}\right) N^{i-j-1} x\right] = N^{j+1} y,$$

where y is defined in an obvious way. Now we can write

$$N^{q-1}x = N^{q-j-1}N^j x = N^{q-j-1}N^{j+1}y = N^q y = 0.$$

We thus have arrived at a contradiction, which proves our result. ■

Next, let us examine the matrix representation of nilpotent transformations.

4.7.29. Theorem. Let V be a q-dimensional vector space, and let $N \in L(V, V)$ be nilpotent of index q. Let $m_0 \in V$ be such that $N^{q-1}m_0 \neq 0$. Then the matrix $\mathbf{N}$ of N with respect to the basis $\{N^{q-1}m_0, N^{q-2}m_0, \ldots, m_0\}$ in V is given by

$$\mathbf{N} = \begin{bmatrix} 0 & 1 & 0 & 0 & \ldots & 0 & 0 \\ 0 & 0 & 1 & 0 & \ldots & 0 & 0 \\ \cdots & \cdots & \cdots & \cdots & \cdots & \cdots & \cdots \\ 0 & 0 & 0 & 0 & \ldots & 0 & 1 \\ 0 & 0 & 0 & 0 & \ldots & 0 & 0 \end{bmatrix}. \tag{4.7.30}$$

Proof. By the previous theorem we know that $\{N^{q-1}m_0, \ldots, m_0\}$ is a linearly independent set. By hypothesis, there are q vectors in the set, and thus $\{N^{q-1}m_0, \ldots, m_0\}$ forms a basis for V. Let $e_i = N^{q-i}m_0$ for $i = 1, \ldots, q$. Then

$$Ne_i = \begin{cases} 0, & i = 1 \\ e_{i-1}, & i = 2, \ldots, q. \end{cases}$$

Hence,

$$\begin{aligned} Ne_1 &= 0 \cdot e_1 + 0 \cdot e_2 + \ldots + 0 \cdot e_{q-1} + 0 \cdot e_q \\ Ne_2 &= 1 \cdot e_1 + 0 \cdot e_2 + \ldots + 0 \cdot e_{q-1} + 0 \cdot e_q \\ &\vdots \\ Ne_q &= 0 \cdot e_1 + 0 \cdot e_2 + \ldots + 1 \cdot e_{q-1} + 0 \cdot e_q. \end{aligned}$$

From Eq. (4.2.2) and Definition 4.2.7, it follows that the representation of N is that given by Eq. (4.7.30). This completes the proof of the theorem. ■

The above theorem establishes the matrix representation of a nilpotent linear transformation of index q on a q-dimensional vector space. We will next determine the representation of a nilpotent operator of index ν on a vector space of dimension m, where $\nu \leq m$. The following lemma shows that we can dismiss the case $\nu > m$.

4.7.31. Lemma. Let $N \in L(V, V)$ be nilpotent of index ν, where dim $V = m$. Then $\nu \leq m$.

Proof. Assume $x \in V$, $N^\nu x = 0$, $N^{\nu-1}x \neq 0$, and $\nu > m$. Then, by Theorem 4.7.28, the vectors $x, Nx, \ldots, N^{\nu-1}x$ are linearly independent, which contradicts the fact that dim $V = m$. ■

To prove the next theorem, we require the following result.

4.7.32. Lemma. Let V be an m-dimensional vector space, let $N \in L(V, V)$, let ν be any positive integer, and let

$$W_1 = \{x: Nx = 0\}, \dim W_1 = l_1,$$
$$W_2 = \{x: N^2x = 0\}, \dim W_2 = l_2,$$
$$\cdots\cdots\cdots\cdots\cdots\cdots,$$
$$W_\nu = \{x: N^\nu x = 0\}, \dim W_\nu = l_\nu.$$

Also, for any i such that $1 < i < \nu$, let $\{e_1, \ldots, e_m\}$ be a basis for V such that $\{e_1, \ldots, e_{l_i}\}$ is a basis for W_i. Then

(i) $W_1 \subset W_2 \subset \ldots \subset W_\nu$; and

(ii) $\{e_1, \ldots, e_{l_{i-1}}, Ne_{l_i+1}, \ldots, Ne_{l_{i+1}}\}$ is a linearly independent set of vectors in W_i.

Proof. To prove the first part, let $x \in W_i$ for any $i < \nu$. Then $N^i x = 0$. Hence, $N^{i+1}x = 0$, which implies $x \in W_{i+1}$.

To prove the second part, let $r = l_{i-1}$ and let $t = l_{i+1} - l_i$. We note that if $x \in W_{i+1}$, then $N^i(Nx) = 0$, and so $Nx \in W_i$. This implies that $Ne_j \in W_i$ for $j = l_i + 1, \ldots, l_{i+1}$. This means that the set of vectors $\{e_1, \ldots, e_r, Ne_{l_i+1}, \ldots, Ne_{l_{i+1}}\}$ is in W_i. We show that this set is linearly independent by contradiction. Assume there are scalars $\alpha_1, \ldots, \alpha_r$ and $\beta_1, \ldots, \beta_t$, not all zero, such that

$$\alpha_1 e_1 + \ldots + \alpha_r e_r + \beta_1 Ne_{l_i+1} + \ldots + \beta_t Ne_{l_{i+1}} = 0.$$

Since $\{e_1, \ldots, e_r\}$ is a linearly independent set, at least one of the β_i must be non-zero. Rearranging the last equation we have

$$\beta_1 Ne_{l_i+1} + \ldots + \beta_t Ne_{l_{i+1}} = -\alpha_1 e_1 - \ldots - \alpha_r e_r \in W_{i-1}.$$

Hence,

$$N^{i-1}(\beta_1 Ne_{l_i+1} + \ldots + \beta_t Ne_{l_{i+1}}) = 0.$$

Thus,

$$N^i(\beta_1 e_{l_t+1} + \ldots + \beta_t e_{l_{i+1}}) = 0,$$

and $(\beta_1 e_{l_i+1} + \ldots + \beta_t e_{l_{i+1}}) \in W_i$. If $\beta_1 e_{l_i+1} + \ldots + \beta_t e_{l_{i+1}} \neq 0$, it can be written as a linear combination of $e_1, \ldots, e_{l_i}$, which contradicts the fact that $\{e_1, \ldots, e_{l_{i+1}}\}$ is a linearly independent set. If $\beta_1 e_{l_i+1} + \ldots + \beta_t e_{l_{i+1}} = 0$, we contradict the fact that $\{e_1, \ldots, e_{l_{i+1}}\}$ is a linearly independent set. Hence, we conclude that $\alpha_i = 0$ for $i = 1, \ldots, r$ and $\beta_i = 0$ for $i = 1, \ldots, t$. This completes the proof of the theorem. ■

We are now in a position to consider the general representation of a nilpotent operator on a finite-dimensional vector space.

4.7.33. Theorem. Let V be an m-dimensional vector space over C, and let $N \in L(V, V)$ be nilpotent of index ν. Let $W_1 = \{x: Nx = 0\}, \ldots, W_\nu = \{x: N^\nu x = 0\}$, and let $l_i = \dim W_i$, $i = 1, \ldots, \nu$. Then there exists a basis for V such that the matrix $\mathbf{N}$ of N is of block diagonal form,

$$\mathbf{N} = \begin{bmatrix} \mathbf{N}_1 & \cdots & \mathbf{0} \\ \cdot & \cdot & \cdot \\ \cdot & \cdot & \cdot \\ \cdot & \cdot & \cdot \\ \mathbf{0} & \cdots & \mathbf{N}_r \end{bmatrix}, \tag{4.7.34}$$

where

$$\mathbf{N}_i = \begin{bmatrix} 0 & 1 & 0 & 0 & \ldots & 0 & 0 \\ 0 & 0 & 1 & 0 & \ldots & 0 & 0 \\ \cdots & \cdots & \cdots & \cdots & \cdots & \cdots & \cdots \\ 0 & 0 & 0 & 0 & \ldots & 0 & 1 \\ 0 & 0 & 0 & 0 & \ldots & 0 & 0 \end{bmatrix}, \tag{4.7.35}$$

$i = 1, \ldots, r$, where $r = l_1$, $\mathbf{N}_i$ is a $(k_i \times k_i)$ matrix, $1 \leq k_i \leq \nu$, and k_i is determined in the following way: there are

$l_\nu - l_{\nu-1}$	$(\nu \times \nu)$ matrices,
$2l_i - l_{i+1} - l_{i-1}$	$(i \times i)$ matrices, $i = 2, \ldots, \nu - 1$, and
$2l_1 - l_2$	(1×1) matrices.

The basis for V consists of strings of vectors of the form

$$N^{k_1-1}x_1, \ldots, x_1, \quad N^{k_2-1}x_2, \quad \ldots, \quad x_2, \quad \ldots, \quad N^{k_r-1}x_r, \quad \ldots, \quad x_r.$$

Proof. By Lemma 4.7.32, $W_1 \subset W_2 \subset \ldots \subset W_\nu$. Let $\{e_1, \ldots, e_m\}$ be a basis for V such that $\{e_1, \ldots, e_{l_i}\}$ is a basis for W_i. We see that $W_\nu = V$. Since N is nilpotent of index ν, $W_{\nu-1} \neq W_\nu$ and $l_{\nu-1} < l_\nu$.

We now proceed to select a new basis for V which yields the desired result. We find it convenient to use double subscripting of vectors. Let $f_{1,\nu} = e_{l_{\nu-1}+1}$,

$\ldots, f_{(l_\nu - l_{\nu-1}),\nu} = e_{l_\nu}$ and let $f_{1,\nu-1} = Nf_{1,\nu}, \ldots, f_{(l_\nu - l_{\nu-1}),\nu-1} = Nf_{(l_\nu - l_{\nu-1}),\nu}$. By Lemma 4.7.32, it follows that $\{e_1, \ldots, e_{l_{\nu-2}}, f_{1,\nu-1}, \ldots, f_{(l_\nu - l_{\nu-1}),\nu-1}\}$ is a linearly independent subset of $W_{\nu-1}$, which may or may not be a basis for $W_{\nu-1}$. If it is not, we adjoin additional elements from $W_{\nu-1}$, denoted by $f_{(l_\nu - l_{\nu-1}+1),\nu-1}, \ldots, f_{(l_{\nu-1}-l_{\nu-2}),\nu-1}$, so as to form a basis for $W_{\nu-1}$. Now let $f_{1,\nu-2} = Nf_{1,\nu-1}, f_{2,\nu-2} = Nf_{2,\nu-1}, \ldots, f_{(l_{\nu-1}-l_{\nu-2}),\nu-2} = Nf_{(l_{\nu-1}-l_{\nu-2}),\nu-1}$. By Lemma 4.7.32 it follows, as before, that $\{e_1, \ldots, e_{l_{\nu-3}}, f_{1,\nu-2}, \ldots, f_{(l_{\nu-1}-l_{\nu-2}),\nu-2}\}$ is a linearly independent set in $W_{\nu-2}$. If this set is not a basis we adjoin vectors from $W_{\nu-2}$ so that we do have a basis. We denote the vectors that we adjoin by $f_{(l_{\nu-1}-l_{\nu-2}+1),\nu-2}, \ldots, f_{(l_{\nu-2}-l_{\nu-3}),\nu-2}$. We continue in this manner until we have formed a basis for V. We express this basis in the manner indicated in Figure C.

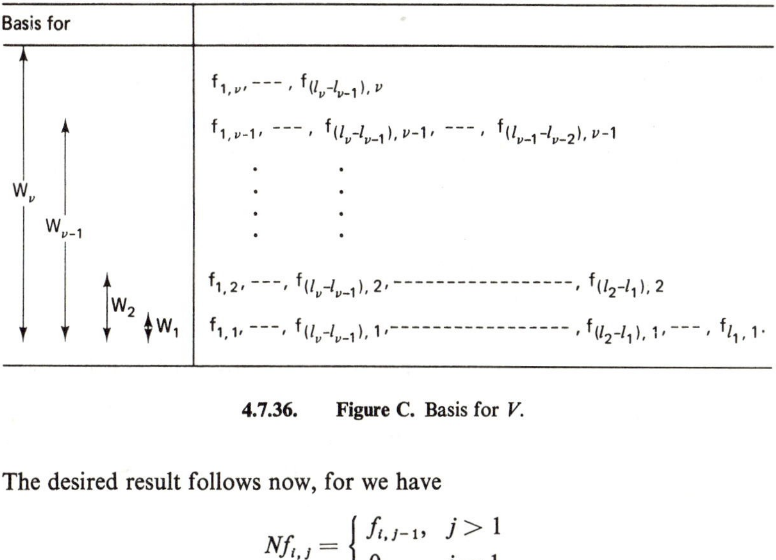

4.7.36. **Figure C.** Basis for V.

The desired result follows now, for we have

$$Nf_{i,j} = \begin{cases} f_{i,j-1}, & j > 1 \\ 0, & j = 1. \end{cases}$$

Hence, if we let $x_1 = f_{1,\nu}$, we see that the first column in Figure C reading bottom to top, is

$$N^{\nu-1}x_1, \quad N^{\nu-2}x_1, \quad \ldots, \quad Nx_1, \quad x_1.$$

We see that each column of Figure C determines a string consisting of k_i entries, where $k_i = \nu$ for $i = 1, \ldots, (l_\nu - l_{\nu-1})$. Note that $(l_\nu - l_{\nu-1}) > 0$, so there is at least one string. In general, the number of strings with j entries is $(l_j - l_{j-1}) - (l_{j+1} - l_j) = 2l_j - l_{j+1} - l_{j-1}$ for $j = 2, \ldots, \nu - 1$. Also, there are $l_1 - (l_2 - l_1) = 2l_1 - l_2$ vectors, or strings with one entry.

Finally, to show that the number of entries, $\mathbf{N}_i$, in $\mathbf{N}$ is l_1, we see that there are a total of $(l_\nu - l_{\nu-1}) + (2l_{\nu-1} - l_\nu - l_{\nu-2}) + \ldots + (2l_2 - l_1 - l_3) + (2l_1 - l_2) = l_1$ columns in the table of Figure C.

This completes the proof of the theorem. ■

The reader should study Figure C to obtain an appreciation of the structure of the basis for the space V.

C. The Jordan Canonical Form

We are finally now in a position to state and prove the result which establishes the Jordan canonical form of matrices.

4.7.37. Theorem. Let X be an n-dimensional vector space over C, and let $A \in L(X, X)$. Let the characteristic polynomial of A be

$$p(\lambda) = (\lambda_1 - \lambda)^{m_1} \ldots (\lambda_p - \lambda)^{m_p},$$

and let the minimal polynomial of A be

$$m(\lambda) = (\lambda - \lambda_1)^{\nu_1} \ldots (\lambda - \lambda_p)^{\nu_p},$$

where $\lambda_1, \ldots, \lambda_p$ are the distinct eigenvalues of A. Let

$$X_i = \{x \in X: (A - \lambda_i I)^{\nu_i} x = 0\}.$$

Then

(i) $X_1, \ldots, X_p$ are invariant subspaces of X under A;
(ii) $X = X_1 \oplus \ldots \oplus X_p$;
(iii) $\dim X_i = m_i$, $i = 1, \ldots, p$; and
(iv) there exists a basis for X such that the matrix $\mathbf{A}$ of A with respect to this basis is of the form

$$\mathbf{A} = \begin{bmatrix} \mathbf{A}_1 & \mathbf{0} & \ldots & \mathbf{0} \\ \mathbf{0} & \mathbf{A}_2 & \ldots & \mathbf{0} \\ \cdots & \cdots & \cdots & \cdots \\ \mathbf{0} & \mathbf{0} & \ldots & \mathbf{A}_p \end{bmatrix}, \tag{4.7.38}$$

where $\mathbf{A}_i$ is an $(m_i \times m_i)$ matrix of the form

$$\mathbf{A}_i = \lambda_i \mathbf{I} + \mathbf{N}_i \tag{4.7.39}$$

and where $\mathbf{N}_i$ is the matrix of the nilpotent operator $(A_i - \lambda_i I)$ of index ν_i on X_i given by Eq. (4.7.34) and Eq. (4.7.35).

Proof. Parts (i)–(iii) are restatements of the primary decomposition theorem (Theorem 4.7.20). From this theorem we also know that $(\lambda - \lambda_i)^{\nu_i}$ is the minimal polynomial of A_i, the restriction of A to X_i. Hence, if we let $N_i = A_i - \lambda_i I$, then N_i is a nilpotent operator of index ν_i on X_i. We are thus able to represent N_i as shown in Eq. (4.7.35).

The completes the proof of the theorem. ■

A little extra work shows that the representation of $A \in L(X, X)$ by a matrix $\mathbf{A}$ of the form given in Eqs. (4.7.38) and (4.7.39) is unique, except for the order in which the block diagonals $\mathbf{A}_1, \ldots, \mathbf{A}_p$ appear in $\mathbf{A}$.

4.7.40. Definition. The matrix $\mathbf{A}$ of $A \in L(X, X)$ given by Eqs. (4.7.38) and (4.7.39) is called the **Jordan canonical form of** A.

We conclude the present section with an example.

4.7.41. Example. Let $X = R^7$, and let $\{u_1, \ldots, u_7\}$ be the natural basis for X (see Example 4.1.15). Let $A \in L(X, X)$ be represented by the matrix

$$\mathbf{A} = \begin{bmatrix} -1 & 0 & -1 & 1 & 1 & 3 & 0 \\ 0 & 1 & 0 & 0 & 0 & 0 & 0 \\ 2 & 1 & 2 & -1 & -1 & -6 & 0 \\ -2 & 0 & -1 & 2 & 1 & 3 & 0 \\ 0 & 0 & 0 & 0 & 1 & 0 & 0 \\ 0 & 0 & 0 & 0 & 0 & 1 & 0 \\ -1 & -1 & 0 & 1 & 2 & 4 & 1 \end{bmatrix}$$

with respect to $\{u_1, \ldots, u_7\}$. Let us find the matrix $\mathbf{A}'$ which represents A in the Jordan canonical form.

We first find that the characteristic polynomial of A is

$$p(\lambda) = (1 - \lambda)^7.$$

This implies that $\lambda_1 = 1$ is the only distinct eigenvalue of A. Its algebraic multiplicity is $m_1 = 7$. In order to find the minimal polynomial of A, let

$$N = A - \lambda_1 I,$$

where I is the identity operator in $L(X, X)$. The representation for N with respect to the natural basis in X is

$$\mathbf{N} = \mathbf{A} - \mathbf{I} = \begin{bmatrix} -2 & 0 & -1 & 1 & 1 & 3 & 0 \\ 0 & 0 & 0 & 0 & 0 & 0 & 0 \\ 2 & 1 & 1 & -1 & -1 & -6 & 0 \\ -2 & 0 & -1 & 1 & 1 & 3 & 0 \\ 0 & 0 & 0 & 0 & 0 & 0 & 0 \\ 0 & 0 & 0 & 0 & 0 & 0 & 0 \\ -1 & -1 & 0 & 1 & 2 & 4 & 0 \end{bmatrix}.$$

We assume the minimal polynomial is of the form $m(\lambda) = (\lambda - 1)^{\nu_1}$ and proceed to find the smallest ν_1 such that $m(\mathbf{A} - \mathbf{I}) = m(\mathbf{N}) = \mathbf{0}$. We first obtain

$$\mathbf{N}^2 = \begin{bmatrix} 0 & -1 & 0 & 0 & 0 & 3 & 0 \\ 0 & 0 & 0 & 0 & 0 & 0 & 0 \\ 0 & 1 & 0 & 0 & 0 & -3 & 0 \\ 0 & -1 & 0 & 0 & 0 & 3 & 0 \\ 0 & 0 & 0 & 0 & 0 & 0 & 0 \\ 0 & 0 & 0 & 0 & 0 & 0 & 0 \\ 0 & 0 & 0 & 0 & 0 & 0 & 0 \end{bmatrix}.$$

Next, we get that

$$\mathbf{N}^3 = \mathbf{0},$$

and so $\nu_1 = 3$. Hence, N is a nilpotent operator of index 3. We see that $X = \mathfrak{N}_1^3$. We will now apply Theorem 4.7.33 to obtain a representation for N in this space.

Using the notation of Theorem 4.7.33, we let $W_1 = \{x: Nx = 0\}$, $W_2 = \{x: N^2x = 0\}$, and $W_3 = \{x: N^3x = 0\}$. We see that $\mathbf{N}$ has three linearly independent rows. This means that the rank of N is 3, and so $\dim(W_1) = l_1 = 4$. Similarly, the rank of N^2 is 1, and so $\dim(W_2) = l_2 = 6$. Clearly, $\dim(W_3) = l_3 = 7$. We can conclude that N will have a representation $\mathbf{N}'$ of the form in Eq. (4.7.34) with $r = 4$. Each of the $\mathbf{N}_i'$ will be of the form in Eq. (4.7.35). There will be $l_3 - l_2 = 1$ (3×3) matrix. There will be $2l_2 - l_3 - l_1 = 1$ (2×2) matrix, and $2l_1 - l_2 = 2$ (1×1) matrices. Hence, there is a basis for X such that N may be represented by the matrix

$$\mathbf{N}' = \left[\begin{array}{ccc|cc|c|c} 0 & 1 & 0 & 0 & 0 & 0 & 0 \\ 0 & 0 & 1 & 0 & 0 & 0 & 0 \\ 0 & 0 & 0 & 0 & 0 & 0 & 0 \\ \hline 0 & 0 & 0 & 0 & 1 & 0 & 0 \\ 0 & 0 & 0 & 0 & 0 & 0 & 0 \\ \hline 0 & 0 & 0 & 0 & 0 & 0 & 0 \\ \hline 0 & 0 & 0 & 0 & 0 & 0 & 0 \end{array}\right].$$

The corresponding basis will consist of strings of vectors of the form

$$N^2x_1, \quad Nx_1, \quad x_1, \quad Nx_2, \quad x_2, \quad x_3, \quad x_4.$$

We will represent the vectors x_1, x_2, x_3, and x_4 by $\mathbf{x}_1, \mathbf{x}_2, \mathbf{x}_3$, and $\mathbf{x}_4$, their coordinate representations, respectively, with respect to the natural basis $\{u_1, \ldots, u_7\}$ in X. We begin by choosing $x_1 \in W_3$ such that $x_1 \notin W_2$; i.e., we find an $\mathbf{x}_1$ such that $\mathbf{N}^3\mathbf{x}_1 = \mathbf{0}$ but $\mathbf{N}^2\mathbf{x}_1 \neq \mathbf{0}$. The vector $\mathbf{x}_1^T = (0,$

1, 0, 0, 0, 0, 0) will do. We see that $(\mathbf{N}\mathbf{x}_1)^T = (0, 0, 1, 0, 0, 0, -1)$ and $(\mathbf{N}^2\mathbf{x}_1)^T = (-1, 0, 1, -1, 0, 0, 0)$. Hence, $Nx_1 \in W_2$ but $Nx_1 \notin W_1$ and $N^2x_1 \in W_1$. We see there will be only one string of length three, and so we next choose $x_2 \in W_2$ such that $x_2 \notin W_1$. Also, the pair $\{Nx_1, x_2\}$ must be linearly independent. The vector $\mathbf{x}_2^T = (1, 0, 0, 0, 0, 0, 0)$ will do. Now $(\mathbf{N}\mathbf{x}_2)^T = (-2, 0, 2, -2, 0, 0, -1)$, and $Nx_2 \in W_1$. We complete the basis for X by selecting two more vectors, $x_3, x_4 \in W_1$, such that $\{N^2x_1, Nx_2, x_3, x_4\}$ are linearly independent. The vectors $\mathbf{x}_3^T = (0, 0, -1, -2, 1, 0, 0)$ and $\mathbf{x}_4^T = (1, 3, 1, 0, 0, 1, 0)$ will suffice.

It follows that the matrix

$$\mathbf{P} = [\mathbf{N}^2\mathbf{x}_1, \mathbf{N}\mathbf{x}_1, \mathbf{x}_1, \mathbf{N}\mathbf{x}_2, \mathbf{x}_2, \mathbf{x}_3, \mathbf{x}_4]$$

is the matrix of the new basis with respect to the natural basis (see Exercise 4.3.9).

The reader can readily show that

$$\mathbf{N}' = \mathbf{P}^{-1}\mathbf{N}\mathbf{P},$$

where

$$\mathbf{P} = \begin{bmatrix} -1 & 0 & 0 & -2 & 1 & 0 & 1 \\ 0 & 0 & 1 & 0 & 0 & 0 & 3 \\ 1 & 1 & 0 & 2 & 0 & -1 & 1 \\ -1 & 0 & 0 & -2 & 0 & -2 & 0 \\ 0 & 0 & 0 & 0 & 0 & 1 & 0 \\ 0 & 0 & 0 & 0 & 0 & 0 & 1 \\ 0 & -1 & 0 & -1 & 0 & 0 & 0 \end{bmatrix}$$

and

$$\mathbf{P}^{-1} = \begin{bmatrix} 0 & 0 & 2 & 1 & 4 & -2 & 2 \\ 0 & 0 & 1 & 1 & 3 & -1 & 0 \\ 0 & 1 & 0 & 0 & 0 & -3 & 0 \\ 0 & 0 & -1 & -1 & -3 & 1 & -1 \\ 1 & 0 & 0 & -1 & -2 & -1 & 0 \\ 0 & 0 & 0 & 0 & 1 & 0 & 0 \\ 0 & 0 & 0 & 0 & 0 & 1 & 0 \end{bmatrix}.$$

Finally the Jordan canonical form for A is given by

$$\mathbf{A}' = \mathbf{N}' + \mathbf{I}.$$

(Recall that the matrix representation for I is the same for any basis in X.) Thus,

$$\mathbf{A}' = \left[\begin{array}{ccc|cc|c|c} 1 & 1 & 0 & 0 & 0 & 0 & 0 \\ 0 & 1 & 1 & 0 & 0 & 0 & 0 \\ 0 & 0 & 1 & 0 & 0 & 0 & 0 \\ \hline 0 & 0 & 0 & 1 & 1 & 0 & 0 \\ 0 & 0 & 0 & 0 & 1 & 0 & 0 \\ \hline 0 & 0 & 0 & 0 & 0 & 1 & 0 \\ \hline 0 & 0 & 0 & 0 & 0 & 0 & 1 \end{array}\right].$$

Again, the reader can show that

$$\mathbf{A}' = \mathbf{P}^{-1}\mathbf{A}\mathbf{P}.$$

In general, it is more convenient as a check to show that $\mathbf{P}\mathbf{A}' = \mathbf{A}\mathbf{P}$. ■

4.7.42. Exercise. Let $X = R^6$, and let $\{u_1, \ldots, u_6\}$ denote the natural basis for X. Let $A \in L(X, X)$ be represented by the matrix

$$\mathbf{A} = \begin{bmatrix} 5 & -1 & 1 & 1 & 0 & 0 \\ 1 & 3 & -1 & -1 & 0 & 0 \\ 0 & 0 & 4 & 0 & 1 & 1 \\ 0 & 0 & 0 & 4 & -1 & -1 \\ 0 & 0 & 0 & 0 & 3 & 1 \\ 0 & 0 & 0 & 0 & 1 & 3 \end{bmatrix}.$$

Show that the Jordan canonical form of A is given by

$$\mathbf{A}' = \left[\begin{array}{ccc|cc|c} 4 & 1 & 0 & 0 & 0 & 0 \\ 0 & 4 & 1 & 0 & 0 & 0 \\ 0 & 0 & 4 & 0 & 0 & 0 \\ \hline 0 & 0 & 0 & 4 & 1 & 0 \\ 0 & 0 & 0 & 0 & 4 & 0 \\ \hline 0 & 0 & 0 & 0 & 0 & 2 \end{array}\right],$$

and find a basis for X for which $\mathbf{A}'$ represents A.

4.8. BILINEAR FUNCTIONALS AND CONGRUENCE

In the present section we consider the representation and some of the properties of bilinear functionals on real finite-dimensional vector spaces. (We will consider bilinear functionals defined on complex vector spaces in Chapter 6.)

Throughout this section X is assumed to be an n-dimensional vector space over the field of real numbers. We recall that if f is a bilinear functional on a real vector space X, then $f\colon X \times X \longrightarrow R$ and

$$f(\alpha x_1 + \beta x_2, y) = \alpha f(x_1, y) + \beta f(x_2, y)$$

and

$$f(x, \alpha y_1 + \beta y_2) = \alpha f(x, y_1) + \beta f(x, y_2)$$

for all $\alpha, \beta \in R$ and for all $x, x_1, x_2, y, y_1, y_2 \in X$. As a consequence of these properties we have, more generally,

$$f\left(\sum_{j=1}^{r} \alpha_j x_j, \sum_{k=1}^{s} \beta_k y_k\right) = \sum_{j=1}^{r} \sum_{k=1}^{s} \alpha_j \beta_k f(x_j, y_k)$$

for all $\alpha_j, \beta_k \in R$ and $x_j, y_k \in X$, $j = 1, \ldots, r$ and $k = 1, \ldots, s$.

4.8.1. Definition. Let $\{e_1, \ldots, e_n\}$ be a basis for the vector space X, and let

$$f_{ij} = f(e_i, e_j), \quad i, j = 1, \ldots, n.$$

The matrix $\mathbf{F} = [f_{ij}]$ is called the **matrix of the bilinear functional f with respect to** $\{e_1, \ldots, e_n\}$.

Our first result provides us with the representation of bilinear functionals on finite-dimensional vector spaces.

4.8.2. Theorem. Let f be a bilinear functional on a vector space X, and let $\{e_1, \ldots, e_n\}$ be a basis for X. Let $\mathbf{F}$ be the matrix of the bilinear functional f with respect to the basis $\{e_1, \ldots, e_n\}$. If x and y are arbitrary vectors in X and if $\mathbf{x}$ and $\mathbf{y}$ are their coordinate representation with respect to the basis $\{e_1, e_2, \ldots, e_n\}$, then

$$f(x, y) = \mathbf{x}^T\mathbf{F}\mathbf{y} = \sum_{i=1}^{n} \sum_{j=1}^{n} f_{ij}\xi_i\eta_j. \tag{4.8.3}$$

Proof. We have $\mathbf{x}^T = (\xi_1, \ldots, \xi_n)$ and $\mathbf{y}^T = (\eta_1, \ldots, \eta_n)$. Also, $x = \xi_1 e_1 + \ldots + \xi_n e_n$ and $y = \eta_1 e_1 + \ldots + \eta_n e_n$. Therefore,

$$f(x, y) = \sum_{i=1}^{n} \sum_{j=1}^{n} \xi_i \eta_j f(e_i, e_j) = \sum_{i=1}^{n} \sum_{j=1}^{n} f_{ij}\xi_i\eta_j = \mathbf{x}^T\mathbf{F}\mathbf{y}$$

which was to be shown. ■

Conversely, if we are given any $(n \times n)$ matrix $\mathbf{F}$, we can use formula (4.8.3) to define the bilinear functional f whose matrix with respect to the given basis $\{e_1, \ldots, e_n\}$ is, in turn, $\mathbf{F}$ again. In general, it therefore follows that on finite-dimensional vector spaces, bilinear functionals correspond in a one-to-one fashion to matrices. The particular one-to-one correspondence depends on the particular basis chosen.

Now recall that if X is a real vector space, then f is said to be symmetric if $f(x, y) = f(y, x)$ for all $x, y \in X$. We also have the following related concept.

4.8.4. Definition. A bilinear functional f on a vector space X is said to be **skew symmetric** if

$$f(x, y) = -f(y, x) \tag{4.8.5}$$

for all $x, y \in X$.

For symmetric and skew symmetric bilinear functionals we have the following result.

4.8.6. Theorem. Let $\{e_1, \ldots, e_n\}$ be a basis for X, and let $\mathbf{F}$ be the matrix for a bilinear functional f with respect to $\{e_1, \ldots, e_n\}$. Then

(i) f is symmetric if and only if $\mathbf{F} = \mathbf{F}^T$;

(ii) f is skew symmetric if and only if $\mathbf{F} = -\mathbf{F}^T$; and

(iii) for every bilinear functional f, there exists a unique symmetric bilinear functional f_1 and a unique skew symmetric bilinear functional f_2 such that

$$f = f_1 + f_2.$$

We call f_1 the **symmetric part** of f and f_2 the **skew symmetric part** of f.

4.8.7. Exercise. Prove Theorem 4.8.6.

The preceding result motivates the following definitions.

4.8.8. Definition. An $(n \times n)$ matrix $\mathbf{F}$ is said to be

(i) **symmetric** if $\mathbf{F} = \mathbf{F}^T$; and

(ii) **skew symmetric** if $\mathbf{F} = -\mathbf{F}^T$.

The next result is easily verified.

4.8.9. Theorem. Let f be a bilinear functional on X, and let f_1 and f_2 be the symmetric and skew symmetric parts of f, respectively. Then

$$f_1(x, y) = \tfrac{1}{2}[f(x, y) + f(y, x)]$$

and

$$f_2(x, y) = \tfrac{1}{2}[f(x, y) - f(y, x)]$$

for all $x, y \in X$.

4.8.10. Exercise. Prove Theorem 4.8.9.

Now let us recall that the quadratic form induced by f was defined as $\hat{f}(x) = f(x, x)$. On a real finite-dimensional vector space X we now have

$$\hat{f}(x) = \mathbf{x}^T\mathbf{F}\mathbf{x} = \sum_{i=1}^{n}\sum_{j=1}^{n} f_{ij}\xi_i\xi_j.$$

For quadratic forms we have the following result.

4.8.11. Theorem. Let f and g be bilinear functionals on X. The quadratic forms induced by f and g are equal if and only if f and g have the same symmetric part. In other words, $\hat{f}(x) = \hat{g}(x)$ for all $x \in X$ if and only if

$$\tfrac{1}{2}[f(x, y) + f(y, x)] = \tfrac{1}{2}[g(x, y) + g(y, x)]$$

for all $x, y \in X$.

Proof. We note that

$$f(x - y, x - y) = f(x, x) - f(x, y) - f(y, x) + f(y, y).$$

From this it follows that

$$\tfrac{1}{2}[f(x, y) + f(y, x)] = \tfrac{1}{2}[f(x, x) + f(y, y) - f(x - y, x - y)].$$

Now if $g(x, x) = f(x, x)$, then

$$\begin{aligned}\tfrac{1}{2}[f(x, y) + f(y, x)] &= \tfrac{1}{2}[g(x, x) + g(y, y) - g(x - y, x - y)]\\ &= \tfrac{1}{2}[g(x, y) + g(y, x)],\end{aligned}$$

so that

$$\tfrac{1}{2}[f(x, y) + f(y, x)] = \tfrac{1}{2}[g(x, y) + g(y, x)]. \tag{4.8.12}$$

Conversely, assume that Eq. (4.8.12) holds for all $x, y \in X$. Then, in particular, if we let $x = y$, we have $f(x, x) = g(x, x)$ for all $x \in X$. This concludes our proof. ■

From Theorem 4.8.11 the following useful result follows: when treating quadratic functionals, it suffices to work with symmetric bilinear functionals.

We leave the proof of the next result as an exercise.

4.8.13. Theorem. A bilinear functional on a vector space X is skew symmetric if and only if $f(x, x) = 0$ for all $x \in X$.

4.8.14. Exercise. Prove Theorem 4.8.13.

The next result enables us to introduce the concept of congruence.

4.8.15. Theorem. Let f be a bilinear functional on a vector space X, let $\{e_1, \ldots, e_n\}$ be a basis for X, and let $\mathbf{F}$ be the matrix of f with respect to this basis. Let $\{e'_1, \ldots, e'_n\}$ be another basis whose matrix with respect to $\{e_1, \ldots, e_n\}$ is $\mathbf{P}$. Then the matrix $\mathbf{F}'$ of f with respect to the basis $\{e'_1, \ldots, e'_n\}$

is given by

$$\mathbf{F}' = \mathbf{P}^T\mathbf{F}\mathbf{P}. \tag{4.8.16}$$

Proof. Let $\mathbf{F}' = [f'_{ij}]$ where, by definition, $f'_{ij} = f(e'_i, e'_j)$. Then $f\left(\sum_{k=1}^{n} p_{ki}e_k, \sum_{l=1}^{n} p_{lj}e_l\right) = \sum_{k=1}^{n}\sum_{l=1}^{n} p_{ki}p_{lj}f(e_k, e_l) = \sum_{k=1}^{n}\sum_{l=1}^{n} p_{ki}f_{kl}p_{lj}$. Hence, $\mathbf{F}' = \mathbf{P}^T\mathbf{F}\mathbf{P}$. ∎

We now have:

4.8.17. Definition. An $(n \times n)$ matrix $\mathbf{F}'$ is said to be **congruent** to an $(n \times n)$ matrix $\mathbf{F}$ if there exists a non-singular matrix $\mathbf{P}$ such that

$$\mathbf{F}' = \mathbf{P}^T\mathbf{F}\mathbf{P}. \tag{4.8.18}$$

We express this congruence by writing $\mathbf{F}' \sim \mathbf{F}$.

Note that congruent matrices are also equivalent matrices. The next theorem shows that $\sim$ in Definition 4.8.17 is reflexive, symmetric, and transitive, and as such it is an equivalence relation.

4.8.19. Theorem. Let $\mathbf{A}$, $\mathbf{B}$, and $\mathbf{C}$ be $(n \times n)$ matrices. Then,

(i) $\mathbf{A}$ is congruent to $\mathbf{A}$;
(ii) if $\mathbf{A}$ is congruent to $\mathbf{B}$, then $\mathbf{B}$ is congruent to $\mathbf{A}$; and
(iii) if $\mathbf{A}$ is congruent to $\mathbf{B}$ and $\mathbf{B}$ is congruent to $\mathbf{C}$, then $\mathbf{A}$ is congruent to $\mathbf{C}$.

Proof. Clearly $\mathbf{A} = \mathbf{I}^T\mathbf{A}\mathbf{I}$, which proves the first part. To prove the second part, let $\mathbf{A} = \mathbf{P}^T\mathbf{B}\mathbf{P}$, where $\mathbf{P}$ is non-singular. Then

$$\mathbf{B} = (\mathbf{P}^T)^{-1}\mathbf{A}\mathbf{P}^{-1} = (\mathbf{P}^{-1})^T\mathbf{A}(\mathbf{P}^{-1}),$$

which proves the second part.

Let $\mathbf{A} = \mathbf{P}^T\mathbf{B}\mathbf{P}$ and $\mathbf{B} = \mathbf{Q}^T\mathbf{C}\mathbf{Q}$, where $\mathbf{P}$ and $\mathbf{Q}$ are non-singular matrices. Then

$$\mathbf{A} = \mathbf{P}^T\mathbf{Q}^T\mathbf{C}\mathbf{Q}\mathbf{P} = (\mathbf{Q}\mathbf{P})^T\mathbf{C}(\mathbf{Q}\mathbf{P}),$$

where $\mathbf{QP}$ is non-singular. This proves the third part. ∎

For practical reasons we are interested in determining the "nicest" (i.e., the simplest) matrix congruent to a given matrix, or what amounts to the same thing, the "nicest" (i.e., the most convenient) basis to use in expressing a given bilinear functional. If, in particular, we confine our interest to quadratic functionals, then it suffices, in view of Theorem 4.8.11, to consider symmetric bilinear functionals.

We come now to the main result of this section, called **Sylvester's theorem.**

4.8.20. Theorem. Let f be any symmetric bilinear functional on a real n-dimensional vector space X. Then there exists a basis $\{e_1, \ldots, e_n\}$ of X such that the matrix of f with respect to this basis is of the form

$$\begin{bmatrix} +1 & & & & & & & & & \\ & \ddots & & & & & \mathbf{0} & & & \\ & & +1 & & & & & & & \\ & & & -1 & & & & & & \\ & & & & \ddots & & & & & \\ & & & & & -1 & & & & \\ & \mathbf{0} & & & & & 0 & & & \\ & & & & & & & \ddots & & \\ & & & & & & & & & 0 \end{bmatrix} \begin{matrix} \} p \\ \} r \\ \} n \end{matrix} \qquad (4.8.21)$$

The integers r and p in the above matrix are uniquely determined by the bilinear form.

Proof. Since the proof of this theorem is somewhat long, it will be carried out in several steps.

Step 1. We first show that there exists a basis $\{v_1, \ldots, v_n\}$ of X such that $f(v_i, v_j) = 0$ for $i \neq j$. The proof of this step is by induction on the dimension of X. The statement is trivial if $\dim X = 1$. Suppose that the assertion is true for $\dim X = n - 1$. Let f be a bilinear functional on X, where $\dim X = n$. Let $v_1 \in X$ be such that $f(v_1, v_1) \neq 0$. There must be such a v_1; otherwise, by Theorem 4.8.13, f would be skew symmetric, and we would conclude that $f(x, y) = 0$ for all x, y. Now let $\mathfrak{M} = \{x \in X: f(v_1, x) = 0\}$. We now show that $\mathfrak{M}$ is a linear subspace of X. Let $x_1, x_2 \in \mathfrak{M}$ so that $f(v_1, x_1) = f(v_1, x_2) = 0$. Then $f(v_1, x_1 + x_2) = f(v_1, x_1) + f(v_1, x_2) = 0 + 0 = 0$. Similarly, $f(v_1, \alpha x_1) = 0$ for all $\alpha \in R$. Therefore, $\mathfrak{M}$ is a linear subspace of X. Furthermore, $\mathfrak{M} \neq X$ because $v_1 \notin \mathfrak{M}$. Hence, $\dim \mathfrak{M} \leq n - 1$. Now let $\dim \mathfrak{M} = q \leq n - 1$. Since f is a bilinear functional on $\mathfrak{M}$, it follows by the induction hypothesis that there is a basis for $\mathfrak{M}$ consisting of a set of q vectors $\{v_2, \ldots, v_{q+1}\}$ such that $f(v_i, v_j) = 0$ for $i \neq j$, $2 \leq i, j \leq q + 1$. Also, $f(v_1, v_j) = 0$ for $j = 2, \ldots, q + 1$, by definition of $\mathfrak{M}$.

Furthermore, $f(v_i, v_1) = f(v_1, v_i)$. Hence, $f(v_i, v_1) = f(v_1, v_i) = 0$ for $i = 2, \ldots, q+1$. It follows that $f(v_i, v_j) = 0$ for $i \neq j$ and $1 \leq i, j \leq q+1$. We now show that $\{v_1, \ldots, v_{q+1}\}$ is a basis for X. Let $x \in X$ and let $x' = x - \alpha_1 v_1$, where $\alpha_1 = f(v_1, x)/f(v_1, v_1)$. Then $f(v_1, x') = f(v_1, x) - \alpha_1 f(v_1, v_1) = f(v_1, v_1) - f(v_1, v_1) = 0$. Thus, $x' \in \mathfrak{M}$. Since $\{v_2, \ldots, v_{q+1}\}$ is a basis for $\mathfrak{M}$, there exist $\alpha_2, \ldots, \alpha_{q+1}$ such that $x' = \alpha_2 v_2 + \ldots + \alpha_{q+1} v_{q+1}$; i.e., $x = \alpha_1 v_1 + \ldots + \alpha_{q+1} v_{q+1}$. Thus, $\{v_1, \ldots, v_{q+1}\}$ spans X. To show that the set $\{v_1, \ldots, v_{q+1}\}$ is linearly independent, assume that $\alpha_1 v_1 + \ldots + \alpha_{q+1} v_{q+1} = 0$. Then $0 = f(v_1, 0) = f(v_1, \alpha_1 v_1 + \ldots + \alpha_{q+1} v_{q+1}) = \alpha_1 f(v_1, v_1) = 0$, which implies that $\alpha_1 = 0$. Hence, $\alpha_2 v_2 + \ldots + \alpha_{q+1} v_{q+1} = 0$. Since the set $\{v_2, \ldots, v_{q+1}\}$ forms a basis for $\mathfrak{M}$, we must have $\alpha_2 = \ldots = \alpha_{q+1} = 0$. Thus, $\{v_1, \ldots, v_{q+1}\}$ forms a basis for X, and we conclude that $q + 1 = n$. This completes the proof of step 1.

Step 2. Let $\{v_1, \ldots, v_n\}$ be a basis for X such that $f(v_i, v_j) = 0$ for $i \neq j$ and let $\beta_i = f(v_i, v_i)$ for $i, j = 1, \ldots, n$. Let $e_i = \gamma_i v_i$ for $i = 1, \ldots, n$, where $\gamma_i = 1/\sqrt{|\beta_i|}$ if $\beta_i \neq 0$ and $\gamma_i = 1$ if $\beta_i = 0$. Now suppose that $\beta_i = f(v_i, v_i) \neq 0$. Then we have $f(e_i, e_i) = f(\gamma_i v_i, \gamma_i v_i) = \gamma_i^2 f(v_i, v_i) = \beta_i/\sqrt{\beta_i^2} = \pm 1$. Also, if $\beta_i = f(v_i, v_i) = 0$, then $f(e_i, e_i) = \gamma_i^2 f(v_i, v_i) = 0$. Finally, we see that $f(e_i, e_j) = f(\gamma_i v_i, \gamma_j v_j) = \gamma_i \gamma_j f(v_i, v_j) = 0$ if $i \neq j$. Thus, we have established a basis for X such that $f_{ij} = f(e_i, e_j) = 0$ if $i \neq j$ and $f_{ii} = f(e_i, e_i) = +1, -1$, or 0.

Step 3. We now show that the integers p and r in matrix (4.8.21) are uniquely determined by f. Let $\{e_1, \ldots, e_n\}$ and $\{e_1', \ldots, e_n'\}$ be bases for X and let $\mathbf{F}$ and $\mathbf{F}'$ be matrices of f with respect to $\{e_1, \ldots, e_n\}$ and $\{e_1', \ldots, e_n'\}$, respectively, where

$$\mathbf{F} = \begin{bmatrix} 1 & & & & & & & & & \\ & \ddots & & & & & & & & \\ & & 1 & & & & & & \mathbf{0} & \\ & & & -1 & & & & & & \\ & & & & \ddots & & & & & \\ & \mathbf{0} & & & & -1 & & & & \\ & & & & & & 0 & & & \\ & & & & & & & \ddots & & \\ & & & & & & & & & 0 \end{bmatrix} \begin{matrix} \left.\vphantom{\begin{matrix}1\\ \ddots\\ 1\end{matrix}}\right\} p \\ \left.\vphantom{\begin{matrix}-1\\ \ddots\\ -1\\ 0\\ \ddots\\ 0\end{matrix}}\right\} n-p \end{matrix}$$

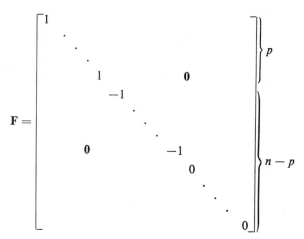

$$
\mathbf{F}' = \left[\begin{array}{ccccccccc}
1 & & & & & & & & \\
 & \cdot & & & & & & & \\
 & & \cdot & & & & & & \\
 & & & 1 & & & \mathbf{0} & & \\
 & & & & -1 & & & & \\
 & & & & & \cdot & & & \\
 & & & & & & \cdot & & \\
 & & & & & & & -1 & \\
 & & & \mathbf{0} & & & 0 & & \\
 & & & & & & & \cdot & \\
 & & & & & & & & \cdot \\
 & & & & & & & & 0
\end{array}\right]
\begin{array}{l}
\left.\vphantom{\begin{array}{c}1\\ \cdot\\ \cdot\\ 1\end{array}}\right\} q \\
\left.\vphantom{\begin{array}{c}-1\\ \cdot\\ \cdot\\ -1\\ 0\\ \cdot\\ \cdot\\ 0\end{array}}\right\} n-q
\end{array}
$$

To prove that $p = q$ we show that $e_1, \ldots, e_p, e'_{q+1}, \ldots, e'_n$ are linearly independent. From this it must follow that $p + (n - q) \leq n$, or $p \leq q$. By the same argument, $q \leq p$, and so $p = q$. Let

$$\gamma_1 e_1 + \ldots + \gamma_p e_p + \gamma'_{q+1} e'_{q+1} + \ldots + \gamma'_n e'_n = 0,$$

where $\gamma_i \in R$, $i = 1, \ldots, p$ and $\gamma'_i \in R$, $i = q + 1, \ldots, n$. Rewriting the above equation we have

$$\gamma_1 e_1 + \ldots + \gamma_p e_p = -(\gamma'_{q+1} e'_{q+1} + \ldots + \gamma'_n e'_n) \triangleq x_0.$$

Then

$$\begin{aligned} f(x_0, x_0) &= f(\gamma_1 e_1 + \ldots + \gamma_p e_p, \gamma_1 e_1 + \ldots + \gamma_p e_p) \\ &= \gamma_1^2 + \ldots + \gamma_p^2 \geq 0, \end{aligned}$$

by choice of $\{e_1, \ldots, e_p\}$. On the other hand,

$$\begin{aligned} f(x_0, x_0) &= f[-(\gamma'_{q+1} e'_{q+1} + \ldots + \gamma'_n e'_n), -(r'_{q+1} e'_{q+1}, \ldots, \gamma'_n e'_n)] \\ &= (-1)^2[-(\gamma'_{q+1})^2 - (\gamma'_{q+2})^2 - \ldots - (\gamma'_n)^2] \leq 0 \end{aligned}$$

by choice of $\{e'_{q+1}, \ldots, e'_{q+n}\}$. From this we conclude that $\gamma_1^2 + \ldots + \gamma_p^2 = 0$; i.e., $\gamma_1 = \ldots = \gamma_p = 0$. Hence, $\gamma'_{q+1} e'_{q+1} + \ldots + \gamma'_n e'_n = 0$. But the set $\{e'_{q+1}, \ldots, e'_n\}$ is linearly independent, and thus $\gamma'_{q+1} = \ldots = \gamma'_n = 0$. Hence, the vectors $e_1, \ldots, e_p, e'_{q+1}, \ldots, e'_n$ are linearly independent, and it follows that $p = q$.

To prove that r is unique, let r be the number of non-zero elements of $\mathbf{F}$ and let r' be the number of non-zero elements of $\mathbf{F}'$. By Theorem 4.8.15, $\mathbf{F}$ and $\mathbf{F}'$ are congruent and hence equivalent. Thus, it follows from Theorem 4.3.16 that $\mathbf{F}$ and $\mathbf{F}'$ must have the same rank, and therefore $r = r'$.

This concludes the proof of the theorem. ■

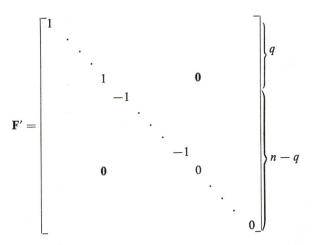

Sylvester's theorem allows the following classification of symmetric bilinear functionals.

4.8.22. Definition. The integer r in Theorem 4.8.20 is called the **rank** of the symmetric bilinear functional f. The integer p is called the **index** of f. The integer n is called the **order** of f. The integer $s = 2p - r$ (i.e., the number of $+1$'s minus the number of -1's) is called the **signature** of f.

Since every real symmetric matrix is congruent to a unique matrix of the form (4.8.21), we define the index, order, and rank of a real symmetric matrix analogously as in Definition 4.8.22.

Now let us recall that a bilinear functional f on a vector space X is said to be positive if $f(x, x) \geq 0$ for all $x \in X$. Also, a bilinear functional f is said to be strictly positive if $f(x, x) > 0$ for all $x \neq 0$, $x \in X$ (it should be noted that $f(x, x) = 0$ for $x = 0$). Our final result of the present section, which is a consequence of Theorem 4.8.20, enables us now to classify symmetric bilinear functionals.

4.8.23. Theorem. Let p, r, and n be defined as in Theorem 4.8.20. A symmetric bilinear functional on a real n-dimensional vector space X is

(i) strictly positive if and only if $p = r = n$; and

(ii) positive if and only if $p = r$.

4.8.24. Exercise. Prove Theorem 4.8.23.

4.9. EUCLIDEAN VECTOR SPACES

A. *Euclidean Spaces: Definition and Properties*

Among the various linear spaces which we will encounter, the so-called Euclidean spaces are so important that we devote the next two sections to them. These spaces will allow us to make many generalizations to facts established in plane geometry, and they will enable us to consider several important special types of linear transformations. In order to characterize these spaces properly, we must make use of two important notions, that of the norm of a vector and that of the inner product of two vectors (refer to Section 3.6). In the real plane, these concepts are related to the length of a vector and to the angle between two vectors, respectively. Before considering the matter on hand, some preliminary remarks are in order.

To begin with, we would like to point out that from a strictly logical point of view Euclidean spaces should actually be treated at a later point of our development. This is so because these spaces are specific examples of metric spaces (to be treated in the next chapter), of normed spaces (to be dealt with in Chapter 6), and of inner product spaces (also to be considered in Chapter 6). However, there are several good reasons for considering Euclidean spaces and their properties at this point. These include: Euclidean spaces are so important in applications that the reader should be exposed to them as early as possible; these spaces and their properties will provide the motivation for subsequent topics treated in this book; and the material covered in the present section and in the next section (dealing with linear transformations defined on Euclidean spaces) constitutes a natural continuation and conclusion of the topics considered thus far in the present chapter.

In order to provide proper motivation for the present section, it is useful to utilize certain facts from plane geometry to indicate the way. To this end let us consider the space R^2 and let $x = (\xi_1, \xi_2)$ and $y = (\eta_1, \eta_2)$ be vectors in R^2. Let $\{u_1, u_2\}$ be the natural basis for R^2. Then the natural coordinate representation of x and y is

$$\mathbf{x} = \begin{bmatrix} \xi_1 \\ \xi_2 \end{bmatrix} \quad \text{and} \quad \mathbf{y} = \begin{bmatrix} \eta_1 \\ \eta_2 \end{bmatrix}, \tag{4.9.1}$$

respectively (see Example 4.1.15). The representation of these vectors in the plane is shown in Figure D. In this figure, $|x|$, $|y|$, and $|x - y|$ denote the

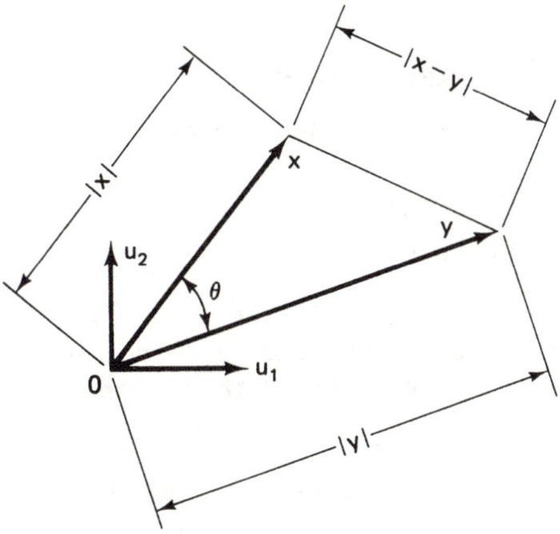

4.9.2. **Figure D.** Length of vectors and angle between vectors.

lengths of vectors x, y, and $(x - y)$, respectively, and θ represents the angle between x and y. The length of vector x is equal to $(\xi_1^2 + \xi_2^2)^{1/2}$, and the length of vector $(x - y)$ is equal to $\{(\xi_1 - \eta_1)^2 + (\xi_2 - \eta_2)^2\}^{1/2}$. By convention, we say in this case that "the distance from x to y" is equal to $\{(\xi_1 - \eta_1)^2 + (\xi_2 - \eta_2)^2\}^{1/2}$, that "the distance from the origin 0 (the null vector) to x" is equal to $(\xi_1^2 + \xi_2^2)^{1/2}$, and the like. Using the notation of the present chapter, we have

$$|x| = \sqrt{\mathbf{x}^T\mathbf{x}} = \sqrt{\xi_1^2 + \xi_2^2} \tag{4.9.3}$$

and

$$|x - y| = \sqrt{(\mathbf{x} - \mathbf{y})^T(\mathbf{x} - \mathbf{y})} = \sqrt{(\mathbf{y} - \mathbf{x})^T(\mathbf{y} - \mathbf{x})} = |y - x|. \tag{4.9.4}$$

The angle θ between vectors x and y can easily be characterized by its cosine, namely,

$$\cos \theta = \frac{(\xi_1\eta_1 + \xi_2\eta_2)}{\sqrt{\xi_1^2 + \xi_2^2}\sqrt{\eta_1^2 + \eta_2^2}}. \tag{4.9.5}$$

Utilizing the notation of the present chapter, we have

$$\cos \theta = \frac{\mathbf{x}^T\mathbf{y}}{\sqrt{\mathbf{x}^T\mathbf{x}}\sqrt{\mathbf{y}^T\mathbf{y}}}. \tag{4.9.6}$$

It turns out that the real-valued function $\mathbf{x}^T\mathbf{y}$, which we used in both Eqs. (4.9.3) and (4.9.6) to characterize the length of any vector x and the angle between any vectors x and y, is of fundamental importance. For this reason we denote it by a special symbol; i.e., we write

$$(x, y) \triangleq \mathbf{x}^T\mathbf{y}. \tag{4.9.7}$$

Now if we let $x = y$ in Eq. (4.9.7), then in view of Eq. (4.9.3) we have

$$|x| = \sqrt{(x, x)}. \tag{4.9.8}$$

By inspection of Eq. (4.9.3) we note that

$$(x, x) > 0 \text{ for all } x \neq 0 \tag{4.9.9}$$

and

$$(x, x) = 0 \text{ for } x = 0. \tag{4.9.10}$$

Also, from Eq. (4.9.7) we have

$$(x, y) = (y, x) \tag{4.9.11}$$

for all x and y. Moreover, for any vectors x, y, and z and for any real scalars α and β we have, in view of Eq. (4.9.7), the relations

$$(x + y, z) = (x, z) + (y, z), \tag{4.9.12}$$

$$(x, y + z) = (x, y) + (x, z), \tag{4.9.13}$$

$$(\alpha x, y) = \alpha(x, y), \tag{4.9.14}$$

and

$$(x, \alpha y) = \alpha(x, y). \tag{4.9.15}$$

In connection with Eq. (4.9.6) we can make several additional observations. First, we note that if $\mathbf{x} = \mathbf{y}$, then $\cos \theta = +1$; if $\mathbf{x} = -\mathbf{y}$, then $\cos \theta = -1$; if $\mathbf{x}^T = (\xi_1, 0)$ and $\mathbf{y}^T = (0, \eta_2)$, then $\cos \theta = 0$; etc. It is easily verified, using Eq. (4.9.6), that $\cos\theta$ assumes all values between $+1$ and -1; i.e., $-1 \leq \cos\theta \leq +1$.

The above formulation agrees, of course, with our notions of length of a vector, distance between two vectors, and angle between two vectors. From Eqs. (4.9.9)–(4.9.15) it is also apparent that relation (4.9.7) satisfies all the axioms of an inner product (see Section 3.6).

Using the above discussion as motivation, let us now begin our treatment of Euclidean vector spaces.

First, we recall the definition of a real inner product: a bilinear functional f on a real vector space X is said to be an **inner product** on X if (i) f is symmetric and (ii) f is strictly positive. We also recall that a real vector space X on which an inner product is defined is called a **real inner product space**. We now have the following important

4.9.16. Definition. A real finite-dimensional vector space on which an inner product is defined is called a **Euclidean space**. A finite-dimensional vector space over the field of complex numbers on which an inner product is defined is called a **unitary space**.

We point out that some authors do not restrict Euclidean spaces to be finite dimensional.

Although many of the results of unitary spaces are essentially identical to those of Euclidean spaces, we postpone our treatment of complex inner product spaces until Chapter 6, where we consider spaces that, in general, may be infinite dimensional.

Throughout the remainder of the present section, X will denote an n-dimensional Euclidean space, unless otherwise specified. Since we will always be concerned with a given bilinear functional on X, we will henceforth write (x, y) in place of $f(x, y)$ to denote the inner product of x and y. Finally, for purposes of completeness, we give a summary of the axioms of a real inner product. We have

(i) $(x, x) > 0$ for all $x \neq 0$ and $(x, x) = 0$ if $x = 0$;
(ii) $(x, y) = (y, x)$ for all $x, y \in X$;
(iii) $(\alpha x + \beta y, z) = \alpha(x, z) + \beta(y, z)$ for all $x, y, z \in X$ and all $\alpha, \beta \in R$; and
(iv) $(x, \alpha y + \beta z) = \alpha(x, y) + \beta(x, z)$ for all $x, y \in X$ and all $\alpha, \beta \in R$.

We note that Eqs. (4.9.9)–(4.9.15) are clearly in agreement with these axioms.

4.9.17. Theorem. The inner product $(x, y) = 0$ for all $x \in X$ if and only if $y = 0$.

Proof. If $y = 0$, then $y = 0 \cdot x$ and $(x, 0) = (x, 0 \cdot x) = 0 \cdot (x, x) = 0$ for all $x \in X$.

On the other hand, let $(x, y) = 0$ for all $x \in X$. Then, in particular, it must be true that $(x, y) = 0$ if $x = y$. We thus have $(y, y) = 0$, which implies that $y = 0$. ■

The reader can prove the next results readily.

4.9.18. Corollary. Let $A \in L(X, X)$. Then $(x, Ay) = 0$ for all $x, y \in X$ if and only if $A = 0$.

4.9.19. Corollary. Let $A, B \in L(X, X)$. If $(x, Ay) = (x, By)$ for all $x, y \in X$, then $A = B$.

4.9.20. Corollary. Let **A** be a real $(n \times n)$ matrix. If $\mathbf{x}^T\mathbf{A}\mathbf{y} = 0$ for all $\mathbf{x}, \mathbf{y} \in R^n$, then $\mathbf{A} = \mathbf{0}$.

4.9.21. Exercise. Prove Corollaries 4.9.18–4.9.20.

Of crucial importance is the notion of norm. We have:

4.9.22. Definition. For each $x \in X$, let

$$|x| = (x, x)^{1/2}.$$

We call $|x|$ the **norm** of x.

Let us consider a specific case.

4.9.23. Example. Let $X = R^n$ and let $x, y \in X$, where $x = (\xi_1, \ldots, \xi_n)$ and $y = (\eta_1, \ldots, \eta_n)$. From Example 3.6.23 it follows that

$$(x, y) = \sum_{i=1}^{n} \xi_i \eta_i \tag{4.9.24}$$

is an inner product on X. The coordinate representation of x and y with respect to the natural basis in R^n is given by

$$\mathbf{x} = \begin{bmatrix} \xi_1 \\ \cdot \\ \cdot \\ \cdot \\ \xi_n \end{bmatrix} \quad \text{and} \quad \mathbf{y} = \begin{bmatrix} \eta_1 \\ \cdot \\ \cdot \\ \cdot \\ \eta_n \end{bmatrix},$$

respectively (see Example 4.1.15). We thus have

$$(x, y) = \mathbf{x}^T\mathbf{y}, \tag{4.9.25}$$

and

$$|x| = \left(\sum_{i=1}^{n} \xi_i^2\right)^{1/2} = (\mathbf{x}^T\mathbf{x})^{1/2}. \quad ■ \tag{4.9.26}$$

The above example gives rise to:

4.9.27. Definition. The vector space R^n with the inner product defined in Eq. (4.9.24) is denoted by E^n. The norm of x given by Eq. (4.9.26) is called the **Euclidean norm** on R^n.

Relation (4.9.29) of the next result is called the **Schwarz inequality**.

4.9.28. Theorem. Let x and y be any elements of X. Then

$$|(x, y)| \leq |x| \cdot |y|, \tag{4.9.29}$$

where in Eq. (4.9.29) $|(x, y)|$ denotes the absolute value of a real scalar and $|x|$ denotes the norm of x.

Proof. For any x and y in X and for any real scalar α we have

$$(x + \alpha y, x + \alpha y) = (x, x) + \alpha(x, y) + \alpha(y, x) + \alpha^2(y, y) \geq 0.$$

Now assume first that $y \neq 0$, and let

$$\alpha = \frac{-(x, y)}{(y, y)}.$$

Then

$$\begin{aligned}(x + \alpha y, x + \alpha y) &= (x, x) + 2\alpha(x, y) + \alpha^2(y, y)\\ &= (x, x) - \frac{2(x, y)}{(y, y)}(x, y) + \frac{(x, y)^2}{(y, y)^2}(y, y)\\ &= (x, x) - \frac{(x, y)^2}{(y, y)} \geq 0,\end{aligned}$$

or

$$(x, x)(y, y) \geq (x, y)^2.$$

Taking the square root of both sides, we have the desired inequality

$$|(x, y)| \leq |x| \cdot |y|.$$

To complete the proof, consider the case $y = 0$. Then $(x, y) = 0$, $|y| = 0$, and in this case the inequality follows trivially. ■

4.9.30. Exercise. For $x, y \in X$, show that

$$|(x, y)| = |x| \cdot |y|$$

if and only if x and y are linearly dependent.

In the next result we establish the **axioms of a norm**.

4.9.31. Theorem. For all x and y in X and for all real scalars α, the following hold:

(i) $|x| > 0$ unless $x = 0$, in which case $|x| = 0$;

(ii) $|\alpha x| = |\alpha| \cdot |x|$, where $|\alpha|$ denotes the absolute value of the scalar α; and

(iii) $|x + y| \leq |x| + |y|$.

Proof. The proof of part (i) follows from the definition of an inner product. To prove part (ii), we note that

$$|\alpha x|^2 = (\alpha x, \alpha x) = \alpha(x, \alpha x) = \alpha^2(x, x) = \alpha^2|x|^2.$$

Taking the square root of both sides we have the desired relation

$$|\alpha x| = |\alpha||x|.$$

To verify the last part of the theorem we note that

$$|x + y|^2 = (x + y, x + y) = (x, x) + 2(x, y) + (y, y)$$
$$= |x|^2 + 2(x, y) + |y|^2.$$

Using the Schwarz inequality we obtain

$$|x + y|^2 \leq |x|^2 + 2|x| \cdot |y| + |y|^2 = (|x| + |y|)^2.$$

Taking the square root of both sides we have

$$|x + y| \leq |x| + |y|,$$

which is the desired result. ■

Part (iii) of Theorem 4.9.31 is called the **triangle inequality**. Part (ii) is called the **homogeneous property** of a norm. In Chapter 6 we will define functions on general vector spaces satisfying axioms (i), (ii), and (iii) of Theorem 4.9.31 without making use of inner products. In such cases we will speak of **normed linear spaces** (Euclidean spaces are examples of normed linear spaces).

Our next result is called the **parallelogram law**. Its meaning in the plane is evident from Figure E.

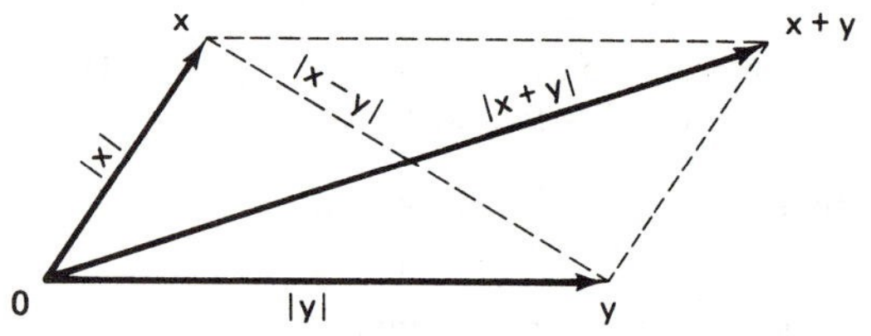

4.9.32. **Figure E.** Interpretation of the parallelogram law.

4.9.33. Theorem. For all $x, y \in X$ the equality

$$|x + y|^2 + |x - y|^2 = 2|x|^2 + 2|y|^2$$

holds.

4.9.34. Exercise. Prove Theorem 4.9.33.

Generalizing Eq. (4.9.4), we define the **distance** between two vectors x

and y of X as

$$\rho(x, y) = |x - y|. \tag{4.9.35}$$

It is not difficult for the reader to prove the next result.

4.9.36. Theorem. For all $x, y, z \in X$, the following hold:

(i) $\rho(x, y) = \rho(y, x)$;
(ii) $\rho(x, y) \geq 0$ and $\rho(x, y) = 0$ if and only if $x = y$; and
(iii) $\rho(x, y) \leq \rho(x, z) + \rho(z, y)$.

A function $\rho(x, y)$ having properties (i), (ii), and (iii) of Theorem 4.9.36 is called a **metric**. Without making use of inner products, we will in Chapter 5 define such functions on non-empty sets (not necessarily linear spaces) and we will in such cases speak of **metric spaces** (Euclidean spaces are examples of metric spaces).

4.9.37. Exercise. Prove Theorem 4.9.36.

B. Orthogonal Bases

Following our discussion at the beginning of the present section further, we now recall the important concept of orthogonality, using inner products. In accordance with Definition 3.6.22, two vectors $x, y \in X$ are said to be **orthogonal** (to one another) if $(x, y) = 0$. We recall that this is written as $x \perp y$. From the discussion at the beginning of this section it is clear that in the plane $x \neq 0$ is orthogonal to $y \neq 0$ if and only if the angle between x and y is some odd multiple of 90°.

The reader has undoubtedly encountered a special case of our next result, known as the **Pythagorean theorem.**

4.9.38. Theorem. Let $x, y \in X$. If $x \perp y$, then

$$|x + y|^2 = |x|^2 + |y|^2.$$

Proof. Since by assumption $x \perp y$, we have $(x, y) = 0$. Thus,

$$|x + y|^2 = (x + y, x + y) = (x, x) + (x, y) + (y, x) + (y, y) = |x|^2 + |y|^2,$$

which is the desired result. ■

4.9.39. Definition. A vector $x \in X$ is said to be a **unit vector** if $|x| = 1$.

Let us choose any vector $y \neq 0$ and let $z = \frac{1}{|y|}y$. Then the norm of z is

$$|z| = \left|\frac{1}{|y|}y\right| = \frac{1}{|y|}|y| = 1,$$

i.e., z is a unit vector. This process is called **normalizing the vector** y.

Next, let $\{f_1, \ldots, f_n\}$ be an arbitrary basis for X and let $\mathbf{F} = [f_{ij}]$ denote the matrix of the inner product with respect to this basis; i.e., $f_{ij} = (f_i, f_j)$ for all i and j. More specifically, $\mathbf{F}$ denotes the matrix of the bilinear functional f that is used in determining the inner product on X with respect to the indicated basis (see Definition 4.8.1). Let $\mathbf{x}$ and $\mathbf{y}$ denote the coordinate representation of x and y, respectively, with respect to $\{f_1, \ldots, f_n\}$. Then we have, by Theorem 4.8.2,

$$(x, y) = \mathbf{x}^T\mathbf{F}\mathbf{y} = \mathbf{y}^T\mathbf{F}\mathbf{x} = \sum_{j=1}^{n} \sum_{i=1}^{n} f_{ij}\xi_j\eta_i.$$

Now by Theorems 4.8.20 and 4.8.23, since the inner product is symmetric and strictly positive, there exists a basis $\{e_1, \ldots, e_n\}$ for X such that the matrix of the inner product with respect to this basis is the $(n \times n)$ identity matrix $\mathbf{I}$, i.e.,

$$(e_i, e_j) = \delta_{ij} = \begin{cases} 0 & \text{if } i \neq j \\ 1 & \text{if } i = j. \end{cases}$$

This motivates the following:

4.9.40. Definition. If $\{e_1, \ldots, e_n\}$ is a basis for X such that $(e_i, e_j) = 0$ for all $i \neq j$, i.e., if $e_i \perp e_j$ for all $i \neq j$, then $\{e_1, \ldots, e_n\}$ is called an **orthogonal basis**. If in addition, $(e_i, e_i) = 1$, i.e., if $|e_i| = 1$ for all i, then $\{e_1, \ldots, e_n\}$ is said to be an **orthonormal basis for** X (thus, $\{e_1, \ldots e_n\}$ is orthonormal if and only if $(e_i, e_j) = \delta_{ij}$).

Using the properties of inner products and the definitions of orthogonal and orthonormal bases, we are now in a position to establish several useful results.

4.9.41. Theorem. Let $\{e_1, \ldots, e_n\}$ be an orthonormal basis for X. Let x and y be arbitrary vectors in X, and let the coordinate representation of x and y with respect to this basis be $\mathbf{x}^T = (\xi_1, \ldots, \xi_n)$ and $\mathbf{y}^T = (\eta_1, \ldots, \eta_n)$, respectively. Then

$$(x, y) = \mathbf{x}^T\mathbf{y} = \mathbf{y}^T\mathbf{x} = \sum_{i=1}^{n} \xi_i\eta_i \tag{4.9.42}$$

and

$$|x| = (\mathbf{x}^T\mathbf{x})^{1/2} = \sqrt{\xi_1^2 + \ldots + \xi_n^2}. \tag{4.9.43}$$

Proof. From the above discussion we have

$$(x, y) = \mathbf{x}^T\mathbf{F}\mathbf{y} = \sum_{i,j=1}^{n} f_{ij}\xi_i\eta_j = \sum_{i,j=1}^{n} \delta_{ij}\xi_i\eta_j = \sum_{i=1}^{n} \xi_i\eta_i.$$

In particular, we have

$$(x, x) = \sum_{i=1}^{n} \xi_i^2. \quad \blacksquare$$

The reader should note that Eqs. (4.9.7) and (4.9.8) introduced at the beginning of this section are, of course, in agreement with Eqs. (4.9.42) and (4.9.43). (See also Example 4.9.23.)

Our next result enables us to determine the coordinates of a vector with respect to a given orthonormal basis.

4.9.44. Theorem. Let $\{e_1, \ldots, e_n\}$ be an orthonormal basis for X and let x be an arbitrary vector. The coordinates of x with respect to $\{e_1, \ldots, e_n\}$ are given by the formulas

$$\xi_1 = (x, e_1), \ldots, \xi_n = (x, e_n).$$

Proof. Since $x = \xi_1 e_1 + \ldots + \xi_n e_n$, we have

$$(x, e_1) = (\xi_1 e_1 + \ldots + \xi_n e_n, e_1) = \xi_1(e_1, e_1) + \ldots + \xi_n(e_n, e_1) = \xi_1.$$

Repeating this procedure for (x, e_i), $i = 2, \ldots, n$, yields the desired result. ■

Let us consider some specific cases.

4.9.45. Example. Let $X = E^2$ (see Definition 4.9.27). Let $x, y \in E^2$, where $x = (\xi_1, \xi_2)$ and $y = (\eta_1, \eta_2)$. Then

$$(x, y) = \xi_1\eta_1 + \xi_2\eta_2.$$

The natural basis for E^2 is given by $u_1 = (1, 0)$ and $u_2 = (0, 1)$. Since $(u_i, u_j) = \delta_{ij}$, it follows that $\{u_1, u_2\}$ is an orthonormal basis for E^2. Furthermore, we have

$$|x| = (\xi_1^2 + \xi_2^2)^{1/2}. \quad ■$$

4.9.46. Example. Let $X = R^2$, and let the inner product on R^2 be defined by

$$(x, y) = \xi_1\eta_1 + 4\xi_2\eta_2. \tag{4.9.47}$$

(The reader may verify that this is indeed an inner product.) Let $\{u_1, u_2\}$ denote the natural basis for R^2; i.e., $u_1 = (1, 0)$ and $u_2 = (0, 1)$. The matrix representation of the bilinear functional, which determines the above inner product with respect to the basis $\{u_1, u_2\}$ is

$$(x, y) = \mathbf{x}^T \begin{bmatrix} 1 & 0 \\ 0 & 4 \end{bmatrix} \mathbf{y},$$

where $\mathbf{x}$ and $\mathbf{y}$ are the coordinate vectors of x and y with respect to $\{u_1, u_2\}$. We see that $(u_1, u_2) = 1 \cdot 0 + 4 \cdot 0 \cdot 1 = 0$; i.e., u_1 and u_2 are orthogonal with respect to the inner product (4.9.47). Note however that $|u_1| = 1$ and $|u_2| = 2$; i.e., the vectors u_1 and u_2 are not orthonormal.

Now let $e_1 = (1, 0)$ and $e_2 = (0, \frac{1}{2})$. Then it is readily verified that $\{e_1, e_2\}$ is an orthonormal basis for X. Furthermore, for $x = \xi_1' e_1 + \xi_2' e_2$, we have

$\xi'_1 = (x, e_1)$ and $\xi'_2 = (x, e_2)$. If we let

$$\mathbf{x}' = \begin{bmatrix} \xi'_1 \\ \xi'_2 \end{bmatrix} \quad \text{and} \quad \mathbf{y}' = \begin{bmatrix} \eta'_1 \\ \eta'_2 \end{bmatrix}$$

denote the coordinate representation of x and y, respectively, with respect to $\{e_1, e_2\}$, then

$$(x, y) = (\mathbf{x}')^T\mathbf{y}'.$$

This illustrates the fact that the norm of a vector must be interpreted with respect to the inner product used in determining the norm. ■

Our next result allows us to represent vectors in X in a convenient way.

4.9.48. Theorem. Let $\{e_1, \ldots, e_n\}$ be an orthogonal basis for X. Then for all $x \in X$ we have

$$x = \frac{(x, e_1)}{(e_1, e_1)}e_1 + \ldots + \frac{(x, e_n)}{(e_n, e_n)}e_n.$$

Proof. Normalizing $e_1, \ldots, e_n$, we obtain the orthonormal basis $\{e'_1, \ldots, e'_n\}$, where $e'_i = \frac{1}{|e_i|}e_i$, $i = 1, \ldots, n$. By Theorem 4.9.44 we have

$$\begin{aligned} x &= (x, e'_1)e'_1 + \ldots + (x, e'_n)e'_n \\ &= \left(x, \frac{1}{|e_1|}e_1\right)\left(\frac{1}{|e_1|}\right)e_1 + \ldots + \left(x, \frac{1}{|e_n|}e_n\right)\left(\frac{1}{|e_n|}\right)e_n \\ &= \frac{(x, e_1)}{|e_1|^2}e_1 + \ldots + \frac{(x, e_n)}{|e_n|^2}e_n \\ &= \frac{(x, e_1)}{(e_1, e_1)}e_1 + \ldots + \frac{(x, e_n)}{(e_n, e_n)}e_n. \quad ■ \end{aligned}$$

We are now in a position to characterize inner products by means of **Parseval's identity**, given in our next result.

4.9.49. Corollary. Let $\{e_1, \ldots, e_n\}$ be an orthogonal basis for X. Then for any $x, y \in X$ we have

$$(x, y) = \sum_{i=1}^{n} \frac{(x, e_i)(y, e_i)}{(e_i, e_i)}.$$

4.9.50. Exercise. Verify Corollary 4.9.49.

Our next result establishes the linear independence of orthogonal vectors. We have:

4.9.51. Theorem. Suppose that $x_1, \ldots, x_k$ are mutually orthogonal non-zero vectors in X; i.e., $x_i \perp x_j$, $i \neq j$. Then $x_1, \ldots, x_k$ are linearly independent.

Proof. Assume that for real scalars $\alpha_1, \ldots, \alpha_k$ we have

$$\alpha_1 x_1 + \ldots + \alpha_k x_k = 0.$$

For arbitrary $i = 1, \ldots, k$, we have

$$0 = (0, x_i) = (\alpha_1 x_1 + \ldots + \alpha_k x_k, x_i) = \alpha_1(x_1, x_i) + \ldots + \alpha_k(x_k, x_i)$$
$$= \alpha_i(x_i, x_i);$$

i.e., $\alpha_i(x_i, x_i) = 0$. This implies that $\alpha_i = 0$ for arbitrary i, which proves the linear independence of $x_1, \ldots, x_k$. ■

Note that the converse to the above theorem is not true. We leave the proofs of the next two results as an exercise.

4.9.52. Corollary. A set of k non-zero mutually orthogonal vectors is a basis for X if and only if $k = \dim X = n$.

4.9.53. Corollary. For X there exist not more than n mutually orthonormal vectors. (In this case we speak of a **complete orthonormal set of vectors.**)

4.9.54. Exercise. Prove Corollaries 4.9.52 and 4.9.53.

Our next result, which is called the **Gram-Schmidt process**, allows us to construct an orthonormal basis from an arbitrary basis.

4.9.55. Theorem. Let $\{f_1, \ldots, f_n\}$ be an arbitrary basis for X. Set

$$\begin{aligned} g_1 &= f_1, & e_1 &= g_1/|g_1|, \\ g_2 &= f_2 - (f_2, e_1)e_1, & e_2 &= g_2/|g_2|, \\ &\ldots\ldots\ldots\ldots, & &\ldots\ldots, \\ g_n &= f_n - \sum_{j=1}^{n-1} (f_n, e_j)e_j, & e_n &= g_n/|g_n|. \end{aligned}$$

Then $\{e_1, \ldots, e_n\}$ is an orthonormal basis for X.

4.9.56. Exercise. Prove Theorem 4.9.55. To accomplish this, show that $(e_i, e_j) = 0$ for $i \neq j$, that $|e_i| = 1$ for $i = 1, \ldots, n$, and that $\{e_1, \ldots, e_n\}$ forms a basis for X.

The next result is a direct consequence of Theorem 4.9.55 and Theorem 3.3.44.

4.9.57. Corollary. If $e_1, \ldots, e_k$, $k < n$, are mutually orthogonal non-zero vectors in X, then we can find a set of vectors $e_{k+1}, \ldots, e_n$ such that the set $\{e_1, \ldots, e_n\}$ forms a basis for X.

Our next result is known as the **Bessel inequality.**

4.9.58. Theorem. If $\{x_1, \ldots, x_k\}$ is an arbitrary set of mutually orthonormal vectors in X, then

$$\sum_{i=1}^{k} |(x, x_i)|^2 \leq |x|^2$$

for all $x \in X$. Moreover, the vector

$$y = x - \sum_{i=1}^{k} (x, x_i)x_i$$

is orthogonal to each x_i, $i = 1, \ldots, k$.

Proof. Let $\alpha_i = (x, x_i)$. We have

$$0 \leq \left| x - \sum_{i=1}^{k} \alpha_i x_i \right|^2 = \left(x - \sum_{i=1}^{k} \alpha_i x_i, x - \sum_{j=1}^{k} \alpha_j x_j \right)$$
$$= (x, x) - \sum_{i=1}^{k} \alpha_i \bar{\alpha}_i - \sum_{j=1}^{k} \bar{\alpha}_j \alpha_j + \sum_{i=1}^{k} \sum_{j=1}^{k} \alpha_i \bar{\alpha}_j (x_i, x_j).$$

Now since the vectors $x_1, \ldots, x_k$ are mutually orthonormal, we have

$$0 \leq |x|^2 - \sum_{i=1}^{k} |\alpha_i|^2 = |x|^2 - \sum_{i=1}^{k} |(x, x_i)|^2,$$

which proves the first part of the theorem.

To prove the second part, we note that

$$(y, x_j) = (x, x_j) - \sum_{i=1}^{k} \alpha_i (x_i, x_j) = \alpha_j - \alpha_j = 0. \quad \blacksquare$$

In Theorem 4.9.58, let U denote the linear subspace of X which is spanned by the set of vectors $\{x_1, \ldots, x_k\}$. Then clearly each vector y defined in this theorem is orthogonal to each vector of U; i.e., $y \perp U$ (see Definition 3.6.22).

Let us next consider:

4.9.59. Theorem. Let Y be a linear subspace of X, and let

$$Y^{\perp} = \{x \in X\colon (x, y) = 0 \text{ for all } y \in Y\}. \tag{4.9.60}$$

(i) Let $\{f_1, \ldots, f_k\}$ span Y. Then $x \in Y^{\perp}$ if and only if $x \perp f_j$ for $j = 1, \ldots, k$.

(ii) $Y^{\perp}$ is a linear subspace of X.

(iii) $n = \dim X = \dim Y + \dim Y^{\perp}$.

(iv) $(Y^{\perp})^{\perp} = Y$.

(v) $X = Y \oplus Y^{\perp}$

(vi) Let $x, y \in X$. If $x = x_1 + x_2$ and $y = y_1 + y_2$, where $x_1, y_1 \in Y$ and $x_2, y_2 \in Y^{\perp}$, then

$$(x, y) = (x_1, y_1) + (x_2, y_2)$$

and

$$|x| = \sqrt{|x_1|^2 + |x_2|^2}.$$

Proof. To prove the first part, note that if $x \in Y^\perp$, then $x \perp f_1, \ldots, x \perp f_k$, since $f_i \in Y$ for $i = 1, \ldots, k$. On the other hand, let $x \perp f_i$, $i = 1, \ldots, k$. Then for any $y \in Y$ there exist scalars η_i, $i = 1, \ldots, k$ such that $y = n_1 f_1 + \ldots + \eta_k f_k$. Hence,

$$(x, y) = \left(x, \sum_{i=1}^{k} \eta_i f_i\right) = \sum_{i=1}^{k} \eta_i (x, f_i) = 0.$$

Thus, $x \in Y^\perp$.

The remaining parts of the theorem are left as an exercise. ∎

4.9.61. Exercise. Prove parts (ii) through (vi) of Theorem 4.9.59.

4.9.62. Definition. Let Y be a linear subspace of X. The subspace $Y^\perp$ defined in Eq. (4.9.60) is called the **orthogonal complement** of Y.

Before closing the present section we state and prove the following important result.

4.9.63. Theorem. Let f be a linear functional on X. There exists a unique $y \in X$ such that

$$f(x) = (x, y) \tag{4.9.64}$$

for all $x \in X$.

Proof. If $f(x) = 0$ for all $x \in X$, then $y = 0$ is the unique vector such that Eq. (4.9.64) is satisfied for all $x \in X$, by Theorem 4.9.17. So let us suppose that $f(x) \neq 0$ for some $x \in X$, and let

$$\mathfrak{N} = \{x \in X : f(x) = 0\}.$$

Then $\mathfrak{N}$ is a linear subspace of X. Let $\mathfrak{N}^\perp$ be the orthogonal complement of $\mathfrak{N}$. Then it follows from Theorem 4.9.59 that $X = \mathfrak{N} \oplus \mathfrak{N}^\perp$. Furthermore, $\mathfrak{N}^\perp$ contains a non-zero vector. Let $y_0 \in \mathfrak{N}^\perp$ and, without loss of generality, let y_0 be chosen in such a fashion that $|y_0| = 1$. Now let $y = f(y_0)y_0$ and for any $x \in X$ let $x_0 = x - \alpha y_0$, where $\alpha = f(x)/f(y_0)$. Then $f(x_0) = 0$, and thus $x_0 \in \mathfrak{N}$. We now have $x = x_0 + \alpha y_0$, and

$$\begin{aligned}(x, y) &= (x_0, f(y_0) \cdot y_0) + (\alpha y_0, f(y_0) \cdot y_0)\\ &= f(y_0) \cdot (x_0, y_0) + \alpha f(y_0) \cdot (y_0, y_0)\\ &= \alpha f(y_0) = f(x);\end{aligned}$$

i.e., for all $x \in X$, $f(x) = (x, y)$.

To show that y is unique, suppose that $(x, y_1) = (x, y_2)$ for all $x \in X$. Then $(x, y_1 - y_2) = 0$ for all $x \in X$. But this implies that $y_1 - y_2 = 0$, or $y_1 = y_2$. This completes the proof of the theorem. ∎

4.10. LINEAR TRANSFORMATIONS ON EUCLIDEAN VECTOR SPACES

A. Orthogonal Transformations

In the present section we concern ourselves with special types of linear transformations defined on Euclidean vector spaces. We will have occasion to reconsider similar types of transformations again in Chapter 7, in a much more general setting. *Unless otherwise specified, X will denote an n-dimensional Euclidean vector space throughout the present section.*

The first special type of linear transformation defined on Euclidean vector spaces which we consider is the so-called "orthogonal transformation." Let $\{e_1, \ldots, e_n\}$ be an orthonormal basis for X, let $e'_i = \sum_{j=1}^{n} p_{ji} e_j$, $i = 1, \ldots, n$, and let $\mathbf{P}$ denote the matrix determined by the real scalars p_{ij}. The following question arises: when is the set $\{e'_1, \ldots, e'_n\}$ also an orthonormal basis for X? To determine the desired properties of $\mathbf{P}$, we consider

$$(e'_i, e'_j) = \left(\sum_{k=1}^{n} p_{ki} e_k, \sum_{l=1}^{n} p_{lj} e_l\right) = \sum_{k,l} p_{ki} p_{lj} (e_k, e_l).$$

In order that $(e'_i, e'_j) = 0$ for $i \neq j$ and $(e'_i, e'_j) = 1$ for $i = j$, we require that

$$(e'_i, e'_j) = \sum_{k,l=1}^{n} p_{ki} p_{lj} \delta_{kl} = \sum_{k=1}^{n} p_{ki} p_{kj} = \delta_{ij},$$

i.e., we require that

$$\mathbf{P}^T \mathbf{P} = \mathbf{I},$$

where, as usual, $\mathbf{I}$ denotes the $n \times n$ identity matrix. We summarize.

4.10.1. Theorem. Let $\{e_1, \ldots, e_n\}$ be an orthonormal basis for X. Let $e'_i = \sum_{j=1}^{n} p_{ji} e_j$, $i = 1, \ldots, n$. Then $\{e'_1, \ldots, e'_n\}$ is an orthonormal basis for X if and only if $\mathbf{P}^T = \mathbf{P}^{-1}$.

This result gives rise to the following:

4.10.2. Definition. A matrix $\mathbf{P}$ such that $\mathbf{P}^T = \mathbf{P}^{-1}$, i.e., such that $\mathbf{P}^T \mathbf{P} = \mathbf{P}^{-1} \mathbf{P} = \mathbf{I}$, is called an **orthogonal matrix**.

4.10.3. Exercise. Show that if $\mathbf{P}$ is an orthogonal matrix, then either $\det \mathbf{P} = 1$ or $\det \mathbf{P} = -1$. Also, show that if $\mathbf{P}$ and $\mathbf{Q}$ are $(n \times n)$ orthogonal matrices, then so is $\mathbf{PQ}$.

The nomenclature used in our next definition will become clear shortly.

4.10.4. Definition. A linear transformation A from X into X is called an **orthogonal linear transformation** if $(Ax, Ay) = (x, y)$ for all $x, y \in X$.

Let us now establish some of the properties of orthogonal transformations.

4.10.5. Theorem. Let $A \in L(X, X)$. Then A is orthogonal if and only if $|Ax| = |x|$ for all $x \in X$.

Proof. If A is orthogonal, then $(Ax, Ax) = (x, x)$ and $|Ax| = |x|$. Conversely, if $|Ax| = |x|$ for all $x \in X$, then

$$\begin{aligned}|A(x + y)|^2 &= (A(x + y), A(x + y)) = (Ax + Ay, Ax + Ay)\\ &= |Ax|^2 + 2(Ax, Ay) + |Ay|^2\\ &= |x|^2 + 2(Ax, Ay) + |y|^2.\end{aligned}$$

Also,

$$|A(x + y)|^2 = |x + y|^2 = (x + y, x + y) = |x|^2 + 2(x, y) + |y|^2,$$

and therefore

$$(Ax, Ay) = (x, y)$$

for all $x, y \in X$. ■

We note that if A is an orthogonal linear transformation, then $x \perp y$ for all $x, y \in X$ if and only if $Ax \perp Ay$. For $(x, y) = 0$ if and only if $(Ax, Ay) = 0$.

4.10.6. Corollary. Every orthogonal linear transformation of X into X is non-singular.

Proof. Let $Ax = 0$. Then $|Ax| = |x| = 0$. Thus, $x = 0$ and A is non-singular. ■

Our next result establishes the link between Definitions 4.10.2 and 4.10.4.

4.10.7. Theorem. Let $\{e_1, \ldots, e_n\}$ be an orthonormal basis for X. Let $A \in L(X, X)$, and let $\mathbf{A}$ be the matrix of A with respect to this basis. Then A is orthogonal if and only if $\mathbf{A}$ is orthogonal.

Proof. Let x and y be arbitrary vectors in X, and let $\mathbf{x}$ and $\mathbf{y}$ denote their coordinate representation, respectively, with respect to the basis $\{e_1, \ldots, e_n\}$. Then $\mathbf{Ax}$ and $\mathbf{Ay}$ denote the coordinate representation of Ax and Ay, respectively, with respect to this basis. Now,

$$(Ax, Ay) = (\mathbf{Ax})^T(\mathbf{Ay}) = \mathbf{x}^T\mathbf{A}^T\mathbf{Ay},$$

and

$$(x, y) = \mathbf{x}^T\mathbf{y} = \mathbf{x}^T\mathbf{I}\mathbf{y}.$$

Now suppose that $\mathbf{A}$ is orthogonal. Then $\mathbf{A}^T\mathbf{A} = \mathbf{I}$ and $(Ax, Ay) = \mathbf{x}^T\mathbf{y} = (x, y)$ for all $x, y \in X$. On the other hand, if A is orthogonal, then $(Ax, Ay) = \mathbf{x}^T\mathbf{A}^T\mathbf{A}\mathbf{y} = \mathbf{x}^T\mathbf{y} = (x, y)$ for all $x, y \in X$. Thus, $\mathbf{x}^T[(\mathbf{A}^T\mathbf{A} - \mathbf{I})\mathbf{y}] = 0$. Since this holds for all $x, y \in X$, we conclude from Corollary 4.9.20 that $\mathbf{A}^T\mathbf{A} - \mathbf{I} = \mathbf{0}$; i.e., $\mathbf{A}^T\mathbf{A} = \mathbf{I}$. ■

The next two results are left as an exercise.

4.10.8. Corollary. Let $A \in L(X, X)$. If A is orthogonal, then $\det A = \pm 1$.

4.10.9. Corollary. Let $A, B \in L(X, X)$. If A and B are orthogonal transformations, then AB is also an orthogonal linear transformation.

4.10.10. Exercise. Prove Corollaries 4.10.8 and 4.10.9.

For reasons that will become apparent later, we introduce the following convention.

4.10.11. Definition. Let $A \in L(X, X)$ be an orthogonal linear transformation. If $\det A = +1$, then A is called a **rotation**. If $\det A = -1$, then A is called a **reflection**.

B. Adjoint Transformations

The next important class of linear transformations on Euclidean spaces which we consider are so-called adjoint linear transformations. Our next result enables us to introduce such transformations in a natural way.

4.10.12. Theorem. Let $G \in L(X, X)$ and define $g: X \times X \longrightarrow R$ by $g(x, y) = (x, Gy)$ for all $x, y \in X$. Then g is a bilinear functional on X. Moreover, if $\{e_1, \ldots, e_n\}$ is an orthonormal basis for X, then the matrix of g with respect to this basis, denoted by $\mathbf{G}$, is the matrix of G with respect to $\{e_1, \ldots, e_n\}$. Conversely, given an arbitrary bilinear functional g defined on X, there exists a unique linear transformation $G \in L(X, X)$ such that $(x, Gy) = g(x, y)$ for all $x, y \in X$.

Proof. Let $G \in L(X, X)$, and let $g(x, y) = (x, Gy)$. Then

$$g(x_1 + x_2, y) = (x_1 + x_2, Gy) = (x_1, Gy) + (x_2, Gy) = g(x_1, y) + g(x_2, y).$$

Also,

$$\begin{aligned} g(x, y_1 + y_2) &= (x, G(y_1 + y_2)) = (x, Gy_1 + Gy_2) = (x, Gy_1) + (x, Gy_2) \\ &= g(x, y_1) + g(x, y_2). \end{aligned}$$

Furthermore,

$$g(\alpha x, y) = (\alpha x, Gy) = \alpha(x, Gy) = \alpha g(x, y),$$

and

$$g(x, \alpha y) = (x, G(\alpha y)) = (x, \alpha G(y)) = \alpha(x, Gy) = \alpha g(x, y),$$

where α is a real scalar. Therefore, g is a bilinear functional.

Next, let $\{e_1, \ldots, e_n\}$ be an orthonormal basis for X. Then the matrix $\mathbf{G}$ of g with respect to this basis is determined by the elements $g_{ij} = g(e_i, e_j)$. Now let $\mathbf{G}' = [g'_{ij}]$ be the matrix of G with respect to $\{e_1, \ldots, e_n\}$. Then $Ge_j = \sum_{k=1}^{n} g'_{kj}e_k$ for $j = 1, \ldots, n$. Hence, $(e_i, Ge_j) = \left(e_i, \sum_{k=1}^{n} g'_{kj}e_k\right) = g'_{ij}$. Since $g_{ij} = g(e_i, e_j) = (e_i, Ge_j) = g'_{ij}$, it follows that $\mathbf{G}' = \mathbf{G}$; i.e., $\mathbf{G}$ is the matrix of G.

To prove the last part of the theorem, choose any orthonormal basis $\{e_1, \ldots, e_n\}$ for X. Given a bilinear functional g defined on X, let $\mathbf{G} = [g_{ij}]$ denote its matrix with respect to this basis, and let G be the linear transformation corresponding to $\mathbf{G}$. Then $(x, Gy) = g(x, y)$ by the identical argument given above. Finally, since the matrix of the bilinear functional and the matrix of the linear transformation were determined independently, this correspondence is unique. ■

It should be noted that the correspondence between bilinear functionals and linear transformations determined by the relation $(x, Gy) = g(x, y)$ for all $x, y \in X$ does not depend on the particular basis chosen for X; however, it does depend on the way the inner product is chosen for X at the outset.

Now let $G \in L(X, X)$, set $g(x, y) = (x, Gy)$, and let $h(x, y) = g(y, x) = (y, Gx) = (Gx, y)$. By Theorem 4.10.12, there exists a unique linear transformation, denote it by G^*, such that $h(x, y) = (x, G^*y)$ for all $x, y \in X$. We call the linear transformation $G^* \in L(X, X)$ the **adjoint** of G.

4.10.13. Theorem

(i) For each $G \in L(X, X)$, there is a unique $G^* \in L(X, X)$ such that $(x, G^*y) = (Gx, y)$ for all $x, y \in X$.

(ii) Let $\{e_1, \ldots, e_n\}$ be an orthonormal basis for X, and let $\mathbf{G}$ be the matrix of the linear transformation $G \in L(X, X)$ with respect to this basis. Let $\mathbf{G}^*$ be the matrix of G^* with respect to $\{e_1, \ldots, e_n\}$. Then $\mathbf{G}^* = \mathbf{G}^T$.

Proof. The proof of the first part follows from the discussion preceding the present theorem.

To prove the second part, let $\{e_1, \ldots, e_n\}$ be an orthonormal basis for X, and let $\mathbf{G}^*$ denote the matrix of G^* with respect to this basis. Let $\mathbf{x}$ and $\mathbf{y}$ be the coordinate representation of x and y, respectively, with respect to this basis. Then

$$(x, G^*y) = \mathbf{x}^T\mathbf{G}^*\mathbf{y} = (Gx, y) = (\mathbf{Gx})^T\mathbf{y} = \mathbf{x}^T\mathbf{G}^T\mathbf{y}.$$

Thus, for all $\mathbf{x}$ and $\mathbf{y}$ we have $\mathbf{x}^T(\mathbf{G}^* - \mathbf{G}^T)\mathbf{y} = 0$. Hence, $\mathbf{G}^* = \mathbf{G}^T$. ■

The above result allows the following equivalent definition of the adjoint linear transformation.

4.10.14. Definition. Let $G \in L(X, X)$. The **adjoint transformation,** G^* is defined by the formula

$$(x, G^*y) = (Gx, y)$$

for all $x, y \in X$.

Although there is obviously great similarity between the adjoint linear transformation and the transpose of a linear transformation, it should be noted that these two transformations constitute different concepts. The differences of these will become more apparent in our subsequent discussion of linear transformations defined on complex vector spaces in Chapter 7.

Our next result includes some of the elementary properties of the adjoint of linear transformations. The reader should compare these with the properties of the transpose of linear transformations.

4.10.15. Theorem. Let $A, B \in L(X, X)$, let A^*, B^* denote their respective adjoints, and let α be a real scalar. Then

(i) $(A^*)^* = A$;
(ii) $(A + B)^* = A^* + B^*$;
(iii) $(\alpha A)^* = \alpha A^*$;
(iv) $(AB)^* = B^*A^*$;
(v) $I^* = I$, where I denotes the identity transformation;
(vi) $0^* = 0$, where 0 denotes the null transformation;
(vii) A is non-singular if and only if A^* is non-singular; and
(viii) if A is non-singular, then $(A^*)^{-1} = (A^{-1})^*$.

4.10.16. Exercise. Prove Theorem 4.10.15.

Our next result enables us to characterize orthogonal transformations in terms of their adjoints.

4.10.17. Theorem. Let $A \in L(X, X)$. Then A is orthogonal if and only if $A^* = A^{-1}$.

Proof. We have $(Ax, Ay) = (A^*Ax, y)$. But A is orthogonal if and only if

$(Ax, Ay) = (x, y)$ for all $x, y \in X$. Therefore,

$$(A^*Ax, y) = (x, y)$$

for all x and y. From this it follows that $A^*A = I$, which implies that $A^* = A^{-1}$. ■

The proof of the next theorem is left as an exercise.

4.10.18. Theorem. Let $A \in L(X, X)$. Then A is orthogonal if and only if A^{-1} is orthogonal, and A^{-1} is orthogonal if and only if A^* is orthogonal.

4.10.19. Exercise. Prove Theorem 4.10.18.

C. *Self-Adjoint Transformations*

Using adjoints, we now introduce two additional important types of linear transformations.

4.10.20. Definition. Let $A \in L(X, X)$. Then A is said to be **self-adjoint** if $A^* = A$, and it is said to be **skew-adjoint** if $A^* = -A$.

Some of the properties of such transformations are as follows.

4.10.21. Theorem. Let $A \in L(X, X)$. Let $\{e_1, \ldots, e_n\}$ be an orthonormal basis for X, and let $\mathbf{A}$ be the matrix of A with respect to this basis. The following are equivalent:

(i) A is self-adjoint;
(ii) $\mathbf{A}$ is symmetric; and
(iii) $(Ax, y) = (x, Ay)$ for all $x, y \in X$.

4.10.22. Theorem. Let $A \in L(X, X)$, and let $\{e_1, \ldots, e_n\}$ be an orthonormal basis for X. Let $\mathbf{A}$ be the matrix of A with respect to this basis. The following are equivalent:

(i) A is skew-adjoint;
(ii) $\mathbf{A}$ is skew-symmetric (see Definition 4.8.8); and
(iii) $(Ax, y) = -(x, Ay)$ for all $x, y \in X$.

4.10.23. Exercise. Prove Theorems 4.10.21 and 4.10.22.

The following corollary follows from part (iii) of Theorem 4.10.22.

4.10.24. Corollary. Let A be as defined in Theorem 4.10.22. Then the following are equivalent:

(i) A is skew-symmetric;
(ii) $(x, Ax) = 0$ for all $x \in X$; and
(iii) $Ax \perp x$ for all $x \in X$.

Our next result enables us to represent arbitrary linear transformations as the sum of self-adjoint and skew-adjoint transformations.

4.10.25. Corollary. Let $A \in L(X, X)$. Then there exist unique $A_1, A_2 \in L(X, X)$ such that $A = A_1 + A_2$, where A_1 is self-adjoint and A_2 is skew-adjoint.

4.10.26. Exercise. Prove Corollaries 4.10.24 and 4.10.25.

4.10.27. Exercise. Show that every real $n \times n$ matrix can be written in one and only one way as the sum of a symmetric and skew-symmetric matrix.

Our next result is applicable to real as well as complex vector spaces.

4.10.28. Theorem. Let X be a complex vector space. Then the eigenvalues of a real symmetric matrix $\mathbf{A}$ are all real. (If all eigenvalues of $\mathbf{A}$ are positive (negative), then $\mathbf{A}$ is called **positive (negative) definite.**)

Proof. Let $\lambda = r + is$ denote an eigenvalue of $\mathbf{A}$, where r and s are real numbers and where $i = \sqrt{-1}$. We must show that $s = 0$.

Since λ is an eigenvalue we know that the matrix $(\mathbf{A} - \lambda\mathbf{I})$ is singular. So is the matrix

$$\begin{aligned}\mathbf{B} &= [\mathbf{A} - (r + is)\mathbf{I}][\mathbf{A} - (r - is)\mathbf{I}] \\ &= \mathbf{A}^2 - (r + is)\mathbf{IA} - (r - is)\mathbf{IA} + (r + is)(r - is)\mathbf{I}^2 \\ &= \mathbf{A}^2 - 2r\mathbf{A} + (r^2 + s^2)\mathbf{I}^2 = (\mathbf{A} - r\mathbf{I})^2 + s^2\mathbf{I}.\end{aligned}$$

Since $\mathbf{B}$ is singular, there exists an $\mathbf{x} \neq \mathbf{0}$ such that $\mathbf{Bx} = \mathbf{0}$. Also,

$$0 = \mathbf{x}^T\mathbf{Bx} = \mathbf{x}^T[(\mathbf{A} - r\mathbf{I})^2 + s^2\mathbf{I}]\mathbf{x} = \mathbf{x}^T(\mathbf{A} - r\mathbf{I})^2\mathbf{x} + s^2\mathbf{x}^T\mathbf{x}.$$

Since $\mathbf{A}$ and $\mathbf{I}$ are symmetric,

$$(\mathbf{A} - r\mathbf{I})^T = \mathbf{A}^T - r\mathbf{I}^T = \mathbf{A} - r\mathbf{I}.$$

Therefore,

$$0 = \mathbf{x}^T(\mathbf{A} - r\mathbf{I})^T(\mathbf{A} - r\mathbf{I})\mathbf{x} + s^2(\mathbf{x}^T\mathbf{x});$$

i.e.,

$$0 = \mathbf{y}^T\mathbf{y} + s^2\mathbf{x}^T\mathbf{x},$$

where $\mathbf{y} = (\mathbf{A} - r\mathbf{I})\mathbf{x}$. Now $\mathbf{y}^T\mathbf{y} = \sum_{i=1}^{n} \eta_i^2 \geq 0$ and $\mathbf{x}^T\mathbf{x} = \sum_{i=1}^{n} \xi_i^2 > 0$, because by assumption $\mathbf{x} \neq \mathbf{0}$. Thus, we have

$$0 = \mathbf{y}^T\mathbf{y} + s^2(\mathbf{x}^T\mathbf{x}) \geq 0 + s^2\mathbf{x}^T\mathbf{x}.$$

The only way that this last relation can hold is if $s = 0$. Therefore, $\lambda = r$, and λ is real. ■

Now let $\mathbf{A}$ be the matrix of the linear transformation $A \in L(X, X)$ with respect to some basis. If $\mathbf{A}$ is symmetric, then all its eigenvalues are real. In this case A is self-adjoint and all its eigenvalues are also real; in fact, the eigenvalues of A and $\mathbf{A}$ are identical. Thus, there exist unique real scalars $\lambda_1, \ldots, \lambda_p$, $p \leq n$, such that

$$\det(A - \lambda I) = \det(\mathbf{A} - \lambda \mathbf{I}) = (\lambda_1 - \lambda)^{m_1}(\lambda_2 - \lambda)^{m_2} \ldots (\lambda_p - \lambda)^{m_p}. \quad (4.10.29)$$

We summarize these observations in the following:

4.10.30. Corollary. Let $A \in L(X, X)$. If A is self-adjoint, then all eigenvalues of A are real and there exist unique real numbers $\lambda_1, \ldots, \lambda_p$, $p \leq n$, such that Eq. (4.10.29) holds.

As in Section 4.5, we say that in Corollary 4.10.30 the eigenvalues λ_i, $i = 1, \ldots, p \leq n$, have **algebraic multiplicities** m_i, $i = 1, \ldots, p$, respectively.

Another direct consequence of Theorem 4.10.28 is the following result.

4.10.31. Corollary. Let $A \in L(X, X)$. If A is self-adjoint, then A has at least one eigenvalue.

4.10.32. Exercise. Prove Corollary 4.10.31.

Let us now examine some of the properties of the eigenvalues and eigenvectors of self-adjoint linear transformations. First, we have:

4.10.33. Theorem. Let $A \in L(X, X)$ be a self-adjoint transformation, and let $\lambda_1, \ldots, \lambda_p$, $p \leq n$, denote the distinct eigenvalues of A. If x_i is an eigenvector for λ_i and if x_j is an eigenvector for λ_j, then $x_i \perp x_j$ for all $i \neq j$.

Proof. Assume that $\lambda_i \neq \lambda_j$ and consider $Ax_i = \lambda_i x_i$ and $Ax_j = \lambda_j x_j$, where $x_i \neq 0$ and $x_j \neq 0$. We have

$$\lambda_i(x_i, x_j) = (\lambda_i x_i, x_j) = (Ax_i, x_j) = (x_i, Ax_j) = (x_i, \lambda_j x_j) = \lambda_j(x_i, x_j).$$

Thus,

$$(\lambda_i - \lambda_j)(x_i, x_j) = 0.$$

Since $\lambda_i \neq \lambda_j$, we have $(x_i, x_j) = 0$, which means $x_i \perp x_j$. ■

Now let $A \in L(X, X)$, and let λ_i be an eigenvalue of A. Recall that $\mathfrak{N}_i$

denotes the **null space** of the linear transformation $A - \lambda_i I$, i.e.,

$$\mathfrak{N}_i = \{x \in X: (A - \lambda_i I)x = 0\}. \tag{4.10.34}$$

Recall also that $\mathfrak{N}_i$ is a linear subspace of X. From Theorem 4.10.33 we now have immediately:

4.10.35. Corollary. Let $A \in L(X, X)$ be a self-adjoint transformation, and let λ_i and λ_j be eigenvalues of A. If $\lambda_i \neq \lambda_j$, then $\mathfrak{N}_i \perp \mathfrak{N}_j$.

4.10.36. Exercise. Prove Corollary 4.10.35.

Making use of Theorem 4.9.59, we now prove the following important result.

4.10.37. Theorem. Let $A \in L(X, X)$ be a self-adjoint transformation, and let $\lambda_1, \ldots, \lambda_p$, $p \leq n$, denote the distinct eigenvalues of A. Then

$$\dim X = n = \dim \mathfrak{N}_1 + \dim \mathfrak{N}_2 + \ldots + \dim \mathfrak{N}_p.$$

Proof. Let $\dim \mathfrak{N}_i = n_i$, and let $\{e_1, \ldots, e_{n_1}\}$ be an orthonormal basis for $\mathfrak{N}_1$. Next, let $\{e_{n_1+1}, \ldots, e_{n_1+n_2}\}$ be an orthonormal basis for $\mathfrak{N}_2$. We continue in this manner, finally letting $\{e_{n_1+\ldots+n_{p-1}+1}, \ldots, e_{n_1+\ldots+n_p}\}$ be an orthonormal basis for $\mathfrak{N}_p$. Let $n_1 + \ldots + n_p = m$. Since $\mathfrak{N}_i \perp \mathfrak{N}_j$, $i \neq j$, it follows that the vectors $e_1, \ldots, e_m$, relabeled in an obvious way, are orthonormal in X. We can conclude, by Corollary 4.9.52, that these vectors are a basis for X, if we can prove that $m = n$.

Let Y be the linear subspace of X generated by the orthonormal vectors $e_1, \ldots, e_m$. Then $\{e_1, \ldots, e_m\}$ is an orthonormal basis for Y and $\dim Y = m$. Since $\dim Y + \dim Y^\perp = \dim X = n$ (see Theorem 4.9.59), we need only prove that $\dim Y^\perp = 0$. To this end let x be an arbitrary vector in $Y^\perp$. Then $(x, e_1) = 0, \ldots, (x, e_m) = 0$; i.e., $x \perp e_1, \ldots, x \perp e_m$, by Theorem 4.9.59. So, in particular, again by Theorem 4.9.59, we have $x \perp \mathfrak{N}_i$, $i = 1, \ldots, p$. Now let y be in $\mathfrak{N}_i$. Then

$$(Ax, y) = (x, Ay) = (x, \lambda_i y) = \lambda_i(x, y) = 0,$$

since A is self-adjoint, since y is in $\mathfrak{N}_i$, and since $x \perp \mathfrak{N}_i$. Thus, $Ax \perp \mathfrak{N}_i$ for $i = 1, \ldots, p$, and again by Theorem 4.9.59, $Ax \perp e_i$, $i = 1, \ldots, m$. Thus, by Theorem 4.9.59, $Ax \perp Y^\perp$. Therefore, for each $x \in Y^\perp$ we also have $Ax \in Y^\perp$. Hence, A induces a linear transformation, say A', from $Y^\perp$ into $Y^\perp$, where $A'x = Ax$ for all $x \in Y^\perp$. Now A' is a self-adjoint linear transformation from $Y^\perp$ into $Y^\perp$, because for all x and y in $Y^\perp$ we have

$$(A'x, y) = (Ax, y) = (x, Ay) = (x, A'y).$$

Assume now that $\dim Y^\perp > 0$. Then by Corollary 4.10.31, A' has an eigenvalue, say λ_0, and a corresponding eigenvector $x_0 \neq 0$. Thus, $x_0 \neq 0$

is in $Y^\perp$ and $A'x_0 = Ax_0 = \lambda_0 x_0$; i.e., λ_0 is also an eigenvector of A, say $\lambda_0 = \lambda_i$. So now it follows that $x_0 \in \mathfrak{N}_i$. But from above, $x_0 \in Y^\perp$, which means $x_0 \perp \mathfrak{N}_i$. This implies that $x_0 \perp x_0$, or $(x_0, x_0) = 0$, which in turn implies that $x_0 = 0$. But this contradicts our earlier assumption that $x_0 \neq 0$. Hence, we have arrived at a contradiction, and it therefore follows that dim $Y^\perp = 0$. This proves the theorem. ■

Our next result is a direct consequence of Theorem 4.10.37.

4.10.38. Corollary. Let $A \in L(X, X)$. If A is self-adjoint, then

(i) there exists an orthonormal basis in X such that the matrix of A with respect to this basis is diagonal; and

(ii) for each eigenvalue λ_i of A we have dim $\mathfrak{N}_i$ = multiplicity of λ_i.

Proof. As in the proof of Theorem 4.10.37 we choose an orthonormal basis $\{e_1, \ldots, e_m\}$, where $m = n$. We have $Ae_1 = \lambda_1 e_1, \ldots, Ae_{n_1} = \lambda_1 e_{n_1}$, $Ae_{n_1+1} = \lambda_2 e_{n_1+1}, \ldots, Ae_{n_1+\ldots+n_p} = \lambda_p e_{n_1+\ldots+n_p}$. Thus, the matrix $\mathbf{A}$ of A with respect to $\{e_1, \ldots, e_n\}$ is

$$\mathbf{A} = \begin{bmatrix} \lambda_1 & & & & & & & & & & 0 \\ & \ddots & & & & & & & & & \\ & & \lambda_1 & & & & & & & & \\ & & & \lambda_2 & & & & & & & \\ & & & & \ddots & & & & & & \\ & & & & & \lambda_2 & & & & & \\ & & & & & & \ddots & & & & \\ & & & & & & & \lambda_p & & & \\ & & & & & & & & \ddots & & \\ 0 & & & & & & & & & & \lambda_p \end{bmatrix} \begin{matrix} \left.\vphantom{\begin{matrix}a\\a\\a\end{matrix}}\right\} n_1 \\ \left.\vphantom{\begin{matrix}a\\a\\a\end{matrix}}\right\} n_2 \\ \vdots \\ \left.\vphantom{\begin{matrix}a\\a\\a\end{matrix}}\right\} n_p \end{matrix}$$

To prove the second part, we note that the characteristic polynomial of A is

$$\det(A - \lambda I) = \det(\mathbf{A} - \lambda \mathbf{I}) = (\lambda_1 - \lambda)^{n_1}(\lambda_2 - \lambda)^{n_2} \ldots (\lambda_p - \lambda)^{n_p},$$

and, hence, $n_i = \dim \mathfrak{N}_i$ = multiplicity of λ_i, $i = 1, \ldots, p$. ■

Another consequence of Theorem 4.10.37 is the following:

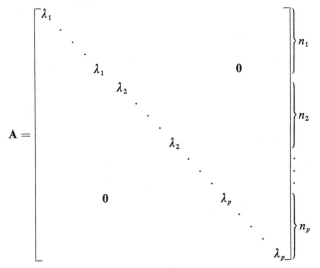

4.10.39. Corollary. Let $\mathbf{A}$ be a real $(n \times n)$ symmetric matrix. Then there exists an orthogonal matrix $\mathbf{P}$ such that the matrix $\mathbf{A}'$ defined by

$$\mathbf{A}' = \mathbf{P}^{-1}\mathbf{A}\mathbf{P} = \mathbf{P}^T\mathbf{A}\mathbf{P}$$

is diagonal.

4.10.40. Exercise. Prove Corollary 4.10.39.

For symmetric bilinear functionals defined on Euclidean vector spaces we have the following result.

4.10.41. Corollary. Let $f(x, y)$ be a symmetric bilinear functional on X. Then there exists an orthonormal basis for X such that the matrix of f with respect to this basis is diagonal.

Proof. By Theorem 4.10.12 there exists an $F \in L(X, X)$ such that $f(x, y) = (x, Fy)$ for all $x, y \in X$. Since f is symmetric, $f(y, x) = f(x, y) = (y, Fx) = (x, Fy) = (Fx, y)$ for all $x, y \in X$, and thus, by Theorem 4.10.21, F is self-adjoint. Hence, by Corollary 4.10.38, there is an orthonormal basis for X such that the matrix of F is diagonal. By Theorem 4.10.12, this matrix is also the representation of f with respect to the same basis. ■

The proof of the next result is left as an exercise.

4.10.42. Corollary. Let $\hat{f}(x)$ be a quadratic form defined on X. Then there exists an orthonormal basis for X such that if $\mathbf{x}^T = (\xi_1, \ldots, \xi_n)$ is the coordinate representation of x with respect to this basis, then $\hat{f}(x) = \alpha_1\xi_1^2 + \ldots + \alpha_n\xi_n^2$ for some real scalars $\alpha_1, \ldots, \alpha_n$.

4.10.43. Exercise. Prove Corollary 4.10.42.

Next, we state and prove the **spectral theorem** for self-adjoint linear transformations. First, we recall that a transformation $P \in L(X, X)$ is a projection on a linear subspace of X if and only if $P^2 = P$ (see Theorem 3.7.4). Also, for any projection P, $X = \mathcal{R}(P) \oplus \mathfrak{N}(P)$, where $\mathcal{R}(P)$ is the range of P and $\mathfrak{N}(P)$ is the null space of P (see Eq. (3.7.8)). Furthermore, recall that a projection P is called an orthogonal projection if $\mathcal{R}(P) \perp \mathfrak{N}(P)$ (see Definition 3.7.16).

4.10.44. Theorem. Let $A \in L(X, X)$ be a self-adjoint transformation, let $\lambda_1, \ldots, \lambda_p$ denote the distinct eigenvalues of A, and let $\mathfrak{N}_i$ be the null space of $A - \lambda_i I$ (see Eq. (4.10.34)). For each $i = 1, \ldots, p$, let P_i denote the projection on $\mathfrak{N}_i$ along $\mathfrak{N}_i^\perp$. Then

(i) P_i is an orthogonal projection for each $i = 1, \ldots, p$;
(ii) $P_iP_j = 0$ for $i \neq j$, $i, j = 1, \ldots, p$;

(iii) $\sum_{j=1}^{p} P_j = I$, where $I \in L(X, X)$ denotes the identity transformation; and

(iv) $A = \sum_{j=1}^{p} \lambda_j P_j$.

Proof. To prove the first part, note that $X = \mathfrak{N}_i \oplus \mathfrak{N}_i^{\perp}, i = 1, \ldots, p$, by Theorem 4.9.59. Thus, by Theorem 3.7.3, $\mathfrak{R}(P_i) = \mathfrak{N}_i$ and $\mathfrak{N}(P_i) = \mathfrak{N}_i^{\perp}$, and hence, P_i is an orthogonal projection.

To prove the second part, let $i \neq j$ and let $x \in X$. Then $P_j x \triangleq x_j \in \mathfrak{N}_j$. Since $\mathfrak{R}(P_i) = \mathfrak{N}_i$ and since $\mathfrak{N}_i \perp \mathfrak{N}_j$, we must have $x_j \in \mathfrak{N}(P_i)$; i.e., $P_i P_j x = 0$ for all $x \in X$.

To prove the third part, let $P = \sum_{i=1}^{p} P_i$. We must show that $P = I$. To do so, we first show that P is a projection. This follows immediately from the fact that for arbitrary $x \in X$, $P^2 x = (P_1 + \ldots + P_p)(P_1 x + \ldots + P_p x) = P_1^2 x + \ldots + P_p^2 x$, because $P_i P_j = 0$ for $i \neq j$. Hence, $P^2 x = (P_1 + \ldots + P_p)x = Px$, and thus P is a projection. Next, we show that $\dim [\mathfrak{R}(P)] = n$. It is straightforward to show that

$$\dim [\mathfrak{R}(P)] = \sum_{i=1}^{p} \dim [\mathfrak{N}_i].$$

But by Theorem 4.10.37, $\sum_{i=1}^{p} \dim [\mathfrak{N}_i] = n$, and thus $\dim [\mathfrak{R}(P)] = n$. Since $X = \mathfrak{R}(P) \oplus \mathfrak{N}(P)$, we conclude that $\mathfrak{R}(P) = X$. Finally, since P is a projection with range X, we conclude that $Px = x$ for all $x \in X$, i.e., $P = I$.

To prove the last part of the theorem, let $x \in X$. From part (iii) we have

$$x = P_1 x + P_2 x + \ldots + P_p x.$$

Let $x_i = P_i x$ for $i = 1, \ldots, p$. Then $x_i \in \mathfrak{N}_i$ and $Ax_i = \lambda_i x_i$. Hence,

$$\begin{aligned} Ax = A(x_1 + \ldots + x_p) &= Ax_1 + \ldots + Ax_p = \lambda_1 x_1 + \ldots + \lambda_p x_p \\ &= \lambda_1 P_1 x + \ldots + \lambda_p P_p x = (\lambda_1 P_1 + \ldots + \lambda_p P_p)x, \end{aligned}$$

which concludes the proof of the theorem. ∎

Any set of linear transformations $\{P_1, \ldots, P_p\}$ satisfying parts (i)–(iii) of Theorem 4.10.44 is said to be a **resolution of the identity** in the setting of a Euclidean space. We shall give a more general definition of this concept in Chapter 7.

D. *Some Examples*

At this point it is appropriate to consider some specific cases.

4.10.45. Example. Let $X = E^2$, let $A \in L(X, X)$, and let $\{e_1, e_2\}$ be an arbitrary basis for X. Suppose that

$$\mathbf{A} = \begin{bmatrix} a_{11} & a_{12} \\ a_{21} & a_{22} \end{bmatrix}$$

is the matrix of A with respect to the basis $\{e_1, e_2\}$. Let $x \in E^2$, and let $\mathbf{x}^T = (\xi_1, \xi_2)$ denote the coordinate representation of x with respect to this basis. Then $\mathbf{Ax}$ is the coordinate representation of Ax with respect to this basis, and we have

$$\mathbf{Ax} = \begin{bmatrix} a_{11}\xi_1 + a_{12}\xi_2 \\ a_{21}\xi_1 + a_{22}\xi_2 \end{bmatrix} \triangleq \begin{bmatrix} \eta_1 \\ \eta_2 \end{bmatrix} = \mathbf{y}.$$

This transformation is depicted pictorially in Figure F.

Now assume that A is a self-adjoint linear transformation. Then there exists an orthonormal basis $\{e'_1, e'_2\}$ such that

$$Ae'_1 = \lambda_1 e'_1, \quad Ae'_2 = \lambda_2 e'_2,$$

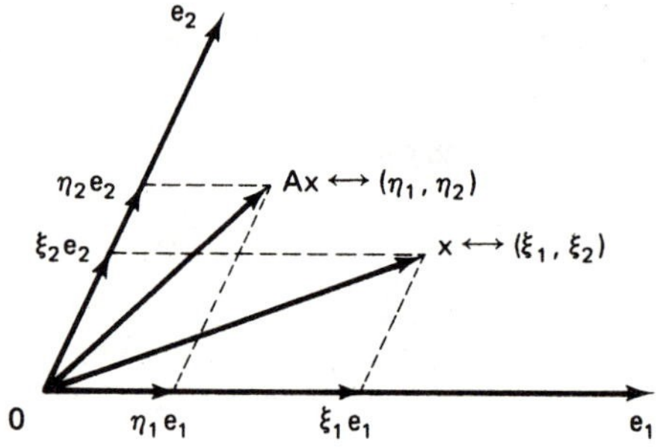

4.10.46. Figure F

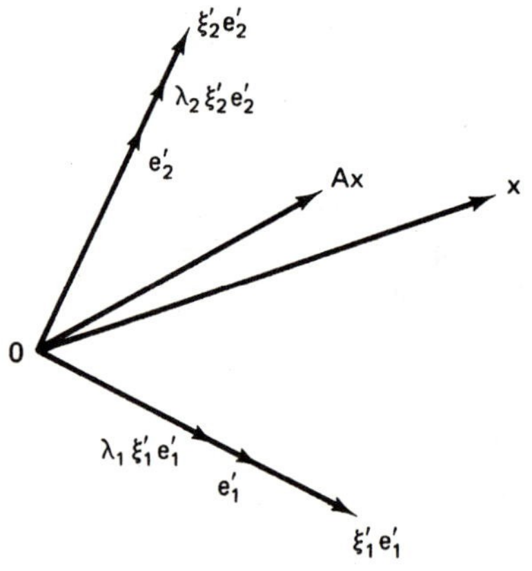

4.10.47. Figure G

where λ_1 and λ_2 denote the eigenvalues of A. Suppose that the coordinates of x with respect to $\{e'_1, e'_2\}$ are ξ'_1 and ξ'_2, respectively. Then

$$Ax = A(\xi'_1 e'_1 + \xi'_2 e'_2) = \xi'_1 Ae'_1 + \xi'_2 Ae'_2 = \xi'_1 \lambda_1 e'_1 + \xi'_2 \lambda_2 e'_2;$$

i.e., the coordinate representation of Ax with respect to $\{e'_1, e'_2\}$ is $(\lambda_1 \xi'_1, \lambda_2 \xi'_2)$. Thus, in order to determine Ax, we merely "stretch" or "compress" the coordinates ξ'_1, ξ'_2 along lines colinear with e'_1 and e'_2, respectively. This is illustrated in Figure G. ■

4.10.48. Example. Consider a transformation R from E^2 into E^2 which rotates vectors as shown in Figure H. By inspection we can characterize R, with respect to the indicated orthonormal basis $\{e_1, e_2\}$, as

$$Re_1 = \cos\theta e_1 + \sin\theta e_2$$
$$Re_2 = -\sin\theta e_1 + \cos\theta e_2.$$

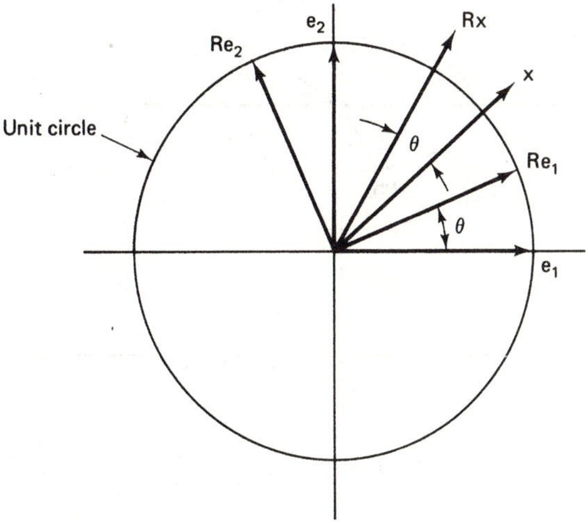

4.10.49. Figure H

The reader can readily verify that R is indeed a linear transformation. The matrix of R with respect to this basis is

$$\mathbf{R}_\theta = \begin{bmatrix} \cos\theta & -\sin\theta \\ \sin\theta & \cos\theta \end{bmatrix}.$$

By direct computation we can verify that

$$\mathbf{R}_\theta^T = \mathbf{R}_\theta^{-1} = \begin{bmatrix} \cos\theta & \sin\theta \\ -\sin\theta & \cos\theta \end{bmatrix},$$

and, moreover, that

$$\det \mathbf{R}_\theta = \cos^2\theta + \sin^2\theta = 1.$$

Thus, R is indeed a rotation as defined in Definition 4.10.11.

For the matrix $\mathbf{R}_\theta$ we also note that $\mathbf{R}_0 = \mathbf{I}$, $\mathbf{R}_\theta^{-1} = \mathbf{R}_{-\theta}$ and $\mathbf{R}_\theta \mathbf{R}_\varphi = \mathbf{R}_{\theta+\varphi}$. ■

4.10.50. Example. Consider now a transformation A from E^3 into E^3, as depicted in Figure J. The vectors e_1, e_2, e_3 form an orthonormal basis for

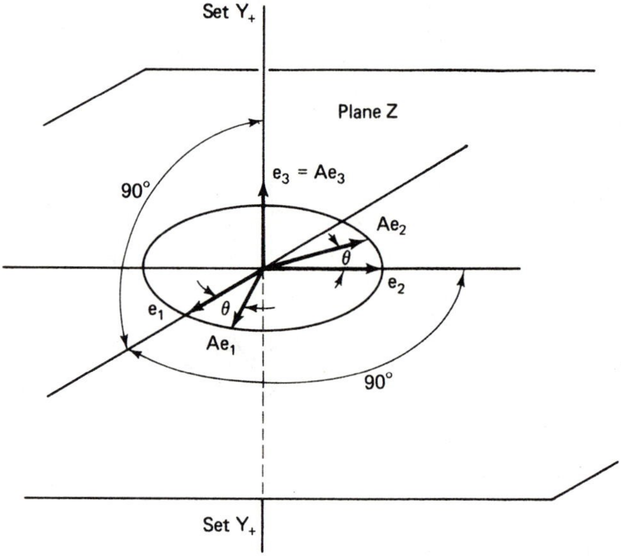

4.10.51. Figure J

E^3. The plane Z is spanned by e_1 and e_2. This transformation accomplishes a rotation about the vector e_3 in the plane Z. By inspection of Figure J it is clear that this transformation is characterized by the set of equations

$$\begin{aligned} Ae_1 &= \cos\theta e_1 + \sin\theta e_2 + 0 \cdot e_3 \\ Ae_2 &= -\sin\theta e_1 + \cos\theta e_2 + 0 \cdot e_3 \\ Ae_3 &= 0 \cdot e_1 + 0 \cdot e_2 + 1 \cdot e_3. \end{aligned}$$

The reader can readily verify that A is a linear transformation. The matrix of A with respect to the basis $\{e_1, e_2, e_3\}$ is

$$\mathbf{A} = \begin{bmatrix} \cos\theta & -\sin\theta & 0 \\ \sin\theta & \cos\theta & 0 \\ 0 & 0 & 1 \end{bmatrix}.$$

For this transformation the following facts are immediately evident (assume $\sin\theta \neq 0$): (a) e_3 is an eigenvector with eigenvalue 1; (b) plane Z is a linear subspace of E^3; (c) $Ax \in Z$ whenever $x \in Z$; (d) the set Y_+ is a linear subspace of E^3; (e) $Ax \in Y_+$ whenever $x \in Y_+$; (f) $Z \perp Y_+$; and (g) $\dim Y_+ = 1$, $\dim Z = 2$, and $\dim Y_+ + \dim Z = \dim E^3$. ■

E. Further Properties of Orthogonal Transformations

The preceding example motivates several of our subsequent results. Let $A \in L(X, X)$. We recall that a linear subspace Y of X is invariant under A if $Ax \in Y$ whenever $x \in Y$. We now prove the following:

4.10.52. Theorem. Let $A \in L(X, X)$ be an orthogonal transformation. Then

(i) the only possible real eigenvalues of A, if there are any, are $+1$ and -1;

(ii) if Y is a linear subspace of X which is invariant under A, then the restriction A' of A to Y is an orthogonal transformation from Y into Y; and

(iii) if Y is a linear subspace of X which is invariant under A, then $Y^\perp$ is also a linear subspace of X which is invariant under A.

Proof. To prove the first part, assume that A has a real eigenvalue, say λ_0. (The definition of eigenvalue of $A \in L(X, X)$ excludes the possibility of complex eigenvalues, since X is a vector space over the field R of real numbers.) Then $Ax = \lambda_0 x$ for $x \neq 0$ and

$$|Ax| = |\lambda_0 x| = |\lambda_0||x|.$$

But $|Ax| = |x|$, because A is by assumption an orthogonal linear transformation. Therefore, $|\lambda_0| = 1$, and we have $\lambda_0 = +1$ or -1.

To prove the second part assume that Y is invariant under A. Then $Ax \in Y$ whenever $x \in Y$, and thus the restriction A' of A to Y, defined by

$$A'x = Ax$$

for all x in Y, is clearly a linear transformation of Y into Y. Now, trivially, for all x in Y we have

$$|A'x| = |Ax| = |x|,$$

since $A \in L(X, X)$ is an orthogonal transformation. Therefore, A' is an orthogonal transformation from Y into Y.

To prove the last part, let Y be an invariant subspace of X under A. Then $x \in Y^\perp$ if and only if $x \perp y$ for all $y \in Y$. Suppose then that $x \in Y^\perp$ and consider Ax. Then for each $y \in Y$ we have

$$(Ax, y) = (x, A^*y) = (x, A^{-1}y),$$

because A is orthogonal. But $A^{-1}y$ is also in Y, for the following reasons. The restriction A' of A to Y is orthogonal on Y by part (ii) and is therefore a non-singular transformation from Y into Y. Hence, $(A')^{-1}$ exists and, moreover, $(A')^{-1}$ must be a transformation from Y into Y. Thus, $(A')^{-1}y = A^{-1}y$ and $A^{-1}y$ is in Y. We finally have

$$(Ax, y) = (x, A^{-1}y) = 0$$

for each y in Y. Thus, $Ax \in Y^\perp$ whenever $x \in Y^\perp$. This proves that $Y^\perp$ is invariant under A. ■

We also have:

4.10.53. Theorem. Let $A \in L(X, X)$ be an orthogonal transformation, let Y_+ denote the set of all $x \in X$ such that $Ax = x$, and let Y_- denote the set of all $x \in X$ such that $Ax = -x$. Then Y_+ and Y_- are linear subspaces of X and $Y_+ \perp Y_-$.

Proof. Since $Y_+ = \mathfrak{N}(A - I)$ and $Y_- = \mathfrak{N}(A + I)$, it follows that Y_+ and Y_- are linear subspaces of X. Now let $x \in Y_+$, and let $y \in Y_-$. Then

$$(x, y) = (Ax, Ay) = (x, -y) = -(x, y),$$

which implies that $(x, y) = 0$. Therefore, $x \perp y$ and $Y_+ \perp Y_-$. ■

Using the above theorem we now can prove the following result.

4.10.54. Corollary. Let A, Y_+ and Y_- be defined as in Theorem 4.10.53, and let Z denote the set of all $x \in X$ such that $x \perp Y_+$ and $x \perp Y_-$. Then Z is a linear subspace of X and $\dim Y_+ + \dim Y_- + \dim Z = \dim X = n$. Furthermore, the restriction of A to Z has no (real) eigenvalues.

Proof. Let $\{e_1, \ldots, e_{n_1}\}$ be an orthonormal basis for Y_+, and let $\{e_{n_1+1}, \ldots, e_{n_1+n_2}\}$ be an orthonormal basis for Y_-, where $\dim Y_+ = n_1$ and $\dim Y_- = n_2$. Then the set $\{e_1, \ldots, e_{n_1+n_2}\}$ is orthonormal. Let Y denote the linear subspace generated by $\{e_1, \ldots, e_{n_1+n_2}\}$. Then $\dim Y = n_1 + n_2$. By the definition of Z and by Theorem 4.9.59 we have $Z = Y^\perp$, and thus Z is a linear subspace of X. Therefore,

$$\begin{aligned} n = \dim X &= \dim Y + \dim Y^\perp = n_1 + n_2 + \dim Z \\ &= \dim Y_+ + \dim Y_- + \dim Z, \end{aligned}$$

which was to be shown.

To prove the second assertion, let A' denote the restriction of A to Z. Suppose there exists a non-zero vector $x \in Z$ such that $A'x = \lambda_0 x$. Since A' is orthogonal by part (ii) of Theorem 4.10.52, we have $\lambda_0 = \pm 1$ by part (i) of Theorem 4.10.52. Thus, x is either in Y_+ or in Y_-. But by assumption, $x \in Z$ and $Z \perp Y_+$ and $Z \perp Y_-$. Therefore, $x = 0$, a contradiction to our earlier assumption. Hence, the restriction A' of A to Z cannot have a real eigenvalue. ■

Our next result is concerned with orthogonal transformations on two-dimensional Euclidean spaces.

4.10.55. Theorem. Let $A \in L(X, X)$ be an orthogonal transformation, where $\dim X = 2$.

(i) If $\det A = +1$ (i.e., A is a rotation), there exists some real θ such that for every orthonormal basis $\{e_1, e_2\}$ the corresponding matrix of A is

$$\mathbf{R}_\theta = \begin{bmatrix} \cos\theta & -\sin\theta \\ \sin\theta & \cos\theta \end{bmatrix}. \tag{4.10.56}$$

(ii) If $\det A = -1$ (i.e., A is a reflection), there exists some orthonormal basis $\{e_1, e_2\}$ such that the matrix of A with respect to this basis is

$$\mathbf{Q} = \begin{bmatrix} 1 & 0 \\ 0 & -1 \end{bmatrix}. \tag{4.10.57}$$

Proof. To prove the first part assume that $\det A = +1$ and choose an arbitrary orthonormal basis $\{e_1, e_2\}$. Let

$$\mathbf{A} = \begin{bmatrix} a_{11} & a_{12} \\ a_{21} & a_{22} \end{bmatrix}$$

denote the matrix of A with respect to this basis. Then, since A is orthogonal, so is $\mathbf{A}$ and we have

$$\mathbf{A}^T\mathbf{A} = \mathbf{I} \tag{4.10.58}$$

and

$$\det A = 1. \tag{4.10.59}$$

Solving Eqs. (4.10.58) and (4.10.59) (we leave the details to the reader) yields $a_{11} = \cos\theta$, $a_{12} = -\sin\theta$, $a_{21} = \sin\theta$, and $a_{22} = \cos\theta$.

To prove the second part assume that A is orthogonal and that $\det A = -1$. Consider the characteristic polynomial of A,

$$p(\lambda) = \lambda^2 + \alpha_1\lambda + \alpha_0.$$

Since $\det A = -1$ we have $\alpha_0 = -1$. Solving for λ_1 and λ_2 we have

$$\lambda_1, \lambda_2 = \frac{-\alpha_1 \pm \sqrt{\alpha_1^2 + 4}}{2},$$

which implies that both λ_1 and λ_2 are real and that $\lambda_1 \neq \lambda_2$. From Theorem 4.10.52 these eigenvalues are $+1$ and -1. Therefore, there exists an orthonormal basis such that the matrix of A with respect to this basis is

$$\begin{bmatrix} \lambda_1 & 0 \\ 0 & \lambda_2 \end{bmatrix} = \begin{bmatrix} 1 & 0 \\ 0 & -1 \end{bmatrix}. \quad \blacksquare$$

In the above proof we have $e_1 \in Y_+$ and $e_2 \in Y_-$, in view of Theorem 4.10.53. Also, from the preceding theorem it is clear that if A is orthogonal and (a) if $\det A = 1$, then $\det(A - \lambda I) = 1 - 2\cos\theta\lambda + \lambda^2$, and (b) if $\det A = -1$, then $\det(A - \lambda I) = \lambda^2 - 1$.

4.10.60. Theorem. Let $A \in L(X, X)$ be an orthogonal transformation having no (real) eigenvalues. Then there exist linear subspaces $Y_1, \ldots, Y_r$ of X such that

(i) $\dim Y_i = 2$, $i = 1, \ldots, r$;
(ii) $Y_i \perp Y_j$ for all $i \neq j$;
(iii) $\dim Y_1 + \ldots + \dim Y_r = \dim X = n$; and
(iv) each subspace Y_i is invariant under A; in fact, the restriction of A to Y_i is a non-trivial rotation (i.e., for the matrix given by Eq. (4.10.56) we have $\theta \neq k\pi$, $k = 0, 1, 2, \ldots$).

Proof. Since by assumption A does not have any (real) eigenvalues, we have

$$\det(A - \lambda I) = (\alpha_1 + \beta_1\lambda + \lambda^2)\ldots(\alpha_r + \beta_r\lambda + \lambda^2),$$

where the α_i, β_i, $i = 1, \ldots, r$ are real (i.e., $\det(A - \lambda I)$ does not have any linear factors $(\lambda_i - \lambda)$, with λ_i real). Solving the first quadratic factor we have

$$\lambda_1 = \frac{-\beta_1 + \sqrt{\beta_1^2 - 4\alpha_1}}{2}$$

and

$$\lambda_2 = \frac{-\beta_1 - \sqrt{\beta_1^2 - 4\alpha_1}}{2},$$

where λ_1 and λ_2 are complex. By Theorem 4.5.33, part (iv), if $f(\cdot)$ is any polynomial function, then $f(\lambda_1)$ will be an eigenvalue of $f(A)$. In particular, if $f(\lambda) = \alpha_1 + \beta_1\lambda + \lambda^2$, we know that one of the eigenvalues of the linear transformation $\alpha_1 I + \beta_1 A + A^2$ will be $\alpha_1 + \beta_1\lambda_1 + \lambda_1^2 = 0$, by choice. Thus, the linear transformation $(\alpha_1 I + \beta_1 A + A^2)$ has 0 as an eigenvalue. Therefore, there exists a vector $f_1 \neq 0$ in X such that

$$(\alpha_1 I + \beta_1 A + A^2)f_1 = 0 \cdot f_1$$

or

$$\alpha_1 f_1 + \beta_1 A f_1 + A^2 f_1 = 0. \tag{4.10.61}$$

Now let $f_2 = Af_1$. We assert that f_1 and f_2 are linearly independent. For if they were not, we would have $f_2 = \gamma f_1 = Af_1$, where γ is a real scalar, and f_1 would be an eigenvector corresponding to a real eigenvalue γ of A, which is impossible by hypothesis. Next, let Y_1 be the linear subspace of X generated by f_1 and f_2. Then Y_1 is two dimensional. We now show that Y_1 is invariant under A. Let $x \in Y_1$. Then

$$x = \xi_1 f_1 + \xi_2 f_2$$

for some ξ_1 and ξ_2, and

$$Ax = \xi_1 Af_1 + \xi_2 Af_2 = \xi_1 Af_1 + \xi_2 A^2 f_1.$$

But from Eq. (4.10.61) it follows that

$$A^2 f_1 = -\alpha_1 f_1 - \beta_1 Af_1.$$

Thus,

$$\begin{aligned} Ax &= \xi_1 Af_1 + \xi_2(-\alpha_1 f_1 - \beta_1 Af_1) = -\xi_2\alpha_1 f_1 + (\xi_1 - \xi_2\beta_1)Af_1 \\ &= -\xi_2\alpha_1 f_1 + (\xi_1 - \xi_2\beta_1)f_2, \end{aligned}$$

which shows that $Ax \in Y_1$ whenever $x \in Y_1$. Thus, Y_1 is invariant under A.

By Theorem 4.10.52, the restriction A' of A to Y_1 is an orthogonal transformation from Y_1 into Y_1. This restriction cannot have any (real) eigenvalues, for then A would also have (real) eigenvalues.

From Theorem 4.10.55, A' cannot be a reflection, for in that case A' would have eigenvalues equal to $+1$ and -1. Moreover, A' cannot be a trivial rotation, for then the eigenvalues of A' would be equal to 1 if $\theta = 0°$ and -1 if $\theta = 180°$. But from Corollary 4.10.8 we know that if A is orthogonal, then $\det A = \pm 1$. Therefore, it follows now from Theorem 4.10.55 that the restriction of A to Y_1 is a non-trivial rotation.

Now let $Z_1 = Y_1^\perp$. Since Y_1 is invariant under A, so is Z_1, by Theorem 4.10.52, part (iii), and $\dim Z_1 = \dim X - 2$. The restriction A_1 of A to Z_1 is an orthogonal transformation from Z_1 into Z_1, and it cannot have any (real) eigenvalues. Applying the argument already given for A and X now to A_1 and Z_1, we can conclude that there exists a two-dimensional linear subspace Y_2 of Z_1 such that the restriction of A_1 to Y_2 is a non-trivial rotation. Now since Y_2 is contained in Z_1 and since by definition $Z_1 = Y_1^\perp$, we have $Y_1 \perp Y_2$.

Next, let Z_2 be the linear subspace which is orthogonal to both Y_1 and Y_2, and let A_2 be the restriction of A to Z_2. Repeating the argument given thus far, we can conclude that there exists a two-dimensional linear subspace Y_3 of Z_2 such that the restriction of A_2 to Y_3 is a non-trivial rotation and such that $Y_2 \perp Y_3$ and $Y_1 \perp Y_3$.

To conclude the proof of the theorem, we continue the above process until we have exhausted the original space X. ■

Combining Theorems 4.10.53 and 4.10.60, we obtain the following:

4.10.62. Corollary. Let $A \in L(X, X)$ be an orthogonal linear transformation. Then there exist linear subspaces $Y_+, Y_-, Y_1, \ldots, Y_r$ of X such that

(i) all of the above linear subspaces are orthogonal to one another;
(ii) $n = \dim X = \dim Y_+ + \dim Y_- + \dim Y_1 + \ldots + \dim Y_r$;
(iii) $x \in Y_+$ if and only if $Ax = x$;
(iv) $x \in Y_-$ if and only if $Ax = -x$; and
(v) the restriction of A to each Y_i, $i = 1, \ldots, r$, is a non-trivial rotation.

Since in the above corollary the dimension of each Y_i, $i = 1, \ldots, r$, is two, we have the following additional result.

4.10.63. Corollary. If in Corollary 4.10.62 $\dim X$ is odd, then A has a real eigenvalue.

We leave the proof of the next result as an exercise.

4.10.64. Theorem. If A is an orthogonal transformation from X into X, then the characteristic polynomial of A is of the form

$$(1 - \lambda)^{n_1}(-1 - \lambda)^{n_2}(1 - 2\cos\theta_1\lambda + \lambda^2) \ldots (1 - 2\cos\theta_r\lambda + \lambda^2) = \det(A - \lambda I).$$

Moreover, there exists an orthonormal basis $\{e_1, \ldots, e_n\}$ of X such that the matrix of A with respect to this basis is of the form

$$\mathbf{A} = \begin{array}{l} Y_1\left\{\begin{array}{l} \\ \\ \end{array}\right. \\ \vdots \\ Y_r\left\{\begin{array}{l} \\ \\ \end{array}\right. \\ Y_-\left\{\begin{array}{l} \\ \\ \\ \end{array}\right. \\ Y_+\left\{\begin{array}{l} \\ \\ \\ \end{array}\right. \end{array} \begin{bmatrix} \cos\theta_1 & -\sin\theta_1 & & & & & & & & \\ \sin\theta_1 & \cos\theta_1 & & & & & & & & \\ & & \ddots & & & & & \mathbf{0} & & \\ & & & \cos\theta_r & -\sin\theta_r & & & & & \\ & & & \sin\theta_r & \cos\theta_r & & & & & \\ & & & & & -1 & & & & \\ & & & & & & \ddots & & & \\ & & \mathbf{0} & & & & & -1 & & \\ & & & & & & & & +1 & \\ & & & & & & & & & \ddots \\ & & & & & & & & & & +1 \end{bmatrix}$$

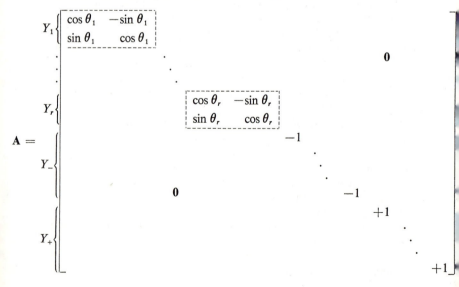

4.10.65. Exercise. Prove Theorem 4.10.64.

In our next result the canonical form of skew-adjoint linear transformations is established.

4.10.66. Theorem. Let A be a skew-adjoint linear transformation from X into X. Then there exists an orthonormal basis $\{e_1, \ldots, e_n\}$ such that the matrix $\mathbf{A}$ of A with respect to this basis is of the form

$$\mathbf{A} = \begin{bmatrix} \begin{array}{cc|} 0 & \lambda_1 \\ -\lambda_1 & 0 \\ \hline \end{array} & & \mathbf{0} \\ & \ddots & \\ \mathbf{0} & & \begin{array}{|cc} \hline 0 & \lambda_r \\ -\lambda_r & 0 \end{array} \end{bmatrix},$$

where the λ_i, $i = 1, \ldots, r$ are real and where some of the λ_i may be zero.

4.10.67. Exercise. Prove Theorem 4.10.66.

Before closing the present section, we briefly introduce so-called "normal transformations." We will have quite a bit more to say about such transformations and their representation in Chapter 7.

4.10.68. Definition. A transformation $A \in L(X, X)$ is said to be a **normal linear transformation** if $A^*A = AA^*$.

Some of the properties of such transformations are as follows.

4.10.69. Theorem. Let $A \in L(X, X)$. Then

(i) if A is a self-adjoint transformation then it is also a normal transformation;

(ii) if A is a skew-adjoint transformation, then it is also a normal transformation;

(iii) if A is an orthogonal transformation, then it is also a normal transformation; and

(iv) if A is a normal linear transformation then there exists an orthonormal basis $\{e_1, \ldots, e_n\}$ of X such that the matrix $\mathbf{A}$ of A with respect to this basis is of the form

$$
\mathbf{A} = \begin{bmatrix}
\begin{array}{cc} \mu_1 & \lambda_1 \\ -\lambda_1 & \mu_1 \end{array} & & & & & & \mathbf{0} \\
& \ddots & & & & & \\
& & \begin{array}{cc} \mu_r & \lambda_r \\ -\lambda_r & \mu_r \end{array} & & & & \\
& & & \nu_1 & & & \\
& \mathbf{0} & & & \ddots & & \\
& & & & & & \nu_{n-2r}
\end{bmatrix}.
$$

The proofs of parts (i)–(iii) follow from the definitions of normal, self-adjoint, skew-adjoint, and orthogonal linear transformations. To prove part (iv), let $A = A_1 + A_2$, where $A_1 = \frac{1}{2}(A + A^*)$ and $A_2 = \frac{1}{2}(A - A^*)$, and note that A_1 is self-adjoint and A_2 is skew adjoint. This representation is unique by Corollary 4.10.25. Making use of Theorem 4.10.66 and Corollary 4.10.38, we obtain the desired result. We leave the details of the proof of this theorem as an exercise.

4.10.70. Exercise. Prove Theorem 4.10.69.

4.11. APPLICATIONS TO ORDINARY DIFFERENTIAL EQUATIONS

In the present section we present applications to the material covered in the present chapter and the preceding chapter. Because of their importance in almost all branches of science and engineering, we consider some topics in ordinary differential equations. Specifically, we concern ourselves with initial-value problems described by ordinary differential equations. The present section is divided into two parts. In subsection A, we define the initial-value problem, while in subsection B we treat linear initial-value problems. At the end of the next chapter, we will continue our discussion of ordinary differential equations.

A. Initial-Value Problem: Definition

Let R denote the set of real numbers, and let $D \subset R^2$ be a domain (i.e., D is an open and connected subset of R^2). We will call R^2 the (t, x) plane. Let f be a real-valued function which is defined and continuous on D, and let $\dot{x} \triangleq dx/dt$ (i.e., $\dot{x}$ denotes the derivative of x with respect to t). We call

$$\dot{x} = f(t, x) \tag{4.11.1}$$

an **ordinary differential equation of the first order**. Let $T = (t_1, t_2) \subset R$ be an open interval which we call a t interval (i.e., $T = (t_1, t_2) = \{t \in R: t_1 < t < t_2\}$). A real differentiable function φ (if it exists) defined on T such that the points $(t, \varphi(t)) \in D$ for all $t \in T$ and such that

$$\dot{\varphi}(t) = f(t, \varphi(t)) \tag{4.11.2}$$

for all $t \in T$ is called a **solution** of the differential equation (4.11.1).

4.11.3. Definition. Let $(\tau, \xi) \in D$. If φ is a solution of the differential equation (4.11.1) and if $\varphi(\tau) = \xi$, then φ is called a **solution of the initial-value problem**

$$\left.\begin{aligned} \dot{x} &= f(t, x) \\ x(\tau) &= \xi \end{aligned}\right\}. \tag{4.11.4}$$

In Figure K a typical solution of an initial-value problem is depicted.

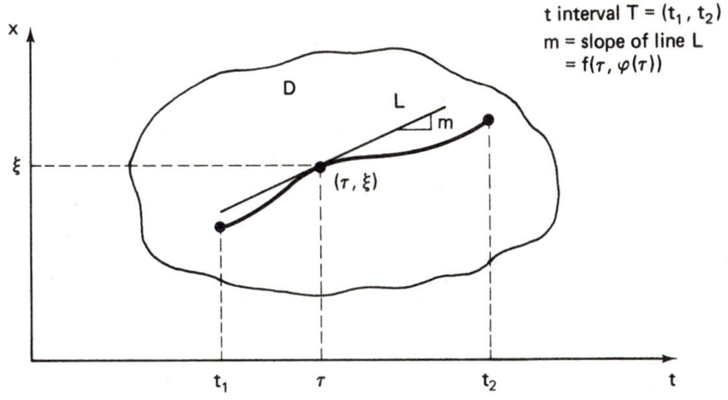

4.11.5. Figure K. Typical solution of an initial-value problem.

We can represent the initial-value problem given in Eq. (4.11.4) equivalently by means of the integral equation

$$\varphi(t) = \xi + \int_\tau^t f(s, \varphi(s))\, ds. \tag{4.11.6}$$

Here we say that two problems are equivalent if they have the same solution.

To prove this equivalence, let φ be a solution of the initial-value problem (4.11.4). Then $\varphi(\tau) = \xi$ and

$$\dot{\varphi}(t) = f(t, \varphi(t))$$

for all $t \in T$. Integrating from τ to t we have

$$\int_\tau^t \dot{\varphi}(s)\, ds = \int_\tau^t f(s, \varphi(s))\, ds$$

or

$$\varphi(t) = \xi + \int_\tau^t f(s, \varphi(s))\, ds.$$

Thus, φ is a solution of the integral equation (4.11.6).

Conversely, let φ be a solution of the integral equation (4.11.6). Then $\varphi(\tau) = \xi$, and differentiating both sides of Eq. (4.11.6) with respect to t we have

$$\dot{\varphi}(t) = f(t, \varphi(t)),$$

and thus φ is a solution of the initial-value problem (4.11.4).

Next, we consider initial-value problems described by means of several first-order ordinary differential equations. Let $D \subset R^{n+1}$ be a domain (i.e., D is an open and connected subset of R^{n+1}). We will call R^{n+1} the $(t, x_1, \ldots, x_n)$ space. Let $f_1, \ldots, f_n$ be n real-valued functions which are defined and continuous on D (i.e., $f_i(t, x_1, \ldots, x_n)$, $i = 1, \ldots, n$, are defined for all points in D and are continuous with respect to all arguments $t, x_1, \ldots, x_n$). We call

$$\dot{x}_i = f_i(t, x_1, \ldots, x_n), \quad i = 1, \ldots, n, \tag{4.11.7}$$

a **system of n ordinary differential equations of the first order.** A set of n real differentiable functions $\{\varphi_1, \ldots, \varphi_n\}$ (if it exists) defined on a real t interval $T = (t_1, t_2) \subset R$ such that the points $(t, \varphi_1(t), \ldots, \varphi_n(t)) \in D$ for all $t \in T$ and such that

$$\dot{\varphi}_i(t) = f_i(t, \varphi_1(t), \ldots, \varphi_n(t)), \quad i = 1, \ldots, n \tag{4.11.8}$$

for all $t \in T$, is called a **solution of the system of ordinary differential equations** (4.11.7).

4.11.9. Definition. Let $(\tau, \xi_1, \ldots, \xi_n) \in D$. If the set $\{\varphi_1, \ldots, \varphi_n\}$ is a solution of the system of equations (4.11.7) and if $(\varphi_1(\tau), \ldots, \varphi_n(\tau)) = (\xi_1, \ldots, \xi_n)$, then the set $\{\varphi_1, \ldots, \varphi_n\}$ is called a **solution of the initial-value problem**

$$\left.\begin{aligned} &\dot{x}_i = f_i(t, x_1, \ldots, x_n), \quad i = 1, \ldots, n \\ &x_i(\tau) = \xi_i, \quad i = 1, \ldots, n \end{aligned}\right\}. \tag{4.11.10}$$

It is convenient to use vector notation to represent Eq. (4.11.10). Let

$$\mathbf{x} = \begin{bmatrix} x_1 \\ \cdot \\ \cdot \\ \cdot \\ x_n \end{bmatrix}, \quad \boldsymbol{\xi} = \begin{bmatrix} \xi_1 \\ \cdot \\ \cdot \\ \cdot \\ \xi_n \end{bmatrix}, \quad \boldsymbol{\varphi} = \begin{bmatrix} \varphi_1 \\ \cdot \\ \cdot \\ \cdot \\ \varphi_n \end{bmatrix}$$

$$\mathbf{f}(t, \mathbf{x}) = \begin{bmatrix} f_1(t, x_1, \ldots, x_n) \\ \cdot \\ \cdot \\ \cdot \\ f_n(t, x_1, \ldots, x_n) \end{bmatrix} = \begin{bmatrix} f_1(t, \mathbf{x}) \\ \cdot \\ \cdot \\ \cdot \\ f_n(t, \mathbf{x}) \end{bmatrix}$$

and *define* $\dot{\mathbf{x}} = d\mathbf{x}/dt$ componentwise; i.e.,

$$\dot{\mathbf{x}} \triangleq \begin{bmatrix} \dot{x}_1 \\ \cdot \\ \cdot \\ \cdot \\ \dot{x}_n \end{bmatrix}.$$

We can express Eq. (4.11.10) equivalently as

$$\left.\begin{aligned} \dot{\mathbf{x}} &= \mathbf{f}(t, \mathbf{x}) \\ \mathbf{x}(\tau) &= \boldsymbol{\xi} \end{aligned}\right\}. \tag{4.11.11}$$

If in Eq. (4.11.11) $\mathbf{f}(t, \mathbf{x})$ does not depend on t (i.e., $\mathbf{f}(t, \mathbf{x}) = \mathbf{f}(\mathbf{x})$ for all $(t, \mathbf{x}) \in D$), then we have

$$\dot{\mathbf{x}} = \mathbf{f}(\mathbf{x}). \tag{4.11.12}$$

In this case we speak of an **autonomous system of first-order ordinary differential equations.**

Of special importance are systems of first-order ordinary differential equations described by

$$\dot{\mathbf{x}} = \mathbf{A}(t)\mathbf{x} + \mathbf{v}(t), \tag{4.11.13}$$

$$\dot{\mathbf{x}} = \mathbf{A}(t)\mathbf{x}, \tag{4.11.14}$$

and

$$\dot{\mathbf{x}} = \mathbf{A}\mathbf{x}, \tag{4.11.15}$$

where $\mathbf{x}$ is a real n-vector, $\mathbf{A}(t) = [a_{ij}(t)]$ is a real $(n \times n)$ matrix with elements $a_{ij}(t)$ that are defined and continuous on a t interval T, $\mathbf{A} = [a_{ij}]$ is an $(n \times n)$ matrix with real constant coefficients, and $\mathbf{v}(t)$ is a real n-vector with components $v_i(t)$, $i = 1, \ldots, n$, which are defined and at least piecewise continuous on T. These equations are clearly a special case of Eq. (4.11.7). For example, if in Eq. (4.11.7) we let

$$f_i(t, x_1, \ldots, x_n) = f_i(t, \mathbf{x}) = \sum_{j=1}^{n} a_{ij}(t)x_j, \quad i = 1, \ldots, n,$$

then Eq. (4.11.14) results. In the case of Eqs. (4.11.14) and (4.11.15), we speak of a **linear homogeneous system** of ordinary differential equations, in the case of Eq. (4.11.13) we have a **linear non-homogeneous system** of ordinary differential equations, and in the case of Eq. (4.11.15) we speak of a linear system of ordinary differential equations with **constant coefficients.**

Next, we consider initial-value problems described by means of nth-order ordinary differential equations. Let f be a real function which is defined and continuous in a domain D of the real $(t, x_1, \ldots, x_n)$ space, and let $x^{(k)} \triangleq d^k x/dt^k$. We call

$$x^{(n)} = f(t, x, x^{(1)}, \ldots, x^{(n-1)}) \tag{4.11.16}$$

an **nth-order ordinary differential equation.** A real function φ (if it exists) which is defined on a t interval $T = (t_1, t_2) \subset R$ and which has n derivatives on T is called a **solution** of Eq. (4.11.16) if $(t, \varphi(t), \ldots, \varphi^{(n)}(t)) \in D$ for all $t \in T$ and if

$$\varphi^{(n)}(t) = f(t, \varphi(t), \ldots, \varphi^{(n-1)}(t)) \tag{4.11.17}$$

for all $t \in T$.

4.11.18. Definition. Let $(\tau, \xi_1, \ldots, \xi_n) \in D$. If φ is a solution of Eq. (4.11.16) and if $\varphi(\tau) = \xi_1, \ldots, \varphi^{(n-1)}(\tau) = \xi_n$, then φ is called a solution of the **initial value problem**

$$\left.\begin{aligned} &x^{(n)} = f(t, x, x^{(1)}, \ldots, x^{(n-1)}) \\ &x(\tau) = \xi_1, \ldots, x^{(n-1)}(\tau) = \xi_n \end{aligned}\right\}. \tag{4.11.19}$$

Of particular interest are nth-order ordinary differential equations

$$a_n(t)x^{(n)} + a_{n-1}(t)x^{(n-1)} + \ldots + a_1(t)x^{(1)} + a_0(t)x = v(t), \tag{4.11.20}$$

$$a_n(t)x^{(n)} + a_{n-1}(t)x^{(n-1)} + \ldots + a_1(t)x^{(1)} + a_0(t)x = 0, \tag{4.11.21}$$

and

$$a_n x^{(n)} + a_{n-1}x^{(n-1)} + \ldots + a_1 x^{(1)} + a_0 x = 0, \tag{4.11.22}$$

where $a_n(t), \ldots, a_0(t)$ are real continuous functions defined on the interval T, where $a_n(t) \neq 0$ for all $t \in T$, where $a_n, \ldots, a_0$ are real constants, where $a_n \neq 0$, and where $v(t)$ is a real function defined and piecewise continuous on T. We call Eq. (4.11.21) a **linear homogeneous ordinary differential equation of order** n, Eq. (4.11.20) a **linear non-homogeneous ordinary differential equation of order** n, and Eq. (4.11.22) a **linear ordinary differential equation of order** n **with constant coefficients.**

We now show that the theory of nth-order ordinary differential equations reduces to the theory of a system of n first-order ordinary differential equations. To this end, let in Eq. (4.11.19) $x = x_1$, and let

$$\left.\begin{aligned} &\dot{x}_1 = x_2 = \dot{x} \\ &\dot{x}_2 = x_3 = x^{(2)} \\ &\cdots\cdots\cdots\cdots\cdots\cdots \\ &\dot{x}_{n-1} = x_n = x^{(n-1)} \\ &\dot{x}_n = f(t, x_1, \ldots, x_n) = x^{(n)} \end{aligned}\right\}. \tag{4.11.23}$$

This system of equations is clearly defined for all $(t, x_1, \ldots, x_n) \in D$. Now assume that the vector $\boldsymbol{\varphi}^T = (\varphi_1, \ldots, \varphi_n)$ is a solution of Eq. (4.11.23) on an interval T. Since $\varphi_2 = \dot{\varphi}_1, \varphi_3 = \dot{\varphi}_2, \ldots, \varphi_n = \varphi_1^{(n-1)}$, and since

$$f(t, \varphi_1(t), \ldots, \varphi_n(t)) = f(t, \varphi_1(t), \ldots, \varphi_1^{(n-1)}(t)) = \varphi_1^{(n)}(t),$$

it follows that the first component φ_1 of the vector $\boldsymbol{\varphi}$ is a solution of Eq. (4.11.16) on the interval T. Conversely, assume that φ_1 is a solution of Eq. (4.11.16) on the interval T. Then the vector $\boldsymbol{\varphi}^T = (\varphi, \varphi^{(1)}, \ldots, \varphi^{(n-1)})$ is clearly a solution of the system of equations (4.11.23). Note that if $\varphi_1(\tau) = \xi_1, \ldots, \varphi_1^{(n-1)}(\tau) = \xi_n$, then the vector $\boldsymbol{\varphi}$ satisfies $\boldsymbol{\varphi}(\tau) = \boldsymbol{\xi}$, where $\boldsymbol{\xi}^T = (\xi_1, \ldots, \xi_n)$. The converse is also true.

Thus far we have concerned ourselves with initial-value problems characterized by **real ordinary differential equations**. It is possible to consider initial-value problems involving **complex ordinary differential equations**. For example, let t be real and let $\mathbf{z}^T = (z_1, \ldots, z_n)$ be a complex vector (i.e., z_k is of the form $u_k + iv_k$, where u_k and v_k are real and $i = \sqrt{-1}$). Let D be a domain in the $(t, \mathbf{z})$ space, and let $f_1, \ldots, f_n$ be n continuous complex-valued functions defined on D. Let $\mathbf{f}^T = (f_1, \ldots, f_n)$, and let $\dot{\mathbf{z}} = d\mathbf{z}/dt$. We call

$$\dot{\mathbf{z}} = \mathbf{f}(t, \mathbf{z}) \tag{4.11.24}$$

a system of n complex ordinary differential equations of the first order. A complex vector $\boldsymbol{\varphi}^T = (\varphi_1, \ldots, \varphi_n)$ which is defined and differentiable on a real t interval $T = (T_1, T_2) \subset R$ such that the points $(t, \varphi_1(t), \ldots, \varphi_n(t)) \in D$ for all $t \in T$ and such that

$$\dot{\boldsymbol{\varphi}}(t) = \mathbf{f}(t, \boldsymbol{\varphi}(t))$$

for all $t \in T$, is called a solution of the system of equations (4.11.24). If in addition, $(\tau, \xi_1, \ldots, \xi_n) \in D$ and if $(\varphi_1(\tau), \ldots, \varphi_n(\tau)) = (\xi_1, \ldots, \xi_n) = \boldsymbol{\xi}^T$, then $\boldsymbol{\varphi}$ is said to be a solution of the initial-value problem

$$\left.\begin{aligned} \dot{\mathbf{z}} &= \mathbf{f}(t, \mathbf{z}) \\ \mathbf{z}(\tau) &= \boldsymbol{\xi} \end{aligned}\right\}. \tag{4.11.25}$$

Of particular interest in applications are initial-value problems characterized by complex linear ordinary differential equations having forms analogous to those given in equations (4.11.13)–(4.11.15).

We can similarly consider initial-value problems described by complex nth-order ordinary differential equations.

Let us look now at some specific examples. The first example demonstrates that the solution to an initial-value problem may not be unique.

4.11.26. Example. Consider the initial-value problem

$$\dot{x} = x^{1/3}$$

$$x(0) = 0.$$

We can readily verify that this problem has *infinitely* many solutions passing through the origin of the (t, x) plane, given by

$$\varphi_\mu(t) = \begin{cases} 0, & 0 \le t \le \mu \\ [\frac{2}{3}(t-\mu)]^{3/2}, & \mu < t \le 1 \end{cases}$$

where μ is any real number such that $0 \le \mu \le 1$. ■

The next example shows that the t interval for which a solution to the initial-value problem exists may be restricted.

4.11.27. Example. Consider the initial-value problem

$$\dot{x} = x^2,$$
$$x(t_1) = \xi,$$

where ξ is any real number. By direct computation we can verify that

$$\varphi(t) = \xi[1 - (t - t_1)\xi]^{-1}$$

is a solution of this problem. We note that if $t = t_1 + 1/\xi$, then the solution $\varphi(t)$ is not defined. Thus, there is a restriction on the t interval for which a solution to the above problem exists. Namely, if $\xi > 0$, the above solution is valid over any interval (t_1, t_2) such that

$$t_1 < t_2 = t_1 + \frac{1}{\xi}.$$

In this case we say the solution fails to exist for $t \ge t_2$. On the other hand, if $\xi \le 0$, the solution given above is valid for any $t \ge t_1$, and we say the solution exists on any interval (t_1, t_2). ■

The preceding examples give rise to several important questions:

When does an initial-value problem possess a solution?
When is a solution unique?
What is the extent of the interval over which such a solution exists?
Is the solution continuously dependent on the initial condition ξ?

At the end of the next chapter we will state and prove results which give answers to these questions.

B. Initial-Value Problem: Linear Systems

In the remainder of the present section we concern ourselves exclusively with initial-value problems described by linear ordinary differential equations. Let again $T = (t_1, t_2)$ be a real t interval, let $\mathbf{x}^T = (x_1, \ldots, x_n)$ denote an n-dimensional vector, let $\mathbf{A} = [a_{ij}]$ be a constant $(n \times n)$ matrix, let $\mathbf{A}(t) = [a_{ij}(t)]$ be an $(n \times n)$ matrix with elements $a_{ij}(t)$ that are defined and continuous on the interval T, and let $\mathbf{v}(t)^T = (v_1(t), \ldots, v_n(t))$ denote an n-

vector with components $v_i(t)$ that are defined and piecewise continuous on T. In the following we consider matrices and vectors with components which may be either real- or complex-valued. In the former case the field for the $\mathbf{x}$ space is the field of real numbers, while in the latter case the field for the $\mathbf{x}$ space is the field of complex numbers. Also, let

$$D = \{(t, \mathbf{x}) \colon t \in T, \mathbf{x} \in R^n(\text{or } C^n)\}. \tag{4.11.28}$$

At first we consider systems of ordinary differential equations given by

$$\dot{\mathbf{x}} = \mathbf{A}(t)\mathbf{x} + \mathbf{v}(t), \tag{4.11.29}$$

$$\dot{\mathbf{x}} = \mathbf{A}(t)\mathbf{x}, \tag{4.11.30}$$

and

$$\dot{\mathbf{x}} = \mathbf{A}\mathbf{x}. \tag{4.11.31}$$

In the applications section of the next chapter we will show that, with the above assumptions, equations (4.11.29)–(4.11.31) possess unique solutions for every $(\tau, \xi) \in D$ which exist over the entire interval $T = (t_1, t_2)$ and which depend continuously on the initial conditions. This is an extremely important result in applications, where we usually require that $T = (-\infty, \infty)$.

4.11.32. Theorem. The set $\mathcal{S}$ of all solutions of Eq. (4.11.30) on T forms an n-dimensional vector space.

Proof. Let $\boldsymbol{\varphi}_1$ and $\boldsymbol{\varphi}_2$ be solutions of Eq. (4.11.30), let F denote the field for the $\mathbf{x}$ space, and let $\alpha_1, \alpha_2 \in F$. Since

$$\begin{aligned}\frac{d}{dt}[\alpha_1\boldsymbol{\varphi}_1(t) + \alpha_2\boldsymbol{\varphi}_2(t)] &= \alpha_1\dot{\boldsymbol{\varphi}}_1(t) + \alpha_2\dot{\boldsymbol{\varphi}}_2(t) \\ &= \alpha_1\mathbf{A}(t)\boldsymbol{\varphi}_1(t) + \alpha_2\mathbf{A}(t)\boldsymbol{\varphi}_2(t) \\ &= \mathbf{A}(t)[\alpha_1\boldsymbol{\varphi}_1(t) + \alpha_2\boldsymbol{\varphi}_2(t)],\end{aligned}$$

it follows that $\alpha_1\boldsymbol{\varphi}_1 + \alpha_2\boldsymbol{\varphi}_2 \in \mathcal{S}$ whenever $\boldsymbol{\varphi}_1, \boldsymbol{\varphi}_2 \in \mathcal{S}$ and whenever $\alpha_1, \alpha_2 \in F$. Furthermore, the **trivial solution** $\boldsymbol{\varphi} = \mathbf{0}$ defined by $\boldsymbol{\varphi}(t) = \mathbf{0}$ for all $t \in T$ is clearly in $\mathcal{S}$, and for every $\boldsymbol{\eta} \in \mathcal{S}$ there exists a $\boldsymbol{\psi} = -\boldsymbol{\eta} \in \mathcal{S}$ such that $\boldsymbol{\eta} + \boldsymbol{\psi} = \mathbf{0}$. It is now an easy matter to verify that all the axioms of a vector space are satisfied for $\mathcal{S}$ (we leave the details to the reader to verify).

Next, we must show that $\mathcal{S}$ is n-dimensional; i.e., we must find a set of solutions $\boldsymbol{\varphi}_1, \ldots, \boldsymbol{\varphi}_n$ which is linearly independent and which spans $\mathcal{S}$. Let $\boldsymbol{\xi}_1, \ldots, \boldsymbol{\xi}_n$ be a set of linearly independent vectors in the n-dimensional $\mathbf{x}$ space. By the existence results which we will prove in the next chapter (and which we will accept here on faith), if $\tau \in T$, there exist n solutions $\boldsymbol{\varphi}_1, \ldots, \boldsymbol{\varphi}_n$ of Eq. (4.11.30) such that $\boldsymbol{\varphi}_i(\tau) = \boldsymbol{\xi}_i$, $i = 1, \ldots, n$. We first show that these solutions are linearly independent. For purposes of contradiction, assume that these solutions are linearly dependent. Then there exist scalars $\alpha_1, \ldots,$

$\alpha_n \in F$, not all zero, such that

$$\sum_{i=1}^{n} \alpha_i \boldsymbol{\varphi}_i(t) = \mathbf{0}$$

for all $t \in T$. This implies that

$$\sum_{i=1}^{n} \alpha_i \boldsymbol{\varphi}_i(\tau) = \sum_{i=1}^{n} \alpha_i \boldsymbol{\xi}_i = \mathbf{0}.$$

But this last equation contradicts the assumption that the $\boldsymbol{\xi}_i$ are linearly independent. Thus, the $\boldsymbol{\varphi}_i$, $i = 1, \ldots, n$, are linearly independent. Finally, to show that these solutions span $\mathcal{S}$, let $\boldsymbol{\varphi}$ be any solution of Eq. (4.11.30) on T, such that $\boldsymbol{\varphi}(\tau) = \boldsymbol{\xi}$. Then there exist unique scalars $\alpha_1, \ldots, \alpha_n \in F$ such that

$$\boldsymbol{\xi} = \sum_{i=1}^{n} \alpha_i \boldsymbol{\xi}_i,$$

because the vectors $\boldsymbol{\xi}_i$, $i = 1, \ldots, n$, form a basis for the $\mathbf{x}$ space. It now follows that

$$\boldsymbol{\psi} = \sum_{i=1}^{n} \alpha_i \boldsymbol{\varphi}_i$$

is a solution of Eq. (4.11.30) on T such that $\boldsymbol{\psi}(\tau) = \boldsymbol{\xi}$. By the uniqueness results which we will prove in the next chapter (and which we accept here on faith),

$$\boldsymbol{\varphi} = \sum_{i=1}^{n} \alpha_i \boldsymbol{\varphi}_i.$$

Since $\boldsymbol{\varphi}$ was chosen arbitrarily, it follows that the solutions $\boldsymbol{\varphi}_i$, $i = 1, \ldots, n$, span $\mathcal{S}$. This concludes the proof. ■

The above result motivates the following two definitions.

4.11.33. Definition. A set of n linearly independent solutions of Eq. (4.11.30) on T is called a **fundamental set** of solutions of (4.11.30). An $(n \times n)$ matrix $\boldsymbol{\Psi}$ whose n columns are linearly independent solutions of Eq. (4.11.30) on T is called a **fundamental matrix**.

Thus, if $\{\boldsymbol{\psi}_1, \ldots, \boldsymbol{\psi}_n\}$ is a set of n linearly independent solutions of Eq. (4.11.30) and if $\boldsymbol{\psi}_i^T = (\psi_{1i}, \ldots, \psi_{ni})$, then

$$\boldsymbol{\Psi} = \begin{bmatrix} \psi_{11} & \psi_{12} & \cdots & \psi_{1n} \\ \psi_{21} & \psi_{22} & \cdots & \psi_{2n} \\ \cdots & \cdots & \cdots & \cdots \\ \psi_{n1} & \psi_{n2} & \cdots & \psi_{nn} \end{bmatrix} = [\boldsymbol{\psi}_1 | \boldsymbol{\psi}_2 | \cdots | \boldsymbol{\psi}_n]$$

is a fundamental matrix.

In our next definition we employ the natural basis for the $\mathbf{x}$ space, given by

$$\mathbf{u}_1 = \begin{bmatrix} 1 \\ 0 \\ \cdot \\ \cdot \\ \cdot \\ 0 \end{bmatrix}, \quad \mathbf{u}_2 = \begin{bmatrix} 0 \\ 1 \\ 0 \\ \cdot \\ \cdot \\ 0 \end{bmatrix}, \quad \ldots, \quad \mathbf{u}_n = \begin{bmatrix} 0 \\ \cdot \\ \cdot \\ \cdot \\ 0 \\ 1 \end{bmatrix}.$$

4.11.34. Definition. A fundamental matrix $\mathbf{\Phi}$ (for Eq. (4.11.30)) whose columns are determined by the linearly independent solutions $\boldsymbol{\varphi}_i$, $i = 1, \ldots, n$, with

$$\boldsymbol{\varphi}_i(\tau) = \mathbf{u}_i, \quad i = 1, \ldots, n,$$

$\tau \in T$, is called the **state transition matrix $\mathbf{\Phi}$** of Eq. (4.11.30).

Let $\mathbf{X} = [x_{ij}]$ be an $(n \times n)$ matrix and *define* differentiation of $\mathbf{X}$ with respect to $t \in T$ componentwise; i.e., $\dot{\mathbf{X}} \triangleq [\dot{x}_{ij}]$. We now have:

4.11.35. Theorem. Let $\mathbf{\Psi}$ be a fundamental matrix of Eq. (4.11.30) and let $\mathbf{X}$ denote an $(n \times n)$ matrix. Then $\mathbf{\Psi}$ satisfies the *matrix equation*

$$\dot{\mathbf{X}} = \mathbf{A}(t)\mathbf{X}, \quad t \in T. \tag{4.11.36}$$

Proof. We have

$$\begin{aligned}\dot{\mathbf{\Psi}} &= [\dot{\boldsymbol{\psi}}_1 \mid \dot{\boldsymbol{\psi}}_2 \mid \ldots \mid \dot{\boldsymbol{\psi}}_n] = [\mathbf{A}(t)\boldsymbol{\psi}_1 \mid \mathbf{A}(t)\boldsymbol{\psi}_2 \mid \ldots \mid \mathbf{A}(t)\boldsymbol{\psi}_n] \\ &= \mathbf{A}(t)[\boldsymbol{\psi}_1 \mid \boldsymbol{\psi}_2 \mid \ldots \mid \boldsymbol{\psi}_n] = \mathbf{A}(t)\mathbf{\Psi}. \quad \blacksquare\end{aligned}$$

We also have:

4.11.37. Theorem. If $\mathbf{\Psi}$ is a solution of the matrix equation (4.11.36) on T and if $t, \tau \in T$, then

$$\det \mathbf{\Psi}(t) = \det \mathbf{\Psi}(\tau) e^{\int_\tau^t \operatorname{tr} \mathbf{A}(s)\, ds}, \quad t \in T. \tag{4.11.38}$$

Proof. Recall that if $\mathbf{C} = [c_{ij}]$ is an $(n \times n)$ matrix, then $\operatorname{tr} \mathbf{C} = \sum_{i=1}^{n} c_{ii}$. Let $\mathbf{\Psi} = [\psi_{ij}]$ and $\mathbf{A}(t) = [a_{ij}(t)]$. Then $\dot{\psi}_{ij} = \sum_{k=1}^{n} a_{ik}(t)\psi_{kj}$. Now

$$\frac{d}{dt}(\det \mathbf{\Psi}) = \begin{vmatrix} \dot{\psi}_{11} & \dot{\psi}_{12} & \cdots & \dot{\psi}_{1n} \\ \psi_{21} & \psi_{22} & \cdots & \psi_{2n} \\ \cdots & \cdots & \cdots & \cdots \\ \psi_{n1} & \psi_{n2} & \cdots & \psi_{nn} \end{vmatrix} + \begin{vmatrix} \psi_{11} & \psi_{12} & \cdots & \psi_{1n} \\ \dot{\psi}_{21} & \dot{\psi}_{22} & \cdots & \dot{\psi}_{2n} \\ \cdots & \cdots & \cdots & \cdots \\ \psi_{n1} & \psi_{n2} & \cdots & \psi_{nn} \end{vmatrix} + \cdots + \begin{vmatrix} \psi_{11} & \psi_{12} & \cdots & \psi_{1n} \\ \psi_{21} & \psi_{22} & \cdots & \psi_{2n} \\ \cdots & \cdots & \cdots & \cdots \\ \dot{\psi}_{n1} & \dot{\psi}_{n2} & \cdots & \dot{\psi}_{nn} \end{vmatrix}. \tag{4.11.39}$$

Also,

$$\begin{vmatrix} \dot{\psi}_{11} & \dot{\psi}_{12} & \cdots & \dot{\psi}_{1n} \\ \psi_{21} & \psi_{22} & \cdots & \psi_{2n} \\ \cdots & \cdots & \cdots & \cdots \\ \psi_{n1} & \psi_{n2} & \cdots & \psi_{nn} \end{vmatrix} = \begin{vmatrix} \sum_{k=1}^{n} a_{1k}(t)\psi_{k1} & \sum_{k=1}^{n} a_{1k}(t)\psi_{k2} & \cdots & \sum_{k=1}^{n} a_{1k}(t)\psi_{kn} \\ \psi_{21} & \psi_{22} & \cdots & \psi_{2n} \\ \cdots & \cdots & \cdots & \cdots \\ \psi_{n1} & \psi_{n2} & \cdots & \psi_{nn} \end{vmatrix}.$$

The last determinant is unchanged if we subtract from the first row a_{12} times the second row plus a_{13} times the third row up to a_{1n} times the nth row. This yields

$$\begin{vmatrix} a_{11}(t)\psi_{11} & a_{11}(t)\psi_{12} & \cdots & a_{11}(t)\psi_{1n} \\ \psi_{21} & \psi_{22} & \cdots & \psi_{2n} \\ \cdots & \cdots & \cdots & \cdots \\ \psi_{n1} & \psi_{n2} & \cdots & \psi_{nn} \end{vmatrix} = a_{11}(t) \det \mathbf{\Psi}.$$

Repeating the above procedure for the remaining determinants we get

$$\begin{aligned} \frac{d}{dt}[\det \mathbf{\Psi}(t)] &= a_{11}(t) \det \mathbf{\Psi}(t) + a_{22}(t) \det \mathbf{\Psi}(t) + \ldots + a_{nn}(t) \det \mathbf{\Psi}(t) \\ &= [\operatorname{tr} \mathbf{A}(t)] \det \mathbf{\Psi}(t). \end{aligned}$$

This now implies

$$\det \mathbf{\Psi}(t) = \det \mathbf{\Psi}(\tau) e^{\int_\tau^t \operatorname{tr} \mathbf{A}(s)\, ds}$$

for all $t \in T$. ■

4.11.40. Exercise. Verify Eq. (4.11.39).

We now prove:

4.11.41. Theorem. A solution $\mathbf{\Psi}$ of the matrix equation (4.11.36) is a fundamental matrix for Eq. (4.11.30) if and only if $\det \mathbf{\Psi}(t) \neq 0$ for all $t \in T$.

Proof. Assume that $\mathbf{\Psi} = [\boldsymbol{\psi}_1 \mid \boldsymbol{\psi}_2 \mid \ldots \mid \boldsymbol{\psi}_n]$ is a fundamental matrix for Eq. (4.11.30), and let $\boldsymbol{\psi}$ be a nontrivial solution for (4.11.30). By Theorem 4.11.32 there exist unique scalars $\alpha_1, \ldots, \alpha_n \in F$, not all zero, such that

$$\boldsymbol{\psi} = \sum_{j=1}^{n} \alpha_j \boldsymbol{\psi}_j$$

or

$$\boldsymbol{\psi} = \mathbf{\Psi}\mathbf{a}, \tag{4.11.42}$$

where $\mathbf{a}^T = (\alpha_1, \ldots, \alpha_n)$. Equation (4.11.42) constitutes a system of n linear equations with unknowns $\alpha_1, \ldots, \alpha_n$ at any $\tau \in T$ and has a unique solution for any choice of $\boldsymbol{\psi}(\tau)$. Hence, we have $\det \mathbf{\Psi}(\tau) \neq 0$, and it now follows from Theorem 4.11.37 that $\det \mathbf{\Psi}(t) \neq 0$ for any $t \in T$.

Conversely, let $\mathbf{\Psi}$ be a solution of the matrix equation (4.11.36) and assume that det $\mathbf{\Psi}(t) \neq 0$ for all $t \in T$. Then the columns of $\mathbf{\Psi}$ are linearly independent for all $t \in T$. ■

The reader can readily prove the next result.

4.11.43. Theorem. Let $\mathbf{\Psi}$ be a fundamental matrix for Eq. (4.11.30), and let $\mathbf{C}$ be an arbitrary $(n \times n)$ non-singular constant matrix. Then $\mathbf{\Psi C}$ is also a fundamental matrix for Eq. (4.11.30). Moreover, if $\mathbf{\Upsilon}$ is any other fundamental matrix for Eq. (4.11.30) then there exists a constant $(n \times n)$ non-singular matrix $\mathbf{P}$ such that $\mathbf{\Upsilon} = \mathbf{\Psi P}$.

4.11.44. Exercise. Prove Theorem 4.11.43.

Now let $\mathbf{R}(t) = [r_{ij}(t)]$ be an arbitrary matrix such that the scalar valued functions $r_{ij}(t)$ are Riemann integrable on T. We *define* integration of $\mathbf{R}(t)$ componentwise, i.e.,

$$\int R(t)dt = \int [r_{ij}(t)]dt = \left[\int r_{ij}(t)dt\right].$$

Integration of vectors is defined similarly.

In the next result we establish some of the properties of the state transition matrix, $\mathbf{\Phi}$. Hereafter, in order to indicate the dependence of $\mathbf{\Phi}$ on τ as well as t, we will write $\mathbf{\Phi}(t, \tau)$. By $\dot{\mathbf{\Phi}}(t, \tau)$, we mean $\partial\mathbf{\Phi}(t, \tau)/\partial t$.

4.11.45. Theorem. Let D be defined by Eq. (4.11.28), let $\tau \in T$, let $\boldsymbol{\varphi}(\tau) = \boldsymbol{\xi}$, let $(\tau, \boldsymbol{\xi}) \in D$, and let $\mathbf{\Phi}(t, \tau)$ denote the state transition matrix for Eq. (4.11.30) for all $t \in T$. Then

(i) $\dot{\mathbf{\Phi}}(t, \tau) = \mathbf{A}(t)\mathbf{\Phi}(t, \tau)$ with $\mathbf{\Phi}(\tau, \tau) = \mathbf{I}$, where $\mathbf{I}$ denotes the $(n \times n)$ identity matrix;

(ii) the unique solution of Eq. (4.11.30) is given by

$$\boldsymbol{\varphi}(t) = \mathbf{\Phi}(t, \tau)\boldsymbol{\xi} \tag{4.11.46}$$

for all $t \in T$;

(iii) $\mathbf{\Phi}(t, \tau)$ is non-singular for all $t \in T$;

(iv) for any $t, \sigma \in T$ we have

$$\mathbf{\Phi}(t, \tau) = \mathbf{\Phi}(t, \sigma)\mathbf{\Phi}(\sigma, \tau);$$

(v) $[\mathbf{\Phi}(t, \tau)]^{-1} \triangleq \mathbf{\Phi}^{-1}(t, \tau) = \mathbf{\Phi}(\tau, t)$ for all $t \in T$; and

(vi) the unique solution of Eq. (4.11.29) is given by

$$\boldsymbol{\varphi}(t) = \mathbf{\Phi}(t, \tau)\boldsymbol{\xi} + \int_\tau^t \mathbf{\Phi}(t, \eta)\mathbf{v}(\eta)d\eta. \tag{4.11.47}$$

Proof. The first part of the theorem follows from the definition of the state transition matrix.

To prove the second part, assume that $\boldsymbol{\varphi}(t) = \boldsymbol{\Phi}(t, \tau)\boldsymbol{\xi}$. Differentiating with respect to t we have

$$\dot{\boldsymbol{\varphi}}(t) = \dot{\boldsymbol{\Phi}}(t, \tau)\boldsymbol{\xi} = \mathbf{A}(t)\boldsymbol{\Phi}(t, \tau)\boldsymbol{\xi} = \mathbf{A}(t)\boldsymbol{\varphi}(t).$$

Furthermore, $\boldsymbol{\varphi}(\tau) = \boldsymbol{\Phi}(\tau, \tau)\boldsymbol{\xi} = \boldsymbol{\xi}$. From the uniqueness results (to be presented in the next chapter) it follows that the specified $\boldsymbol{\varphi}$ is indeed the solution of Eq. (4.11.30).

The third part of the theorem is a consequence of Theorem 4.11.41.

To prove the fourth part of the theorem we note that $\boldsymbol{\varphi}(t) = \boldsymbol{\Phi}(t, \tau)\boldsymbol{\xi}$ is the unique solution of Eq. (4.11.30) satisfying $\boldsymbol{\varphi}(\tau) = \boldsymbol{\xi}$, and also that $\boldsymbol{\varphi}(\sigma) = \boldsymbol{\Phi}(\sigma, \tau)\boldsymbol{\xi}$, $\sigma \in T$. Now consider the solution of Eq. (4.11.30) with initial condition given at σ in place of τ; i.e., $\boldsymbol{\varphi}(t) = \boldsymbol{\Phi}(t, \sigma)\boldsymbol{\varphi}(\sigma)$. Then

$$\boldsymbol{\varphi}(t) = \boldsymbol{\Phi}(t, \tau)\boldsymbol{\xi} = \boldsymbol{\Phi}(t, \sigma)\boldsymbol{\Phi}(\sigma, \tau)\boldsymbol{\xi}.$$

Since this equation holds for arbitrary $\boldsymbol{\xi}$ in the $\mathbf{x}$ space, we have

$$\boldsymbol{\Phi}(t, \tau) = \boldsymbol{\Phi}(t, \sigma)\boldsymbol{\Phi}(\sigma, \tau).$$

To prove the fifth part of the theorem we note that $\boldsymbol{\Phi}^{-1}(t, \tau)$ exists by part (iii). From part (iv) it now follows that

$$\mathbf{I} = \boldsymbol{\Phi}(t, \tau)\boldsymbol{\Phi}(\tau, t),$$

where $\mathbf{I}$ denotes the $(n \times n)$ identity matrix. Thus,

$$\boldsymbol{\Phi}^{-1}(t, \tau) = \boldsymbol{\Phi}(\tau, t)$$

for all $t \in T$.

In the next chapter we will show that under the present assumptions, Eq. (4.11.29) possesses a unique solution for every $(\tau, \boldsymbol{\xi}) \in D$, where $\boldsymbol{\varphi}(\tau) = \boldsymbol{\xi}$. Thus, to prove the last part of the theorem, we must show that the function (4.11.47) is this solution. Differentiating with respect to t we have

$$\begin{aligned}\dot{\boldsymbol{\varphi}}(t) &= \dot{\boldsymbol{\Phi}}(t, \tau)\boldsymbol{\xi} + \boldsymbol{\Phi}(t, t)\mathbf{v}(t) + \int_\tau^t \dot{\boldsymbol{\Phi}}(t, \eta)\mathbf{v}(\eta)\, d\eta \\ &= \mathbf{A}(t)\boldsymbol{\Phi}(t, \tau)\boldsymbol{\xi} + \mathbf{v}(t) + \int_\tau^t \mathbf{A}(t)\boldsymbol{\Phi}(t, \eta)\mathbf{v}(\eta)\, d\eta \\ &= \mathbf{A}(t)[\boldsymbol{\Phi}(t, \tau)\boldsymbol{\xi} + \int_\tau^t \boldsymbol{\Phi}(t, \eta)\mathbf{v}(\eta)\, d\eta] + \mathbf{v}(t) \\ &= \mathbf{A}(t)\boldsymbol{\varphi}(t) + \mathbf{v}(t).\end{aligned}$$

Also, $\boldsymbol{\varphi}(\tau) = \boldsymbol{\xi}$. Therefore, $\boldsymbol{\varphi}$ is the unique solution of Eq. (4.11.29). ■

In engineering and physics, $\boldsymbol{\varphi}$ is interpreted as representing the "state" of a physical system described by appropriate ordinary differential equations. In Eq. (4.11.46), the matrix $\boldsymbol{\Phi}(t, \tau)$ relates the "states" of the system at the points $t \in T$ and $\tau \in T$. Hence, the name "state transition matrix."

Next, we wish to examine the properties of linear ordinary differential equations with constant coefficients given by Eq. (4.11.31). We require the following preliminary result.

4.11.48. Theorem. Let $\mathbf{A}$ be a constant $(n \times n)$ matrix ($\mathbf{A}$ may be real or complex). Let $\mathbf{S}_N(t)$ denote the matrix

$$\mathbf{S}_N(t) = \mathbf{I} + \sum_{k=1}^{N} \frac{t^k}{k!}\mathbf{A}^k.$$

Then each element of matrix $\mathbf{S}_N(t)$ converges absolutely and uniformly on any finite interval $(-t_1, t_1)$, $t_1 > 0$, as $N \longrightarrow \infty$.

Proof. Let $a_{ij}^{(k)}$ denote the (i, j)th element of matrix $\mathbf{A}^k$, where $i, j = 1, \ldots, n$, and $k = 0, 1, 2, \ldots$. Then the (i, j)th element of $\mathbf{S}_N(t)$ is equal to

$$\delta_{ij} + \sum_{k=1}^{N} a_{ij}^{(k)} \frac{t^k}{k!},$$

where δ_{ij} is the Kronecker delta. We now show that

$$\delta_{ij} + \sum_{k=1}^{\infty} \left| a_{ij}^{(k)} \frac{t^k}{k!} \right| < \infty \text{ for all } i, j.$$

Let $m = \max\limits_{1 \le i \le n} \left(\sum_{j=1}^{n} |a_{ij}| \right)$. Then m is a constant which depends on the elements of the matrix $\mathbf{A}$. Since $\mathbf{A}^{k+1} = \mathbf{A}\mathbf{A}^k$, we have $\max\limits_{i,j} |a_{ij}^{(k+1)}| = \max\limits_{i,j} \left| \sum_{p=1}^{n} a_{ip} a_{pj}^{(k)} \right|$ $\le \max\limits_{i,j} \left(\sum_{p=1}^{n} |a_{ip}| \cdot |a_{pj}^{(k)}| \right) \le \left(\max\limits_{i} \sum_{p=1}^{n} |a_{ip}| \right) \left(\max\limits_{p,j} |a_{pj}^{(k)}| \right)$. Therefore, $\max\limits_{i,j} |a_{ij}^{(k+1)}| \le m \cdot \max\limits_{i,j} |a_{ij}^{(k)}|$. When $k = 1$, we have $\max\limits_{i,j} |a_{ij}| \le m$, and by induction it follows that $\max\limits_{i,j} |a_{ij}^{(k)}| \le m^k$. Now let $M_k = (mt_1)^k / k!$. Then we have for any $t \in (-t_1, t_1)$, $t_1 > 0$, and for any i, j,

$$\left| a_{ij}^{(k)} \frac{t^k}{k!} \right| \le M_k.$$

Since $1 + \sum_{k=1}^{\infty} M_k = e^{mt_1}$, we now have

$$\delta_{ij} + \sum_{k=1}^{\infty} a_{ij}^{(k)} \frac{t^k}{k!}$$

is an absolutely and uniformly convergent series for each i, j over the interval $(-t_1, t_1)$ by the Weierstrass M-test. ■

We are now in a position to consider the following:

4.11.49. Definition. Let $\mathbf{A}$ be a constant $(n \times n)$ matrix. We define $e^{\mathbf{A}t}$ to be the matrix

$$e^{\mathbf{A}t} = \mathbf{I} + \sum_{k=1}^{\infty} \frac{t^k}{k!}\mathbf{A}^k$$

for any $-\infty < t < \infty$.

We note immediately that $e^{\mathbf{A}t}|_{t=0} = \mathbf{I}$.

We now prove:

4.11.50. Theorem. Let $T = (-\infty, \infty)$, let $\tau \in T$, and let $\mathbf{A}$ be a constant $(n \times n)$ matrix. Then

(i) the state transition matrix for Eq. (4.11.31) is given by

$$\mathbf{\Phi}(t, \tau) = e^{\mathbf{A}(t-\tau)}$$

for all $t \in T$;

(ii) the matrix $e^{\mathbf{A}t}$ is non-singular for all $t \in T$;

(iii) $e^{\mathbf{A}t_1}e^{\mathbf{A}t_2} = e^{\mathbf{A}(t_1+t_2)}$ for all $t_1, t_2 \in T$;

(iv) $\mathbf{A}e^{\mathbf{A}t} = e^{\mathbf{A}t}\mathbf{A}$ for all $t \in T$; and

(v) $(e^{\mathbf{A}t})^{-1} = e^{-\mathbf{A}t}$ for all $t \in T$.

Proof. To prove the first part we must show that $\mathbf{\Phi}(t, \tau)$ satisfies the matrix equation

$$\dot{\mathbf{\Phi}}(t, \tau) = \mathbf{A}\mathbf{\Phi}(t, \tau)$$

for all $t \in T$, with $\mathbf{\Phi}(\tau, \tau) = \mathbf{I}$. Now, by definition,

$$\mathbf{\Phi}(t, \tau) = e^{\mathbf{A}(t-\tau)} = \mathbf{I} + \sum_{k=1}^{\infty} \frac{(t-\tau)^k}{k!}\mathbf{A}^k.$$

In view of Theorem 4.11.48 we may differentiate the above series term by term. In doing so we obtain

$$\frac{d}{dt}[e^{\mathbf{A}(t-\tau)}] = \mathbf{A} + \sum_{k=1}^{\infty} \frac{(t-\tau)^k}{k!}\mathbf{A}^{k+1} = \mathbf{A}\left[\mathbf{I} + \sum_{k=1}^{\infty} \frac{(t-\tau)^k}{k!}\mathbf{A}^k\right]$$
$$= \mathbf{A}e^{\mathbf{A}(t-\tau)},$$

and thus we have

$$\dot{\mathbf{\Phi}}(t, \tau) = \mathbf{A}\mathbf{\Phi}(t, \tau)$$

for all $t \in T$, with $\mathbf{\Phi}(\tau, \tau) = e^{\mathbf{A}(\tau-\tau)} = \mathbf{I}$. Therefore, $e^{\mathbf{A}(t-\tau)}$ is the state transition matrix for Eq. (4.11.31).

The second part of the theorem is obvious.

To prove the third part of the theorem, we note that for any $t_1, t_2 \in T$, we have

$$\mathbf{\Phi}(t_1, -t_2) = \mathbf{\Phi}(t_1, 0)\mathbf{\Phi}(0, -t_2).$$

Now $\mathbf{\Phi}(t_1, -t_2) = e^{\mathbf{A}(t_1+t_2)}$, $\mathbf{\Phi}(t_1, 0) = e^{\mathbf{A}t_1}$, and $\mathbf{\Phi}(0, -t_2) = e^{\mathbf{A}t_2}$, which yields the desired result.

To prove the fourth part of the theorem we note that for all $t \in T$,

$$\mathbf{A}\left(\mathbf{I} + \sum_{k=1}^{\infty} \frac{t^k}{k!}\mathbf{A}^k\right) = \mathbf{A} + \sum_{k=1}^{\infty} \frac{t^k}{k!}\mathbf{A}^{k+1} = \left(\mathbf{I} + \sum_{k=1}^{\infty} \frac{t^k}{k!}\mathbf{A}^k\right)\mathbf{A}.$$

Finally, to prove the last part of the theorem, note that for all $t \in T$,

$$e^{\mathbf{A}t} \cdot e^{\mathbf{A}(-t)} = e^{\mathbf{A}(t-t)} = \mathbf{I}.$$

Therefore, $(e^{\mathbf{A}t})^{-1} = e^{-\mathbf{A}t}$. ■

The following natural question arises: can we find an expression similar to $e^{\mathbf{A}t}$ for the case when $\mathbf{A} = \mathbf{A}(t)$, $t \in T$. The answer is, in general, no. However, there is a special case when such a generalization is valid.

4.11.51. Theorem. If for Eq. (4.11.30) $\mathbf{A}(t_1)\mathbf{A}(t_2) = \mathbf{A}(t_2)\mathbf{A}(t_1)$ for all $t_1, t_2 \in T$, then the state transition matrix $\mathbf{\Phi}(t, \tau)$ is given by

$$\mathbf{\Phi}(t, \tau) = e^{\int_\tau^t \mathbf{A}(\eta)\, d\eta} = e^{\mathbf{B}(t,\tau)} = \mathbf{I} + \sum_{k=1}^{\infty} \frac{1}{k!}\mathbf{B}^k(t, \tau),$$

where $\mathbf{B}(t, \tau) \triangleq \int_\tau^t \mathbf{A}(\eta)\, d\eta$.

4.11.52. Exercise. Prove Theorem 4.11.51.

We note that a sufficient condition for $\mathbf{A}(t_1)$ to commute with $\mathbf{A}(t_2)$ for all $t_1, t_2 \in T$ is that $\mathbf{A}(t)$ be a diagonal matrix.

4.11.53. Exercise. Find the state transition matrix for $\dot{\mathbf{x}} = \mathbf{A}(t)\mathbf{x}$, where

$$\mathbf{A}(t) = \begin{bmatrix} 1 & 0 \\ t & 1 \end{bmatrix}.$$

The reader will find it instructive to verify the following additional results.

4.11.54. Exercise. Let $\mathbf{\Lambda}$ denote the $(n \times n)$ diagonal matrix

$$\mathbf{\Lambda} = \begin{bmatrix} \lambda_1 & & & \mathbf{0} \\ & \cdot & & \\ & & \cdot & \\ & & & \cdot \\ \mathbf{0} & & & \lambda_n \end{bmatrix}.$$

Show that

$$e^{\mathbf{\Lambda}t} = \begin{bmatrix} e^{\lambda_1 t} & & & \mathbf{0} \\ & \cdot & & \\ & & \cdot & \\ & & & \cdot \\ \mathbf{0} & & & e^{\lambda_n t} \end{bmatrix}$$

for all $t \in T = (-\infty, \infty)$.

4.11.55. Exercise. Let $t \in T = (-\infty, \infty)$, let $\tau \in T$, and let $\boldsymbol{\xi} \in R^n$ (or C^n). Let $\mathbf{A}$ be the $(n \times n)$ matrix for Eq. (4.11.31), and let $\boldsymbol{\varphi}$ denote the unique solution of Eq. (4.11.31) with $\boldsymbol{\varphi}(\tau) = \boldsymbol{\xi}$. Let $\mathbf{P}$ be a similarity transformation for $\mathbf{A}$, and let $\mathbf{B} = \mathbf{P}^{-1}\mathbf{A}\mathbf{P}$.

(a) Show that $e^{\mathbf{A}t} = \mathbf{P}e^{\mathbf{B}t}\mathbf{P}^{-1}$ for all $t \in T$.

(b) Show that the unique solution of Eq. (4.11.31) is given by

$$\boldsymbol{\varphi} = \mathbf{P}\boldsymbol{\psi},$$

where $\boldsymbol{\psi}$ is the unique solution of the initial-value problem

$$\dot{\mathbf{y}} = \mathbf{B}\mathbf{y}$$

with

$$\boldsymbol{\psi}(\tau) = \mathbf{P}^{-1}\boldsymbol{\varphi}(\tau) = \mathbf{P}^{-1}\boldsymbol{\xi}.$$

4.11.56. Exercise. Let D be defined by (4.11.28). In Eq. (4.11.29), let $\mathbf{A}(t) = \mathbf{A}$ for all $t \in T$; i.e.,

$$\dot{\mathbf{x}} = \mathbf{A}\mathbf{x} + \mathbf{v}(t). \tag{4.11.57}$$

Let $\tau \in T$, and let $\boldsymbol{\varphi}$ denote the unique solution of Eq. (4.11.57) with $\boldsymbol{\varphi}(\tau) = \boldsymbol{\xi}$. Let $\mathbf{P}$ be a similarity transformation for $\mathbf{A}$, and let $\mathbf{B} = \mathbf{P}^{-1}\mathbf{A}\mathbf{P}$. Show that the unique solution of Eq. (4.11.57) is given by

$$\boldsymbol{\varphi} = \mathbf{P}\boldsymbol{\psi},$$

where $\boldsymbol{\psi}$ is the unique solution of the initial-value problem

$$\dot{\mathbf{y}} = \mathbf{B}\mathbf{y} + \mathbf{P}^{-1}\mathbf{v}(t)$$

with

$$\boldsymbol{\psi}(\tau) = \mathbf{P}^{-1}\boldsymbol{\varphi}(\tau) = \mathbf{P}^{-1}\boldsymbol{\xi},$$

$(\tau, \boldsymbol{\psi}(\tau)) \in D$, $t \in T$.

4.11.58. Exercise. Let $\mathbf{J}$ denote the Jordan canonical form of the $(n \times n)$ matrix $\mathbf{A}$ of Eq. (4.11.31), and let $\mathbf{M}$ denote the non-singular $(n \times n)$ matrix which transforms $\mathbf{A}$ into $\mathbf{J}$; i.e., $\mathbf{J} = \mathbf{M}^{-1}\mathbf{A}\mathbf{M}$. Then

$$\mathbf{J} = \begin{bmatrix} \mathbf{J}_0 & & & & & \mathbf{0} \\ & \mathbf{J}_1 & & & & \\ & & \cdot & & & \\ & & & \cdot & & \\ & & & & \cdot & \\ \mathbf{0} & & & & & \mathbf{J}_p \end{bmatrix},$$

where

$$\mathbf{J}_0 = \begin{bmatrix} \lambda_1 & & & & & 0 \\ & \lambda_2 & & & & \\ & & \cdot & & & \\ & & & \cdot & & \\ & & & & \cdot & \\ 0 & & & & & \lambda_k \end{bmatrix},$$

where

$$\mathbf{J}_m = \begin{bmatrix} \lambda_{k+m} & 1 & 0 & \cdots & 0 & 0 \\ 0 & \lambda_{k+m} & 1 & \cdots & 0 & 0 \\ \cdot & & \cdot & \cdot & & \\ \cdot & & & \cdot & \cdot & \\ \cdot & & & & \cdot & \cdot \\ \cdot & & & & \cdot & 1 \\ 0 & 0 & \cdots\cdots & & \lambda_{k+m} & 1 \\ 0 & 0 & \cdots\cdots & & 0 & \lambda_{k+m} \end{bmatrix},$$

$m = 1, \ldots, p$, and where $\lambda_1, \ldots, \lambda_k, \lambda_{k+1}, \ldots, \lambda_{k+p}$ denote the (not necessarily distinct) eigenvalues of $\mathbf{A}$. Show that

$$e^{\mathbf{J}t} = \begin{bmatrix} e^{\mathbf{J}_0 t} & & & & \mathbf{0} \\ & e^{\mathbf{J}_1 t} & & & \\ & & \cdot & & \\ & & & \cdot & \\ & & & & \cdot \\ \mathbf{0} & & & & e^{\mathbf{J}_p t} \end{bmatrix},$$

where

$$e^{\mathbf{J}_0 t} = \begin{bmatrix} e^{\lambda_1 t} & & & \mathbf{0} \\ & \cdot & & \\ & & \cdot & \\ & & & \cdot \\ \mathbf{0} & & & e^{\lambda_k t} \end{bmatrix}$$

and

$$e^{\mathbf{J}_m t} = e^{t\lambda_{k+m}} \begin{bmatrix} 1 & t & \dfrac{t^2}{2!} & \cdots & \dfrac{t^{\nu_m - 1}}{(\nu_m - 1)!} \\ 0 & 1 & t & \cdots & \dfrac{t^{\nu_m - 2}}{(\nu_m - 2)!} \\ \cdots & \cdots & \cdots & \cdots & \cdots \\ 0 & 0 & 0 & \cdots & 1 \end{bmatrix},$$

where $\mathbf{J}_m$ is a $\nu_m \times \nu_m$ matrix and $k + \nu_1 + \ldots + \nu_p = n$.

Next, we consider initial-value problems characterized by linear nth-order ordinary differential equations given by

$$a_n(t)x^{(n)} + a_{n-1}(t)x^{(n-1)} + \ldots + a_1(t)x^{(1)} + a_0(t)x = v(t), \quad (4.11.59)$$

$$a_n(t)x^{(n)} + a_{n-1}(t)x^{(n-1)} + \ldots + a_1(t)x^{(1)} + a_0(t)x = 0, \quad (4.11.60)$$

and

$$a_n x^{(n)} + a_{n-1}x^{(n-1)} + \ldots + a_1 x^{(1)} + a_0 x = 0. \quad (4.11.61)$$

In Eqs. (4.11.59) and (4.11.60), $v(t)$ and $a_i(t)$, $i = 0, \ldots, n$, are functions which are defined and continuous on a real t interval T, and in Eq. (4.11.61), the a_i, $i = 0, \ldots, n$, are constant coefficients. We assume that $a_n \neq 0$, that $a_n(t) \neq 0$ for any $t \in T$, and that $v(t)$ is not identically zero. Furthermore, the coefficients a_i, $a_i(t)$, $i = 0, \ldots, n$, may be either real or complex.

In accordance with Eq. (4.11.23), we can reduce the study of Eq. (4.11.60) to the study of the system of n first-order ordinary differential equations

$$\dot{\mathbf{x}} = \mathbf{A}(t)\mathbf{x}, \tag{4.11.62}$$

where

$$\mathbf{A}(t) = \begin{bmatrix} 0 & 1 & 0 & \cdots & 0 \\ 0 & 0 & 1 & \cdots & 0 \\ \cdots & \cdots & \cdots & \cdots & \cdots \\ 0 & 0 & 0 & \cdots & 1 \\ \dfrac{-a_0(t)}{a_n(t)} & \dfrac{-a_1(t)}{a_n(t)} & \dfrac{-a_2(t)}{a_n(t)} & \cdots & \dfrac{-a_{n-1}(t)}{a_n(t)} \end{bmatrix}. \tag{4.11.63}$$

In this case the matrix $\mathbf{A}(t)$ is said to be in **companion form**. Since $\mathbf{A}(t)$ is continuous on T, there exists for all $t \in T$ a unique solution $\boldsymbol{\mu}$ to the initial-value problem

$$\left.\begin{aligned} \dot{\mathbf{x}} &= \mathbf{A}(t)\mathbf{x} \\ \mathbf{x}(\tau) &= \boldsymbol{\xi} = (\xi_1, \ldots, \xi_n)^T \end{aligned}\right\}, \tag{4.11.64}$$

where $\tau \in T$ and $\boldsymbol{\xi} \in R^n$ (or C^n) (this will be proved in the next chapter). Moreover, the first component of $\boldsymbol{\mu}$, μ_1, is the solution of Eq. (4.11.60) satisfying

$$\mu_1(\tau) = \mu(\tau) = \xi_1, \quad \mu^{(1)}(\tau) = \xi_2, \quad \ldots, \quad \mu^{(n-1)}(\tau) = \xi_n.$$

Now let $\psi_1, \ldots, \psi_n$ be solutions of Eq. (4.11.60). Then we can readily verify that the matrix

$$\mathbf{\Psi} = \begin{bmatrix} \psi_1 & \psi_2 & \cdots & \psi_n \\ \psi_1^{(1)} & \psi_2^{(1)} & \cdots & \psi_n^{(1)} \\ \cdots & \cdots & \cdots & \cdots \\ \psi_1^{(n-1)} & \psi_2^{(n-1)} & \cdots & \psi_n^{(n-1)} \end{bmatrix} \tag{4.11.65}$$

is a solution of the matrix equation

$$\dot{\mathbf{\Psi}} = \mathbf{A}(t)\mathbf{\Psi}, \tag{4.11.66}$$

where $\mathbf{A}(t)$ is defined by Eq. (4.11.63). We call the determinant of $\mathbf{\Psi}$ the **Wronskian** of Eq. (4.11.60) with respect to the solutions $\psi_1, \ldots, \psi_n$, and we denote it by

$$\det \mathbf{\Psi} = W(\psi_1, \ldots, \psi_n). \tag{4.11.67}$$

Note that for a fixed set of solutions $\psi_1, \ldots, \psi_n$ (and considering τ fixed), the Wronskian is a function of t. To indicate this, we write $W(\psi_1, \ldots, \psi_n)(t)$.

In view of Theorem 4.11.37 we have for all $t \in T$,

$$W(\psi_1, \ldots, \psi_n)(t) = \det \boldsymbol{\Psi}(t) = \det \boldsymbol{\Psi}(\tau)e^{\int_\tau^t \operatorname{tr} \mathbf{A}(s)\, ds}$$
$$= W(\psi_1, \ldots, \psi_n)(\tau)e^{\int_\tau^t -[a_{n-1}(\eta)/a_n(\eta)]\, d\eta}. \quad (4.11.68)$$

4.11.69. Example. Consider the second-order ordinary differential equation

$$t^2 x^{(2)} + t x^{(1)} - x = 0, \quad 0 < t < \infty. \quad (4.11.70)$$

The functions $\psi_1(t) = t$ and $\psi_2(t) = 1/t$ are clearly solutions of Eq. (4.11.70). Consider now the matrix

$$\boldsymbol{\Psi}(t) = \begin{bmatrix} t & \dfrac{1}{t} \\ 1 & -\dfrac{1}{t^2} \end{bmatrix} = \begin{bmatrix} \psi_1 & \psi_2 \\ \psi_1^{(1)} & \psi_2^{(1)} \end{bmatrix}.$$

Then

$$W(\psi_1, \psi_2)(t) = \det \boldsymbol{\Psi}(t) = -\frac{2}{t}, \quad t > 0.$$

Using the notation of Eq. (4.11.63), we have in the present case $a_1(t)/a_2(t) = 1/t$. From Eq. (4.11.68) we have, for any $\tau > 0$,

$$W(\psi_1, \psi_2)(t) = \det \boldsymbol{\Psi}(t) = W(\psi_1, \psi_2)(\tau)e^{\int_\tau^t -[a_1(\eta)/a_2(\eta)]\, d\eta}$$
$$= -\frac{2}{\tau}e^{\ln(\tau/t)} = -\frac{2}{t}, \quad t > 0,$$

which checks. ■

The reader will have no difficulty in proving the following:

4.11.71. Theorem. A set of n solutions of Eq. (4.11.60), $\psi_1, \ldots, \psi_n$, is linearly independent on a t interval T if and only if $W(\psi_1, \ldots, \psi_n)(t) \neq 0$ for all $t \in T$. Moreover, every solution of Eq. (4.11.60) is a linear combination of any set of n linearly independent solutions.

4.11.72. Exercise. Prove Theorem 4.11.71.

We call a set of n solutions of Eq. (4.11.60), $\psi_1, \ldots, \psi_n$, which is linearly independent on T a **fundamental set** for Eq. (4.11.60).

Let us next turn our attention to the non-homogeneous linear nth-order ordinary differential equation (4.11.59). Without loss of generality, let us assume that $a_n(t) = 1$ for all $t \in T$; i.e., let us consider

$$x^{(n)} + a_{n-1}(t)x^{(n-1)} + \ldots + a_1(t)x^{(1)} + a_0(t)x = v(t). \quad (4.11.73)$$

The study of this equation reduces to the study of the system of n first-order ordinary differential equations

$$\dot{\mathbf{x}} = \mathbf{A}(t)\mathbf{x} + \mathbf{b}(t), \tag{4.11.74}$$

where

$$\mathbf{A}(t) = \begin{bmatrix} 0 & 1 & 0 & \cdots & 0 \\ 0 & 0 & 1 & \cdots & 0 \\ \cdots & \cdots & \cdots & \cdots & \cdots \\ 0 & 0 & 0 & \cdots & 1 \\ -a_0(t) & -a_1(t) & -a_2(t) & \cdots & -a_{n-1}(t) \end{bmatrix}, \quad \mathbf{b}(t) = \begin{bmatrix} 0 \\ \cdot \\ \cdot \\ \cdot \\ 0 \\ v(t) \end{bmatrix}. \tag{4.11.75}$$

In the next chapter we will show that for all $t \in T$ there exists a unique solution $\boldsymbol{\zeta}$ to the initial-value problem

$$\left.\begin{aligned} \dot{\mathbf{x}} &= \mathbf{A}(t)\mathbf{x} + \mathbf{b}(t) \\ \mathbf{x}(\tau) &= \boldsymbol{\xi} = (\xi_1, \ldots, \xi_n)^T \end{aligned}\right\}, \tag{4.11.76}$$

where $\tau \in T$ and $\boldsymbol{\xi} \in R^n$ (or C^n). The first component of $\boldsymbol{\zeta}$, ζ_1, is the solution of Eq. (4.11.59), with $a_n(t) = 1$ for all $t \in T$, satisfying

$$\zeta_1(\tau) = \xi_1, \quad \zeta^{(1)}(\tau) = \xi_2, \quad \ldots, \quad \zeta^{(n-1)}(\tau) = \xi_n.$$

We now have:

4.11.77. Theorem. Let $\{\psi_1, \ldots, \psi_n\}$ be a fundamental set for the equation

$$x^{(n)} + a_{n-1}(t)x^{(n-1)} + \ldots + a_1(t)x^{(1)} + a_0(t)x = 0. \tag{4.11.78}$$

Then the solution ζ of the equation

$$x^{(n)} + a_{n-1}(t)x^{(n-1)} + \ldots + a_1(t)x^{(1)} + a_0(t)x = v(t), \tag{4.11.79}$$

satisfying $\boldsymbol{\zeta}(\tau) = \boldsymbol{\xi} = (\zeta(\tau), \zeta^{(1)}(\tau), \ldots, \zeta^{(n-1)}(\tau))^T = (\xi_1, \ldots, \xi_n)^T$, $\tau \in T$, $\boldsymbol{\xi} \in R^n$ (or C^n) is given by the expression

$$\zeta(t) = \zeta_h(t) + \sum_{i=1}^{n} \psi_i(t) \int_\tau^t \left\{ \frac{W_i(\psi_1, \ldots, \psi_n)(s)}{W(\psi_1, \ldots, \psi_n)(s)} \right\} v(s)\, ds, \tag{4.11.80}$$

where ζ_h is the solution of Eq. (4.11.78) with $\zeta_h(\tau) = \xi_1$, and where $W_i(\psi_1, \ldots, \psi_n)(t)$ is obtained from $W(\psi_1, \ldots, \psi_n)(t)$ by replacing the ith column of $W(\psi_1, \ldots, \psi_n)(t)$ by $(0, 0, \ldots, 1)^T$.

4.11.81. Exercise. Prove Theorem 4.11.77.

Let us consider a specific case.

4.11.82. Example. Consider the second-order ordinary differential equation

$$t^2 x^{(2)} + t x^{(1)} - x = b(t), \quad t > 0, \tag{4.11.83}$$

where $b(t)$ is a real continuous function for all $t > 0$. This equation is equivalent to

$$x^{(2)} + \frac{1}{t}x^{(1)} - \frac{1}{t^2}x = v(t), \tag{4.11.84}$$

where $v(t) = b(t)/t^2$. From Example 4.11.69 we have $\psi_1(t) = t$, $\psi_2(t) = 1/t$, and $W(\psi_1, \psi_2)(t) = -2/t$. Also,

$$W_1(\psi_1, \psi_2)(t) = \begin{vmatrix} 0 & \frac{1}{t} \\ 1 & \frac{-1}{t^2} \end{vmatrix} = -\frac{1}{t}, \qquad W_2(\psi_1, \psi_2)(t) = \begin{vmatrix} t & 0 \\ 1 & 1 \end{vmatrix} = t.$$

Finally, in view of Theorem 4.11.77 we have

$$\zeta(t) = \zeta_h(t) + \psi_1(t)\int_\tau^t \frac{W_1(\psi_1, \psi_2)(s)}{W(\psi_1, \psi_2)(s)}v(s)\,ds + \psi_2(t)\int_\tau^t \frac{W_2(\psi_1, \psi_2)(s)}{W(\psi_1, \psi_2)(s)}v(s)\,ds$$

$$= \zeta_h(t) + \frac{t}{2}\int_\tau^t v(s)\,ds - \frac{1}{2t}\int_\tau^t s^2 v(s)\,ds. \quad \blacksquare$$

Let us next focus our attention on linear nth-order ordinary differential equations with constant coefficients. Without loss of generality, let us assume that, in Eq. (4.11.61), $a_n = 1$. We have

$$x^{(n)} + a_{n-1}x^{(n-1)} + \ldots + a_1 x^{(1)} + a_0 x = 0. \tag{4.11.85}$$

We call the algebraic equation

$$p(\lambda) = \lambda^n + a_{n-1}\lambda^{n-1} + \ldots + a_1\lambda + a_0 = 0 \tag{4.11.86}$$

the **characteristic equation** of the differential equation (4.11.85).

As was done before, we see that the study of Eq. (4.11.85) reduces to the study of the system of first-order ordinary differential equations given by

$$\dot{\mathbf{x}} = \mathbf{A}\mathbf{x}, \tag{4.11.87}$$

where

$$\mathbf{A} = \begin{bmatrix} 0 & 1 & 0 & 0 & \cdots & 0 \\ 0 & 0 & 1 & 0 & \cdots & 0 \\ \cdots & \cdots & \cdots & \cdots & \cdots & \cdots \\ -a_0 & -a_1 & -a_2 & -a_3 & \cdots & -a_{n-1} \end{bmatrix}. \tag{4.11.88}$$

We now show that the eigenvalues of matrix $\mathbf{A}$ of Eq. (4.11.88) are precisely the roots of the characteristic equation (4.11.86). First we consider

$$\det(\mathbf{A} - \lambda\mathbf{I}) = \begin{vmatrix} -\lambda & 1 & 0 & \cdots & 0 & 0 \\ 0 & -\lambda & 0 & \cdots & 0 & 0 \\ \cdots & \cdots & \cdots & \cdots & \cdots & \cdots \\ 0 & 0 & 0 & \cdots & -\lambda & 1 \\ -a_0 & -a_1 & -a_2 & \cdots & -a_{n-2} & -(\lambda + a_{n-1}) \end{vmatrix}$$

$$= -\lambda \begin{vmatrix} -\lambda & 1 & 0 & \cdots & 0 & 0 \\ 0 & -\lambda & 1 & \cdots & 0 & 0 \\ \cdots & \cdots & \cdots & \cdots & \cdots & \cdots \\ 0 & 0 & 0 & \cdots & -\lambda & 1 \\ -a_1 & -a_2 & -a_3 & \cdots & -a_{n-2} & -(\lambda + a_{n-1}) \end{vmatrix}$$

$$+ (-1)^{n+1}(-a_0) \begin{vmatrix} 1 & 0 & \cdots & 0 \\ -\lambda & 1 & \cdots & 0 \\ 0 & -\lambda & \cdots & 0 \\ \cdots & \cdots & \cdots & \cdots \\ 0 & 0 & \cdots & 1 \end{vmatrix}.$$

Using induction we arrive at the expression

$$\det (\mathbf{A} - \lambda \mathbf{I}) = (-1)^n \{\lambda^n + a_{n-1}\lambda^{n-1} + \ldots + a_1\lambda + a_0\}. \quad (4.11.89)$$

It follows from Eq. (4.11.89) that λ is an eigenvalue of $\mathbf{A}$ if and only if λ is a root of the characteristic equation (4.11.86).

4.11.90. Exercise. Assume that the eigenvalues of matrix $\mathbf{A}$ given in Eq. (4.11.88) are all real and distinct. Let $\mathbf{\Lambda}$ denote the diagonal matrix

$$\mathbf{\Lambda} = \begin{bmatrix} \lambda_1 & & & \mathbf{0} \\ & \lambda_2 & & \\ & & \ddots & \\ \mathbf{0} & & & \lambda_n \end{bmatrix}, \quad (4.11.91)$$

where $\lambda_1, \ldots, \lambda_n$ denote the eigenvalues of matrix $\mathbf{A}$. Let $\mathbf{V}$ denote the **Vandermonde matrix** given by

$$\mathbf{V} = \begin{bmatrix} 1 & 1 & \cdots & 1 \\ \lambda_1 & \lambda_2 & \cdots & \lambda_n \\ \lambda_1^2 & \lambda_2^2 & \cdots & \lambda_n^2 \\ \cdots & \cdots & \cdots & \cdots \\ \lambda_1^{n-1} & \lambda_2^{n-1} & \cdots & \lambda_n^{n-1} \end{bmatrix}.$$

(a) Show that $\mathbf{V}$ is non-singular.
(b) Show that $\mathbf{\Lambda} = \mathbf{V}^{-1}\mathbf{A}\mathbf{V}$.

Before closing the present section, let us consider so-called "adjoint systems." To this end let us consider once more Eq. (4.11.30); i.e.,

$$\dot{\mathbf{x}} = \mathbf{A}(t)\mathbf{x}. \quad (4.11.92)$$

Let $\mathbf{A}^*(t)$ denote the conjugate transpose of $\mathbf{A}(t)$. (That is, if $\mathbf{A}(t) = [a_{ij}(t)]$, then $\mathbf{A}^*(t) = [\bar{a}_{ij}(t)]^T = [\bar{a}_{ji}(t)]$, where $\bar{a}_{ij}(t)$ denotes the complex conjugate of $\bar{a}_{ij}(t)$.) We call the system of linear first-order ordinary differential equations

$$\dot{\mathbf{y}} = -\mathbf{A}^*(t)\mathbf{y} \tag{4.11.93}$$

the **adjoint system** to (4.11.92).

4.11.94. Exercise. Let $\mathbf{\Psi}$ be a fundamental matrix of Eq. (4.11.92). Show that $\mathbf{\Upsilon}$ is a fundamental matrix for Eq. (4.11.93) if and only if

$$\mathbf{\Upsilon}^*\mathbf{\Psi} = \mathbf{C},$$

where $\mathbf{C}$ is a constant non-singular matrix, and where $\mathbf{\Upsilon}^*$ denotes the conjugate transpose of $\mathbf{\Upsilon}$.

It is also possible to consider adjoint equations for linear nth-order ordinary differential equations. Let us for example consider Eq. (4.11.85), the study of which can be reduced to that of Eq. (4.11.87), with $\mathbf{A}$ specified by Eq. (4.11.88). Now consider the adjoint system to Eq. (4.11.87), given by

$$\dot{\mathbf{y}} = -\mathbf{A}^*\mathbf{y}, \tag{4.11.95}$$

where

$$-\mathbf{A}^* = \begin{bmatrix} 0 & 0 & \cdots & 0 & \bar{a}_0 \\ -1 & 0 & \cdots & 0 & \bar{a}_1 \\ 0 & -1 & \cdots & 0 & \bar{a}_2 \\ \cdots & \cdots & \cdots & \cdots & \cdots \\ 0 & 0 & \cdots & -1 & \bar{a}_{n-1} \end{bmatrix}, \tag{4.11.96}$$

where $\bar{a}_i$ denotes the complex conjugate of a_i, $i = 0, \ldots, n-1$. Equation (4.11.95) represents the system of equations

$$\begin{aligned} \dot{y}_1 &= \bar{a}_0 y_n, \\ \dot{y}_2 &= -y_1 + \bar{a}_1 y_n, \\ &\cdots\cdots\cdots\cdots, \\ \dot{y}_n &= -y_{n-1} + \bar{a}_{n-1} y_n. \end{aligned} \tag{4.11.97}$$

Differentiating the last expression in Eq. (4.11.97) $(n-1)$ times, eliminating $y_1, \ldots, y_{n-1}$, and letting $y_n = y$, we obtain

$$(-1)^n y^{(n)} + (-1)^{n-1}\bar{a}_{n-1}y^{(n-1)} + \ldots + (-1)\bar{a}_1 y^{(1)} + \bar{a}_0 y = 0. \tag{4.11.98}$$

Equation (4.11.98) is called the **adjoint** of Eq. (4.11.85).

4.12. NOTES AND REFERENCES

There are many excellent texts on finite-dimensional vector spaces and matrices that can be used to supplement this chapter (see e.g., [4.1], [4.2], [4.4], and [4.6]–[4.10]). References [4.1], [4.2], [4.6], and [4.10] include applications. (In particular, consult the references in [4.10] for a list of diversified areas of applications.)

Excellent references on ordinary differential equations include [4.3], [4.5], and [4.11].

REFERENCES

[4.1] N. R. AMUNDSON, *Mathematical Methods in Chemical Engineering: Matrices and Their Applications.* Englewood Cliffs, N.J.: Prentice-Hall, Inc., 1966.

[4.2] R. E. BELLMAN, *Introduction to Matrix Algebra.* New York: McGraw-Hill Book Company, Inc., 1970.

[4.3] F. BRAUER and J. A. NOHEL, *Qualitative Theory of Ordinary Differential Equations: An Introduction.* New York: W. A. Benjamin, Inc., 1969. *

[4.4] E. T. BROWNE, *Introduction to the Theory of Determinants and Matrices.* Chapel Hill, N.C.: The University of North Carolina Press, 1958.

[4.5] E. A. CODDINGTON and N. LEVINSON, *Theory of Ordinary Differential Equations.* New York: McGraw-Hill Book Company, Inc., 1955.

[4.6] F. R. GANTMACHER, *Theory of Matrices.* Vols. I, II. New York: Chelsea Publishing Company, 1959.

[4.7] P. R. HALMOS, *Finite Dimensional Vector Spaces.* Princeton, N.J.: D. Van Nostrand Company, Inc., 1958.

[4.8] K. HOFFMAN and R. KUNZE, *Linear Algebra.* Englewood Cliffs, N.J.: Prentice-Hall, Inc., 1961.

[4.9] S. LIPSCHUTZ, *Linear Algebra.* New York: McGraw-Hill Book Company, 1968.

[4.10] B. NOBLE, *Applied Linear Algebra.* Englewood Cliffs, N.J.: Prentice-Hall, Inc., 1969.

[4.11] L. S. PONTRIAGIN, *Ordinary Differential Equations.* Reading, Mass.: Addison-Wesley Publishing Co., Inc., 1962.

*Reprinted by Dover Publications, Inc., New York, 1989.

5

METRIC SPACES

Up to this point in our development we have concerned ourselves primarily with algebraic structure of mathematical systems. In the present chapter we focus our attention on topological structure. In doing so, we introduce the concepts of "distance" and "closeness." In the final two chapters we will consider mathematical systems endowed with algebraic as well as topological structure.

A generalization of the concept of "distance" is the notion of metric. Using the terminology from geometry, we will refer to elements of an *arbitrary* set X as **points** and we will characterize metric as a real-valued, non-negative function on $X \times X$ satisfying the properties of "distance" between two points of X. We will refer to a mathematical system consisting of a basic set X and a metric defined on it as a **metric space**. We emphasize that in the present chapter the underlying space X need not be a linear space.

In the first nine sections of the present chapter we establish several basic facts from the theory of metric spaces, while in the last section of the present chapter, which consists of two parts, we consider some applications to the material of the present chapter.

5.1. DEFINITION OF METRIC SPACE

We begin with the following definition of metric and metric space.

5.1.1. Definition. Let X be an arbitrary non-empty set, and let ρ be a real-valued function on $X \times X$, i.e., $\rho: X \times X \longrightarrow R$, where ρ has the following properties:

(i) $\rho(x, y) \geq 0$ for all $x, y \in X$ and $\rho(x, y) = 0$ if and only if $x = y$;
(ii) $\rho(x, y) = \rho(y, x)$ for all $x, y \in X$; and
(iii) $\rho(x, y) \leq \rho(x, z) + \rho(z, y)$ for all $x, y, z \in X$.

The function ρ is called a **metric** on X, and the mathematical system consisting of ρ and X, $\{X; \rho\}$, is called a **metric space.**

The set X is often called the **underlying set** of the metric space, the elements of X are often called **points**, and $\rho(x, y)$ is frequently called the **distance** from a point $x \in X$ to a point $y \in X$. In view of axiom (i) the distance between two different points is a unique positive number and is equal to zero if and only if two points coincide. Axiom (ii) indicates that the distance between points x and y is equal to the distance between points y and x. Axiom (iii) represents the well-known **triangle inequality** encountered, for example, in plane geometry. Clearly, if ρ is a metric for X and if α is any real positive number, then the function $\alpha\rho(x, y)$ is also a metric for X. We are thus in a position to define infinitely many metrics on X.

The above definition of metric was motivated by our notion of distance. Our next result enables us to define metric in an equivalent (and often convenient) way.

5.1.2. Theorem. Let $\rho: X \times X \longrightarrow R$. Then ρ is a metric if and only if

(i) $\rho(x, y) = 0$ if and only if $x = y$; and
(ii) $\rho(y, z) \leq \rho(x, y) + \rho(x, z)$ for all $x, y, z \in X$.

Proof. The necessity is obvious. To prove sufficiency, let $x, y, z \in X$ with $y = z$. Then $0 = \rho(y, y) \leq 2\rho(x, y)$. Hence, $\rho(x, y) \geq 0$ for all $x, y \in X$. Next, let $z = x$. Then $\rho(y, x) \leq \rho(x, y)$. Since x and y are arbitrary, we can reverse their role and conclude $\rho(x, y) \leq \rho(y, x)$. Therefore, $\rho(x, y) = \rho(y, x)$ for all $x, y \in X$. This proves that ρ is a metric. ■

Different metrics defined on the same underlying set X yield different metric spaces. In applications, the choice of a specific metric is often dictated by the particular problem on hand. If in a particular situation the metric ρ is understood, then we simply write X in place of $\{X; \rho\}$ to denote the particular metric space under consideration.

Let us now consider a few examples of metric spaces.

5.1.3. Example. Let X be the set of real numbers R, and let the function ρ on $R \times R$ be defined as

$$\rho(x, y) = |x - y| \tag{5.1.4}$$

for all $x, y \in R$, where $|x|$ denotes the absolute value of x. Now clearly $\rho(x, y) = |x - y| = 0$ if and only if $x = y$. Also, for all $x, y, z \in R$, we have $\rho(y, z) = |y - z| = |(y - x) + (x - z)| \leq |x - y| + |x - z| = \rho(x, y) + \rho(x, z)$. Therefore, by Theorem 5.1.2, ρ is a metric and $\{R; \rho\}$ is a metric space. We call $\rho(x, y)$ defined by Eq. (5.1.4) the **usual metric** on R, and we call the metric space $\{R; \rho\}$ the **real line.** ■

5.1.5. Example. Let X be the set of all complex numbers C. If $z \in C$, then $z = a + ib$, where $i = \sqrt{-1}$, and where a, b are real numbers. Let $\bar{z} = a - ib$ and define ρ as

$$\rho(z_1, z_2) = [(z_1 - z_2)\overline{(z_1 - z_2)}]^{1/2}. \tag{5.1.6}$$

It can readily be shown that $\{C; \rho\}$ is a metric space. We call (5.1.6) the **usual metric** for C. ■

5.1.7. Example. Let X be an arbitrary non-empty set and define the function ρ on $X \times X$ as

$$\rho(x, y) = \begin{cases} 0 & \text{if } x = y \\ 1 & \text{if } x \neq y. \end{cases} \tag{5.1.8}$$

Clearly $\rho(x, y) \geq 0$ for all $x, y \in X$, $\rho(x, x) = 0$ for all $x \in X$, and $\rho(x, y) \leq \rho(x, z) + \rho(z, y)$ for all $x, y, z \in X$. Therefore, (5.1.8) is a metric on X. The function defined in Eq. (5.1.8) is called the **discrete metric** and is important in analysis because it can be used to **metrize** any set X. ■

We distinguish between bounded and unbounded metric spaces.

5.1.9. Definition. Let $\{X; \rho\}$ be a metric space. If there exists a positive number r such that $\rho(x, y) \leq r$ for all $x, y \in X$, we say $\{X; \rho\}$ is a **bounded metric space.** If $\{X; \rho\}$ is not bounded, we say $\{X; \rho\}$ is an **unbounded metric space.**

If $\{X; \rho\}$ is an unbounded metric space, then ρ takes on arbitrarily large values. The metric spaces in Examples 5.1.3 and 5.1.5 are unbounded, whereas the metric space in Example 5.1.7 is clearly bounded.

5.1.10. Exercise. Let $\{X; \rho\}$ be an arbitrary metric space. Define the function $\rho_1: X \times X \rightarrow R$ by

$$\rho_1(x, y) = \frac{\rho(x, y)}{1 + \rho(x, y)}. \tag{5.1.11}$$

Show that $\rho_1(x, y)$ is a metric. Show that $\{X; \rho_1\}$ is a bounded metric space, even though $\{X; \rho\}$ may not be bounded. Thus, the function (5.1.11) can be used to generate a bounded metric space from any unbounded metric space. (Hint: Show that if $\varphi: R \longrightarrow R$ is given by $\varphi(t) = t/(1 + t)$, then $\varphi(t_1) \leq \varphi(t_2)$ for all t_1, t_2 such that $0 \leq t_1 \leq t_2$.)

Subsequently, we will call

$$R^* = R \cup \{-\infty\} \cup \{+\infty\}$$

the **extended real numbers**. In the following exercise, we define a useful metric on R^*. This metric is, of course, not the only metric possible.

5.1.12. Exercise. Let $X = R^*$ and define the function $f: R^* \longrightarrow R$ as

$$f(x) = \begin{cases} \dfrac{x}{1 + |x|}, & x \in R \\ +1, & x = +\infty \\ -1, & x = -\infty \end{cases}$$

Let $\rho_*: R^* \times R^* \longrightarrow R$ be defined by $\rho_*(x, y) = |f(x) - f(y)|$ for all $x, y \in R^*$. Show that $\{R^*; \rho_*\}$ is a bounded metric space. The function ρ_* is called the **usual metric** for R^*, and $\{R^*; \rho_*\}$ is called the **extended real line**.

We will have occasion to use the next result.

5.1.13. Theorem. Let $\{X; \rho\}$ be a metric space, and let x, y, and z be any elements of X. Then

$$|\rho(x, z) - \rho(y, z)| \leq \rho(x, y) \tag{5.1.14}$$

for all $x, y, z \in X$.

Proof. From axiom (iii) of Definition 5.1.1 it follows that

$$\rho(x, z) \leq \rho(x, y) + \rho(y, z) \tag{5.1.15}$$

and

$$\rho(y, z) \leq \rho(y, x) + \rho(x, z). \tag{5.1.16}$$

From (5.1.15) we have

$$\rho(x, z) - \rho(y, z) \leq \rho(x, y) \tag{5.1.17}$$

and from (5.1.16) we have

$$-\rho(y, x) \leq \rho(x, z) - \rho(y, z). \tag{5.1.18}$$

In view of axiom (ii) of Definition 5.1.1 we have $\rho(x, y) = \rho(y, x)$, and thus relations (5.1.17) and (5.1.18) imply

$$-\rho(x, y) \leq \rho(x, z) - \rho(y, z) \leq \rho(x, y).$$

This proves that $|\rho(x, z) - \rho(y, z)| \leq \rho(x, y)$ for all $x, y, z \in X$. ■

The notion of metric makes it possible to consider various geometric concepts. We have:

5.1.19. Definition. Let $\{X; \rho\}$ be a metric space, and let Y be a non-void subset of X. If $\rho(x, y)$ is bounded for all $x, y \in Y$, we define the **diameter of set** Y, denoted $\delta(Y)$ or diam (Y), as

$$\delta(Y) = \sup \{\rho(x, y): x, y \in Y\}.$$

If $\rho(x, y)$ is unbounded, we write $\delta(Y) = +\infty$ and we say that Y has infinite diameter, or Y is **unbounded**. If Y is empty, we define $\delta(Y) = 0$.

5.1.20. Exercise. Show that if $Y \subset Z \subset X$, where $\{X; \rho\}$ is a metric space, then $\delta(Y) \leq \delta(Z)$. Also, show that if Z is non-empty, then $\delta(Z) = 0$ if and only if Z is a singleton.

We also have:

5.1.21. Definition. Let $\{X; \rho\}$ be a metric space, and let Y and Z be two non-void subsets of X. We define the **distance between sets** Y **and** Z as

$$d(Y, Z) = \inf \{\rho(y, z): y \in Y, z \in Z\}.$$

Let $p \in X$ and define

$$d(p, Z) = \inf \{\rho(p, z): z \in Z\}.$$

We call $d(p, Z)$ the **distance between point** p **and set** Z.

Since $\rho(y, z) = \rho(z, y)$ for all $y \in Y$ and $z \in Z$, it follows that $d(Y, Z) = d(Z, Y)$. We note that, in general, $d(Y, Z) = 0$ does not imply that Y and Z have points in common. For example, let X be the real line with the usual metric ρ. If $Y = \{x \in X: 0 < x < 1\}$ and $Z = \{x \in X: 1 < x < 2\}$, then clearly $d(Y, Z) = 0$, even though $Y \cap Z = \varnothing$. Similarly, $d(p, Z) = 0$ does not imply that $p \in Z$.

5.1.22. Theorem. Let $\{X; \rho\}$ be a metric space, and let Y be any non-void subset of X. If ρ' denotes the restriction of ρ to $Y \times Y$, i.e., if

$$\rho'(x, y) = \rho(x, y) \text{ for all } x, y \in Y,$$

then $\{Y; \rho'\}$ is a metric space.

5.1.23. Exercise. Prove Theorem 5.1.22.

We call ρ' the **metric induced** by ρ on Y, and we say that $\{Y; \rho'\}$ is a **metric subspace** of $\{X; \rho\}$ or simply a **subspace** of X. Since usually there is no room for confusion, we drop the prime from ρ' and simply denote the metric subspace by $\{Y; \rho\}$. We emphasize that *any* non-void subset of a metric space can be made into a metric subspace. This is not so in the case of linear subspaces. If $Y \neq X$, then we speak of a **proper subspace**.

5.2. SOME INEQUALITIES

In order to present some of the important metric spaces that arise in applications, we first need to establish some important inequalities. These are summarized and proved in the following:

5.2.1. Theorem. Let R denote the set of real numbers, and let C denote the set of complex numbers.

(i) Let $p, q \in R$ such that $1 < p < \infty$ and such that $\frac{1}{p} + \frac{1}{q} = 1$. Then for all $\alpha, \beta \in R$ such that $\alpha \geq 0$ and $\beta \geq 0$, we have

$$\alpha\beta \leq \frac{\alpha^p}{p} + \frac{\beta^q}{q}. \tag{5.2.2}$$

(ii) **(Hölder's inequality)** Let $p, q \in R$ be such that $1 < p < \infty$, and $\frac{1}{p} + \frac{1}{q} = 1$.

(a) *Finite Sums.* Let n be any positive integer and, let $\xi_1, \ldots, \xi_n$ and $\eta_1, \ldots, \eta_n$ belong either to R or to C. Then

$$\sum_{i=1}^{n} |\xi_i \eta_i| \leq \left[\sum_{i=1}^{n} |\xi_i|^p\right]^{1/p} \left[\sum_{i=1}^{n} |\eta_i|^q\right]^{1/q}. \tag{5.2.3}$$

(b) *Infinite Sums.* Let $\{\xi_i\}$ and $\{\eta_i\}$ be infinite sequences in either R or C. If $\sum_{i=1}^{\infty} |\xi_i|^p < \infty$ and $\sum_{i=1}^{\infty} |\eta_i|^q < \infty$, then

$$\sum_{i=1}^{\infty} |\xi_i \eta_i| \leq \left[\sum_{i=1}^{\infty} |\xi_i|^p\right]^{1/p} \left[\sum_{i=1}^{\infty} |\eta_i|^q\right]^{1/q}. \tag{5.2.4}$$

(c) *Integrals.* Let $[a, b]$ be an interval on the real line, and let $f, g: [a, b] \to R$. If $\int_a^b |f(t)|^p\, dt < \infty$ and $\int_a^b |g(t)|^q\, dt < \infty$ (integration is in the Riemann sense), then

$$\int_a^b |f(t)g(t)|\, dt \leq \left[\int_a^b |f(t)|^p\, dt\right]^{1/p} \left[\int_a^b |g(t)|^q\, dt\right]^{1/q}. \tag{5.2.5}$$

(iii) **(Minkowski's inequality)** Let $p \in R$, where $1 \leq p < \infty$.

(a) *Finite Sums.* Let n be any positive integer, and let $\xi_1, \ldots, \xi_n$ and $\eta_1, \ldots, \eta_n$ belong either to R or to C. Then

$$\left[\sum_{i=1}^{n} |\xi_i \pm \eta_i|^p\right]^{1/p} \leq \left[\sum_{i=1}^{n} |\xi_i|^p\right]^{1/p} + \left[\sum_{i=1}^{n} |\eta_i|^p\right]^{1/p}. \tag{5.2.6}$$

(b) *Infinite Sums.* Let $\{\xi_i\}$ and $\{\eta_i\}$ be infinite sequences in either R or C. If $\sum_{i=1}^{\infty} |\xi_i|^p < \infty$ and $\sum_{i=1}^{\infty} |\eta_i|^p < \infty$, then

$$\left[\sum_{i=1}^{\infty} |\xi_i \pm \eta_i|^p\right]^{1/p} \leq \left[\sum_{i=1}^{\infty} |\xi_i|^p\right]^{1/p} + \left[\sum_{i=1}^{\infty} |\eta_i|^p\right]^{1/p}. \quad (5.2.7)$$

(c) *Integrals.* Let $[a, b]$ be an interval on the real line, and let $f, g: [a, b] \longrightarrow R$. If $\int_a^b |f(t)|^p dt < \infty$ and $\int_a^b |g(t)|^p\, dt < \infty$, then

$$\left[\int_a^b |f(t) \pm g(t)|^p\, dt\right]^{1/p} \leq \left[\int_a^b |f(t)|^p\, dt\right]^{1/p} + \left[\int_a^b |g(t)|^p\, dt\right]^{1/p}. \quad (5.2.8)$$

Proof. To prove part (i), consider the graph of $\eta = \xi^{p-1}$ in the (ξ, η) plane, depicted in Figure A. Let $\sigma_1 = \int_0^\alpha \xi^{p-1}\, d\xi$ and $\sigma_2 = \int_0^\beta \eta^{q-1}\, d\eta$. We have $\sigma_1 = \alpha^p/p$ and $\sigma_2 = \beta^q/q$. From Figure A it is clear that $\sigma_1 + \sigma_2 \geq \alpha\beta$ for any choice of $\alpha, \beta \geq 0$, and hence relation (5.2.2) follows.

To prove part (iia) we first note that if $\left(\sum_{i=1}^{n} |\xi_i|^p\right)^{1/p} = 0$ or if $\left(\sum_{i=1}^{n} |\eta_i|^q\right)^{1/q} = 0$, then inequality (5.2.3) follows trivially. Therefore, we assume that $\left(\sum_{i=1}^{n} |\xi_i|^p\right)^{1/p} \neq 0$ and $\left(\sum_{i=1}^{n} |\eta_i|^q\right)^{1/q} \neq 0$. From (5.2.2) we now have

$$\frac{|\xi_i|}{\left(\sum_{i=1}^{n} |\xi_i|^p\right)^{1/p}} \cdot \frac{|\eta_i|}{\left(\sum_{i=1}^{n} |\eta_i|^q\right)^{1/q}} \leq \frac{1}{p} \cdot \frac{|\xi_i|^p}{\left(\sum_{i=1}^{n} |\xi_i|^p\right)} + \frac{1}{q} \cdot \frac{|\eta_i|^q}{\left(\sum_{i=1}^{n} |\eta_i|^q\right)}.$$

Hence,

$$\sum_{i=1}^{n} \frac{|\xi_i|}{\left(\sum_{i=1}^{n} |\xi_i|^p\right)^{1/p}} \cdot \frac{|\eta_i|}{\left(\sum_{i=1}^{n} |\eta_i|^q\right)^{1/q}} \leq \frac{1}{p\left(\sum_{i=1}^{n} |\xi_i|^p\right)} \sum_{i=1}^{n} |\xi_i|^p$$

$$+ \frac{1}{q\left(\sum_{i=1}^{n} |\eta_i|^q\right)} \sum_{i=1}^{n} |\eta_i|^q = \frac{1}{p} + \frac{1}{q} = 1.$$

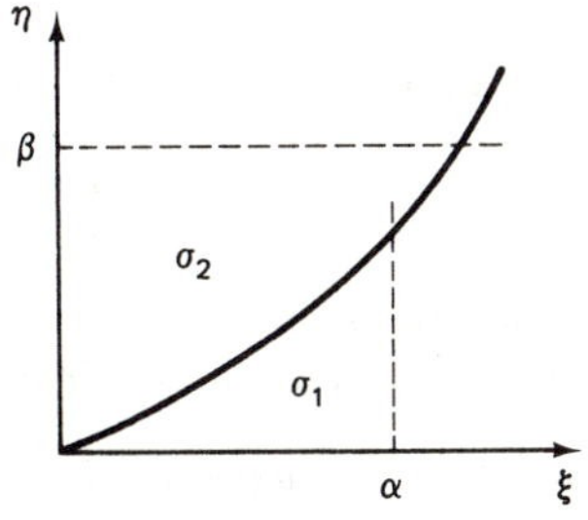

5.2.9. Figure A.

It now follows that

$$\sum_{i=1}^{n} |\xi_i \eta_i| = \sum_{i=1}^{n} |\xi_i||\eta_i| \leq \left(\sum_{i=1}^{n} |\xi_i|^p\right)^{1/p} \left(\sum_{i=1}^{n} |\eta_i|^q\right)^{1/q},$$

which was to be proved.

To prove part (iib), we note that for any positive integer n,

$$\sum_{i=1}^{n} |\xi_i \eta_i| \leq \left(\sum_{i=1}^{n} |\xi_i|^p\right)^{1/p} \left(\sum_{i=1}^{n} |\eta_i|^q\right)^{1/q} \leq \left(\sum_{i=1}^{\infty} |\xi_i|^p\right)^{1/p} \left(\sum_{i=1}^{\infty} |\eta_i|^q\right)^{1/q}.$$

If we let $n \longrightarrow \infty$ in the above inequality, then (5.2.4) follows.

The proof of part (iic) is established in a similar fashion. We leave the details of the proof to the reader.

To prove part (iiia), we first note that if $p = 1$, then inequality (5.2.6) follows trivially. It therefore suffices to consider the case $1 < p < \infty$. We observe that for any ξ_i and η_i, we have

$$(|\xi_i| + |\eta_i|)^p = (|\xi_i| + |\eta_i|)^{p-1}|\xi_i| + (|\xi_i| + |\eta_i|)^{p-1}|\eta_i|.$$

Summing the above identity with respect to i from 1 to n, we now have

$$\sum_{i=1}^{n} |\xi_i \pm \eta_i|^p \leq \sum_{i=1}^{n} [|\xi_i| + |\eta_i|]^p$$
$$= \sum_{i=1}^{n} [|\xi_i| + |\eta_i|]^{p-1}|\xi_i| + \sum_{i=1}^{n} [|\xi_i| + |\eta_i|]^{p-1}|\eta_i|.$$

Applying the Hölder inequality (5.2.3) to each of the sums on the right side of the above relation and noting that $(p - 1)q = p$, we now obtain

$$\sum_{i=1}^{n} [|\xi_i| + |\eta_i|]^p \leq \left[\sum_{i=1}^{n} (|\xi_i| + |\eta_i|)^p\right]^{1/q} \left[\sum_{i=1}^{n} |\xi_i|^p\right]^{1/p} + \left[\sum_{i=1}^{n} (|\xi_i| + |\eta_i|)^p\right]^{1/q} \left[\sum_{i=1}^{n} |\eta_i|^p\right]^{1/p}.$$

If we assume that $\left[\sum_{i=1}^{n} (|\xi_i| + |\eta_i|)^p\right]^{1/q} \neq 0$ and divide both sides of the above inequality by this term, we have

$$\left[\sum_{i=1}^{n} (|\xi_i| + |\eta_i|)^p\right]^{1/p} \leq \left[\sum_{i=1}^{n} |\xi_i|^p\right]^{1/p} + \left[\sum_{i=1}^{n} |\eta_i|^p\right]^{1/p}.$$

Since $\left[\sum_{i=1}^{n} |\xi_i \pm \eta_i|^p\right]^{1/p} \leq \left[\sum_{i=1}^{n} (|\xi_i| + |\eta_i|)^p\right]^{1/p}$, the desired result follows. We note that in case $\left[\sum_{i=1}^{n} (|\xi_i| + |\eta_i|)^p\right]^{1/q} = 0$, inequality (5.2.6) follows trivially.

Applying the same reasoning as above, the reader can now prove the Minkowski inequality for infinite sums and for integrals. ■

If in (5.2.3), (5.2.4), or (5.2.5) we let $p = q = \frac{1}{2}$, then we speak of the **Schwarz inequality** for finite sums, infinite sums, and integrals, respectively.

5.2.10. Exercise. Prove Hölder's inequality for integrals (5.2.5), Minkowski's inequality for infinite sums (5.2.7), and Minkowski's inequality for integrals (5.2.8).

5.3. EXAMPLES OF IMPORTANT METRIC SPACES

In the present section we consider specific examples of metric spaces which are very important in applications. It turns out that all of the spaces of this section are also vector spaces.

As in Chapter 4, we denote elements $x, y \in R^n$ (elements $x, y \in C^n$) by $x = (\xi_1, \ldots, \xi_n)$ and $y = (\eta_1, \ldots, \eta_n)$, respectively, where $\xi_i, \eta_i \in R$ for $i = 1, \ldots, n$ (where $\xi_i, \eta_i \in C$ for $i = 1, \ldots, n$). Similarly, elements x, $y \in R^\infty$ (elements $x, y \in C^\infty$) are denoted by $x = (\xi_1, \xi_2, \ldots)$ and $y = (\eta_1, \eta_2, \ldots)$, respectively, where $\xi_i, \eta_i \in R$ for all i (where $\xi_i, \eta_i \in C$ for all i).

5.3.1. Example. Let $X = R^n$ (let $X = C^n$), let $1 \leq p < \infty$, and let

$$\rho_p(x, y) = \left[\sum_{i=1}^{n} |\xi_i - \eta_i|^p\right]^{1/p}. \tag{5.3.2}$$

We now show that $\{R^n; \rho_p\}(\{C^n; \rho_p\})$ is a metric space.

Axioms (i) and (ii) of Definition 5.1.1 are readily verified. To show that axiom (iii) is satisfied, let $a, b, d \in R^n$ (let $a, b, d \in C^n$), where $a = (\alpha_1, \ldots, \alpha_n)$, $b = (\beta_1, \ldots, \beta_n)$, and $d = (\delta_1, \ldots, \delta_n)$. If $x = a - b$ and $y = b - d$, then we have from inequality (5.2.6),

$$\rho_p(a, d) = \left\{\sum_{i=1}^{n} |\alpha_i - \delta_i|^p\right\}^{1/p} = \left\{\sum_{i=1}^{n} |\alpha_i - \beta_i + \beta_i - \delta_i|^p\right\}^{1/p}$$
$$\leq \left\{\sum_{i=1}^{n} |\alpha_i - \beta_i|^p\right\}^{1/p} + \left\{\sum_{i=1}^{n} |\beta_i - \delta_i|^p\right\}^{1/p} = \rho(a, b) + \rho(b, d),$$

the triangle inequality. It thus follows that $\{R^n; \rho_p\}(\{C^n; \rho_p\})$ is a metric space; in fact, it is an unbounded metric space.

We frequently abbreviate $\{R^n; \rho_p\}$ by R^n_p and $\{C^n; \rho_p\}$ by C^n_p. For the case $p = 2$, we call ρ_2 the **Euclidean metric** or the **usual metric** on R^n. ■

5.3.3. Example. For $x, y \in R^n$ (for $x, y \in C^n$), let

$$\rho_\infty(x, y) = \max\{|\xi_1 - \eta_1|, \ldots, |\xi_n - \eta_n|\}. \tag{5.3.4}$$

It is readily shown that $\{R^n; \rho_\infty\}(\{C^n; \rho_\infty\})$ is a metric space. ■

5.3.5. Example. Let $1 \leq p < \infty$, let $X = R^\infty$ (or $X = C^\infty$), and define

$$l_p = \left\{x \in X: \sum_{i=1}^{\infty} |\xi_i|^p < \infty\right\}. \tag{5.3.6}$$

For $x, y \in l_p$, let

$$\rho_p(x, y) = \left[\sum_{i=1}^{\infty} |\xi_i - \eta_i|^p\right]^{1/p}. \tag{5.3.7}$$

We can readily verify that $\{l_p; \rho_p\}$ is a metric space. ■

5.3.8. Example. Let $X = R^\infty$ (or $X = C^\infty$), and let

$$l_\infty = \{x \in X: \sup_i \{|\xi_i|\} < \infty\}. \tag{5.3.9}$$

For $x, y \in l_\infty$, define

$$\rho_\infty(x, y) = \sup_i \{|\xi_i - \eta_i|\}. \tag{5.3.10}$$

We can easily show that $\{l_\infty; \rho_\infty\}$ is a metric space. ■

5.3.11. Exercise. Use the inequalities of Section 5.2 to show that the spaces of Examples 5.3.3, 5.3.5, and 5.3.8 are metric spaces.

5.3.12. Example. Let $[a, b]$, $a < b$, be an interval on the real line, and let $\mathcal{C}[a, b]$ be the set of all real-valued continuous functions defined on $[a, b]$. Let $1 \leq p < \infty$ and for $x, y \in \mathcal{C}[a, b]$, define

$$\rho_p(x, y) = \left[\int_a^b |x(t) - y(t)|^p \, dt\right]^{1/p}. \tag{5.3.13}$$

We now show that $\{\mathcal{C}[a, b]; \rho_p\}$ is a metric space.

Clearly, $\rho_p(x, y) = \rho_p(y, x)$, and $\rho_p(x, y) \geq 0$ for all $x, y \in \mathcal{C}[a, b]$. If $x(t) = y(t)$ for all $t \in [a, b]$, then $\rho_p(x, y) = 0$. To prove the converse of this statement, suppose that $x(t) \neq y(t)$ for some $t \in [a, b]$. Since $x, y \in \mathcal{C}[a, b]$, $x - y \in \mathcal{C}[a, b]$, and there is some interval in $[a, b]$, i.e., a subinterval of $[a, b]$, such that $|x(t) - y(t)| > 0$ for all t in that subinterval. Hence,

$$\left[\int_a^b |x(t) - y(t)|^p \, dt\right]^{1/p} > 0.$$

Therefore, $\rho_p(x, y) = 0$ if and only if $x(t) = y(t)$ for *all* $t \in [a, b]$.

To show that the triangle inequality holds, let $u, v, w \in \mathcal{C}[a, b]$, and let $x = u - v$ and $y = v - w$. Then we have, from inequality (5.2.8),

$$\begin{aligned}
\rho_p(u, w) &= \left\{\int_a^b |u(t) - w(t)|^p \, dt\right\}^{1/p} \\
&= \left\{\int_a^b |u(t) - v(t) + v(t) - w(t)|^p \, dt\right\}^{1/p} \\
&\leq \left\{\int_a^b |u(t) - v(t)|^p \, dt\right\}^{1/p} + \left\{\int_a^b |v(t) - w(t)|^p \, dt\right\}^{1/p} \\
&= \rho_p(u, v) + \rho_p(v, w),
\end{aligned}$$

the triangle inequality. It now follows that $\{\mathcal{C}[a, b]; \rho_p\}$ is a metric space. It is easy to see that this space is an unbounded metric space. ■

5.3.14. Example. Let $\mathcal{C}[a, b]$ be defined as in the preceding example. For $x, y \in \mathcal{C}[a, b]$, let

$$\rho_\infty(x, y) = \sup_{a \leq t \leq b} |x(t) - y(t)|. \tag{5.3.15}$$

To show that $\{\mathcal{C}[a, b]; \rho_\infty\}$ is a metric space we first note that $\rho_\infty(x, y) = \rho_\infty(y, x)$, that $\rho(x, y) \geq 0$ for all x, y, and that $\rho(x, y) = 0$ if and only if $x(t) = y(t)$ for all $t \in [a, b]$. To show that ρ_∞ satisfies the triangle inequality we note that

$$\begin{aligned}\rho_\infty(x, y) &= \sup_{a \leq t \leq b} |x(t) - y(t)| \\ &= \sup_{a \leq t \leq b} |x(t) - z(t) + z(t) - y(t)|\end{aligned}$$

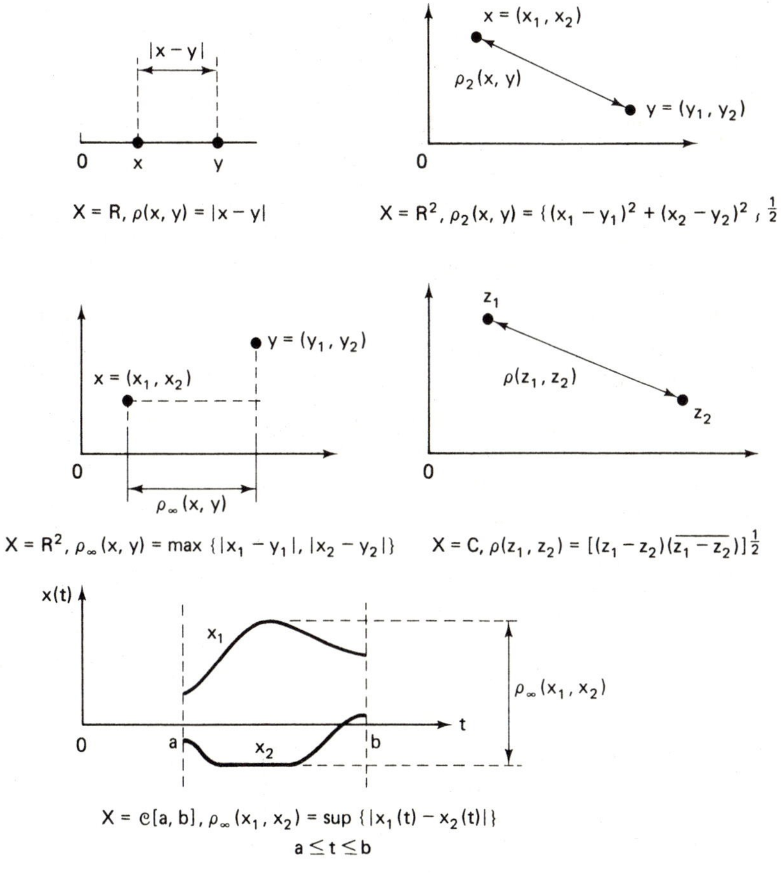

5.3.16. **Figure B.** Illustration of various metrics.

$$\leq \sup_{a \leq t \leq b} \{|x(t) - z(t)| + |z(t) - y(t)|\}$$
$$\leq \sup_{a \leq t \leq b} |x(t) - z(t)| + \sup_{a \leq t \leq b} |z(t) - y(t)|$$
$$= \rho_\infty(x, z) + \rho_\infty(z, y).$$

It thus follows that $\{\mathcal{C}[a, b]; \rho_\infty\}$ is a metric space. ■

In Figure B, several metrics considered in Section 5.1 and in the present section are depicted pictorially.

5.3.17. Exercise. Show that the metric defined in Eq. (5.3.4) is equivalent to

$$\rho_\infty(x, y) = \lim_{p \to \infty} \left[\sum_{i=1}^{n} |\xi_i - \eta_i|^p\right]^{1/p}.$$

5.3.18. Exercise. Let $X = R$ denote the set of real numbers, and define $d(x, y) = (x - y)^2$ for all $x, y \in R$. Show that the function d is not a metric. This illustrates the necessity for the exponent $1/p$ in Eq. (5.3.2).

We conclude the present section by considering Cartesian products of metric spaces. Let $\{X; \rho_x\}$ and $\{Y; \rho_y\}$ be two metric spaces, and let $Z = X \times Y$. Utilizing the metrics ρ_x and ρ_y we can define metrics on Z in an infinite variety of ways. Some of the more interesting cases are given in the following:

5.3.19. Theorem. Let $\{X; \rho_x\}$ and $\{Y; \rho_y\}$ be metric spaces, and let $Z = X \times Y$. Let $z_1 = (x_1, y_1)$ and $z_2 = (x_2, y_2)$ be two points of $Z = X \times Y$. Define the functions

$$\rho_p(z_1, z_2) = \{[\rho_x(x_1, x_2)]^p + [\rho_y(y_1, y_2)]^p\}^{1/p}, \quad 1 \leq p < \infty$$

and

$$\rho_\infty(z_1, z_2) = \max\{\rho_x(x_1, x_2), \rho_y(y_1, y_2)\}.$$

Then $\{Z; \rho_p\}$ and $\{Z; \rho_\infty\}$ are metric spaces.

The spaces $\{Z; \rho_p\}$ and $\{Z; \rho_\infty\}$ are examples of **product (metric) spaces.**

5.3.20. Exercise. Prove Theorem 5.3.19.

We can extend the above concept to the product of n metric spaces. We have:

5.3.21. Theorem. Let $\{X_1; \rho_1\}, \ldots, \{X_n; \rho_n\}$ be n metric spaces, and let $X = X_1 \times \ldots \times X_n = \prod_{i=1}^{n} X_i$. For $x = (x_1, \ldots, x_n) \in X$, $y = (y_1, \ldots, y_n) \in X$, define the functions

$$\rho'(x, y) = \sum_{i=1}^{n} \rho_i(x_i, y_i)$$

and

$$\rho''(x, y) = \left(\sum_{i=1}^{n} [\rho_i(x_i, y_i)]^2\right)^{1/2}.$$

Then $\{X; \rho'\}$ and $\{X; \rho''\}$ are metric spaces.

5.3.22. Exercise. Prove Theorem 5.3.21.

5.4. OPEN AND CLOSED SETS

Having introduced the notion of metric, we are now in a position to consider several important fundamental concepts which we will need throughout the remainder of this book.

In the present section $\{X; \rho\}$ will denote an arbitrary metric space.

5.4.1. Definition. Let $x_0 \in X$ and let $r \in R, r > 0$. An **open sphere** or **open ball**, denoted by $S(x_0; r)$, is defined as the set

$$S(x_0; r) = \{x \in X: \rho(x, x_0) < r\}.$$

We call the fixed point x_0 the **center** and the number r the **radius** of $S(x_0; r)$. For simplicity, we often call an open sphere simply a **sphere**.

The radius of a sphere is always positive and finite. In place of the terms ball or sphere we also use the term **spherical neighborhood** of x_0.

In Figure C, spheres in several types of metric spaces considered in the previous sections are depicted. Note that in these figures the indicated spheres do not include boundaries.

5.4.3. Exercise. Describe the open sphere in R^2 as a function of r if the metric is the discrete metric of Example 5.1.7.

We can now categorize the points or elements of a metric space in several ways.

5.4.4. Definition. Let Y be a subset of X. A point $x \in X$ is called a **contact point** or **adherent point** of set Y if every open sphere with center x contains at least one point of Y. The set of all adherent points of Y is called the **closure** of Y and is denoted by $\bar{Y}$.

We note that every point of Y is an adherent point of Y; however, there may be points not in Y which are also adherent points of Y.

5.4.5. Definition. Let Y be a subset of X, and let $x \in X$ be an adherent point of Y. Then x is called an **isolated point** if there is a sphere with center x

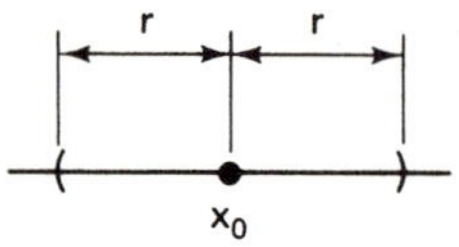

Sphere $S(x_0; r)$, where $X = R$ and $\rho(x, y) = |x - y|$

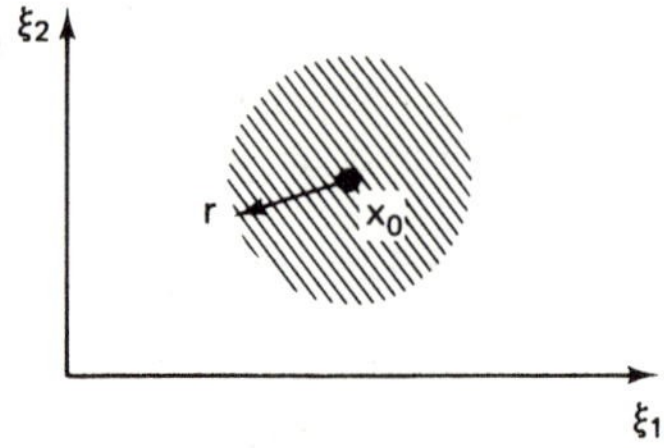

Sphere $S(x_0; r)$, where $X = R^2$ and $\rho(x, y) = \rho_2(x, y) = [(\xi_1 - \eta_1)^2 + (\xi_2 - \eta_2)^2]^{\frac{1}{2}}$

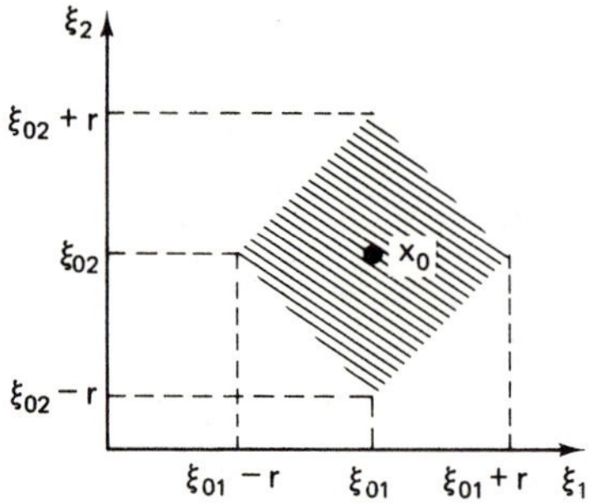

Sphere $S(x_0; r)$, where $X = R^2$ and $\rho(x, y) = \rho_1(x, y) = |\xi_1 - \eta_1| + |\xi_2 - \eta_2|$

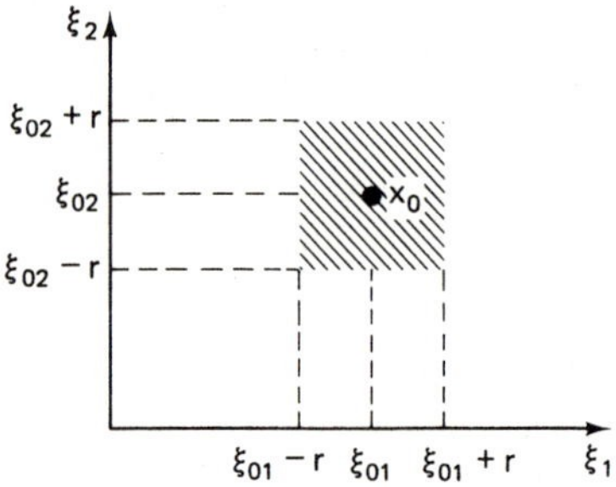

Sphere $S(x_0; r)$ where $X = R^2$ and $\rho(x, y) = \rho_\infty(x, y) = \max \{|\xi_1 - \eta_1|, |\xi_2 - \eta_2|\}$

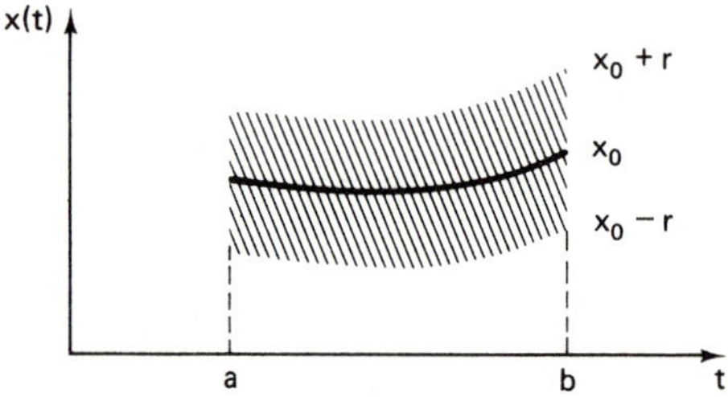

Sphere $S(x_0; r)$, where $X = \mathcal{C}[a, b]$ and $\rho(x, y) = \rho_\infty(x, y) = \sup_{a \leq t \leq b} |x(t) - y(t)|$

5.4.2. Figure C. Spheres in various metric spaces.

which contains no point of Y other than x itself. The point x is called a **limit point** or **point of accumulation** of set Y if every sphere with center at x contains an infinite number of points of Y. The set of all limit points of Y is called the **derived set** of Y and is denoted by Y'.

Our next result shows that adherent points are either limit points or isolated points.

5.4.6. Theorem. Let Y be a subset of X and let $x \in X$. If x is an adherent point of Y, then x is either a limit point or an isolated point.

Proof. We prove the theorem by assuming that x is an adherent point of Y but not an isolated point. We must then show that x is a limit point of Y. To do so, consider the family of spheres $S(x; 1/n)$ for $n = 1, 2, \ldots$. Let $x_n \in S(x; 1/n)$ be such that $x_n \in Y$ but $x_n \neq x$ for each n. Now suppose there are only a finite number of distinct such points x_n, say, $\{x_1, \ldots, x_k\}$. If we let $d = \min_{1 \leq i \leq k} \rho(x, x_i)$, then $d > 0$. But this contradicts the fact that there is an $x_n \in S(x; 1/n)$ for every $n = 1, 2, 3, \ldots$. Hence, there are infinitely many x_n and thus x is a limit point of Y. ■

We can now categorize adherent points of $Y \subset X$ into the following three classes: (a) isolated points of Y, which always belong to Y; (b) points of accumulation which belong to Y; and (c) points of accumulation which do not belong to Y.

5.4.7. Example. Let $X = R$, let ρ be the usual metric, and let $Y = \{x \in R: 0 < x < 1, x = 2\}$, as depicted in Figure D. The element $x = 2$ is an isolated point of Y, the elements 0 and 1 are adherent points of Y which do not belong to Y, and each point of the set $\{x \in R: 0 < x < 1\}$ is a limit point of Y belonging to Y. ■

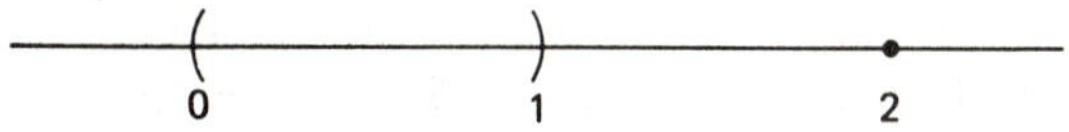

5.4.8. **Figure D.** Set $Y = \{x \in R: 0 < x < 1, x = 2\}$ of Example 5.4.7.

5.4.9. Example. Let $\{R; \rho\}$ be the real line with the usual metric, and let Q be the set of rational numbers in R. For every $x \in R$, any open sphere $S(x; r)$ contains a point in Q. Thus, every point in R is an adherent point of Q; i.e., $R \subset \bar{Q}$. Since $\bar{Q} \subset R$, it follows that $R = \bar{Q}$. Clearly, there are no isolated points in Q. Also, for any $x \in R$, every sphere $S(x; r)$ contains an infinite number of points in Q. Therefore, every point in R is a limit point of Q; i.e., $R \subset Q'$. This implies that $Q' = R$. ■

Let us now consider the following basic results.

5.4.10. Theorem. Let Y and Z be subsets of X, and let $\bar{Y}$ and $\bar{Z}$ denote the closures of Y and Z, respectively. Let $\bar{\bar{Y}}$ denote the closure of $\bar{Y}$, and let Y' be the derived set of Y. Then

(i) $Y \subset \bar{Y}$;
(ii) $\bar{\bar{Y}} = \bar{Y}$;
(iii) if $Y \subset Z$, then $\bar{Y} \subset \bar{Z}$;
(iv) $\overline{Y \cup Z} = \bar{Y} \cup \bar{Z}$;
(v) $\overline{Y \cap Z} \subset \bar{Y} \cap \bar{Z}$; and
(vi) $\bar{Y} = Y \cup Y'$.

Proof. To prove the first part, let $x \in Y$. Then $x \in S(x; r)$ for every $r > 0$. Hence, $x \in \bar{Y}$. Therefore, $Y \subset \bar{Y}$.

To prove the second part, let $x \in \bar{\bar{Y}}$, and let $r > 0$. Then there is an $x_1 \in \bar{Y}$ such that $x_1 \in S(x; r)$, and hence $\rho(x, x_1) = r_1 < r$. Let $r_0 = r - r_1 > 0$. We now wish to show that $S(x_1; r_0) \subset S(x; r)$. In doing so, let $y \in S(x_1; r_0)$. Then $\rho(y, x_1) < r_0$. By the triangle inequality we have $\rho(x, y) \leq \rho(x, x_1) + \rho(x_1, y) < r_1 + (r - r_1) = r$, and hence $y \in S(x; r)$. Since $x_1 \in \bar{Y}$, the sphere $S(x_1; r_0)$ contains a point $x_2 \in Y$. Thus, $x_2 \in S(x; r)$. Since $S(x; r)$ is an arbitrary spherical neighborhood of x, we have $x \in \bar{Y}$. This proves that $\bar{\bar{Y}} \subset \bar{Y}$. Also, in view of part (i), we have $\bar{Y} \subset \bar{\bar{Y}}$. Therefore, it follows that $\bar{\bar{Y}} = \bar{Y}$.

To prove the third part of the theorem, let $r > 0$ and let $x \in \bar{Y}$. Then there is a $y \in Y$ such that $y \in S(x; r)$. Since $Y \subset Z$, $y \in Z$ and thus x is an adherent point of Z.

To prove the fourth part, note that $Y \subset Y \cup Z$ and $Z \subset Y \cup Z$. From part (iii) it now follows that $\bar{Y} \subset \overline{Y \cup Z}$ and $\bar{Z} \subset \overline{Y \cup Z}$. Thus, $\bar{Y} \cup \bar{Z} \subset \overline{Y \cup Z}$. To show that $\overline{Y \cup Z} \subset \bar{Y} \cup \bar{Z}$, let $x \in \overline{Y \cup Z}$ and suppose that $x \notin \bar{Y} \cup \bar{Z}$. Then there exist spheres $S(x; r_1)$ and $S(x; r_2)$ such that $S(x; r_1) \cap Y = \varnothing$ and $S(x; r_2) \cap Z = \varnothing$. Let $r = \min\{r_1, r_2\}$. Then $S(x; r) \cap [Y \cup Z] = \varnothing$. But this is impossible since $x \in \overline{Y \cup Z}$. Hence, $x \in \bar{Y} \cup \bar{Z}$, and thus $\overline{Y \cup Z} \subset \bar{Y} \cup \bar{Z}$.

The proof of the remainder of the theorem is left as an exercise. ■

5.4.11. Exercise. Prove parts (v) and (vi) of Theorem 5.4.10.

We can further classify points and subsets of metric spaces.

5.4.12. Definition. Let Y be a subset of X and let $Y^{\sim}$ denote the complement of Y. A point $x \in X$ is called an **interior point** of the set Y if there exists a sphere $S(x; r)$ such that $S(x; r) \subset Y$. The set of all interior points of set Y is called the **interior of** Y and is denoted by Y°. A point $x \in X$ is an **exterior point** of Y if it is an interior point of the complement of Y. The **exterior of** Y is the set of all exterior points of set Y. The set of all points $x \in X$ such that $x \in \bar{Y} \cap \overline{(Y^{\sim})}$ is called the **frontier** of set Y. The **boundary** of a set Y is the set of all points in the frontier of Y which belong to Y.

5.4.13. Example. Let $\{R; \rho\}$ be the real line with the usual metric, and let $Y = \{y \in R: 0 < y \leq 1\} = (0, 1]$. The interior of Y is the set $(0, 1) = \{y \in R: 0 < y < 1\}$. The exterior of Y is the set $(-\infty, 0) \cup (1, +\infty)$, $\bar{Y} = \{y \in R: 0 \leq y \leq 1\} = [0, 1]$ and $\overline{Y^{\sim}} = (-\infty, 0] \cup [1, +\infty)$. Thus, the frontier of Y is the set $\{0, 1\}$, and the boundary of Y is the singleton $\{1\}$. ■

We now introduce the following important concepts.

5.4.14. Definition. A subset Y of X is said to be an **open subset** of X if every point of Y is an interior point of Y; i.e., $Y = Y^{\circ}$. A subset Z of X is said to be a **closed subset** of X if $Z = \bar{Z}$.

When there is no room for confusion, we usually call Y an **open set** and Z a **closed set**. On occasions when we want to be very explicit, we will say that Y is **open relative to** $\{X; \rho\}$ or **with respect to** $\{X; \rho\}$.

In our next result we establish some of the important properties of open sets.

5.4.15. Theorem.

(i) X and $\varnothing$ are open sets.

(ii) If $\{Y_\alpha\}_{\alpha \in A}$ is an arbitrary family of open subsets of X, then $\bigcup_{\alpha \in A} Y_\alpha$ is an open set.

(iii) The intersection of a *finite* number of open sets of X is open.

Proof. To prove the first part, note that for every $x \in X$, any sphere $S(x; r) \subset X$. Hence, every point in X is an interior point. Thus, X is open. Also, observe that $\varnothing$ has no points and therefore every point of $\varnothing$ is an interior point of $\varnothing$. Hence, $\varnothing$ is an open subset of X.

To prove the second part, let $\{Y_\alpha\}_{\alpha \in A}$ be a family of open sets in X, and let $Y = \bigcup_{\alpha \in A} Y_\alpha$. If Y_α is empty for every $\alpha \in A$, then $Y = \varnothing$ is an open subset of X. Now suppose that $Y \neq \varnothing$ and let $x \in Y$. Then $x \in Y_\alpha$ for some $\alpha \in A$. Since Y_α is an open set, there is a sphere $S(x; r)$ such that $S(x; r) \subset Y_\alpha$. Hence, $S(x; r) \subset Y$, and thus x is an interior point of Y. Therefore, Y is an open set.

To prove the third part, let Y_1 and Y_2 be open subsets of X. If $Y_1 \cap Y_2 = \varnothing$, then $Y_1 \cap Y_2$ is open. So let us assume that $Y_1 \cap Y_2 \neq \varnothing$, and let

$x \in Y = Y_1 \cap Y_2$. Since $x \in Y_1$, there is an $r_1 > 0$ such that $x \in S(x; r_1) \subset Y_1$. Similarly, there is an $r_2 > 0$ such that $x \in S(x; r_2) \subset Y_2$. Let $r = \min\{r_1, r_2\}$. Then $x \in S(x; r)$, where $S(x; r) \subset S(x; r_1)$ and $S(x; r) \subset S(x; r_2)$. Thus, $S(x; r) \subset Y_1 \cap Y_2$, and x is an interior point of $Y_1 \cap Y_2$. Hence, $Y_1 \cap Y_2$ is an open subset of X. By induction, we can show that the intersection of any finite number of open subsets of X is open. ■

We now make the following

5.4.16. Definition. Let $\{X; \rho\}$ be a metric space. The **topology of X determined by ρ** is defined to be the family of all open subsets of X.

In our next result we establish a connection between open and closed subsets of X.

5.4.17. Theorem.

(i) X and $\varnothing$ are closed sets.
(ii) If Y is an open subset of X, then $Y^{\sim}$ is closed.
(iii) If Z is a closed subset of X, then $Z^{\sim}$ is open.

Proof. The first part of this theorem follows immediately from the definitions of X, $\varnothing$, and closed set.

To prove the second part, let Y be any open subset of X. We may assume that $Y \neq \varnothing$ and $Y \neq X$. Let x be any adherent point of $Y^{\sim}$. Then x cannot belong to Y, for if it did, then there would exist a sphere $S(x; r) \subset Y$, which is impossible. Therefore, every adherent point of $Y^{\sim}$ belongs to $Y^{\sim}$, and thus $Y^{\sim}$ is closed if Y is open.

To prove the third part, let Z be any closed subset of X. Again, we may assume that $Z \neq \varnothing$ and $Z \neq X$. Let $x \in Z^{\sim}$. Then there exists a sphere $S(x; r)$ which contains no point of Z. This is so because if every such sphere would contain a point of Z, then x would be an adherent point of Z and consequently would belong to Z, since Z is closed. Thus, there is a sphere $S(x; r) \subset Z^{\sim}$; i.e., x is an interior point of $Z^{\sim}$. Since this holds for arbitrary $x \in Z^{\sim}$, $Z^{\sim}$ is an open set. ■

In the next result we present additional important properties of open sets.

5.4.18. Theorem.

(i) Every open sphere in X is an open set.
(ii) If Y is an open subset of X, then there is a family of open spheres, $\{S_\alpha\}_{\alpha \in A}$, such that $Y = \bigcup_{\alpha \in A} S_\alpha$.
(iii) The interior of any subset Y of X is the largest open set contained in Y.

Proof. To prove the first part, let $S(x; r)$ be any open sphere in X. Let $x_1 \in S(x; r)$, and let $\rho(x, x_1) = r_1$. If we let $r_0 = r - r_1$, then according to the proof of part (ii) of Theorem 5.4.10 we have $S(x_1; r_0) \subset S(x; r)$. Hence, x_1 is an interior point of $S(x; r)$. Since this is true for any $x_1 \in S(x; r)$, it follows that $S(x; r)$ is an open subset of X.

To prove the second part of the theorem, we first note that if $Y = \varnothing$, then Y is open and is the union of an empty family of spheres. So assume that $Y \neq \varnothing$ and that Y is open. Then each point $x \in Y$ is the center of a sphere $S(x; r) \subset Y$, and moreover Y is the union of the family of all such spheres.

The proof of the last part of the theorem is left as an exercise. ■

5.4.19. Exercise. Prove part (iii) of Theorem 5.4.18.

Let $\{Y; \rho\}$ be a subspace of a metric space $\{X; \rho\}$, and suppose that V is a subset of Y. It can happen that V may be an open subset of Y and at the same time not be an open subset of X. Thus, when a set is described as open, it is important to know in what space it is open. We have:

5.4.20. Theorem. Let $\{Y; \rho\}$ be a metric subspace of $\{X; \rho\}$.

(i) A subset $V \subset Y$ is open relative to $\{Y; \rho\}$ if and only if there is a subset $U \subset X$ such that U is open relative to $\{X; \rho\}$ and $V = Y \cap U$.

(ii) A subset $G \subset Y$ is closed relative to $\{Y; \rho\}$ if and only if there is a subset F of X such that F is closed relative to $\{X; \rho\}$ and $G = F \cap Y$.

Proof. Let $S(x_0; r) = \{x \in X: \rho(x, x_0) < r\}$ and $S'(x_0; r) = \{x \in Y: \rho(x, x_0) < r\}$. Then $S'(x_0; r) = Y \cap S(x_0; r)$.

To prove the necessity of part (i), let V be an open set relative to $\{Y; \rho\}$, and let $x \in V$. Then there is a sphere $S'(x; r) \subset V$ (r may depend on x). Now

$$V = \bigcup_{x \in V} S'(x; r) = \bigcup_{x \in V} S(x; r) \cap Y.$$

By part (ii) of Theorem 5.4.15, $\bigcup_{x \in V} S(x; r) = U$ is an open set in $\{X; \rho\}$.

To prove the sufficiency of part (i), let $V = Y \cap U$, where U is an open subset of X. Let $x \in V$. Then $x \in U$, and hence there is a sphere $S(x; r) \subset U$. Thus, $S'(x; r) = Y \cap S(x; r) \subset Y \cap U = V$. This proves that x is an interior point of V and that V is an open subset of Y.

The proof of part (ii) of the theorem is left as an exercise. ■

5.4.21. Exercise. Prove part (ii) of Theorem 5.4.20.

The first part of the preceding theorem may be stated in another equivalent way. Let $\mathfrak{I}$ and $\mathfrak{I}'$ be the topology of $\{X; \rho\}$ and $\{Y; \rho\}$, respectively, generated by ρ. Then $\mathfrak{I}' = \{Y \cap U: U \in \mathfrak{I}\}$.

Let us now consider some specific examples.

5.4.22. Example. Let $X = R$, and let ρ be the usual metric on R; i.e., $\rho(x, y) = |x - y|$. Any set $Y = (a, b) = \{x: a < x < b\}$ is an open subset of X. We call (a, b) an **open interval** on R. ■

5.4.23. Example. We now show that the word "finite" is crucial in part (iii) of Theorem 5.4.15. Let $\{R; \rho\}$ denote again the real line with the usual metric, and let $a < b$. If $Y_n = \{x \in R: a < x < b + 1/n\}$, then for each positive integer n, Y_n is an open subset of the real line. However, the set

$$\bigcap_{n=1}^{\infty} Y_n = \{x \in R: a < x \leq b\} = (a, b]$$

is not an open subset of R. (This can readily be verified, since every sphere $S(b; r)$ contains a point greater than b and hence is not in $\bigcap_{n=1}^{\infty} Y_n$.) ■

In the above example, let $Y = (a, b]$. We saw that Y is not an open subset of R; i.e., b is not an interior point of Y. However, if we were to consider $\{Y; \rho\}$ as a metric space *by itself*, then Y is an open set.

5.4.24. Example. Let $\{\mathcal{C}[a, b]; \rho_\infty\}$ denote the metric space of Example 5.3.14. Let λ be an arbitrary finite positive number. Then the set of continuous functions satisfying the condition $|x(t)| < \lambda$ for all $a \leq t \leq b$ is an open subset of the metric space $\{\mathcal{C}[a, b]; \rho_\infty\}$. ■

Theorems 5.4.15 and 5.4.17 tell us that the sets X and $\varnothing$ are both open and closed in any metric space. In some metric spaces there may be *proper subsets* of X which are both open and closed, as illustrated in the following example.

5.4.25. Example. Let X be the set of real numbers given by $X = (-2, -1) \cup (+1, +2)$, and let $\rho(x, y) = |x - y|$ for $x, y \in X$. Then $\{X; \rho\}$ is clearly a metric space. Let $Y = (-2, -1) \subset X$ and $Z = (+1, +2) \subset X$. Note that both Y and Z are open subsets of X. However, $Y^\sim = Z$, $Z^\sim = Y$, and thus Y and Z are also closed subsets of X. Therefore, Y and Z are proper subsets of the metric space $\{X; \rho\}$ which are both open and closed. (Note that in the preceding we are not viewing X as a subset of R. As such X would be open. Considering $\{X; \rho\}$ as our metric space, X is both open and closed.) ■

5.4.26. Exercise. Let $\{X; \rho\}$ be a metric space with ρ the discrete metric defined in Example 5.1.7. Show that every subset of X is both open and closed.

In our next result we summarize several important properties of closed sets.

5.4.27. Theorem.

(i) Every subset of X consisting of a finite number of elements is closed.

(ii) Let $x_0 \in X$, let $r > 0$, and let $K(x_0; r) = \{x \in X: \rho(x, x_0) \leq r\}$. Then $K(x_0; r)$ is closed.

(iii) A subset $Y \subset X$ is closed if and only if $\bar{Y} \subset Y$.

(iv) A subset $Y \subset X$ is closed if and only if $Y' \subset Y$.

(v) Let $\{Y_\alpha\}_{\alpha \in A}$ be any family of closed sets in X. Then $\bigcap_{\alpha \in A} Y_\alpha$ is closed.

(vi) The union of a *finite* number of closed sets in X is closed.

(vii) The closure of a subset Y of X is the intersection of all closed sets containing Y.

Proof. Only the proof of part (v) is given. Let $\{Y_\alpha\}_{\alpha \in A}$ be any family of closed subsets of X. Then $\{Y_\alpha^\sim\}_{\alpha \in A}$ is a family of open sets. Now $(\bigcap_{\alpha \in A} Y_\alpha)^\sim = \bigcup_{\alpha \in A} Y_\alpha^\sim$ is an open set, and hence $\bigcap_{\alpha \in A} Y_\alpha$ is a closed subset of X. ■

5.4.28. Exercise. Prove parts (i) to (iv), (vi), and (vii) of Theorem 5.4.27.

We now consider several specific examples of closed sets.

5.4.29. Example. Let $X = R$, and let ρ be the usual metric, $\rho(x, y) = |x - y|$. Any set $Y = \{x \in R: a \leq x \leq b\}$, where $a \leq b$ is a closed subset of R. We call Y a **closed interval** on R and denote it by $[a, b]$. ■

5.4.30. Example. We now show that the word "finite" is essential in part (vi) of Theorem 5.4.27. Let $\{R; \rho\}$ denote the real line with the usual metric, and let $a > 0$. If $Y_n = \{x \in R: 1/n \leq x \leq a\}$ for each positive integer n, then Y_n is a closed subset of the real line. However, the set

$$\bigcup_{n=1}^{\infty} Y_n = \{x \in R: 0 < x \leq a\} = (0, a]$$

is not a closed subset of the real line, as can readily be verified since 0 is an adherent point of $(0, a]$. ■

5.4.31. Exercise. The set $K(x_0; r)$ defined in part (ii) of Theorem 5.4.27 is sometimes called a **closed sphere**. It need not coincide with $\bar{S}(x_0; r)$, i.e., the closure of the open sphere $S(x_0; r)$.

(i) Show that $\bar{S}(x_0; r) \subset K(x_0; r)$.

(ii) Let $\{X; \rho\}$ be the discrete metric space defined in Example 5.1.7.

Describe the sets $S(x; 1)$, $\bar{S}(x; 1)$, and $K(x; 1)$ for any $x \in X$ and conclude that, in general, $\bar{S}(x; 1) \neq K(x; 1)$ if X contains more than one point.

(iii) Let $X = (-\infty, 0) \cup J$, where J denotes the set of positive integers, and let $\rho(x, y) = |x - y|$. Describe $S(0; 1)$, $\bar{S}(0; 1)$, and $K(0; 1)$ and conclude that $\bar{S}(0; 1) \neq K(0; 1)$.

We are now in a position to introduce certain additional concepts which are important in analysis and applications.

5.4.32. Definition. Let Y and Z be subsets of X. The set Y is said to be **dense** in Z (or dense with respect to Z) if $\bar{Y} \supset Z$. The set Y is said to be **everywhere dense** in $\{X; \rho\}$ (or simply, everywhere dense in X) if $\bar{Y} = X$. If the exterior of Y is everywhere dense in X, then Y is said to be **nowhere dense** in X. A subset Y of X is said to be **dense-in-itself** if every point of Y is a limit point of Y. A subset Y of X which is both closed and dense-in-itself is called a **perfect set**.

5.4.33. Definition. A metric space $\{X; \rho\}$ is said to be **separable** if there is a countable subset Y in X which is everywhere dense in X.

The following result enables us to characterize separable metric spaces in an equivalent way. We have:

5.4.34. Theorem. A metric space $\{X; \rho\}$ is separable if and only if there is a countable set $S = \{x_1, x_2, \ldots\} \subset X$ such that for every $x \in X$, for given $\epsilon > 0$ there is an $x_n \in S$ such that $\rho(x, x_n) < \epsilon$.

5.4.35. Exercise. Prove Theorem 5.4.34.

Let us now consider some specific cases.

5.4.36. Example. The real line with the usual metric is a separable space. As we saw in Example 5.4.9, if Q is the set of rational numbers, then $\bar{Q} = R$. ■

5.4.37. Example. Let $\{R^n; \rho_p\}$ be the metric space defined in Example 5.3.1 (recall that $1 \leq p < \infty$). The set of vectors $x = (\xi_1, \ldots, \xi_n)$ with rational coordinates (i.e., ξ_i is a rational real number, $i = 1, \ldots, n$) is a denumerable everywhere dense set in R^n and, therefore, $\{R^n; \rho_p\}$ is a separable metric space. ■

5.4.38. Example. Let $\{l_p; \rho_p\}$ be the metric space defined in Example 5.3.5 (recall that $1 \leq p < \infty$). We can show that this space is separable in the following manner. Let

$$Y = \{y_n \in l_p : y_n = (\eta_1, \ldots, \eta_n, 0, 0, \ldots) \text{ for some } n,$$
$$\text{where } \eta_i \text{ is a rational real number, } i = 1, \ldots, n\}.$$

Then Y is a countable subset of l_p. To show that it is everywhere dense, let $\epsilon > 0$ and let $x \in l_p$, where $x = (\xi_1, \xi_2, \ldots)$. Choose n sufficiently large so that

$$\sum_{k=n+1}^{\infty} |\xi_k|^p < \frac{\epsilon^p}{2}.$$

We can now find a $y_n \in Y$ such that

$$\sum_{k=1}^{n} |\xi_k - \eta_k|^p < \frac{\epsilon^p}{2}.$$

Hence,

$$[\rho_p(x, y_n)]^p = \sum_{k=1}^{n} |\xi_k - \eta_k|^p + \sum_{k=n+1}^{\infty} |\xi_k|^p < \epsilon^p;$$

i.e., $\rho_p(x, y_n) < \epsilon$. By Theorem 5.4.34, $\{l_p; \rho_p\}$ is separable. ■

In order to establish the separability of the space of continuous functions, it is necessary to use the **Weierstrass approximation theorem**, which we state without proof.

5.4.39. Theorem. Let $\mathcal{C}[a, b]$ be the space of real continuous functions on the interval $[a, b]$, and let $\mathcal{P}(t)$ be the family of all polynomials (defined on $[a, b]$). Let $\epsilon > 0$, and let $x \in \mathcal{C}[a, b]$. Then there is a $p \in \mathcal{P}(t)$ such that $\sup_{a \le t \le b} |x(t) - p(t)| < \epsilon$.

5.4.40. Exercise. Using the Weierstrass approximation theorem, show that the metric spaces $\{\mathcal{C}[a, b]; \rho_p\}$, defined in Example 5.3.12, and $\{\mathcal{C}[a, b]; \rho_\infty\}$, defined in Example 5.3.14, are separable.

5.4.41. Exercise. Show that the metric space $\{X; \rho\}$, where ρ is the discrete metric defined in Example 5.1.7, is separable if and only if X is a countable set.

We conclude the present section by considering an example of a metric space which is not separable.

5.4.42. Example. Let $\{l_\infty; \rho_\infty\}$ be the metric space defined in Example 5.3.8. Let $Y \subset R^\infty$ denote the set

$$Y = \{y \in R^\infty : y = (\eta_1, \eta_2, \ldots), \text{ where } \eta_i = 0 \text{ or } 1\}.$$

Clearly then $Y \subset l_\infty$. Now for every real number $\alpha \in [0, 1]$, there is a $y \in Y$ such that $\alpha = \sum_{n=1}^{\infty} \eta_n (\frac{1}{2})^n$, where $y = (\eta_1, \eta_2, \ldots)$. Thus, Y is an uncountable set. Notice now that for every $y_1, y_2 \in Y$, $\rho_\infty(y_1, y_2) = 0$ or 1. That is, ρ_∞ restricted to Y is the discrete metric. It follows from Exercise 5.4.41 that Y cannot be separable and, consequently, $\{l_\infty; \rho_\infty\}$ is not separable. ■

5.5. COMPLETE METRIC SPACES

The set of real numbers R with the usual metric ρ defined on it has many remarkable properties, several of which are attributable to the so-called "completeness property" of this space. For this reason we speak of $\{R; \rho\}$ as being a complete metric space. In the present section we consider general complete metric spaces.

Throughout this section $\{X; \rho\}$ is our underlying metric space, and J denotes the set of positive integers. Before considering the completeness of metric spaces we need to consider a few facts about sequences on metric spaces (cf. Definition 1.1.25).

5.5.1. Definition. A **sequence** $\{x_n\}$ in a set $Y \subset X$ is a function $f: J \longrightarrow Y$. Thus, if $\{x_n\}$ is a sequence in Y, then $f(n) = x_n$ for each $n \in J$.

5.5.2. Definition. Let $\{x_n\}$ be a sequence of points in X, and let x be a point of X. The sequence $\{x_n\}$ is said to **converge** to x if for every $\epsilon > 0$, there is an integer N such that for all $n \geq N$, $\rho(x, x_n) < \epsilon$ (i.e., $x_n \in S(x; \epsilon)$ for all $n \geq N$). In general, N depends on ϵ; i.e., $N = N(\epsilon)$. We call x the **limit** of $\{x_n\}$, and we usually write

$$\lim_n x_n = x,$$

or $x_n \longrightarrow x$ as $n \longrightarrow \infty$. If there is no $x \in X$ to which the sequence converges, then we say that $\{x_n\}$ **diverges**.

Thus, $x_n \longrightarrow x$ if and only if the sequence of real numbers $\{\rho(x_n, x)\}$ converges to zero. In view of the above definition we note that for every $\epsilon > 0$ there is a *finite* number N such that all terms of $\{x_n\}$ except the first $(N - 1)$ terms must lie in the sphere with center x and radius ϵ. Hence, the convergence of a sequence depends on the infinite number of terms $\{x_{N+1}, x_{N+2}, \ldots\}$, and no amount of alteration of a finite number of terms of a divergent sequence can make it converge. Moreover, if a convergent sequence is changed by omitting or adding a finite number of terms, then the resulting sequence is still convergent to the same limit as the original sequence.

Note that in Definition 5.5.2 we called x *the* limit of the sequence $\{x_n\}$. We will show that if $\{x_n\}$ has a limit in X, then that limit is unique.

5.5.3. Definition. Let $\{x_n\}$ be a sequence of points in X, where $f(n) \triangleq x_n$ for each $n \in J$. If the range of f is bounded, then $\{x_n\}$ is said to be a **bounded sequence**.

The range of f in the above definition may consist of a finite number of points or of an infinite number of points. Specifically, if the range of f

consists of one point, then we speak of a **constant sequence**. Clearly, all constant sequences are convergent.

5.5.4. Example. Let $\{R; \rho\}$ denote the set of real numbers with the usual metric. If $n \in J$, then the sequence $\{n^2\}$ diverges and is unbounded, and the range of this sequence is an infinite set. The sequence $\{(-1)^n\}$ diverges, is bounded, and its range is a finite set. The sequence $\left\{a + \frac{(-1)^n}{n}\right\}$ converges to a, is bounded, and its range is an infinite set. ■

5.5.5. Definition. Let $\{x_n\}$ be a sequence in X. Let $n_1, n_2, \ldots, n_k, \ldots$ be a sequence of positive integers which is strictly increasing; i.e., $n_j > n_k$ for all $j > k$. Then the sequence $\{x_{n_k}\}$ is called a **subsequence** of $\{x_n\}$. If the subsequence $\{x_{n_k}\}$ converges, then its limit is called a **subsequential limit** of $\{x_n\}$.

It turns out that many of the important properties of convergence on R can be extended to the setting of arbitrary metric spaces. In the next result several of these properties are summarized.

5.5.6. Theorem. Let $\{x_n\}$ be a sequence in X. Then

(i) there is at most one point $x \in X$ such that $\lim_n x_n = x$;

(ii) if $\{x_n\}$ is convergent, then it is bounded;

(iii) $\{x_n\}$ converges to a point $x \in X$ if and only if every sphere about x contains all but a finite number of terms in $\{x_n\}$;

(iv) $\{x_n\}$ converges to a point $x \in X$ if and only if every subsequence of $\{x_n\}$ converges to x;

(v) if $\{x_n\}$ converges to $x \in X$ and if $y \in X$, then $\lim_n \rho(x_n, y) = \rho(x, y)$;

(vi) if $\{x_n\}$ converges to $x \in X$ and if the sequence $\{y_n\}$ of X converges to $y \in X$, then $\lim_n \rho(x_n, y_n) = \rho(x, y)$; and

(vii) if $\{x_n\}$ converges to $x \in X$, and if there is a $y \in X$ and a $\gamma > 0$ such that $\rho(x_n, y) \leq \gamma$ for all $n \in J$, then $\rho(x, y) \leq \gamma$.

Proof. To prove part (i), assume that $x, y \in X$ and that $\lim_n x_n = x$ and $\lim_n x_n = y$. Then for every $\epsilon > 0$ there are positive integers N_x and N_y such that $\rho(x_n, x) < \epsilon/2$ whenever $n \geq N_x$ and $\rho(x_n, y) < \epsilon/2$ whenever $n \geq N_y$. If we let $N = \max(N_x, N_y)$, then it follows that

$$\rho(x, y) \leq \rho(x_n, x) + \rho(x_n, y) < \frac{\epsilon}{2} + \frac{\epsilon}{2} = \epsilon.$$

Now ϵ is *any* positive number. Since the only non-negative number which is less than every positive number is zero, it follows that $\rho(x, y) = 0$ and therefore $x = y$.

To prove part (iii), assume that $\lim_n x_n = x$ and let $S(x; \epsilon)$ be any sphere about x. Then there is a positive integer N such that the only terms of the sequence $\{x_n\}$ which are possibly not in $S(x; \epsilon)$ are the terms $x_1, x_2, \ldots, x_{N-1}$. Conversely, assume that every sphere about x contains all but a finite number of terms from the sequence $\{x_n\}$. With $\epsilon > 0$ specified, let $M = \max\{n \in J: x_n \notin S(x; \epsilon)\}$. If we set $N = M + 1$, then $x_n \in S(x; \epsilon)$ for all $n \geq N$, which was to be shown.

To prove part (v), we note from Theorem 5.1.13 that

$$|\rho(y, x) - \rho(y, x_n)| \leq \rho(x, x_n).$$

By hypothesis, $\lim_n x_n = x$. Therefore, $\lim_n \rho(x, x_n) = 0$ and so $\lim_n |\rho(y, x) - \rho(y, x_n)| = 0$; i.e., $\lim_n \rho(y, x_n) = \rho(y, x)$.

Finally, to prove part (vii), suppose to the contrary that $\rho(x, y) > \gamma$. Then $\delta = \rho(x, y) - \gamma > 0$. Now $\gamma - \rho(x_n, y) \geq 0$ for all $n \in J$, and thus

$$0 < \delta \leq \rho(x, y) - \rho(x_n, y) \leq \rho(x, x_n)$$

for all $n \in J$. But this is impossible, since $\lim_n x_n = x$. Thus, $\rho(x, y) \leq \gamma$.

We leave the proofs of the remaining parts as an exercise. ■

5.5.7. Exercise. Prove parts (ii), (iv), and (vi) of Theorem 5.5.6.

In Definition 5.4.5, we introduced the concept of limit point of a set $Y \subset X$. In Definition 5.5.2, we defined the limit of a sequence of points, $\{x_n\}$, in X. These two concepts are closely related; however, the reader should carefully note the distinction between the two. The limit point of a set is strictly a property of the set itself. On the other hand, a sequence is not a set. Furthermore, the elements of a sequence are ordered and not necessarily distinct, while the elements of a set are not ordered but are distinct. However, the range of a sequence is a subset of X. We now give a result relating these concepts.

5.5.8. Theorem. Let Y be a subset of X. Then

(i) $x \in X$ is an adherent point of Y if and only if there is a sequence $\{y_n\}$ in Y (i.e., $y_n \in Y$ for all n) such that $\lim_n y_n = x$;

(ii) $x \in X$ is a limit point of the set Y if and only if there is a sequence $\{y_n\}$ of distinct points in Y such that $\lim_n y_n = x$; and

(iii) Y is closed if and only if for every convergent sequence $\{y_n\}$, such that $y_n \in Y$ for all n, $\lim_n y_n = x \in Y$.

Proof. To prove part (i), assume that $\lim_n y_n = x$. Then every sphere about x contains at least one term of the sequence $\{y_n\}$ and, since every term of $\{y_n\}$ is a point of Y, it follows that x is an adherent point of Y. Conversely, assume that x is an adherent point of Y. Then every sphere about x contains at least one point of Y. Now let us choose for each positive integer n a point $y_n \in Y$ such that $y_n \in S(x; 1/n)$. Then it follows readily that the sequence $\{y_n\}$ chosen in this fashion converges to x. Specifically, if $\epsilon > 0$ is given, then we choose a positive integer N such that $1/N < \epsilon$. Then for every $n \geq N$ we have $y_n \in S(x; 1/n) \subset S(x; \epsilon)$. This concludes the proof of part (i).

To prove part (ii), assume that x is a limit point of the set Y. Then every sphere $S(x; 1/n)$ contains an infinite number of points, and so we can choose a $y_n \in S(x; 1/n)$ such that $y_n \neq y_m$ for all $m < n$. The sequence $\{y_n\}$ consists of distinct points and converges to x. Conversely, if $\{y_n\}$ is a sequence of distinct points convergent to x and if $S(x; \epsilon)$ is any sphere with center at x, then by definition of convergence there is an N such that for all $n \geq N$, $y_n \in S(x; \epsilon)$. That is, there are infinitely many points of Y in $S(x; \epsilon)$.

To prove part (iii), assume that Y is closed and let $\{y_n\}$ be a convergent sequence with $y_n \in Y$ for all n and $\lim_n y_n = x$. We want to show that $x \in Y$. By part (i), x must be an adherent point of Y. Since Y is closed, $x \in Y$. Next, we prove the converse. Let x be an adherent point of Y. Then by part (i), there is a sequence $\{y_n\}$ in Y such that $\lim_n y_n = x$. By hypothesis, we must have $x \in Y$. Since Y contains all of its adherent points, it must be closed. ■

Statement (iii) of Theorem 5.5.8 is often used as an alternate way of defining a closed set.

The next theorem provides us with conditions under which a sequence is convergent in a product metric space.

5.5.9. Theorem. Let $\{X; \rho_x\}$ and $\{Y; \rho_y\}$ be two metric spaces, let $Z = X \times Y$, let ρ be any of the metrics defined on Z in Theorem 5.3.19, and let $\{Z; \rho\}$ denote the product metric space of $\{X; \rho_x\}$ and $\{Y; \rho_y\}$. If $z \in Z = X \times Y$, then $z = (x, y)$, where $x \in X$ and $y \in Y$. Let $\{x_n\}$ be a sequence in X, and let $\{y_n\}$ be a sequence in Y. Then,

(i) the sequence $\{(x_n, y_n)\}$ converges in Z if and only if $\{x_n\}$ converges in X and $\{y_n\}$ converges in Y; and

(ii) $\lim_n (x_n, y_n) = (\lim_n x_n, \lim_n y_n)$ whenever this limit exists.

5.5.10. Exercise. Prove Theorem 5.5.9.

In many situations the limit to which a given sequence may converge is unknown. The following concept enables us to consider the convergence of a sequence without knowing the limit to which the sequence may converge.

5.5.11. Definition. A sequence $\{x_n\}$ of points in a metric space $\{X; \rho\}$ is said to be a **Cauchy sequence** or a **fundamental sequence** if for every $\epsilon > 0$ there is an integer N such that $\rho(x_n, x_m) < \epsilon$ whenever $m, n \geq N$.

The next result follows directly from the triangle inequality.

5.5.12. Theorem. Every convergent sequence in a metric space $\{X; \rho\}$ is a Cauchy sequence.

Proof. Assume that $\lim_n x_n = x$. Then for arbitrary $\epsilon > 0$ we can find an integer N such that $\rho(x_n, x) < \epsilon/2$ and $\rho(x_m, x) < \epsilon/2$ whenever $m, n \geq N$. In view of the triangle inequality we now have

$$\rho(x_n, x_m) \leq \rho(x_n, x) + \rho(x_m, x) < \epsilon$$

whenever $m, n \geq N$. This proves the theorem. ■

We emphasize that in an arbitrary metric space $\{X; \rho\}$ a Cauchy sequence is not necessarily convergent.

5.5.13. Theorem. Let $\{x_n\}$ be a Cauchy sequence. Then $\{x_n\}$ is a bounded sequence.

Proof. We need to show that there is a constant γ such that $0 < \gamma < \infty$ and such that $\rho(x_m, x_n) < \gamma$ for all $m, n \in J$.

Letting $\epsilon = 1$, we can find N such that $\rho(x_m, x_n) < 1$ whenever $m, n \geq N$. Now let $\lambda = \max\{\rho(x_1, x_2), \rho(x_1, x_3), \ldots, \rho(x_1, x_N)\}$. Then, by the triangle inequality,

$$\rho(x_1, x_n) \leq \rho(x_1, x_N) + \rho(x_N, x_n) \leq (\lambda + 1)$$

if $n \geq N$. Thus, for all $n \in J$, $\rho(x_1, x_n) \leq \lambda + 1$. Again, by the triangle inequality,

$$\rho(x_m, x_n) \leq \rho(x_m, x_1) + \rho(x_1, x_n)$$

for all $m, n \in J$. Thus, $\rho(x_m, x_n) \leq 2(\lambda + 1)$ and $\{x_n\}$ is a bounded sequence. ■

We also have:

5.5.14. Theorem. If a Cauchy sequence $\{x_n\}$ contains a convergent subsequence $\{x_{n_k}\}$, then the sequence $\{x_n\}$ is convergent.

5.5.15. Exercise. Prove Theorem 5.5.14.

We now give the definition of complete metric space.

5.5.16. Definition. If every Cauchy sequence in a metric space $\{X; \rho\}$ converges to an element in X, then $\{X; \rho\}$ is said to be a **complete metric space**.

Complete metric spaces are of utmost importance in analysis and applications. We will have occasion to make extensive use of the properties of such spaces in the remainder of this book.

5.5.17. Example. Let $X = (0, 1)$, and let $\rho(x, y) = |x - y|$ for all $x, y \in X$. Let $x_n = 1/n$ for $n \in J$. Then the sequence $\{x_n\}$ is Cauchy (i.e., it is a Cauchy sequence), since $|x_n - x_m| < 1/N$ for all $n \geq m \geq N$. Since there is no $x \in X$ to which $\{x_n\}$ converges, the metric space $\{X; \rho\}$ is not complete. ■

5.5.18. Example. Let $X = Q$, the set of rational numbers, and let $\rho(x, y) = |x - y|$. Let $x_n = 1 + \frac{1}{2!} + \ldots + \frac{1}{n!}$ for $n \in J$. The sequence $\{x_n\}$ is Cauchy. Since there is no limit in Q to which $\{x_n\}$ converges, the metric space $\{Q; \rho\}$ is not complete. ■

5.5.19. Example. Let $R^\# = R - \{0\}$, and let $\rho(x, y) = |x - y|$ for all $x, y \in R^\#$. Let $x_n = 1/n$, $n \in J$. The sequence $\{x_n\}$ is Cauchy; however, it does not converge to a limit in $R^\#$. Thus, $\{R^\#; \rho\}$ is not complete. Some further comments are in order here. If we view $R^\#$ as a subset of R in the metric space $\{R; \rho\}$ (ρ denotes the usual metric on R), then the sequence $\{x_n\}$ converges to zero; i.e., $\lim_n x_n = 0$. By Theorem 5.5.8, $R^\#$ cannot be a closed subset of R. However, $R^\#$ is a closed subset of the metric space $\{R^\#; \rho\}$, since it is the whole space. There is no contradiction here to Theorem 5.5.8, for the sequence $\{x_n\}$ does not converge to a limit in $R^\#$. Specifically, Theorem 5.5.8 states that *if* a sequence does converge to a limit, then the limit must belong to the space. The requirement for completeness is that *every* Cauchy sequence must converge to an element in the space. ■

We now consider several specific examples of important complete metric spaces.

5.5.20. Example. Let ρ denote the usual metric on R, the set of real numbers. The completeness of $\{R; \rho\}$ is one of the fundamental results of analysis. ■

5.5.21. Example. Let $\{X; \rho_x\}$ and $\{Y; \rho_y\}$ be arbitrary complete metric spaces. If $Z = X \times Y$ and if $z \in Z$, then $z = (x, y)$, where $x \in X$ and $y \in Y$ (see Theorem 5.3.19). Define

$$\begin{aligned} \rho_2(z_1, z_2) &= \rho_2((x_1, y_1), (x_2, y_2)) \\ &= \sqrt{[\rho_x(x_1, x_2)]^2 + [\rho_y(y_1, y_2)]^2}. \end{aligned}$$

It can readily be shown that the metric space $\{Z; \rho_2\}$ is complete. ■

5.5.22. Exercise. Verify the completeness of $\{Z; \rho_2\}$ in the above example.

5.5.23. Example. Let ρ be the usual metric defined on C, the set of complex numbers. Utilizing Example 5.5.21 along with the completeness of $\{R; \rho\}$ (see Example 5.5.20), we can readily show that $\{C; \rho\}$ is a complete metric space. ■

5.5.24. Exercise. Verify the completeness of $\{C; \rho\}$.

5.5.25. Exercise. Let $X = R^n$ (let $X = C^n$) denote the set of all real (of all complex) ordered n-tuples $x = (\xi_1, \ldots, \xi_n)$. Let $y = (\eta_1, \ldots, \eta_n)$, let

$$\rho_p(x, y) = \left[\sum_{j=1}^{n} |\xi_j - \eta_j|^p\right]^{1/p}, \quad 1 \leq p < \infty,$$

and let

$$\rho_\infty(x, y) = \max\{|\xi_1 - \eta_1|, \ldots, |\xi_n - \eta_n|\}, \quad \text{i.e., } p = \infty.$$

Utilizing the completeness of the real line (of the complex plane), show that $\{R^n; \rho_p\} = R^n_p(\{C^n; \rho_p\} = C^n_p)$ is a complete metric space for $1 \leq p \leq \infty$. In particular, show that if $\{x_k\}$ is a Cauchy sequence in R^n_p (in C^n_p), where $x_k = (\xi_{1k}, \ldots, \xi_{nk})$, then $\{\xi_{jk}\}$ is a Cauchy sequence in R (in C) for $j = 1, \ldots, n$, and $\{x_k\}$ converges to x, where $x = (\xi_1, \ldots, \xi_n)$ and $\xi_j = \lim_k \xi_{jk}$ for $j = 1, \ldots, n$.

5.5.26. Example. Let $\{l_p; \rho_p\}$ be the metric space defined in Example 5.3.5. We now show that this space is a complete metric space.

Let $\{x_k\}$ be a Cauchy sequence in l_p, where $x_k = (\xi_{1k}, \xi_{2k}, \ldots, \xi_{nk}, \ldots)$. Let $\epsilon > 0$. Then there is an $N \in J$ such that

$$\rho_p(x_j, x_k) = \left[\sum_{m=1}^{\infty} |\xi_{mj} - \xi_{mk}|^p\right]^{1/p} \leq \epsilon$$

for all $k, j \geq N$. This implies that $|\xi_{mj} - \xi_{mk}| \leq \epsilon$ for every $m \in J$ and all $k, j \geq N$. Thus, $\{\xi_{mk}\}$ is a Cauchy sequence in R for every $m \in J$, and hence $\{\xi_{mk}\}$ is convergent to some limit, say $\lim_k \xi_{mk} = \xi_m$ for $m \in J$. Now let $x = (\xi_1, \xi_2, \ldots, \xi_n, \ldots)$. We want to show that (i) $x \in l_p$ and (ii) $\lim_k x_k = x$. Since $\{x_k\}$ is a Cauchy sequence, we know by Theorem 5.5.13 that there exists a $\gamma > 0$ such that

$$\rho_p(0, x_k) = \left[\sum_{m=1}^{\infty} |\xi_{mk}|^p\right]^{1/p} \leq \gamma$$

for all $k \in J$. Now let n be any positive integer, let ρ'_p be the metric on R^n defined in Exercise 5.5.25, and let $x'_k = \{\xi_{1k}, \ldots, \xi_{nk}\}$. Then $\rho'_p(x'_k, x'_j) \leq \rho_p(x_k, x_j)$, and thus $\{x'_k\}$ is a Cauchy sequence in R^n_p. It also follows that $\rho'_p(0, x'_k) \leq \gamma$ for all $k \in J$. Now by Exercise 5.5.25, $\{x'_k\}$ converges to x', where $x' = (\xi_1, \ldots, \xi_n)$. It follows from Theorem 5.5.6, part (vii), that $\rho'_p(0, x') \leq \gamma$; i.e., $\left[\sum_{i=1}^{n} |\xi_i|^p\right]^{1/p} \leq \gamma$. Since this must hold for all $n \in J$, it follows that $x \in l_p$. To show that $\lim_k x_k = x$, let $\epsilon > 0$. Then there is an integer N such that $\rho_p(x_j, x_k) \leq \epsilon$ for all $k, j \geq N$. Again, let n be any positive integer. Then we have $\rho'_p(x'_j, x'_k) \leq \epsilon$ for all $j, k \geq N$. For fixed n, we conclude from Theorem 5.5.6, part (vii), that $\rho'_p(x', x'_k) \leq \epsilon$ for all $k \geq N$. Hence, $\left[\sum_{m=1}^{n} |\xi_m - \xi_{mk}|^p\right]^{1/p} \leq \epsilon$ for all $k \geq N$, where N depends only on ϵ (and not on n). Since this must hold for all $n \in J$, we conclude that $\rho(x, x_k) \leq \epsilon$ for all $k \geq N$. This implies that $\lim_k x_k = x$. ■

5.5.27. Exercise. Show that the discrete metric space of Example 5.1.7 is complete.

5.5.28. Example. Let $\{\mathcal{C}[a, b]; \rho_\infty\}$ be the metric space defined in Example 5.3.14. Thus, $\mathcal{C}[a, b]$ is the set of all continuous functions on $[a, b]$ and

$$\rho_\infty(x, y) = \sup_{a \leq t \leq b} |x(t) - y(t)|.$$

We now show that $\{\mathcal{C}[a, b]; \rho_\infty\}$ is a complete metric space. If $\{x_n\}$ is a Cauchy sequence in $\mathcal{C}[a, b]$, then for each $\epsilon > 0$ there is an N such that $|x_n(t) - x_m(t)| < \epsilon$ whenever $m, n \geq N$ for all $t \in [a, b]$. Thus, for fixed t, the sequence $\{x_n(t)\}$ converges to, say, $x_0(t)$. Since t is arbitrary, the sequence of functions $\{x_n(\cdot)\}$ converges pointwise to a function $x_0(\cdot)$. Also, since $N = N(\epsilon)$ is independent of t, the sequence $\{x_n(\cdot)\}$ converges uniformly to $x_0(\cdot)$. Now from the calculus we know that if a sequence of continuous functions $\{x_n(\cdot)\}$ converges uniformly to a function $x_0(\cdot)$, then $x_0(\cdot)$ is continuous. Therefore, every Cauchy sequence in $\{\mathcal{C}[a, b]; \rho_\infty\}$ converges to an element in this space in the sense of the metric ρ_∞. Therefore, the metric space $\{\mathcal{C}[a, b]; \rho_\infty\}$ is complete. ■

5.5.29. Example. Let $\{\mathcal{C}[a, b]; \rho_2\}$ be the metric space defined in Example 5.3.12, with $p = 2$; i.e.,

$$\rho_2(x, y) = \left\{\int_a^b [x(t) - y(t)]^2 \, dt\right\}^{1/2}.$$

We now show that this metric space is not complete. Without loss of generality let the closed interval be $[-1, 1]$. In particular, consider the sequence $\{x_n\}$ of continuous functions defined by

$$x_n(t) = \left\{\begin{array}{ll} 0, & -1 \leq t \leq 0 \\ nt, & 0 \leq t \leq 1/n \\ 1, & 1/n \leq t \leq 1 \end{array}\right\},$$

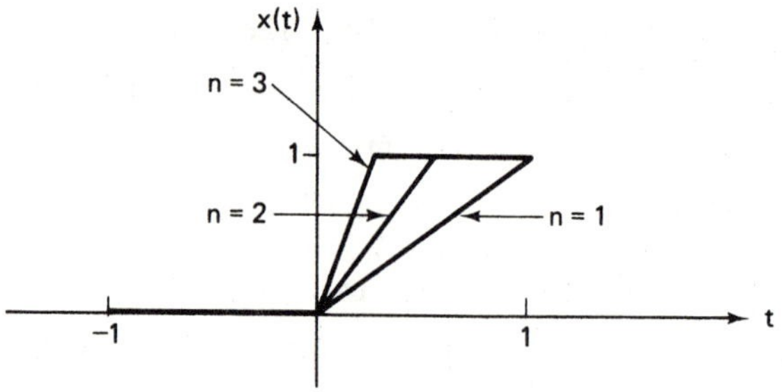

5.5.30. **Figure F.** Sequence $\{x_n\}$ for $\{\mathcal{C}[a, b]; \rho_2\}$.

$n = 1, 2, \ldots$. This sequence is depicted pictorially in Figure F. Now let $m > n$ and note that

$$\{\rho_2(x_m, x_n)\}^2 = (m - n)^2 \int_0^{1/m} t^2\, dt + \int_{1/m}^{1/n} (1 - nt)^2\, dt$$
$$= \frac{(m-n)^2}{3m^2 n} < \frac{1}{3n} < \epsilon$$

whenever $n > 1/(3\epsilon)$. Therefore, $\{x_n\}$ is a Cauchy sequence.

For purposes of contradiction, let us now assume that $\{x_n\}$ converges to a continuous function x, where convergence is taken with respect to the metric ρ_2. In other words, assume that

$$\int_{-1}^{1} |x_n(t) - x(t)|^2\, dt \longrightarrow 0 \text{ as } n \longrightarrow \infty.$$

This implies that the above integral with any limits between $+1$ and -1 also approaches zero as $n \longrightarrow \infty$. Since $x_n(t) = 0$ whenever $t \in [-1, 0]$, we have

$$\int_{-1}^{0} |x_n(t) - x(t)|^2\, dt = 0$$

independent of n. From this it follows that the continuous function x is such that

$$\int_{-1}^{0} |x(t)|^2\, dt = 0,$$

and $x(t) = 0$ whenever $t \in [-1, 0]$. Now if $0 < a \leq 1$, then

$$\int_{a}^{1} |x_n(t) - x(t)|^2\, dt \longrightarrow 0 \text{ as } n \longrightarrow \infty.$$

Choosing $n > 1/a$, we have

$$\int_{a}^{1} |1 - x(t)|^2\, dt \longrightarrow 0 \text{ as } n \longrightarrow \infty.$$

Since this integral is independent of n it vanishes. Also, since x is continuous it follows that $x(t) = 1$ for $t \geq a$. Since a can be chosen arbitrarily close to zero, we end up with a function x such that

$$x(t) = \begin{Bmatrix} 0, & t \in [-1, 0] \\ 1, & t \in (0, 1] \end{Bmatrix}.$$

Therefore, the Cauchy sequence $\{x_n\}$ does not converge to a point in $\mathcal{C}[a, b]$, and the metric space is not complete. ■

The completeness property of certain metric spaces is an essential and important property which we will use and encounter frequently in the remainder of this book. The preceding example demonstrates that not all metric spaces are complete. However, this space $\{\mathcal{C}[a, b]; \rho_2\}$ is a subspace of a larger metric space which is complete. To discuss this complete metric space (i.e., the **completion** of $\{\mathcal{C}[a, b]; \rho_2\}$), it is necessary to make use of the Lebesgue theory of measure and integration. For a thorough treatment of this theory, we refer the reader to the texts by Royden [5.9] and Taylor [5.10]. Although knowledge of this theory is not an essential requirement in the development of the subsequent results in this book, we will want to make reference to certain examples of important metric spaces which are defined in terms of the Lebesgue integral. For this reason, we provide the following *heuristic* comments for those readers who are unfamiliar with this subject.

The **Lebesgue measure** space on the real numbers, R, consists of the triple $\{R, \mathfrak{M}, \mu\}$, where $\mathfrak{M}$ is a certain family of subsets of R, called the **Lebesgue measurable sets** in R, and μ is a mapping, $\mu\colon \mathfrak{M} \longrightarrow R^*$, called **Lebesgue measure**, which may be viewed as a generalization of the concept of length in R. While it is not possible to characterize $\mathfrak{M}$ without providing additional details concerning the Lebesgue theory, it is quite simple to enumerate several important examples of elements in $\mathfrak{M}$. For instance, $\mathfrak{M}$ contains all intervals of the form $(a, b) = \{x \in R\colon a < x < b\}$, $[c, d) = \{x \in R\colon c \leq x < d\}$, $(e, f] = \{x \in R\colon e < x \leq f\}$, $[g, h] = \{x \in R\colon g \leq x \leq h\}$, as well as all countable unions and intersections of such intervals. It is emphasized that $\mathfrak{M}$ does *not* include all subsets of R. Now if $A \in \mathfrak{M}$ is an interval, then the measure of A, $\mu(A)$, is the length of A. For example, if $A = [a, b]$, then $\mu(A) = b - a$. Also, if B is a countable union of disjoint intervals, then $\mu(B)$ is the sum of the lengths of the disjoint intervals (this sum may be infinite). Of particular interest are subsets of R having measure zero. Essentially, this means it is possible to "cover" the set with an arbitrarily small subset of R. Thus, every subset of R containing at most a countable number of points has Lebesgue measure equal to zero. For example, the set of rational numbers has Lebesgue measure zero. (There are also uncountable subsets of R having Lebesgue measure zero.)

In connection with the above discussion, we say that a proposition $P(x)$ is true *almost everywhere* (abbreviated a.e.) if the set $S = \{x \in R\colon P(x)$ is not true} has Lebesgue measure zero. For example, two functions f, g: $R \longrightarrow R$ are said to be *equal a.e.* if the set $S = \{x \in R: f(x) \neq g(x)\} \in \mathfrak{M}$ and if $\mu(S) = 0$.

Let us now consider the integral of real-valued functions defined on the interval $[a, b] \subset R$. It can be shown that a bounded function $f: [a, b] \longrightarrow R$ is **Riemann integrable** (where the Riemann integral is denoted, as usual, by $\int_a^b f(x)dx$) if and only if f is continuous almost everywhere on $[a, b]$. The class of Riemann integrable functions with a metric defined in the same manner as in Example 5.5.29 (for continuous functions on $[a, b]$) is *not* a complete metric space. However, as pointed out before, it is possible to generalize the concept of integral and make it applicable to a class of functions significantly larger than the class of functions which are continuous a.e. In doing so, we must consider the class of measurable functions. Specifically, a function $f: R \longrightarrow R$ is said to be a **Lebesgue measurable function** if $f^{-1}(\mathfrak{U}) \in \mathfrak{M}$ for every open set $\mathfrak{U} \subset R$. Now let f be a Lebesgue measurable function which is bounded on the interval $[a, b]$, and let $M = \sup\{f(x) = y: x \in [a, b]\}$, and let $m = \inf\{f(x) = y: x \in [a, b]\}$. In the Lebesgue approach to integration, the range of f is partitioned into intervals. (This is in contrast with the Riemann approach, where the domain of f is partitioned in developing the integral.) Specifically, let us divide the range of f into the n parts specified by $m = y_0 \leq y_1 \leq \cdots \leq y_{n-1} \leq y_n = M$, let $E_k = \{x \in R: y_{k-1} \leq x \leq y_k\}$ for $k = 1, \ldots, n$, and let ξ_k be such that $y_{k-1} \leq \xi_k \leq y_k$ for $k = 1, \ldots, n$. The sum $\sum_{k=1}^{n} \xi_k \mu(E_k)$ approximates the area under the graph of f, and it can serve as the definition of the integral of f between a and b, after an appropriate limiting process has been performed. Provided that this limit exists, it is called the **Lebesgue integral** of f over $[a, b]$, and it is denoted by $\int_{[a,b]} f d\mu$. It can be shown that any bounded function f which is Riemann integrable over $[a, b]$ is Lebesgue integrable over $[a, b]$, and furthermore $\int_{[a,b]} f d\mu = \int_a^b f(x)dx$. On the other hand, there are functions which are Lebesgue integrable but not Riemann integrable over $[a, b]$. For example, consider the function $f: [a, b] \longrightarrow R$ defined by $f(x) = 0$ if x is rational and $f(x) = 1$ if x is irrational. This function is so erratic that the Riemann integral does not exist in this case. However, since the interval $[a, b] = A \cup B$, where $A = \{x: f(x) = 1\}$ and $B = \{x: f(x) = 0\}$, it follows from the preceding characterization of Lebesgue integral that $\int_{[a,b]} f d\mu = 1.\mu(A) + 0.\mu(B) = b - a$.

Let us now consider an important class of complete metric spaces, given in the next example.

5.5.31. Example. Let $p \geq 1$ (p not necessarily an integer), let $(R, \mathfrak{M}, \mu)$ denote the Lebesgue measure space on the real numbers, and let $[a, b]$ be a subset of R. Let $\mathcal{L}_p[a, b]$ denote the family of functions $f: R \longrightarrow R$ which are Lebesgue measurable and such that $\int_{[a,b]} |f|^p \, d\mu$ exists and is finite.

We define an equivalence relation $\sim$ on $\mathcal{L}_p[a, b]$ by saying that $f \sim g$ if $f(x) = g(x)$ except on a subset of $[a, b]$ having Lebesgue measure zero. Now denote the family of equivalence classes into which $\mathcal{L}_p[a, b]$ is divided by $L_p[a, b]$. Specifically, let us denote the equivalence class $[f] = \{g \in \mathcal{L}_p[a, b]: g \sim f\}$ for $f \in \mathcal{L}_p[a, b]$. Then $L_p[a, b] = \{[f]: f \in \mathcal{L}_p[a, b]\}$.

Now let $X = L_p[a, b]$ and define $\rho_p: X \times X \longrightarrow R$ by

$$\rho_p([f], [g]) = \left[\int_{[a,b]} |f - g|^p \, d\mu\right]^{1/p}. \tag{5.5.32}$$

It can be shown that the value of $\rho([f], [g])$ defined by Eq. (5.5.32) is the same for any f and g in the equivalence classes $[f]$ and $[g]$, respectively. Furthermore, ρ satisfies all the axioms of a metric, and as such $\{L_p[a, b]; \rho_p\}$ is a metric space. One of the important results of the Lebesgue theory is that this space is complete.

It is important to note that the right-hand side of Eq. (5.5.32) cannot be used to define a metric on $\mathcal{L}_p[a, b]$, since there are functions $f \neq g$ such that $\int_{[a,b]} |f - g|^p \, d\mu = 0$; however, in the literature the distinction between $L_p[a, b]$ and $\mathcal{L}_p[a, b]$ is usually suppressed. That is, we usually write $f \in L_p[a, b]$ instead of $[f] \in L_p[a, b]$, where $f \in \mathcal{L}_p[a, b]$.

Finally, in the particular case when $p = 2$, the space $\{\mathcal{C}[a, b]; \rho_2\}$ of Example 5.5.29 is a subspace of the space $\{L_2; \rho_2\}$. ■

Before closing the present section we consider some important general properties of complete metric spaces.

5.5.33. Theorem. Let $\{X; \rho\}$ be a complete metric space, and let $\{Y; \rho\}$ be a metric subspace of $\{X; \rho\}$. Then $\{Y; \rho\}$ is complete if and only if Y is a closed subset of X.

Proof. Assume that $\{Y; \rho\}$ is complete. To show that Y is a closed subset of X we must show that Y contains all of its adherent points. Let y be an adherent point of Y; i.e., let $y \in \bar{Y}$. Then each open sphere $S(y; 1/n)$, $n = 1, 2, \ldots$, contains at least one point y_n in Y. Since $\rho(y_n, y) < 1/n$ it follows that the sequence $\{y_n\}$ converges to y. Since $\{y_n\}$ is a Cauchy sequence in the complete space $\{Y; \rho\}$ we have $\{y_n\}$ converging to a point $y' \in Y$. But the limit of a sequence of points in a metric space is unique by Theorem 5.5.6. Therefore, $y' = y$; i.e., $y \in Y$ and y is closed.

Conversely, assume that Y is a closed subset of X. To show that the space $\{Y; \rho\}$ is complete, let $\{y_n\}$ be an arbitrary Cauchy sequence in $\{Y; \rho\}$. Then $\{y_n\}$ is a Cauchy sequence in the complete metric space $\{X; \rho\}$ and as such it has a limit $y \in X$. However, in view of Theorem 5.5.8, part (iii), the closed subset Y of X contains all its adherent points. Therefore, $\{Y; \rho\}$ is complete. ∎

We emphasize that completeness and closure are not necessarily equivalent in arbitrary metric spaces. For example, a metric space is always closed, yet it is not necessarily complete.

Before characterizing a complete metric space in an alternate way, we need to introduce the following concept.

5.5.34. Definition. A sequence $\{S_i\}$ of subsets of a metric space $\{X; \rho\}$ is called a **nested sequence of sets** if

$$S_1 \supset S_2 \supset S_3 \supset \ldots .$$

We leave the proof of the last result of the present section as an exercise.

5.5.35. Theorem. Let $\{X; \rho\}$ be a metric space. Then,

(i) $\{X; \rho\}$ is complete if and only if every sequence of closed nested spheres in $\{X; \rho\}$ with radii tending to zero have non-void interesection; and

(ii) if $\{X; \rho\}$ is complete, if $\{S_i\}$ is a nested sequence of non-empty closed subsets of X, and if $\lim_n \text{diam}(S_n) = 0$, then the intersection $\bigcap_{n=1}^{\infty} S_n$ is not empty; in fact, it consists of a single point.

5.5.36. Exercise. Prove Theorem 5.5.35.

5.6. COMPACTNESS

We recall the **Bolzano-Weierstrass theorem** from the calculus: Every bounded, infinite subset of the real line (i.e., the set of real numbers with the usual metric) has at least one point of accumulation. Thus, if Y is an arbitrary bounded infinite subset of R, then in view of this theorem we know that any sequence formed from elements of Y has a convergent subsequence. For example, let $Y = [0, 2]$, and let $\{x_n\}$ be the sequence of real numbers given by

$$x_n = \frac{1 - (-1)^n}{2} + \frac{1}{n}, \quad n = 1, 2, \ldots .$$

Then the range of this sequence lies in Y and is thus bounded. Hence, the range has at least one accumulation point. It, in fact, has two.

A theorem from the calculus which is closely related to the Bolzano-Weierstrass theorem is the **Heine-Borel theorem**. We need the following terminology.

5.6.1. Definition. Let Y be a set in a metric space $\{X; \rho\}$, and let A be an index set. A collection of sets $\{Y_\alpha : \alpha \in A\}$ in $\{X; \rho\}$ is called a **covering** of Y if $Y \subset \bigcup_{\alpha \in A} Y_\alpha$. A subcollection $\{Y_\beta : \beta \in B\}$ of the covering $\{Y_\alpha : \alpha \in A\}$, i.e., $B \subset A$ such that $Y \subset \bigcup_{\beta \in B} Y_\beta$ is called a **subcovering** of $\{Y_\alpha; \alpha \in A\}$. If all the members Y_α and Y_β are open sets, then we speak of an **open covering** and **open subcovering**. If A is a finite set, then we speak of a **finite covering**. In general, A may be an uncountable set.

We now recall the Heine-Borel theorem as it applies to subsets of the real line (i.e., of R): Let Y be a closed and bounded subset of R. If $\{Y_\alpha : \alpha \in A\}$ is any family of open sets on the real line which covers Y, then it is possible to find a *finite* subcovering of sets from $\{Y_\alpha : \alpha \in A\}$.

Many important properties of the real line follow from the Bolzano-Weierstrass theorem and from the Heine-Borel theorem. In general, these properties cannot be carried over directly to *arbitrary* metric spaces. The concept of compactness, to be introduced in the present section, will enable us to isolate those metric spaces which possess the Heine-Borel and Bolzano-Weierstrass property.

Because of its close relationship to compactness, we first introduce the concept of total boundedness.

5.6.2. Definition. Let Y be any set in a metric space $\{X; \rho\}$, and let ϵ be an arbitrary positive number. A set S_ϵ in X is said to be an **ϵ-net** for Y if for any point $y \in Y$ there exists at least one point $s \in S_\epsilon$ such that $\rho(s,y) < \epsilon$. The ϵ-net, S_ϵ, is said to be **finite** if S_ϵ contains a finite number of points. A subset Y of X is said to be **totally bounded** if X contains a finite ϵ-net for Y for *every* $\epsilon > 0$.

Some authors use the terminology **ϵ-dense set** for ϵ-net and **precompact** for totally bounded sets.

An obvious equivalent characterization of total boundedness is contained in the following result.

5.6.3. Theorem. A subset $Y \subset X$ is totally bounded if and only if Y can be covered by a finite number of spheres of radius ϵ for any $\epsilon > 0$.

5.6.4. Exercise. Prove Theorem 5.6.3.

In Figure G a pictorial demonstration of the preceding concepts is given. If in this figure the size of ϵ would be decreased, then correspondingly, the

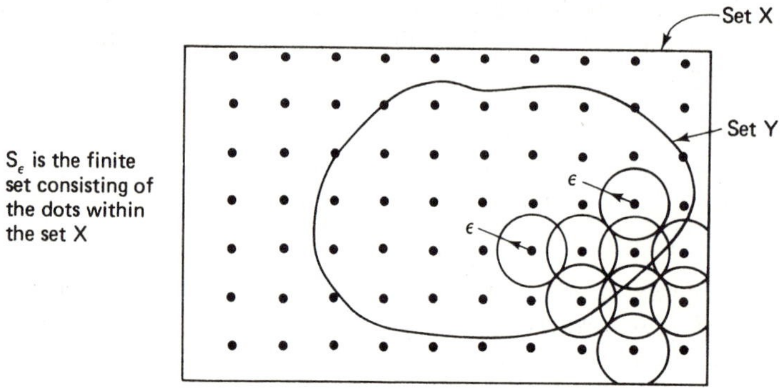

5.6.5. Figure G. Total boundedness of a set Y.

number of elements in S_ϵ would increase. If for arbitrarily small ϵ the number of elements in S_ϵ remains finite, then we have a totally bounded set Y.

Total boundedness is a stronger property than boundedness. We leave the proof of the next result as an exercise.

5.6.6. Theorem. Let $\{X; \rho\}$ be a metric space, and let Y be a subset of X. Then,

(i) if Y is totally bounded, then it is bounded;
(ii) if Y is totally bounded, then its closure $\bar{Y}$ is totally bounded; and
(iii) if the metric space $\{X; \rho\}$ is totally bounded, then it is separable.

5.6.7. Exercise. Prove Theorem 5.6.6.

We note, for example, that all finite sets (including the empty set) are totally bounded. Whereas all totally bounded sets are also bounded, the converse does, in general, not hold. We demonstrate this by means of the following example.

5.6.8. Example. Let $\{l_2; \rho_2\}$ be the metric space defined in Example 5.3.5. Consider the subset $Y \subset l_2$ defined by

$$Y = \{y \in l_2 : \sum_{i=1}^{\infty} |\eta_i|^2 \leq 1\}.$$

We show that Y is bounded but not totally bounded. For any $x, y \in Y$, we have by the Minkowski inequality (5.2.7),

$$\rho_2(x, y) = \left[\sum_{i=1}^{\infty} |\xi_i - \eta_i|^2\right]^{1/2} \leq \left[\sum_{i=1}^{\infty} |\xi_i|^2\right]^{1/2} + \left[\sum_{i=1}^{\infty} |\eta_i|^2\right]^{1/2} \leq 2.$$

Thus, Y is bounded. To show that Y is not totally bounded, consider the set of points $E = \{e_1, e_2, \ldots\} \subset Y$, where $e_1 = (1, 0, 0, \ldots)$, $e_2 = (0, 1,$

$0, \ldots)$, etc. Then $\rho_2(e_i, e_j) = \sqrt{2}$ for $i \neq j$. Now suppose there is a finite ϵ-net for Y for say $\epsilon = \frac{1}{2}$. Let $\{s_1, \ldots, s_n\}$ be the net S_ϵ. Now if e_j is such that $\rho(e_j, s_i) < \frac{1}{2}$ for some i, then $\rho(e_k, s_i) \geq \rho(e_k, e_j) - \rho(e_j, s_i) > \frac{1}{2}$ for $k \neq j$. Hence, there can be at most one element of the set E in each sphere $S(s_i; \frac{1}{2})$ for $i = 1, \ldots, n$. Since there are infinitely many points in E and only a finite number of spheres $S(s_i; \frac{1}{2})$, this contradicts the fact that S_ϵ is an ϵ-net. Hence, there is no finite ϵ-net for $\epsilon = \frac{1}{2}$, and Y is not totally bounded. ■

Let us now consider an example of a totally bounded set.

5.6.9. Example. Let $\{R^n; \rho_2\}$ be the metric space defined in Example 5.3.1, and let Y be the subset of R^n defined by $Y = \{y \in R^n: \sum_{i=1}^{n} \eta_i^2 \leq 1\}$. Clearly, Y is bounded. To show that Y is totally bounded, we construct an ϵ-net for Y for an arbitrary $\epsilon > 0$. To this end, let N be a positive integer such that $N\epsilon \geq \sqrt{n}$, and let S_ϵ be the set of all n-tuples given by

$$S_\epsilon = \{s = (\sigma_1, \ldots, \sigma_n) \in Y: \sigma_i = m_i/N, \text{ some integer } m_i,$$
$$\text{where } -N \leq m_i \leq N, i = 1, \ldots, n\}.$$

Then clearly $S_\epsilon \subset Y$ and S_ϵ is finite. Now for any $y = (\eta_1, \ldots, \eta_n) \in Y$, there is an $s \in S_\epsilon$ such that $|\sigma_i - \eta_i| \leq 1/N$ for $i = 1, \ldots, n$. Thus, $\rho_2(y, s) \leq \left[\sum_{i=1}^{n} (1/N)^2\right]^{1/2} = \sqrt{n}/N \leq \epsilon$. Therefore, S_ϵ is a finite ϵ-net. Since ϵ is arbitrary, Y is totally bounded. ■

In general, any bounded subset of $R_2^n = \{R^n; \rho_2\}$ is totally bounded.

5.6.10. Exercise. Let $\{l_2; \rho_2\}$ be the metric space defined in Example 5.3.5, and let $Y \subset l_2$ be the subset defined by

$$Y = \{y \in l_2: |\eta_1| \leq 1, |\eta_2| \leq \tfrac{1}{2}, \ldots, |\eta_n| \leq (\tfrac{1}{2})^{n-1}, \ldots\}.$$

Show that Y is totally bounded.

In studying compactness of metric spaces, we will find it convenient to introduce the following concept.

5.6.11. Definition. A metric space $\{X; \rho\}$ is said to be **sequentially compact** if every sequence of elements in X contains a subsequence which converges to some element $x \in X$. A set Y in the metric space $\{X; \rho\}$ is said to be **sequentially compact** if the subspace $\{Y; \rho\}$ is sequentially compact; i.e., every sequence in Y contains a subsequence which converges to a point in Y.

5.6.12. Example. Let $X = (0, 1]$, and let ρ be the usual metric on the real line R. Consider the sequence $\{x_n\}$, where $x_n = 1/n, n = 1, 2, \ldots$. This sequence has no subsequence which converges to a point in X, and thus $\{X; \rho\}$ is not sequentially compact. ■

We now define compactness.

5.6.13. Definition. A metric space $\{X; \rho\}$ is said to be **compact**, or to possess the **Heine-Borel property**, if every open covering of $\{X; \rho\}$ contains a *finite* open subcovering. A set Y in a metric space $\{X; \rho\}$ is said to be **compact** if the subspace $\{Y; \rho\}$ is compact.

Some authors use the term *bicompact* for Heine-Borel compactness and the term *compact* for what we call *sequentially compact*. As we shall see shortly, in the case of metric spaces, compactness and sequential compactness are equivalent, so no confusion should arise.

We will also show that compact metric spaces can equivalently be characterized by means of the Bolzano-Weierstrass property, given by the following.

5.6.14. Definition. A metric space $\{X; \rho\}$ possesses the **Bolzano-Weierstrass property** if every infinite subset of X has at least one point of accumulation. A set Y in X possesses the **Bolzano-Weierstrass property** if the subspace $\{Y; \rho\}$ possesses the Bolzano-Weierstrass property.

Before setting out on proving the assertions made above, i.e., the equivalence of compactness, sequential compactness, and the Bolzano-Weierstrass property, in metric spaces, a few comments concerning some of these concepts may be of benefit.

Informally, we may view a sequentially compact metric space as having such an abundance of elements that no matter how we choose a sequence, there will always be a clustering of an infinite number of points around at least one point in the metric space. A similar interpretation can be made concerning metric spaces which possess the Bolzano-Weierstrass property.

Utilizing the concepts of sequential compactness and total boundedness, we first state and prove the following result.

5.6.15. Theorem. Let $\{X; \rho\}$ be a metric space, and let Y be a subset of X. The following properties hold:

(i) if Y is sequentially compact, then Y is bounded;
(ii) if Y is sequentially compact, then Y is closed;
(iii) if $\{X; \rho\}$ is sequentially compact, then $\{X; \rho\}$ is totally bounded;
(iv) if $\{X; \rho\}$ is sequentially compact, then $\{X; \rho\}$ is complete; and
(v) if $\{X; \rho\}$ is totally bounded and complete, then it is sequentially compact.

Proof. To prove (i), assume that Y is a sequentially compact subset of X and assume, for purposes of contradiction, that Y is unbounded. Then we can construct a sequence $\{y_n\}$ with elements arbitrarily far apart. Specifically, let $y_1 \in Y$ and choose $y_2 \in Y$ such that $\rho(y_1, y_2) > 1$. Next, choose $y_3 \in Y$ such that $\rho(y_1, y_3) > 1 + \rho(y_1, y_2)$. Continuing this process, choose $y_n \in Y$ such that $\rho(y_1, y_n) > 1 + \rho(y_1, y_{n-1})$. If $m > n$, then $\rho(y_1, y_m) > 1 + \rho(y_1, y_n)$ and $\rho(y_m, y_n) \geq |\rho(y_1, y_m) - \rho(y_1, y_n)| > 1$. But this implies that $\{y_n\}$ contains no convergent subsequence. However, we assumed that Y is sequentially compact; i.e., every sequence in Y contains a convergent subsequence. Therefore, we have arrived at a contradiction. Hence, Y must be bounded. In the above argument we assumed that Y is an infinite set. We note that if Y is a finite set then there is nothing to prove.

To prove part (ii), let $\bar{Y}$ denote the closure of Y and assume that $y \in \bar{Y}$. Then there is a sequence of points $\{y_n\}$ in Y which converges to y, and every subsequence of $\{y_n\}$ converges to y, by Theorem 5.5.6, part (iv). But, by hypothesis, Y is sequentially compact. Thus, the sequence $\{y_n\}$ in Y contains a subsequence which converges to some element in Y. Therefore, $Y = \bar{Y}$ and Y is closed.

We now prove part (iii). Let $\{X; \rho\}$ be a sequentially compact metric space, and let $x_1 \in X$. With $\epsilon > 0$ fixed we choose if possible $x_2 \in X$ such that $\rho(x_1, x_2) \geq \epsilon$. Next, if possible choose $x_3 \in X$ such that $\rho(x_1, x_2) \geq \epsilon$ and $\rho(x_1, x_3) \geq \epsilon$. Continuing this process we have, for every n, $\rho(x_n, x_1) \geq \epsilon$, $\rho(x_n, x_2) \geq \epsilon, \ldots, \rho(x_n, x_{n-1}) \geq \epsilon$. We now show that this process must ultimately terminate. Clearly, if $\{X; \rho\}$ is a bounded metric space then we can pick ϵ sufficiently large to terminate the process after the first step; i.e., there is no point $x \in X$ such that $\rho(x_1, x) \geq \epsilon$. Now suppose that, in general, the process does not terminate. Then we have constructed a sequence $\{x_n\}$ such that for any two members x_i, x_j of this sequence, we have $\rho(x_i, x_j) \geq \epsilon$. But, by hypothesis, $\{X; \rho\}$ is sequentially compact, and thus $\{x_n\}$ contains a subsequence which is convergent to an element in X. Hence, we have arrived at a contradiction and the process must terminate. Using this procedure we now have for arbitrary $\epsilon > 0$ a *finite* set of points $\{x_1, x_2, \ldots, x_l\}$ such that the spheres, $S(x_n; \epsilon)$, $n = 1, \ldots, l$, cover X; i.e., for any $\epsilon > 0$, X contains a finite ϵ-net. Therefore, the metric space $\{X; \rho\}$ is totally bounded.

We now prove part (iv) of the theorem. Let $\{x_n\}$ be a Cauchy sequence. Then for every $\epsilon > 0$ there is an integer l such that $\rho(x_m, x_n) < \epsilon$ whenever $m > n \geq l$. Since $\{X; \rho\}$ is sequentially compact, the sequence $\{x_n\}$ contains a subsequence $\{x_{l_n}\}$ convergent to a point $x \in X$ so that $\lim_{n\to\infty} \rho(x_{l_n}, x) = 0$. The sequence $\{l_n\}$ is an increasing sequence and $l_m \geq m$. It now follows that

$$0 \leq \rho(x_n, x) \leq \rho(x_n, x_{l_m}) + \rho(x_{l_m}, x) < \epsilon + \rho(x_{l_m}, x)$$

whenever $m > n \geq l$. Letting $m \to \infty$, we have $0 \leq \rho(x_n, x) \leq \epsilon$, whenever $n \geq l$. Hence, the Cauchy sequence $\{x_n\}$ converges to $x \in X$. Therefore, X is complete.

In connection with parts (iv) and (v) we note that a totally bounded metric space is not necessarily sequentially compact. We leave the proof of part (v) as an exercise. ■

5.6.16. Exercise. Prove part (v) of Theorem 5.6.15.

Parts (iii), (iv) and (v) of the above theorem allow us to define a sequentially compact metric space equivalently as a metric space which is complete and totally bounded. We now show that a metric space is sequentially compact if and only if it satisfies the Bolzano-Weierstrass property.

5.6.17. Theorem. A metric space $\{X; \rho\}$ is sequentially compact if and only if every infinite subset of X has at least one point of accumulation.

Proof. Assume that Y is an infinite subset of a sequentially compact metric space $\{X; \rho\}$. If $\{y_n\}$ is any sequence of distinct points in Y, then $\{y_n\}$ contains a convergent subsequence $\{y_{l_n}\}$, because $\{X; \rho\}$ is sequentially compact. The limit y of the subsequence is a point of accumulation of Y.

Conversely, assume that $\{X; \rho\}$ is a metric space such that every infinite subset Y of X has a point of accumulation. Let $\{y_n\}$ be any sequence of points in Y. If a point occurs an infinite number of times in $\{y_n\}$, then this sequence contains a convergent subsequence, a constant subsequence, and we are finished. If this is not the case, then we can assume that all elements of $\{y_n\}$ are distinct. Let Z denote the set of all points y_n, $n = 1, 2, \ldots$. By hypothesis, the infinite set Z has at least one point of accumulation. If $z \in Z$ is such a point of accumulation then we can choose a sequence of points of Z which converges to z (see Theorem 5.5.8, part (i)) and this sequence is a subsequence $\{y_{l_n}\}$ of $\{y_n\}$. Therefore, $\{X; \rho\}$ is sequentially compact. This concludes the proof. ■

Our next objective is to show that in metric spaces the concepts of compactness and sequential compactness are equivalent. In doing so we employ the following lemma, the proof of which is left as an exercise.

5.6.18. Lemma. Let $\{X; \rho\}$ be a sequentially compact metric space. If $\{Y_\alpha : \alpha \in A\}$ is an infinite open covering of $\{X; \rho\}$, then there exists a number $\epsilon > 0$ such that every sphere in X of radius ϵ is contained in at least one of the open sets Y_α.

5.6.19. Exercise. Prove Lemma 5.6.18.

5.6.20. Theorem. A metric space $\{X; \rho\}$ is compact if and only if it is sequentially compact.

Proof. From Theorem 5.6.17, a metric space is sequentially compact if and only if it has the Bolzano-Weierstrass property. Therefore, we first show that every infinite subset of a compact metric space has a point of accumulation.

Let $\{X; \rho\}$ be a compact metric space, and let Y be an infinite subset of X. For purposes of contradiction, assume that Y has no point of accumulation. Then each $x \in X$ is the center of a sphere which contains no point of Y, except possibly x itself. These spheres form an infinite open covering of X. But, by hypothesis, $\{X; \rho\}$ is compact, and therefore we can choose from this infinite covering a finite number of spheres which also cover X. Now each sphere from this finite subcovering contains at most one point of Y, and therefore Y is finite. But this is contrary to our original assumption, and we have arrived at a contradiction. Therefore, Y has at least one point of accumulation, and $\{X; \rho\}$ is sequentially compact.

Conversely, assume that $\{X; \rho\}$ is a sequentially compact metric space, and let $\{Y_\alpha; \alpha \in A\}$ be an arbitrary infinite open covering of X. From Lemma 5.6.18 there exists an $\epsilon > 0$ such that every sphere in X of radius ϵ is contained in at least one of the open sets Y_α. Now, by hypothesis, $\{X; \rho\}$ is sequentially compact and is therefore totally bounded by part (iii) of Theorem 5.6.15. Thus, with arbitrary ϵ fixed we can find a finite ϵ-net, $\{x_1, x_2, \ldots, x_l\}$, such that $X \subset \bigcup_{i=1}^{l} S(x_i; \epsilon)$. Now in view of Lemma 5.6.18, $S(x_i; \epsilon) \subset Y_{\alpha i}$, $i = 1, \ldots, l$, where the sets Y_{α_i} are from the family $\{Y_\alpha; \alpha \in A\}$. Hence,

$$X \subset \bigcup_{i=1}^{l} Y_{\alpha_i},$$

and X has a finite open subcovering chosen from the infinite open covering $\{Y_\alpha; \alpha \in A\}$. Therefore, the metric space $\{X; \rho\}$ is compact. This proves the theorem. ■

There is yet another way of characterizing a compact metric space. Before doing so, we give the following definition.

5.6.21. Definition. Let $\{F_\alpha : \alpha \in A\}$ be an infinite family of closed sets. The family $\{F_\alpha : \alpha \in A\}$ is said to have the **finite intersection property** if for every finite set $B \subset A$ the set $\bigcap_{\alpha \in B} F_\alpha$ is not empty.

5.6.22. Theorem. A metric space $\{X; \rho\}$ is compact if and only if every infinite family $\{F_\alpha : \alpha \in A\}$ of closed sets in X with the finite intersection property has a nonvoid intersection; i.e., $\bigcap_{\alpha \in A} F_\alpha \neq \varnothing$.

5.6.23. Exercise. Prove Theorem 5.6.22.

We now summarize the above results as follows.

5.6.24. Theorem. In a metric space $\{X; \rho\}$ the following are equivalent:

(i) $\{X; \rho\}$ is compact;
(ii) $\{X; \rho\}$ is sequentially compact;
(iii) $\{X; \rho\}$ possesses the Bolzano-Weierstrass property;
(iv) $\{X; \rho\}$ is complete and totally bounded; and
(v) every infinite family of closed sets in $\{X; \rho\}$ with the finite intersection property has a nonvoid intersection.

Concerning product spaces we offer the following exercise.

5.6.25. Exercise. Let $\{X_1; \rho_1\}, \{X_2; \rho_2\}, \ldots, \{X_n; \rho_n\}$ be n compact metric spaces. Let $X = X_1 \times X_2 \times \ldots \times X_n$, and let

$$\rho(x, y) = \rho_1(x_1, y_1) + \ldots + \rho_n(x_n, y_n), \tag{5.6.26}$$

where $x_i, y_i \in X_i$, $i = 1, \ldots, n$, and where $x, y \in X$. Show that the product space $\{X; \rho\}$ is also a compact metric space.

The next result constitutes an important characterization of compact sets in the spaces R^n and C^n.

5.6.27. Theorem. Let $\{R^n; \rho_2\}$ (let $\{C^n; \rho_2\}$) be the metric space defined in Example 5.3.1. A set $Y \subset R^n$ (a set $Y \subset C^n$) is compact if and only if it is closed and bounded.

5.6.28. Exercise. Prove Theorem 5.6.27.

Recall that every non-void compact set in the real line R contains its infimum and its supremum.

In general, it is not an easy task to apply the results of Theorem 5.6.24 to specific spaces in order to establish necessary and sufficient conditions for compactness. From the point of view of applications, criteria such as those established in Theorem 5.6.27 are much more desirable.

We now give a condition which tells us when a subset of a metric space is compact. We have:

5.6.29. Theorem. Let $\{X; \rho\}$ be a compact metric space, and let $Y \subset X$. If Y is closed, then Y is compact.

Proof. Let $\{Y_\alpha; \alpha \in A\}$ be any open covering of Y; i.e., each Y_α is open relative to $\{Y; \rho\}$. Then, by Theorem 5.4.20, for each Y_α there is a U_α which is open relative to $\{X; \rho\}$ such that $Y_\alpha = Y \cap U_\alpha$. Since Y is closed, $Y^\sim$ is an open set in $\{X; \rho\}$. Also, since $X = Y \cup Y^\sim$, $Y^\sim \cup \{U_\alpha : \alpha \in A\}$ is an open covering of X. Since X is compact, it is possible to find a finite sub-covering from this family; i.e., there is a finite set $B \subset A$ such that $X = Y^\sim$

$\cup [\bigcup_{\alpha \in B} U_\alpha]$. Since $Y \subset \bigcup_{\alpha \in B} U_\alpha$, $Y = \bigcup_{\alpha \in B} Y \cap U_\alpha$; i.e., $\{Y_\alpha; \alpha \in B\}$ covers Y. This implies that Y is compact. ■

We close the present section by introducing the concept of relative compactness.

5.6.30. Definition. Let $\{X; \rho\}$ be a metric space and let $Y \subset X$. The subset Y is said to be **relatively compact** in X if $\bar{Y}$ is a compact subset of X.

One of the essential features of a relatively compact set is that every sequence has a convergent subsequence, just as in the case of compact subsets; however, the limit of the subsequence need not be in the subset. Thus, we have the following result.

5.6.31. Theorem. Let $\{X; \rho\}$ be a metric space and let $Y \subset X$. Then Y is relatively compact in X if and only if every sequence of elements in Y contains a subsequence which converges to some $x \in X$.

Proof. Let Y be relatively compact in X, and let $\{y_n\}$ be any sequence in Y. Then $\{y_n\}$ belongs to $\bar{Y}$ also and hence has a convergent subsequence in $\bar{Y}$, since $\bar{Y}$ is sequentially compact. Hence, $\{y_n\}$ contains a subsequence which converges to an element $x \in \bar{Y} \subset X$.

Conversely, let $\{y_n\}$ be a sequence in $\bar{Y}$. Then for each $n = 1, 2, \ldots$, there is an $x_n \in Y$ such that $\rho(x_n, y_n) < 1/n$. Since $\{x_n\}$ is a sequence in Y, it contains a convergent subsequence, say $\{x_{n_k}\}$, which converges to some $x \in X$. Since $\{x_{n_k}\}$ is also in $\bar{Y}$, it follows from part (iii) of Theorem 5.5.8 that $x \in \bar{Y}$. Hence, $\bar{Y}$ is sequentially compact, and so Y is relatively compact in X. ■

5.7. CONTINUOUS FUNCTIONS

Having introduced the concept of metric space, we are in a position to give a generalization of the concept of continuity of functions encountered in calculus.

5.7.1. Definition. Let $\{X; \rho_x\}$ and $\{Y; \rho_y\}$ be two metric spaces, and let $f\colon X \longrightarrow Y$ be a mapping of X into Y. The mapping f is said to be **continuous at the point** $x_0 \in X$ if for every $\epsilon > 0$ there is a $\delta > 0$ such that

$$\rho_y[f(x), f(x_0)] < \epsilon$$

whenever $\rho_x(x, x_0) < \delta$. The mapping f is said to be **continuous on** X or simply **continuous** if it is continuous at each point $x \in X$.

We note that in the above definition the δ is dependent on the choice of x_0 and ϵ; i.e., $\delta = \delta(\epsilon, x_0)$. Now if for each $\epsilon > 0$ there exists a $\delta = \delta(\epsilon) > 0$ such that for any x_0 we have $\rho_y[f(x), f(x_0)] < \epsilon$ whenever $\rho_x(x, x_0) < \delta$, then we say that the function f is **uniformly continuous on** X. Henceforth, if we simply say f is continuous, we mean f is continuous on X.

5.7.2. Example. Let $\{X; \rho_x\} = R_2^n$, and let $\{Y; \rho_y\} = R_2^m$ (see Example 5.3.1). Let $\mathbf{A}$ denote the real matrix

$$\mathbf{A} = \begin{bmatrix} a_{11} & a_{12} & \cdots & a_{1n} \\ a_{21} & a_{22} & \cdots & a_{2n} \\ \cdots & \cdots & \cdots & \cdots \\ a_{m1} & a_{m2} & \cdots & a_{mn} \end{bmatrix}.$$

We denote $\mathbf{x} \in R^n$ and $\mathbf{y} \in R^m$ by

$$\mathbf{x} = \begin{bmatrix} \xi_1 \\ \cdot \\ \cdot \\ \cdot \\ \xi_n \end{bmatrix} \quad \text{and} \quad \mathbf{y} = \begin{bmatrix} \eta_1 \\ \cdot \\ \cdot \\ \cdot \\ \eta_m \end{bmatrix}.$$

Let us define the function $\mathbf{f}: R^n \longrightarrow R^m$ by

$$\mathbf{f}(\mathbf{x}) = \mathbf{A}\mathbf{x}$$

for each $\mathbf{x} \in R^n$. We now show that $\mathbf{f}$ is continuous on R^n. If $\mathbf{x}, \mathbf{x}_0 \in R^n$ and $\mathbf{y}, \mathbf{y}_0 \in R^m$ are such that $\mathbf{y} = \mathbf{f}(\mathbf{x})$ and $\mathbf{y}_0 = \mathbf{f}(\mathbf{x}_0)$, then we have

$$\begin{bmatrix} \eta_1 \\ \cdot \\ \cdot \\ \cdot \\ \eta_m \end{bmatrix} = \begin{bmatrix} a_{11} & \cdots & a_{1n} \\ \cdot & & \cdot \\ \cdot & & \cdot \\ \cdot & & \cdot \\ a_{m1} & \cdots & a_{mn} \end{bmatrix} \begin{bmatrix} \xi_1 \\ \cdot \\ \cdot \\ \cdot \\ \xi_n \end{bmatrix}$$

and

$$[\rho_y(\mathbf{y}, \mathbf{y}_0)]^2 = \sum_{i=1}^{m} \left[\sum_{j=1}^{n} a_{ij}(\xi_j - \xi_{0j}) \right]^2.$$

Using the Schwarz inequality, it follows that

$$[\rho_y(\mathbf{y}, \mathbf{y}_0)]^2 \leq \left[\sum_{i=1}^{m} \left(\sum_{j=1}^{n} a_{ij}^2 \right) \right] \left[\sum_{j=1}^{n} (\xi_j - \xi_{0j})^2 \right].$$

Now let $M = \left\{ \sum_{i=1}^{m} \sum_{j=1}^{n} a_{ij}^2 \right\}^{1/2} \neq 0$ (if $M = 0$ then we are done). Given any $\epsilon > 0$ and choosing $\delta = \epsilon / M$, it follows that $\rho_y(\mathbf{y}, \mathbf{y}_0) < \epsilon$ whenever $\rho_x(\mathbf{x}, \mathbf{x}_0) < \delta$ and any mapping $\mathbf{f}: R^n \longrightarrow R^m$ which is represented by a real, constant $(m \times n)$ matrix $\mathbf{A}$ is continuous on R^n. ■

5.7.3. Example. Let $\{X; \rho_x\} = \{Y; \rho_y\} = \{\mathcal{C}[a, b]; \rho_2\}$, the metric space defined in Example 5.3.12, and let us define a function $f: X \longrightarrow Y$ in the following way. For $x \in X$, $y = f(x)$ is given by

$$y(t) = \int_a^t k(t, s)x(s)ds, \quad t \in [a, b],$$

where $k: R^2 \longrightarrow R$ is continuous in the usual sense, i.e., with respect to the metric spaces R_2^2 and R_2^1. We now show that f is continuous on X. Let x, $x_0 \in X$ and $y, y_0 \in Y$ be such that $y = f(x)$ and $y_0 = f(x_0)$. Then

$$[\rho_y(y, y_0)]^2 = \int_a^b \left\{\int_a^t k(t, s)[x(s) - x_0(s)]ds\right\}^2 dt.$$

It follows from Hölder's inequality for integrals (5.2.5) that

$$\rho_y(y, y_0) \leq M\rho_x(x, x_0),$$

where $M = \left[\int_a^b \int_a^b k^2(t, s)\, dsdt\right]^{1/2}$. Hence, for any $\epsilon > 0$, $\rho_y(y, y_0) < \epsilon$ whenever $\rho_x(x, x_0) < \delta$, where $\delta = \epsilon/M$. ■

5.7.4. Example. Consider the metric space $\{\mathcal{C}[a, b]; \rho_\infty\}$ defined in Example 5.3.14. Let $\mathcal{C}^1[a, b]$ be the subset of $\mathcal{C}[a, b]$ of all functions having continuous first derivatives (on (a, b)), and let $\{X; \rho_x\}$ be the metric subspace $\{\mathcal{C}^1[a, b]; \rho_\infty\}$. Let $\{Y; \rho_y\} = \{\mathcal{C}[a, b]; \rho_\infty\}$ and define the function $f: X \longrightarrow Y$ as follows. For $x \in X$, $y = f(x)$ is given by

$$y(t) = \frac{dx(t)}{dt}.$$

To show that f *is not continuous*, we show that for any $\delta > 0$ there is a pair $x, x_0 \in X$ such that $\rho_x(x, x_0) < \delta$ but $\rho_y(f(x), f(x_0)) \geq 1$. Let $x_0(t) = 0$ for all $t \in [a, b]$, and let $x(t) = \alpha \sin \omega t$, $\alpha > 0$, $\omega > 0$. Then $\rho(x_0, x) \leq \alpha$. Now if $y_0 = f(x_0)$ and $y = f(x)$, then $y_0(t) = 0$ for all $t \in [a, b]$ and $y(t) = \alpha\omega \cos \omega t$. Hence, $\rho(y_0, y) = \alpha\omega$, provided that ω is sufficiently large, i.e., so that $\cos \omega t = \pm 1$ for some $t \in [a, b]$. Now no matter what value of δ we choose, there is an $x \in X$ such that $\rho(x, x_0) < \delta$ if we pick $\alpha < \delta$. However, $\rho(y, y_0) = 1$ if we let $\omega = 1/\alpha$. Therefore, f is not continuous on X. ■

We can interpret the notion of continuity of functions in the following equivalent way.

5.7.5. Theorem. Let $\{X; \rho_x\}$ and $\{Y; \rho_y\}$ be metric spaces, and let f: $X \longrightarrow Y$. Then f is continuous at a point $x_0 \in X$ if and only if for every $\epsilon > 0$, there exists a $\delta > 0$ such that

$$f(S(x_0; \delta)) \subset S(f(x_0); \epsilon).$$

5.7.6. Exercise. Prove Theorem 5.7.5.

Intuitively, Theorem 5.7.5 tells us that f is continuous at x_0 if $f(x)$ is arbitrarily close to $f(x_0)$ when x is sufficiently close to x_0. The concept of continuity is depicted in Figure H for the case where $\{X; \rho_x\} = \{Y; \rho_y\} = R_2^2$.

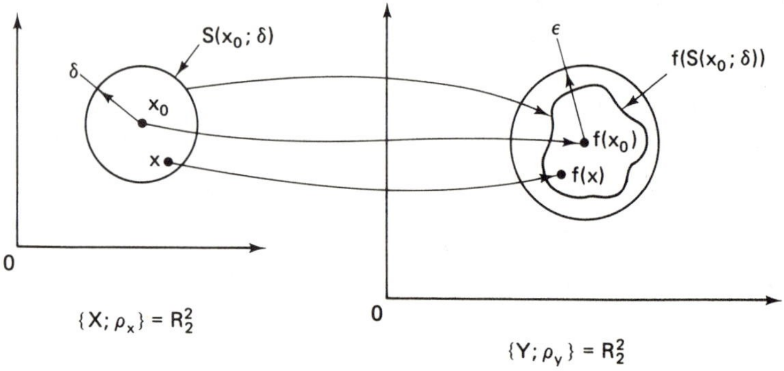

5.7.7. Figure H. Illustration of continuity.

As we did in Chapter 1, we distinguish between mappings on metric spaces which are injective, surjective, or bijective.

It turns out that the concepts of continuity and convergence of sequences are related. Our next result yields a connection between convergence and continuity.

5.7.8. Theorem. Let $\{X; \rho_x\}$ and $\{Y; \rho_y\}$ be two metric spaces. A function $f: X \longrightarrow Y$ is continuous at a point $x_0 \in X$ if and only if for every sequence $\{x_n\}$ of points in X which converges to a point x_0 the corresponding sequence $\{f(x_n)\}$ converges to the point $f(x_0)$ in Y; i.e.,

$$\lim_{n\to\infty} f(x_n) = f(\lim_{n\to\infty} x_n) = f(x_0)$$

whenever $\lim_{n\to\infty} x_n = x_0$.

Proof. Assume that f is continuous at a point $x_0 \in X$, and let $\{x_n\}$ be a sequence such that $\lim_{n\to\infty} x_n = x_0$. Then for every $\epsilon > 0$ there is a $\delta > 0$ such that $\rho_y(f(x), f(x_0)) < \epsilon$ whenever $\rho_x(x, x_0) < \delta$. Also, there is an N such that $\rho_x(x_n, x_0) < \delta$ whenever $n \geq N$. Hence, $\rho_y(f(x_n), f(x_0)) < \epsilon$ whenever $n \geq N$. Thus, if f is continuous at x_0 and if $\lim_{n\to\infty} x_n = x_0$, then $\lim_{n\to\infty} f(x_n) = f(x_0)$.

Conversely, assume that $f(x_n) \longrightarrow f(x_0)$ whenever $x_n \longrightarrow x_0$. For purposes of contradiction, assume that f is not continuous at x_0. Then there exists an $\epsilon > 0$ such that for each $\delta > 0$ there is an x with the property that $\rho_x(x, x_0) < \delta$ and $\rho_y(f(x), f(x_0)) \geq \epsilon$. This implies that for each positive integer n there is an x_n such that $\rho_x(x_n, x_0) < 1/n$ and $\rho_y(f(x_n), f(x_0)) \geq \epsilon$ for all n; i.e., $x_n \longrightarrow x_0$ but $\{f(x_n)\}$ does not converge to $f(x_0)$. But we assumed that $f(x_n) \longrightarrow f(x_0)$ whenever $x_n \longrightarrow x_0$. Hence, we have arrived at a contradiction, and f must be continuous at x_0. This concludes the proof of our theorem. ■

Continuous mappings on metric spaces possess the following important properties.

5.7.9. Theorem. Let $\{X; \rho_x\}$ and $\{Y; \rho_y\}$ be two metric spaces, and let f be a mapping of X into Y. Then

(i) f is continuous on X if and only if the inverse image of each open subset of $\{Y; \rho_y\}$ is open in $\{X; \rho_x\}$; and

(ii) f is continuous on X if and only if the inverse image of each closed subset of $\{Y; \rho_y\}$ is closed in $\{X; \rho_x\}$.

Proof. Let f be continuous on X, and let $V \neq \varnothing$ be an open subset of $\{Y; \rho_y\}$. Let $U = f^{-1}(V)$. Clearly, $U \neq \varnothing$. Now let $x \in U$. Then there exists a unique $y = f(x) \in V$. Since V is open, there is a sphere $S(y; \epsilon)$ which is entirely contained in V. Since f is continuous at x, there is a sphere $S(x; \delta)$ such that its image $f(S(x; \delta))$ is entirely contained in $S(y; \epsilon)$ and therefore in V. But from this it follows that $S(x; \delta) \subset U$. Hence, every $x \in U$ is the center of a sphere which is contained in U. Therefore, U is open.

Conversely, assume that the inverse image of each non-empty open subset of Y is open. For arbitrary $x \in X$ we have $y = f(x)$. Since $S(y; \epsilon) \subset Y$ is open, the set $f^{-1}(S(y; \epsilon))$ is open for every $\epsilon > 0$ and $x \in f^{-1}(S(y; \epsilon))$. Hence, there is a sphere $S(x; \delta)$ such that $S(x; \delta) \subset f^{-1}(S(y; \epsilon))$. From this it follows that for every $\epsilon > 0$ there is a $\delta > 0$ such that $f(S(x; \delta)) \subset S(y; \epsilon)$. Therefore, f is continuous at x. But $x \in X$ was arbitrarily chosen. Hence, f is continuous on X. This concludes the proof of part (i).

To prove part (ii) we utilize part (i) and take complements of open sets. ■

The reader is cautioned that the image of an open subset of X under a continuous mapping $f: X \longrightarrow Y$ is not necessarily an open subset of Y. For example, let $f: R \longrightarrow R$ be defined by $f(x) = x^2$ for every $x \in R$. Clearly, f is continuous on R. Yet the image of the open interval $(-1, 1)$ is the interval $[0, 1)$. But the interval $[0, 1)$ is not open.

We leave the proof of the next result as an exercise to the reader.

5.7.10. Theorem. Let $\{X; \rho_x\}$, $\{Y; \rho_y\}$, and $\{Z; \rho_z\}$ be metric spaces, let f be a mapping of X into Y, and let g be a mapping of Y into Z. If f is continuous on X and g is continuous on Y, then the composite mapping $h = g \circ f$ of X into Z is continuous on X.

5.7.11. Exercise. Prove Theorem 5.7.10.

For continuous mappings on compact spaces we state and prove the following result.

5.7.12. Theorem. Let $\{X; \rho_x\}$ and $\{Y; \rho_y\}$ be two metric spaces, and let $f: X \longrightarrow Y$ be continuous on X.

(i) If $\{X; \rho_x\}$ is compact, then $f(X)$ is a compact subset of $\{Y; \rho_y\}$.
(ii) If U is a compact subset of the metric space $\{X; \rho_x\}$, then $f(U)$ is a compact subset of the metric space $\{Y; \rho_y\}$.
(iii) If $\{X; \rho_x\}$ is compact and if U is a closed subset of X, then $f(U)$ is a closed subset of $\{Y; \rho_y\}$.
(iv) If $\{X; \rho_x\}$ is compact, then f is uniformly continuous on X.

Proof. To prove part (i) let $\{y_n\}$ be a sequence in $f(X)$. Then there are points $\{x_n\}$ in X such that $y_n = f(x_n)$. Since $\{X; \rho_x\}$ is compact we can find a subsequence $\{x_{l_n}\}$ of $\{x_n\}$ which converges to a point in X; i.e., $x_{l_n} \longrightarrow x$. In view of Theorem 5.7.8 we have, since f is continuous at x, $f(x_{l_n}) \longrightarrow f(x) \in f(X)$. From this it follows that the sequence $\{y_n\}$ has a convergent subsequence and $f(X)$ is compact.

To prove part (ii), let U be a compact subset of X. Then $\{U; \rho_x\}$ is a compact metric space. In view of part (i) it now follows that $f(U)$ is also a compact subset of the metric space $\{Y; \rho_y\}$.

To prove part (iii), we first observe that a closed subset U of a compact metric space $\{X; \rho_x\}$ is itself compact and $\{U; \rho_x\}$ is itself a compact metric space. In view of part (ii), $f(U)$ is a compact subset of the metric space $\{Y; \rho_y\}$ and as such is bounded and closed.

To prove part (iv), let $\epsilon > 0$. For every $x \in X$, there is some positive number, $\eta(x)$, such that $f(S(x; 2\eta(x))) \subset S(f(x); \epsilon/2)$. Now the family $\{S(x; \eta(x)): x \in X\}$ is an open covering of X. Since X is compact, there is a finite set, say $F \subset X$, such that $\{S(x; \eta(x)): x \in F\}$ is a covering of X. Now let

$$\delta = \min \{\eta(x): x \in F\}.$$

Since F is a finite set, δ is some positive number. Now let $x, y \in X$ be such that $\rho(x, y) < \delta$. Choose $z \in F$ such that $x \in S(z; \eta(z))$. Since $\delta \leq \eta(z)$, $y \in S(z; 2\eta(z))$. Since $f(S(z; 2\eta(z))) \subset S(f(z); \epsilon/2)$, it follows that $f(x)$ and $f(y)$ are in $S(f(z); \epsilon/2)$. Hence, $\rho_y(f(x), f(y)) < \epsilon$. Since δ does not depend on $x \in X$, f is uniformly continuous on X. This completes the proof of the theorem. ■

Let us next consider some additional generalizations of concepts encountered in the calculus.

5.7.13. Definition. Let $\{X; \rho_x\}$ and $\{Y; \rho_y\}$ be metric spaces, and let $\{f_n\}$ be a sequence of functions from X into Y. If $\{f_n(x)\}$ converges at each $x \in X$, then we say that $\{f_n\}$ is **pointwise convergent**. In this case we write $\lim_n f_n = f$, where f is defined for every $x \in X$.

Equivalently, we say that the sequence $\{f_n\}$ is pointwise convergent to a function f if for every $\epsilon > 0$ and for every $x \in X$ there is an integer $N = N(\epsilon, x)$ such that

$$\rho_y(f_n(x), f(x)) < \epsilon$$

whenever $n \geq N(\epsilon, x)$. In general, $N(\epsilon, x)$ is not necessarily bounded. However, if $N(\epsilon, x)$ is bounded for all $x \in X$, then we say that the sequence $\{f_n\}$ **converges to f uniformly on** X. Let $M(\epsilon) = \sup_{x \in X} N(\epsilon, x) < \infty$. Equivalently, we say that the sequence $\{f_n\}$ converges uniformly to f on X if for every $\epsilon > 0$ there is an $M(\epsilon)$ such that

$$\rho_y(f_n(x), f(x)) < \epsilon$$

whenever $n \geq M(\epsilon)$ for all $x \in X$.

In the next result a connection between uniform convergence of functions and continuity is established. (We used a special case of this result in the proof of Example 5.5.28.)

5.7.14. Theorem. Let $\{X; \rho_x\}$ and $\{Y; \rho_y\}$ be two metric spaces, and let $\{f_n\}$ be a sequence of functions from X into Y such that f_n is continuous on X for each n. If the sequence $\{f_n\}$ converges uniformly to f on X, then f is continuous on X.

Proof. Assume that the sequence $\{f_n\}$ converges uniformly to f on X. Then for every $\epsilon > 0$ there is an N such that $\rho_y(f_n(x), f(x)) < \epsilon$ whenever $n \geq N$ for all $x \in X$. If $M \geq N$ is a fixed integer then f_M is continuous on X. Letting $x_0 \in X$ be fixed, we can find a $\delta > 0$ such that $\rho_y(f_M(x), f_M(x_0)) < \epsilon$ whenever $\rho_x(x, x_0) < \delta$. Therefore, we have

$$\rho_y(f(x), f(x_0)) \leq \rho_y(f(x), f_M(x)) + \rho_y(f_M(x), f_M(x_0)) + \rho_y(f_M(x_0), f(x_0)) < 3\epsilon,$$

whenever $\rho_x(x, x_0) < \delta$. From this it follows that f is continuous at x_0. Since x_0 was arbitrarily chosen, f is continuous at all $x \in X$. This proves the theorem. ■

The reader will recognize in the last result of the present section several generalizations from the calculus to real-valued functions defined on metric spaces.

5.7.15. Theorem. Let $\{X; \rho_x\}$ be a metric space, and let $\{R; \rho\}$ denote the real line R with the usual metric. Let $f: X \longrightarrow R$, and let $U \subset X$. If f is continuous on X and if U is a compact subset of $\{X; \rho_x\}$, then

(i) f is uniformly continuous on U;
(ii) f is bounded on U; and
(iii) if $U \neq \varnothing$, f attains its infimum and supremum on U; i.e., there exist $x_0, x_1 \in U$ such that $f(x_0) = \inf\{f(x): x \in U\}$ and $f(x_1) = \sup\{f(x): x \in U\}$.

Proof. Part (i) follows from part (iv) of Theorem 5.7.12. Since U is a compact subset of X it follows that $f(U)$ is a compact subset of R. Thus, $f(U)$ is bounded and closed. From this it follows that f is bounded. To prove part (iii), note that if U is a non-empty compact subset of $\{X; \rho_x\}$, then $f(U)$ is a non-empty compact subset of R. This implies that f attains its infimum and supremum on U. ■

5.8. SOME IMPORTANT RESULTS IN APPLICATIONS

In this section we present two results which are used widely in applications. The first of these is called the **fixed point principle** while the second is known as the **Ascoli-Arzela theorem**. Both of these results are widely utilized, for example, in establishing existence and uniqueness of solutions of various types of equations (ordinary differential equations, integral equations, algebraic equations, functional differential equations, and the like).

We begin by considering a special class of continuous mappings on metric spaces, so-called contraction mappings.

5.8.1. Definition. Let $\{X; \rho\}$ be a metric space and let $f: X \longrightarrow X$. The function f is said to be a **contraction mapping** if there exists a real number c such that $0 < c < 1$ and

$$\rho(f(x), f(y)) \leq c\rho(x, y) \tag{5.8.2}$$

for all $x, y \in X$.

The reader can readily verify the following result.

5.8.3. Theorem. Every contraction mapping is uniformly continuous on X.

5.8.4. Exercise. Prove Theorem 5.8.3.

The following result is known as the **fixed point principle** or the **principle of contraction mappings.**

5.8.5. Theorem. Let $\{X; \rho\}$ be a complete metric space, and let f be a contraction mapping of X into X. Then

(i) there exists a unique point $x_0 \in X$ such that

$$f(x_0) = x_0, \tag{5.8.6}$$

and

(ii) for any $x_1 \in X$, the sequence $\{x_n\}$ in X defined by

$$x_{n+1} = f(x_n), \quad n = 1, 2, \ldots \tag{5.8.7}$$

converges to the unique element x_0 given in (5.8.6).

The unique point x_0 satisfying Eq. (5.8.6) is called a **fixed point** of f. In this case we say that x_0 is obtained by the **method of successive approximations.**

Proof. We first show that if there is an $x_0 \in X$ satisfying (5.8.6), then it must be unique. Suppose that x_0 and y_0 satisfy (5.8.6). Then by inequality (5.8.2), we have $\rho(x_0, y_0) \leq c\rho(x_0, y_0)$. Since $0 < c < 1$, it follows that $\rho(x_0, y_0) = 0$ and therefore $x_0 = y_0$.

Now let x_1 be any point in X. We want to show that the sequence $\{x_n\}$ generated by Eq. (5.8.7) is a Cauchy sequence. For any $n > 1$, we have $\rho(x_{n+1}, x_n) \leq c\rho(x_n, x_{n-1})$. By induction we see that $\rho(x_{n+1}, x_n) \leq c^{n-1}\rho(x_2, x_1)$ for $n = 1, 2, \ldots$. Thus, for any $m > n$ we have

$$\rho(x_m, x_n) \leq \sum_{k=n}^{m-1} \rho(x_{k+1}, x_k) \leq c^{n-1}\rho(x_2, x_1)[1 + c + \ldots + c^{m-n-1}]$$
$$\leq \frac{c^{n-1}\rho(x_2, x_1)}{1 - c}.$$

Since $0 < c < 1$, the right-hand side of the above inequality can be made arbitrarily small by choosing n sufficiently large. Thus, $\{x_n\}$ is a Cauchy sequence.

Next, since $\{X; \rho\}$ is complete, it follows that $\{x_n\}$ converges; i.e., $\lim_n x_n$ exists. Let $\lim_n x_n = x$. Now since f is continuous on X, we have

$$\lim_n f(x_n) = f(\lim_n x_n).$$

But $f(\lim_n x_n) = f(x)$ and $\lim_n f(x_n) = \lim_n x_{n+1} = x$. Thus, $f(x) = x$ and we have proven the existence of a fixed point of f. Since we have already proven uniqueness, the proof is complete. ■

It may turn out that the composite function $f^{(n)} \triangleq f \circ f \circ \ldots \circ f$ is a contraction mapping, whereas f is not. The following result shows that such a mapping still has a unique fixed point.

5.8.8. Corollary. Let $\{X; \rho\}$ be a complete metric space, and let $f: X \longrightarrow X$ be continuous on X. If the composite function $f^{(n)} = f \circ f \circ \ldots \circ f$ is a contraction mapping, then there is a unique point $x_0 \in X$ such that

$$f(x_0) = x_0. \tag{5.8.9}$$

Moreover, the fixed point can be determined by the method of successive approximations (see Theorem 5.8.5).

5.8.10. Exercise. Prove Corollary 5.8.8.

We will consider several applications of the above results in the last section of this chapter.

Before we can consider the Arzela-Ascoli theorem, we need to introduce the following concept.

5.8.11. Definition. Let $\mathcal{C}[a, b]$ denote the set of all continuous real-valued functions defined on the interval $[a, b]$ of the real line R. A subset Y of $\mathcal{C}[a, b]$ is said to be **equicontinuous on** $[a, b]$ if for every $\epsilon > 0$ there exists a $\delta > 0$ such that $|x(t) - x(t_0)| < \epsilon$ for all $x \in Y$ and all t, t_0 such that $|t - t_0| < \delta$.

Note that in this definition δ depends only on ϵ and not on x or t and t_0. We now state and prove the **Arzela-Ascoli theorem.**

5.8.12. Theorem. Let $\{\mathcal{C}[a, b]; \rho_\infty\}$ be the metric space defined in Example 5.3.14. Let Y be a bounded subset of $\mathcal{C}[a, b]$. If Y is equicontinuous on $[a, b]$, then Y is relatively compact in $\mathcal{C}[a, b]$.

Proof. For each positive integer k, let us divide the interval $[a, b]$ into k equal parts by the set of points $V_k = \{t_{0k}, t_{1k}, \ldots, t_{kk}\} \subset [a, b]$. That is, $a = t_{0k} < t_{1k} < \ldots < t_{kk} = b$, where $t_{ik} = a + (i/k)(b - a)$, $i = 0, 1, \ldots, k$, and $[a, b] = \bigcup_{i=1}^{k} [t_{(i-1)k}, t_{ik}]$ for all $k = 1, 2, \ldots$. Since each V_k is a finite set, $\bigcup_{k=1}^{\infty} V_k$ is a countable set. For convenience of notation, let us denote this set by $\{\tau_1, \tau_2, \ldots\}$. The ordering of this set is immaterial. Next, since Y is bounded, there is a $\gamma > 0$ such that $\rho_\infty(x, y) \leq \gamma$ for all $x, y \in Y$. Let x_0 be held fixed in Y, and let $y \in Y$ be arbitrary. Let $0 \in \mathcal{C}[a, b]$ be the function which is zero for all $t \in [a, b]$. Then $\rho_\infty(y, 0) \leq \rho_\infty(y, x_0) + \rho_\infty(x_0, 0)$. Hence, $\rho_\infty(y, 0) \leq M$ for all $y \in Y$, where $M = \gamma + \rho_\infty(x_0, 0)$. This implies that $\sup_{t \in [a,b]} |y(t)| \leq M$ for all $y \in Y$. Now, let $\{y_n\}$ be an arbitrary sequence in Y. We want to show that $\{y_n\}$ contains a convergent subsequence. Since $|y_n(\tau_1)| \leq M$ for all n, the sequence of real numbers $\{y_n(\tau_1)\}$ contains a convergent subsequence which we shall call $\{y_{1n}(\tau_1)\}$. Again, since $|y_{1n}(\tau_2)| \leq M$ for all n, the sequence of real numbers $\{y_{1n}(\tau_2)\}$ contains a convergent subsequence which we shall call $\{y_{2n}(\tau_2)\}$. We see that $\{y_{2n}(\tau_1)\}$ is a subsequence of $\{y_{1n}(\tau_1)\}$, and hence it is convergent. Proceeding in a similar fashion, we obtain sequences $\{y_{1n}\}, \{y_{2n}\}, \ldots$ such that $\{y_{kn}\}$ is a subsequence of $\{y_{jn}\}$ for all $k > j$. Furthermore, each sequence is such that $\lim_n y_{kn}(\tau_i)$ exists for each i such that $1 \leq i \leq k$. Now let $\{x_n\}$ be the sequence $\{y_{nn}\}$. Then $\{x_n\}$ is a subsequence of $\{y_{kn}\}$ and $\lim_n x_n(\tau_i)$ exists for $i = 1, 2, \ldots$. We now wish to show that $\{x_n\}$ is a Cauchy sequence in $\{\mathcal{C}[a, b]; \rho_\infty\}$. Let $\epsilon > 0$ be given. Since Y is equicontinuous on $[a, b]$, we can find a positive number k such that $|x_n(t) - x_n(t')| < \epsilon/3$ for every n whenever $|t - t'| < 1/k$. Since $\{x_n(\tau_i)\}$ is a convergent sequence of real numbers, there exists a positive integer N such that $|x_n(\tau_i) - x_m(\tau_i)| < \epsilon/3$ whenever $m \geq N$ and $n \geq N$ for all $\tau_i \in V_k$. Now, if $t \in [a, b]$, there is some $\tau_i \in V_k$ such that $|t - \tau_i| < 1/k$.

Hence, for all $m \geq N$ and $n \geq N$, we have

$$|x_n(t) - x_m(t)| \leq |x_n(t) - x_n(t_i)| + |x_n(t_i) - x_m(t_i)| + |x_m(t_i) - x_m(t)| < \epsilon.$$

This implies that $\rho_\infty(x_m, x_n) < \epsilon$ for all $m, n \geq N$. Therefore, $\{x_n\}$ is a Cauchy sequence in $\mathcal{C}[a, b]$. Since $\{\mathcal{C}[a, b]; \rho_\infty\}$ is a complete metric space (see Example 5.5.28), $\{x_n\}$ converges to some point in $\mathcal{C}[a, b]$. This implies that $\{y_n\}$ has a subsequence which converges to a point in $\mathcal{C}[a, b]$ and so, by Theorem 5.6.31, Y is relatively compact in $\mathcal{C}[a, b]$. This completes the proof of the theorem. ■

Our next result follows directly from Theorem 5.8.12. It is sometimes referred to as **Ascoli's lemma.**

5.8.13. Corollary. Let $\{\varphi_n\}$ be a sequence of functions in $\{\mathcal{C}[a, b]; \rho_\infty\}$. If $\{\varphi_n\}$ is equicontinuous on $[a, b]$ and uniformly bounded on $[a, b]$ (i.e., there exists an $M > 0$ such that $\sup_{a \leq t \leq b} |\varphi_n(t)| \leq M$ for all n), then there exists a $\varphi \in \mathcal{C}[a, b]$ and a subsequence $\{\varphi_{n_k}\}$ of $\{\varphi_n\}$ such that $\{\varphi_{n_k}\}$ converges to φ uniformly on $[a, b]$.

5.8.14. Exercise. Prove Corollary 5.8.13.

We close the present section with the following converse to Theorem 5.8.12.

5.8.15. Theorem. Let Y be a subset of $\mathcal{C}[a, b]$ which is relatively compact in the metric space $\{\mathcal{C}[a, b]; \rho_\infty\}$. Then Y is a bounded set and is equicontinuous on $[a, b]$.

5.8.16. Exercise. Prove Theorem 5.8.15.

5.9. EQUIVALENT AND HOMEOMORPHIC METRIC SPACES. TOPOLOGICAL SPACES

It is possible that seemingly different metric spaces may exhibit properties which are very similar with regard to such concepts as open sets, limits of sequences, and continuity of functions. For example, for each p, $1 \leq p \leq \infty$, the spaces R_p^n (see Examples 5.3.1, 5.3.3) are different metric spaces. However, it turns out that the family of all open sets is the same in all of these metric spaces for $1 \leq p \leq \infty$ (e.g., the family of open sets in R_1^n is the same as the family of open sets in R_2^n, which is the same as the family of open sets in R_3^n, etc.). Furthermore, metric spaces which are not even defined on the same underlying set (e.g., the metric spaces $\{X; \rho_x\}$ and $\{Y; \rho_y\}$, where $X \neq Y$) may have many similar properties of the type mentioned above.

We begin with equivalence of metric spaces defined on the same underlying set.

5.9.1. Definition. Let $\{X; \rho_1\}$ and $\{X; \rho_2\}$ be two metric spaces defined on the same underlying set X. Let $\mathfrak{J}_1$ and $\mathfrak{J}_2$ be the topology of X determined by ρ_1 and ρ_2, respectively. Then the metrics ρ_1 and ρ_2 are said to be **equivalent metrics** if $\mathfrak{J}_1 = \mathfrak{J}_2$.

Throughout the present section we use the notation

$$f\colon \{X; \rho_1\} \longrightarrow \{Y; \rho_2\}$$

to indicate a mapping from X into Y, where the metric on X is ρ_1 and the metric on Y is ρ_2. This distinction becomes important in the case where $X = Y$, i.e. in the case $f\colon \{X; \rho_1\} \longrightarrow \{X; \rho_2\}$.

Let us denote by i the identity mapping from X onto X; i.e., $i(x) = x$ for all $x \in X$. Clearly, i is a bijective mapping, and the inverse is simply i itself. However, since the domain and range of i may have different metrics associated with them, we shall write

$$i\colon \{X; \rho_1\} \longrightarrow \{X; \rho_2\}$$

and

$$i^{-1}\colon \{X; \rho_2\} \longrightarrow \{X; \rho_1\}.$$

With the foregoing statements in mind, we provide in the following theorem a number of equivalent statements to characterize equivalent metrics.

5.9.2. Theorem. Let $\{X; \rho_1\}$, $\{X; \rho_2\}$, and $\{Y; \rho_3\}$ be metric spaces. Then the following statements are equivalent:

(i) ρ_1 and ρ_2 are equivalent metrics;
(ii) for any mapping $f\colon X \longrightarrow Y$, $f\colon \{X; \rho_1\} \longrightarrow \{Y; \rho_3\}$ is continuous on X if and only if $f\colon \{X; \rho_2\} \longrightarrow \{Y; \rho_3\}$ is continuous on X;
(iii) the mapping $i\colon \{X; \rho_1\} \longrightarrow \{X; \rho_2\}$ is continuous on X, and the mapping $i^{-1}\colon \{X; \rho_2\} \longrightarrow \{X; \rho_1\}$ is continuous on X; and
(iv) for any sequence $\{x_n\}$ in X, $\{x_n\}$ converges to a point x in $\{X; \rho_1\}$ if and only if $\{x_n\}$ converges to x in $\{X; \rho_2\}$.

Proof. To prove this theorem we show that statement (i) implies statement (ii); that statement (ii) implies statement (iii); that statement (iii) implies statement (iv); and that statement (iv) implies statement (i).

To show that (i) implies (ii), assume that ρ_1 and ρ_2 are equivalent metrics, and let f be any continuous mapping from $\{X; \rho_1\}$ into $\{Y; \rho_3\}$. Let U be any open set in $\{Y; \rho_3\}$. Since f is continuous, $f^{-1}(U)$ is an open set in $\{X; \rho_1\}$. Since ρ_1 and ρ_2 are equivalent metrics, $f^{-1}(U)$ is also an open set in $\{X; \rho_2\}$. Hence, the mapping $f\colon \{X; \rho_2\} \longrightarrow \{Y; \rho_3\}$ is continuous. The proof of the converse in statement (ii) is identical.

We now show that (ii) implies (iii). Clearly, the mapping $i: \{X; \rho_2\} \longrightarrow \{X; \rho_2\}$ is continuous. Now assume the validity of statement (ii), and let $\{Y; \rho_3\} = \{X; \rho_2\}$. Then $i: \{X; \rho_1\} \longrightarrow \{X; \rho_2\}$ is continuous. Again, it is clear that $i^{-1}: \{X; \rho_1\} \longrightarrow \{X; \rho_1\}$ is continuous. Letting $\{Y; \rho_3\} = \{X; \rho_1\}$ in statement (ii), it follows that $i^{-1}: \{X; \rho_2\} \longrightarrow \{X; \rho_1\}$ is continuous.

Next, we show that (iii) implies (iv). Let $i: \{X; \rho_1\} \longrightarrow \{X; \rho_2\}$ be continuous, and let the sequence $\{x_n\}$ in metric space $\{X; \rho_1\}$ converge to x. By Theorem 5.7.8, $\lim_n i(x_n) = i(x)$; i.e., $\lim_n x_n = x$ in $\{X; \rho_2\}$. The converse is proven in the same manner.

Finally, we show that (iv) implies (i). Let U be an open set in $\{X; \rho_1\}$. Then $U^{\sim}$ is closed in $\{X; \rho_1\}$. Now let $\{x_n\}$ be a sequence in $U^{\sim}$ which converges to x in $\{X; \rho_1\}$. Then $x \in U^{\sim}$ by part (iii) of Theorem 5.5.8. By assumption, $\{x_n\}$ converges to x in $\{X; \rho_2\}$ also. Furthermore, since $x \in U^{\sim}$, $U^{\sim}$ is closed in $\{X; \rho_2\}$, by part (iii) of Theorem 5.5.8. Hence, U is open in $\{X; \rho_2\}$. Letting U be an open set in $\{X; \rho_2\}$, by the same reasoning we conclude that U is open in $\{X; \rho_1\}$. Thus, ρ_1 and ρ_2 are equivalent metrics.

This concludes the proof of the theorem. ■

The next result establishes sufficient conditions for two metrics to be equivalent. These conditions are not necessary, however.

5.9.3. Theorem. Let $\{X; \rho_1\}$ and $\{X; \rho_2\}$ be two metric spaces. If there exist two positive real numbers, γ and λ, such that

$$\lambda \rho_2(x, y) \leq \rho_1(x, y) \leq \gamma \rho_2(x, y)$$

for all $x, y \in X$, then ρ_1 and ρ_2 are equivalent metrics.

5.9.4. Exercise. Prove Theorem 5.9.3.

Let us now consider some specific examples of equivalent metric spaces.

5.9.5. Exercise. Let $\{X; \rho\}$ be any metric space. For the example of Exercise 5.1.10 the reader showed that $\{X; \rho_1\}$ is a metric space, where

$$\rho_1(x, y) = \frac{\rho(x, y)}{1 + \rho(x, y)}$$

for all $x, y \in X$. Show that ρ and ρ_1 are equivalent metrics.

5.9.6. Theorem. Let $\{R^n; \rho_1\} = R_1^n$ and $\{R^n; \rho_2\} = R_2^n$ be the metric spaces defined in Example 5.3.1, and let $\{R^n; \rho_\infty\}$ be the metric space defined in Example 5.3.3. Then

(i) $\rho_\infty(x, y) \leq \rho_2(x, y) \leq \sqrt{n}\, \rho_\infty(x, y)$ for all $x, y \in R^n$;
(ii) $\rho_\infty(x, y) \leq \rho_1(x, y) \leq n\rho_\infty(x, y)$ for all $x, y \in R^n$; and
(iii) ρ_1, ρ_2, and ρ_∞ are equivalent metrics.

5.9.7. Exercise. Prove Theorem 5.9.6.

It can be shown that for the metric spaces $\{R^n; \rho_p\}$ and $\{R^n; \rho_q\}$, ρ_p and ρ_q are equivalent metrics for any p, q such that $1 \leq p \leq \infty$, $1 \leq q \leq \infty$.

In Example 5.1.12, we defined a metric ρ_*, called the **usual metric** for R^*. Up until now, it has not been apparent that there is any meaningful connection between ρ_* and the usual metric for R. The following result shows that when ρ_* is restricted to R, it is equivalent to the usual metric on R.

5.9.8. Theorem. Let $\{R; \rho\}$ denote the real line with the usual metric, and let $\{R^*; \rho_*\}$ denote the extended real line (see Exercise 5.1.12). Consider $\{R; \rho_*\}$ which is a metric subspace of $\{R^*; \rho_*\}$. Then

(i) for the metric spaces $\{R; \rho\}$ and $\{R; \rho_*\}$, ρ and ρ_* are equivalent metrics;
(ii) if $U \subset R$, then U is open in $\{R; \rho\}$ if and only if U is open in $\{R^*; \rho_*\}$; and
(iii) if U is open in $\{R^*; \rho_*\}$, then $U \cap R$, $U - \{+\infty\}$, and $U - \{-\infty\}$ are open in $\{R^*; \rho_*\}$.

5.9.9. Exercise. Prove Theorem 5.9.8. (Hint: Use part (iii) of Theorem 5.9.2 to prove part (i) of this theorem.)

Our next example shows that i^{-1} need not be continuous, even though i is continuous.

5.9.10. Example. Let X be any non-empty set, and let ρ_1 be the discrete metric on X (see Example 5.1.7). In Exercise 5.4.26 the reader was asked to show that every subset of X is open in $\{X; \rho_1\}$. Now let $\{X; \rho\}$ be an arbitrary metric space with the same underlying set X. Clearly, $i: \{X; \rho_1\} \longrightarrow \{X; \rho\}$ is continuous. However, $i^{-1}: \{X; \rho\} \longrightarrow \{X; \rho_1\}$ is not continuous unless every subset of $\{X; \rho\}$ is open. Since this is usually not true, i^{-1} need not be continuous. ■

Next, we introduce the concepts of homeomorphism and homeomorphic metric spaces.

5.9.11. Definition. Two metric spaces $\{X; \rho_x\}$ and $\{Y; \rho_y\}$ are said to be **homeomorphic** if there exists a mapping $\varphi: \{X; \rho_x\} \longrightarrow \{Y; \rho_y\}$ such that (i) φ is a bijective mapping of X onto Y, and (ii) $E \subset X$ is open in $\{X; \rho_x\}$ if and only if $\varphi(E)$ is open in $\{Y; \rho_y\}$. The mapping φ is called a **homeomorphism**.

We immediately have the following generalization of Theorem 5.9.2.

5.9.12. Theorem. Let $\{X; \rho_x\}$, $\{Y; \rho_y\}$, and $\{Z; \rho_z\}$ be metric spaces, and let φ be a bijective mapping of $\{X; \rho_x\}$ onto $\{Y; \rho_y\}$. Then the following statements are equivalent.

(i) φ is a homeomorphism;

(ii) for any mapping $f: X \longrightarrow Z$, $f: \{X; \rho_x\} \longrightarrow \{Z; \rho_z\}$ is continuous on X if and only if $f \circ \varphi^{-1}: \{Y; \rho_y\} \longrightarrow \{Z; \rho_z\}$ is continuous on Y;

(iii) $\varphi: \{X; \rho_x\} \longrightarrow \{Y; \rho_y\}$ is continuous and $\varphi^{-1}: \{Y; \rho_y\} \longrightarrow \{X; \rho_x\}$ is continuous; and

(iv) for any sequence $\{x_n\}$ in X, $\{x_n\}$ converges to a point x in $\{X; \rho_x\}$ if and only if $\{\varphi(x_n)\}$ converges to $\varphi(x)$ in $\{Y; \rho_y\}$.

5.9.13. Exercise. Prove Theorem 5.9.12.

The connection between homeomorphic metric spaces defined on the same underlying set and equivalent metrics is provided by the next result.

5.9.14. Theorem. Let $\{X; \rho_1\}$ and $\{X; \rho_2\}$ be two metric spaces with the same underlying set X. Then ρ_1 and ρ_2 are equivalent if and only if the identity mapping $i: \{X; \rho_1\} \longrightarrow \{X; \rho_2\}$ is a homeomorphism.

5.9.15. Exercise. Prove Theorem 5.9.14.

It is possible for $\{X; \rho_1\}$ and $\{X; \rho_2\}$ to be homeomorphic, even though ρ_1 and ρ_2 may not be equivalent.

There are important cases for which the metric relations between the elements of two distinct metric spaces are the same. In such cases only the nature of the elements of the metric spaces differ. Since this difference may be of no importance, such spaces may often be viewed as being essentially identical. Such metric spaces are said to be isometric. Specifically, we have:

5.9.16. Definition. Let $\{X; \rho_x\}$ and $\{Y; \rho_y\}$ be two metric spaces, and let $\varphi: \{X; \rho_x\} \longrightarrow \{Y; \rho_y\}$ be a bijective mapping of X onto Y. The mapping φ is said to be an **isometry** if

$$\rho_x(x, y) = \rho_y(\varphi(x), \varphi(y))$$

for all $x, y \in X$. If such an isometry exists, then the metric spaces $\{X; \rho_x\}$ and $\{Y; \rho_y\}$ are said to be **isometric**.

5.9.17. Theorem. Let φ be an isometry. Then φ is a homeomorphism.

5.9.18. Exercise. Prove Theorem 5.9.17.

We close the present section by introducing the concept of topological space. It turns out that metric spaces are special cases of such spaces.

In Theorem 5.4.15 we showed that, in the case of a metric space $\{X; \rho\}$, (i) the empty set $\varnothing$ and the entire space X are open; (ii) the union of an arbitrary collection of open sets is open; and (iii) the intersection of a finite collection of open sets is open. Examining the various proofs of the present chapter, we note that a great deal of the development of metric spaces is not a consequence of the metric but, rather, depends only on the properties of certain open and closed sets. Taking the notion of open set as basic (instead of the concept of distance, as in the case of metric spaces) and taking the aforementioned properties of open sets as postulates, we can form a mathematical structure which is much more general than the metric space.

5.9.19. Definition. Let X be a non-void set of points, and let $\mathfrak{I}$ be a family of subsets which we will call **open**. We call the pair $\{X; \mathfrak{I}\}$ a **topological space** if the following hold:

(i) $X \in \mathfrak{I}$, $\varnothing \in \mathfrak{I}$;
(ii) if $U_1 \in \mathfrak{I}$ and $U_2 \in \mathfrak{I}$, then $U_1 \cap U_2 \in \mathfrak{I}$; and
(iii) for any index set A, if $\alpha \in A$, and $U_\alpha \in \mathfrak{I}$, then $\bigcup_{\alpha \in A} U_\alpha \in \mathfrak{I}$.

The family $\mathfrak{I}$ is called the **topology** for the set X. The complement of an open set $U \in \mathfrak{I}$ with respect to X is called a **closed set**.

The reader can readily verify the following results:

5.9.20. Theorem. Let $\{X; \mathfrak{I}\}$ be a topological space. Then

(i) $\varnothing$ is closed;
(ii) X is closed;
(iii) the union of a *finite* number of closed sets is closed; and
(iv) the intersection of an *arbitrary* collection of closed sets is closed.

5.9.21. Exercise. Prove Theorem 5.9.20.

We close the present section by citing several specific examples of topological spaces.

5.9.22. Example. In view of Theorem 5.4.15, every metric space is a topological space. ■

5.9.23. Example. Let $X = \{x, y\}$, and let the open sets in X be the void set $\varnothing$, the set X itself, and the set $\{x\}$. If $\mathfrak{I}$ is defined in this way, then $\{X; \mathfrak{I}\}$ is a topological space. In this case the closed sets are $\varnothing$, X, and $\{y\}$. ■

5.9.24. Example. Although many fundamental concepts carry over from metric spaces to topological spaces, it turns out that the concept of topological space is often too general. Therefore, it is convenient to suppose that certain topological spaces satisfy some additional conditions which are also true in metric spaces. These conditions, called the **separation axioms**, are imposed on topological spaces $\{X; \mathfrak{I}\}$ to form the following important special cases:

T_1-spaces: A topological space $\{X; \mathfrak{J}\}$ is called a **T_1-space** if every set consisting of a single point is closed. Equivalently, a space is called a T_1-space, provided that if x and y are distinct points there is an open set containing y but not x. Clearly, metric spaces satisfy the T_1-axiom.

T_2-spaces: A topological space $\{X; \mathfrak{J}\}$ is called a **T_2-space** if for all distinct points $x, y \in X$ there are disjoint open sets U_x and U_y such that $x \in U_x$ and $y \in U_y$. T_2-spaces are also called **Hausdorff spaces**. All metric spaces are Hausdorff spaces. Also, all T_2-spaces are T_1-spaces. However, there are T_1-spaces which do not satisfy the T_2-separation axiom.

T_3-spaces: A topological space $\{X; \mathfrak{J}\}$ is called a **T_3-space** if (i) it is a T_1-space, and (ii) if given a closed set Y and a point x not in Y there are disjoint open sets U_1 and U_2 such that $x \in U_1$ and $Y \subset U_2$. T_3-spaces are also called **regular topological spaces**. All metric spaces are T_3-spaces. All T_3-spaces are T_2-spaces; however, not all T_2-spaces are T_3-spaces.

T_4-spaces: A topological space $\{X; \mathfrak{J}\}$ is called a **T_4-space** if (i) it is a T_1-space, and (ii) if for each pair of disjoint closed sets Y_1, Y_2 in X there exists a pair of disjoint open sets U_1, U_2 such that $Y_1 \subset U_1$ and $Y_2 \subset U_2$. T_4-spaces are also called **normal topological spaces**. Such spaces are clearly T_3-spaces. However, there are T_3-spaces which are not normal topological spaces. On the other hand, all metric spaces are T_4-spaces. ■

5.10. APPLICATIONS

The present section consists of two parts (subsections A and B).

In the first part we make extensive use of the contraction mapping principle to establish existence and uniqueness results for various types of equations. This part consists essentially of some specific examples.

In the second part, we continue the discussion of Section 4.11, dealing with ordinary differential equations. Specifically, we will apply Ascoli's lemma, and we will answer the questions raised at the end of subsection 4.11A.

A. Applications of the Contraction Mapping Principle

In our first example we consider a scalar algebraic equation which may be linear or nonlinear.

5.10.1. Example. Consider the equation

$$x = f(x), \tag{5.10.2}$$

where $f: [a, b] \longrightarrow [a, b]$ and where $[a, b]$ is a closed interval of R. Let $L > 0$, and assume that f satisfies the condition

$$|f(x_2) - f(x_1)| \leq L|x_2 - x_1| \tag{5.10.3}$$

for all $x_1, x_2 \in [a, b]$. In this case f is said to satisfy a **Lipschitz condition**, and L is called a **Lipschitz constant**.

Now consider the complete metric space $\{R; \rho\}$, where ρ denotes the usual metric on the real line. Then $\{[a, b]; \rho\}$ is a complete metric subspace of $\{R; \rho\}$ (see Theorem 5.5.33). If in (5.10.3) we assume that $L < 1$, then f is clearly a contraction mapping, and Theorem 5.8.5 applies. It follows that if $L < 1$, then Eq. (5.10.2) possesses a unique solution. Specifically, if $x_0 \in [a, b]$, then the sequence $\{x_n\}$, $n = 1, 2, \ldots$ determined by $x_n = f(x_{n-1})$ converges to the unique solution of Eq. (5.10.2).

Note that if $|df(x)/dx| = |f'(x)| \leq c < 1$ on the interval $[a, b]$ (in this case $f'(a)$ denotes the right-hand derivative of f at a, and $f'(b)$ denotes the left-hand derivative of f at b), then f is clearly a contraction.

In Figures J and K the applicability of the contraction mapping principle

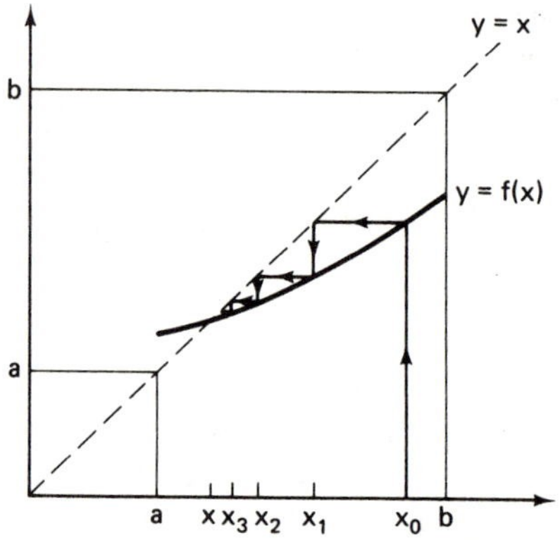

5.10.4. **Figure J.** Successive approximations (convergent case).

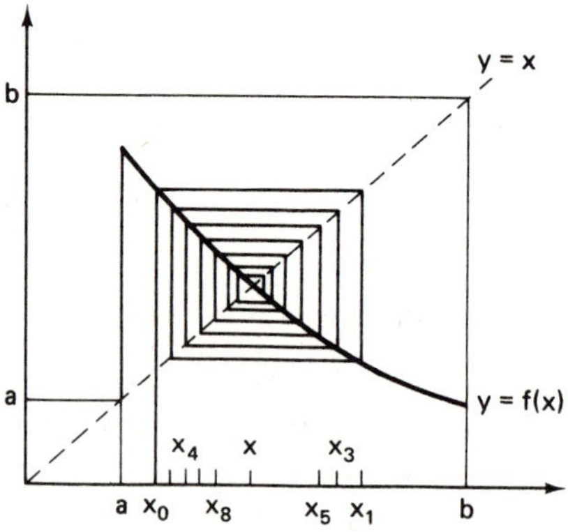

5.10.5. **Figure K.** Successive approximations (convergent case).

is demonstrated pictorially. As indicated, the sequence $\{x_n\}$ determined by successive approximations converges to the fixed point x. ■

In our next example we consider a system of linear equations.

5.10.6. Example. Consider the system of n linear equations

$$\xi_i = \sum_{j=1}^{n} a_{ij}\xi_j + \beta_i, \quad i = 1, \ldots, n. \tag{5.10.7}$$

Assume that $x = (\xi_1, \ldots, \xi_n) \in R^n$, $b = (\beta_1, \ldots, \beta_n) \in R^n$, and $a_{ij} \in R$. Here the constants a_{ij}, β_i are known and the ξ_i are unknown. In the following we use the contraction mapping principle to determine conditions for the existence and uniqueness of solutions of Eq. (5.10.7). In doing so we consider different metric spaces. In all cases we let

$$y = f(x)$$

denote the mapping determined by the system of linear equations

$$\eta_i = \sum_{j=1}^{n} a_{ij}\xi_j + \beta_i, \quad i = 1, \ldots, n,$$

where $y = (\eta_1, \ldots, \eta_n) \in R^n$.

First we consider the complete space $\{R^n; \rho_1\} = R_1^n$. Let $y' = f(x')$, $y'' = f(x'')$, $x' = (\xi_1', \ldots, \xi_n')$ and $x'' = (\xi_1'', \ldots, \xi_n'')$. We have

$$\begin{aligned}\rho_1(y', y'') = \rho_1(f(x'), f(x'')) &= \sum_{i=1}^{n} \left| \sum_{j=1}^{n} a_{ij}\xi_j' + \beta_i - a_{ij}\xi_j'' - \beta_i \right| \\ &= \sum_{i=1}^{n} \left| \sum_{j=1}^{n} a_{ij}(\xi_j' - \xi_j'') \right| \leq \sum_{i=1}^{n} \sum_{j=1}^{n} |a_{ij}||\xi_j' - \xi_j''| \\ &\leq \max_j \left\{ \sum_{i=1}^{n} |a_{ij}| \right\} \rho_1(x', x''),\end{aligned}$$

where in the preceding the Hölder inequality for finite sums was used (see Theorem 5.2.1). Clearly, f is a contraction if the inequality

$$\sum_{i=1}^{n} |a_{ij}| \leq k < 1 \tag{5.10.8}$$

holds. Thus, Eq. (5.10.7) possesses a unique solution if (5.10.8) holds for all j.

Next, we consider the complete space $\{R^n; \rho_2\} = R_2^n$. We have

$$\begin{aligned}\rho_2^2(y', y'') = \rho_2^2(f(x'), f(x'')) &= \sum_{i=1}^{n} \left\{ \sum_{j=1}^{n} a_{ij}\xi_j' + \beta_i - a_{ij}\xi_j'' - \beta_i \right\}^2 \\ &= \sum_{i=1}^{n} \left\{ \sum_{j=1}^{n} a_{ij}(\xi_j' - \xi_j'') \right\}^2 \leq \left\{ \sum_{i=1}^{n} \sum_{j=1}^{n} a_{ij}^2 \right\} \rho_2^2(x', x''),\end{aligned}$$

where, in the preceding, the Schwarz inequality for finite sums was employed (see Theorem 5.2.1). It follows that f is a contraction, provided that the inequality

$$\sum_{i=1}^{n} \sum_{j=1}^{n} a_{ij}^2 \triangleq k^2 < 1 \tag{5.10.9}$$

holds. Therefore, Eq. (5.10.7) possesses a unique solution, if (5.10.9) is satisfied.

Lastly, let us consider the complete metric space $\{R^n; \rho_\infty\} = R^n_\infty$. We have

$$\rho_\infty(y', y'') = \rho_\infty(f(x'), f(x'')) = \max_i \left| \sum_{j=1}^{n} a_{ij}(\xi'_j - \xi''_j) \right|$$

$$\leq \max_i \sum_j |a_{ij}||\xi'_j - \xi''_j| \leq \max_i \sum_{j=1}^{n} |a_{ij}| \max_j |\xi'_j - \xi''_j|$$

$$= \{\max_i \sum_{j=1}^{n} |a_{ij}|\}\rho_\infty(x', x'').$$

Thus, f is a contraction if

$$\{\max_i \sum_{j=1}^{n} |a_{ij}|\} \triangleq k < 1. \tag{5.10.10}$$

Hence, if (5.10.10) holds, then Eq. (5.10.7) has a unique solution.

In summary, if any one of the conditions (5.10.8), (5.10.9), or (5.10.10) holds, then Eq. (5.10.7) possesses a unique solution, namely x. This solution can be determined by the successive approximation

$$\xi_i^{(k)} = \sum_{j=1}^{n} a_{ij}\xi_j^{(k-1)} + b_i, \quad k = 1, 2, \ldots, \tag{5.10.11}$$

for all $i = 1, \ldots, n$, with starting point $x^{(0)} = (\xi_1^{(0)}, \ldots, \xi_n^{(0)})$. ■

Next, let us consider an integral equation.

5.10.12. Example. Let $\varphi \in \mathfrak{C}[a, b]$ and let $K(s, t)$ be a real-valued function which is continuous on the square $[a, b] \times [a, b]$. Let $\lambda \in R$. We call

$$x(s) = \varphi(s) + \lambda \int_a^b K(s, t)x(t)dt \tag{5.10.13}$$

a **Fredholm non-homogeneous linear integral equation of the second kind.** In this equation x is the unknown, $K(s, t)$ and φ are specified, and λ is regarded as an arbitrary parameter.

We now show that for all $|\lambda|$ sufficiently small, Eq. (5.10.13) has a unique solution which is continuous on $[a, b]$. To this end, consider the complete metric space $\{\mathfrak{C}[a, b]; \rho_\infty\}$, and let $y = f(x)$ denote the mapping determined by

$$y(s) = \varphi(s) + \lambda \int_a^b K(s, t)x(t)dt.$$

Clearly $y \in \mathfrak{C}[a, b]$. We thus have $f: \mathfrak{C}[a, b] \longrightarrow \mathfrak{C}[a, b]$. Now let $M = \sup_{\substack{a \leq t \leq b \\ a \leq s \leq b}} |K(s, t)|$. Then

$$\rho_\infty(f(x_1), f(x_2)) \leq |\lambda| M(b - a)\rho_\infty(x_1, x_2).$$

Therefore, if we choose λ so that

$$|\lambda| = < \frac{1}{M(b - a)} \tag{5.10.14}$$

then f is a contraction mapping. From Theorem 5.8.5 it now follows that Eq. (5.10.13) possesses a unique solution $x \in \mathcal{C}[a, b]$, if (5.10.14) holds. Starting at $x_0 \in \mathcal{C}[a, b]$, successive approximations to this solution are given by

$$x_n(s) = \varphi(s) + \lambda \int_a^b K(s, t)x_{n-1}(t)dt, \quad n = 1, 2, 3, \ldots. \quad \blacksquare \qquad (5.10.15)$$

Next, we consider yet another type of integral equation.

5.10.16. Example. Let $\varphi \in \mathcal{C}[a, b]$, let $K(s, t)$ be a real continuous function on the triangle $a \leq t \leq s \leq b$, and let $\lambda \in R$. We call

$$x(s) = \varphi(s) + \lambda \int_a^s K(s, t)x(t)dt, \quad a \leq s \leq b, \qquad (5.10.17)$$

a **linear Volterra integral equation**. Here x is unknown, $K(s, t)$ and φ are specified, and λ is an arbitrary parameter.

We now show that, for all λ, Eq. (5.10.17) possesses a unique continuous solution. We consider again the complete metric space $\{\mathcal{C}[a, b]; \rho_\infty\}$, and we let $y = f(x)$ be the mapping determined by

$$y(s) = \varphi(s) + \lambda \int_a^s K(s, t)x(t)dt.$$

Since the right-hand side of this expression is continuous, it follows that $f: \mathcal{C}[a, b] \longrightarrow \mathcal{C}[a, b]$. Moreover, since K is continuous, there is an M such that $|K(s, t)| \leq M$. Let $y_1 = f(x_1)$, and let $y_2 = f(x_2)$. As in the preceding example, we have

$$\rho_\infty(f(x_1), f(x_2)) = \rho_\infty(y_1, y_2) \leq |\lambda| M(b - a)\rho_\infty(x_1, x_2).$$

Now let $f^{(n)}$ denote the composite mapping $f \circ f \circ \ldots \circ f$, and let $f^{(n)}(x) = y^{(n)}$. A little bit of algebra yields

$$\rho_\infty(f^{(n)}(x_1), f^{(n)}(x_2)) = \rho_\infty(y_1^{(n)}, y_2^{(n)}) \leq \frac{1}{n!}|\lambda|^n M^n(b - a)^n \rho_\infty(x_1, x_2). \qquad (5.10.18)$$

However, $\frac{1}{n!}|\lambda|^n M^n(b - a)^n \longrightarrow 0$ as $n \longrightarrow \infty$. Thus, for an arbitrary value of λ, n can be chosen so large that

$$k \triangleq \frac{1}{n!}|\lambda|^n M^n(b - a)^n < 1.$$

Hence, we have

$$\rho_\infty(f^{(n)}(x_1), f^{(n)}(x_2)) \leq k\rho_\infty(x_1, x_2), \quad 0 < k < 1.$$

Therefore, the composite mapping $f^{(n)}$ is a contraction mapping. It follows from Corollary 5.8.8 that Eq. (5.10.17) possesses a unique continuous solution for arbitrary λ. This solution can be determined by the method of successive approximations. $\blacksquare$

5.10.19. Exercise. Verify inequality (5.10.18).

Next we consider initial-value problems characterized by scalar ordinary differential equations.

5.10.20. Example. Consider the initial-value problem

$$\left.\begin{aligned} \dot{x} &= f(t, x) \\ x(\tau) &= \xi \end{aligned}\right\} \tag{5.10.21}$$

discussed in Section 4.11. We would like to determine conditions for the existence and uniqueness of a solution $\varphi(t)$ of (5.10.21) for $\tau \leq t \leq T$.

Let $k > 0$, and assume that f satisfies the condition

$$|f(t, x_1) - f(t, x_2)| \leq k|x_1 - x_2|$$

for all $t \in [\tau, T]$ and for all $x_1, x_2 \in R$. In this case we say that f satisfies a **Lipschitz condition in** x and we call k a **Lipschitz constant.**

As was pointed out in Section 4.11, Eq. (5.10.21) is equivalent to the integral equation

$$\varphi(t) = \xi + \int_\tau^t f(s, \varphi(s))ds. \tag{5.10.22}$$

Consider now the complete metric space $\{\mathcal{C}[\tau, T]; \rho_\infty\}$, and let

$$F(\varphi) = \xi + \int_\tau^t f(s, \varphi(s))ds, \quad \tau \leq t \leq T.$$

Then clearly $F: \mathcal{C}[\tau, T] \rightarrow \mathcal{C}[\tau, T]$. Now

$$\begin{aligned} \rho_\infty(F(\varphi_1), F(\varphi_2)) &= \sup_{\tau \leq t \leq T} \left| \int_\tau^t [f(s, \varphi_1(s)) - f(s, \varphi_2(s))]ds \right| \\ &\leq \sup_{\tau \leq t \leq T} \int_\tau^t k|\varphi_1(s) - \varphi_2(s)|\, ds \leq k(T - \tau)\rho_\infty(\varphi_1, \varphi_2). \end{aligned}$$

Thus, F is a contraction if $k < 1/(T - \tau)$.

Next, let $F^{(n)}$ denote the composite mapping $F \circ F \circ \ldots \circ F$. Similarly, as in (5.10.18), the reader can verify that

$$\rho_\infty(F^{(n)}(\varphi_1), F^{(n)}(\varphi_2)) \leq \frac{1}{n!} k^n(T - \tau)^n \rho_\infty(\varphi_1, \varphi_2). \tag{5.10.23}$$

Since $\frac{1}{n!} k^n(T - \tau)^n \rightarrow 0$ as $n \rightarrow \infty$, it follows that for sufficiently large n, $\frac{1}{n!} k^n(T - \tau)^n < 1$. Therefore, $F^{(n)}$ is a contraction. It now follows from Corollary 5.8.8 that Eq. (5.10.21) possesses a unique solution for $[\tau, T]$. Furthermore, this solution can be obtained by the method of successive approximations. ■

5.10.24. Exercise. Generalize Example 5.10.20 to the initial-value problem

$$\dot{x}_i = f_i(t, x_1, \ldots, x_n),$$
$$x_i(\tau) = \xi_i, \quad i = 1, \ldots, n,$$

which is discussed in Section 4.11.

B. Further Applications to Ordinary Differential Equations

At the end of Section 4.11A we raised the following questions: (i) When does an initial-value problem possess solutions? (ii) When are these solutions unique? (iii) What is the extent of the interval over which such solutions exist? (iv) Are these solutions continuously dependent on initial conditions?

In Example 5.10.20 we have already given a partial answer to the first two questions. In the remainder of the present section we refine the type of result given in Example 5.10.20, and we give an answer to the remaining items raised above.

As in the beginning of Section 4.11A, we call R^2 the (t, x) plane, we let $D \subset R^2$ denote an open connected set (i.e., D is a domain), we assume that f is a real-valued function which is defined and continuous on D, we call $T = (t_1, t_2) \subset R$ a t interval, and we let φ denote a solution of the differential equation

$$\dot{x} = f(t, x). \tag{5.10.25}$$

The reader should refer to Section 4.11A for the definition of solution φ.

We first concern ourselves with the initial-value problem

$$\dot{x} = f(t, x), \qquad x(\tau) = \xi \tag{5.10.26}$$

characterized in Definition 4.11.3. Our first result is concerned with existence of solutions of this problem. It is convenient to establish this result in two stages, using the notion of ϵ-approximate solution of Eq. (5.10.25).

5.10.27. Definition. A function φ defined and continuous on a t interval T is called an **ϵ-approximate solution** of Eq. (5.10.25) on T if

(i) $(t, \varphi(t)) \in D$ for all $t \in T$;

(ii) φ has a continuous derivative on T except possibly on a finite set S of points in T, where there are jump discontinuities allowed; and

(iii) $|\dot{\varphi}(t) - f(t, \varphi(t))| < \epsilon$ for all $t \in T - S$.

If S is not empty, φ is said to have **piecewise continuous derivatives on** T.

We now prove:

5.10.28. Theorem. In Eq. (5.10.25), let f be continuous on the rectangle

$$D_0 = \{(t, x): |t - \tau| \leq a, \quad |x - \xi| \leq b\}.$$

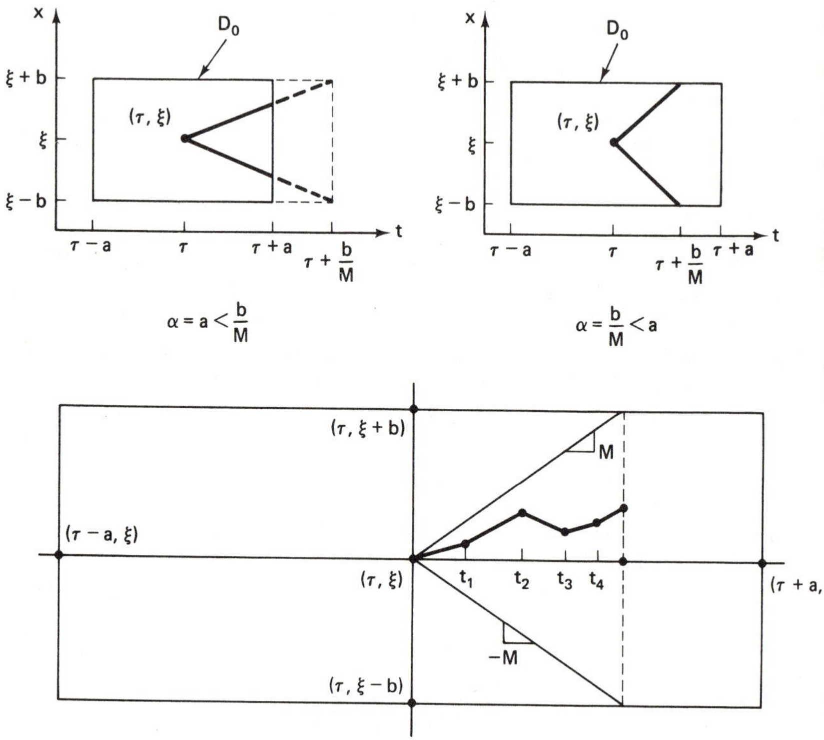

5.10.29. Figure L. Construction of an ϵ-approximate solution.

Given any $\epsilon > 0$, there exists an ϵ-approximate solution φ of Eq. (5.10.25) on an interval $|t - \tau| \leq \alpha \leq a$ such that $\varphi(\tau) = \xi$.

Proof. Let $M = \max_{(t, x) \in D_0} |f(t, x)|$, and let $\alpha = \min(a, b/M)$. Note that $\alpha = a$ if $a < b/M$ and $\alpha = b/M$ if $a > b/M$ (refer to Figure L). We will show that an ϵ-approximate solution exists on the interval $[\tau, \tau + \alpha]$. The proof is similar for the interval $[\tau - \alpha, \tau]$. In our proof we will construct an ϵ-approximate solution starting at (τ, ξ), consisting of a finite number of straight line segments joined end to end (see Figure L).

Since f is continuous on the compact set D_0, it is uniformly continuous on D_0 (see Theorem 5.7.12). Hence, given $\epsilon > 0$, there exists $\delta = \delta(\epsilon) > 0$ such that $|f(t, x) - f(t', x')| < \epsilon$ whenever $(t, x), (t', x') \in D_0$, $|t - t'| < \delta$ and $|x - x'| < \delta$. Now let $\tau = t_0$ and $\tau + \alpha = t_n$. We divide the half-open interval $(t_0, t_n]$ into n half-open subintervals $(t_0, t_1], (t_1, t_2], \ldots, (t_{n-1}, t_n]$ in such a fashion that

$$\max |t_i - t_{i-1}| < \min\left(\delta, \frac{\delta}{M}\right), \tag{5.10.30}$$

Next, we construct a polygonal path consisting of n straight lines joined end to end, starting at the point $(\tau, \xi) \triangleq (t_0, \xi_0)$ and having slopes equal to $m_{i-1} = f(t_{i-1}, \xi_{i-1})$ over the intervals $(t_{i-1}, t_i]$, $i = 1, \ldots, n$, respectively, where $\xi_i = \xi_{i-1} + m_{i-1}|t_i - t_{i-1}|$. A typical polygonal path is shown in Figure L. Note that the graph of this path is confined to the triangular region in Figure L. Let us denote the polygonal path constructed in this way by φ. Note that φ is continuous on the interval $[\tau, \tau + \alpha]$, that φ is a piecewise linear function, and that φ is piecewise continuously differentiable. Indeed, we have $\varphi(\tau) = \xi_0 = \xi$ and

$$\varphi(t) = \varphi(t_{i-1}) + f(t_{i-1}, \varphi(t_{i-1}))(t - t_{i-1}), \quad t_{i-1} < t \leq t_i, \quad i = 1, \ldots, n. \tag{5.10.31}$$

Also note that

$$|\varphi(t) - \varphi(t')| \leq M|t - t'| \tag{5.10.32}$$

for all $t, t' \in [\tau, \tau + \alpha]$. We now show that φ is an ϵ-approximate solution. Let $t \in (t_{i-1}, t_i)$. Then it follows from (5.10.30) and (5.10.32) that $|\varphi(t) - \varphi(t_{i-1})| \leq \delta$. Now since $|f(t, x) - f(t', x')| < \epsilon$ whenever $(t, x), (t', x') \in D_0$, $|t - t'| < \delta$, and $|x - x'| < \delta$, it follows from Eq. (5.10.31) that

$$|\dot{\varphi}(t) - f(t, \varphi))| = |f(t_{i-1}, \varphi(t_{i-1})) - f(t, \varphi(t))| < \epsilon.$$

Therefore, the function φ is an ϵ-approximate solution on the interval $|t - \tau| \leq \alpha \leq a$. ■

We are now in a position to establish conditions for the existence of solutions of the initial-value problem (5.10.26).

5.10.33. Theorem. In Eq. (5.10.25), let f be continuous on the rectangle $D_0 = \{(t, x): |t - \tau| \leq a, |x - \xi| \leq b\}$. Then the initial-value problem (5.10.26) has a solution on some t interval given by $|t - \tau| \leq \alpha \leq a$.

Proof. Let $\epsilon_n > 0$, $\epsilon_{n+1} < \epsilon_n$ and $\lim_{n \to \infty} \epsilon_n = 0$ (i.e., let $\{\epsilon_n\}$, $n = 1, 2, \ldots$, be a monotone decreasing sequence of positive numbers tending to zero). By Theorem 5.10.28, there exists for every ϵ_n an ϵ_n-approximate solution of Eq. (5.10.25), call it φ_n, on some interval $|t - \tau| \leq \alpha$ such that $\varphi_n(\tau) = \xi$. Now for each φ_n it is true, by construction of φ_n, that

$$|\varphi_n(t) - \varphi_n(t')| \leq M|t - t'|. \tag{5.10.34}$$

This shows that $\{\varphi_n\}$ is an equicontinuous set of functions (see Definition 5.8.11). Letting $t' = \tau$ in (5.10.34), we have $|\varphi_n(t) - \xi| \leq M|t - \tau| \leq M\alpha$, and thus $|\varphi_n(t)| \leq |\xi| + M\alpha$ for all n and for all $t \in [\tau, \tau + \alpha]$. Thus, the sequence $\{\varphi_n\}$ is uniformly bounded. In view of the Ascoli lemma (see Corollary 5.8.13) there exists a subsequence $\{\varphi_{n_k}\}$, $k = 1, 2, \ldots$ of the sequence $\{\varphi_n\}$ which converges uniformly on the interval $[\tau - \alpha, \tau + \alpha]$ to a limit function φ; i.e.,

$$\lim_{k \to \infty} \varphi_{n_k} = \varphi.$$

This function is continuous (see Theorem 5.7.14) and, in addition, $|\varphi(t) - \varphi(t')| \leq M|t - t'|$.

To complete the proof, we must show that φ is a solution of Eq. (5.10.26) or, equivalently, that φ satisfies the integral equation

$$\varphi(t) = \xi + \int_\tau^t f(s, \varphi(s))ds. \tag{5.10.35}$$

Let φ_{n_k} be an ϵ_{n_k}-approximate solution, let $\Delta_{n_k}(t) = \dot{\varphi}_{n_k}(t) - f(t, \varphi_{n_k}(t))$ at those points where φ_{n_k} is differentiable, and let $\Delta_{n_k}(t) = 0$ at the points where φ_{n_k} is not differentiable. Then φ_{n_k} can be expressed in integral form as

$$\varphi_{n_k}(t) = \xi + \int_\tau^t [f(s, \varphi_{n_k}(s)) + \Delta_{n_k}(s)]ds. \tag{5.10.36}$$

Since φ_{n_k} is an ϵ-approximate solution, we have $|\Delta_{n_k}(t)| < \epsilon_{n_k}$. Also, since f is uniformly continuous on D_0 and since $\varphi_{n_k} \longrightarrow \varphi$ uniformly on $[\tau - \alpha, \tau + \alpha]$, as $k \longrightarrow \infty$, it follows that $|f(t, \varphi_{n_k}(t)) - f(t, \varphi(t))| < \epsilon$ on the interval $[\tau - \alpha, \tau + \alpha]$ whenever k is so large that $|\varphi_{n_k}(t) - \varphi(t)| < \delta$ on $[\tau - \alpha, \tau + \alpha]$. Using Eq. (5.10.36) we now have

$$\left|\int_\tau^t [f(s, \varphi_{n_k}(s)) - f(s, \varphi(s)) + \Delta_{n_k}(s)]ds\right| \leq \left|\int_\tau^t \left|f(s, \varphi_{n_k}(s)) - f(s, \varphi(s))\right| ds\right|$$
$$+ \left|\int_\tau^t \left|\Delta_{n_k}(s)\right| ds\right| < \alpha(\epsilon_{n_k} + \epsilon) \triangleq \epsilon'.$$

Therefore, $\lim_{k\to\infty} \int_\tau^t [f(s, \varphi_{n_k}(s)) + \Delta_{n_k}(s)]ds = \int_\tau^t f(s, \varphi(s))ds$. It now follows that

$$\varphi(t) = \xi + \int_\tau^t f(s, \varphi(s))ds,$$

which completes the proof. ■

Using Theorem 5.10.33, the reader can readily prove the next result.

5.10.37. Corollary. In Eq. (5.10.25), let f be continuous on a domain D of the (t, x) plane, and let $(\tau, \xi) \in D$. Then the initial-value problem (5.10.26) has a solution φ on some t interval containing τ.

5.10.38. Exercise. Prove Theorem 5.10.37.

Theorem 5.10.33 (along with Corollary 5.10.37) is known in the literature as the **Cauchy-Peano existence theorem**. Note that in these results the solution φ is not guaranteed to be unique.

Next, we seek conditions under which uniqueness of solutions is assured. We require the following preliminary result, called the **Gronwall inequality**.

5.10.39. Theorem. Let r and k be real continuous functions on an interval $[a, b]$. Suppose $r(t) \geq 0$ and $k(t) \geq 0$ for all $t \in [a, b]$, and let $\delta \geq 0$ be a given non-negative constant. If

$$r(t) \leq \delta + \int_a^t k(s)r(s)ds \tag{5.10.40}$$

for all $t \in [a, b]$, then

$$r(t) \leq \delta e^{\int_a^t k(s)ds} \tag{5.10.41}$$

for all $t \in [a, b]$.

Proof. Let $R(t) = \delta + \int_a^t k(s)r(s)ds$. Then $r(t) \leq R(t)$, $R(a) = \delta$, $\dot{R}(t) = k(t)r(t) \leq k(t)R(t)$, and

$$\dot{R}(t) - k(t)R(t) \leq 0 \tag{5.10.42}$$

for all $t \in [a, b]$. Let $K(t) = e^{-\int_a^t k(s)ds}$. Then

$$\dot{K}(t) = -k(t)e^{-\int_a^t k(s)ds} = -K(t)k(t).$$

Multiplying both sides of (5.10.42) by $K(t)$ we have

$$K(t)\dot{R}(t) - K(t)k(t)R(t) \leq 0$$

or

$$K(t)\dot{R}(t) + \dot{K}(t)R(t) \leq 0$$

or

$$\frac{d}{dt}[K(t)R(t)] \leq 0.$$

Integrating this last expression from a to t we obtain

$$K(t)R(t) - K(a)R(a) \leq 0$$

or

$$K(t)R(t) - \delta \leq 0$$

or

$$e^{-\int_a^t k(s)ds} R(t) - \delta \leq 0$$

or

$$r(t) \leq R(t) \leq \delta e^{\int_a^t k(s)ds},$$

which is the desired inequality. ■

In our next result we will require that the function f in Eq. (5.10.25) satisfy a Lipschitz condition

$$|f(t, x') - f(t, x'')| < k|x' - x''|$$

for all $(t, x'), (t, x'') \in D$.

5.10.43. Theorem. In Eq. (5.10.25), let f be continuous on a domain D of the (t, x) plane, and let f satisfy a Lipschitz condition with respect to x on D. Let $(\tau, \xi) \in D$. Then the initial-value problem (5.10.26) has a unique solution on some t interval containing τ (i.e., if φ_1 and φ_2 are two solutions of Eq. (5.10.25) on an interval (a, b), if $\tau \in (a, b)$, and if $\varphi_1(\tau) = \varphi_2(\tau) = \xi$, then $\varphi_1 = \varphi_2$).

Proof. By Corollary 5.10.37, at least one solution exists on some interval (a, b), $\tau \in (a, b)$. Now suppose there is more than one solution, say φ_1 and φ_2, to the initial-value problem (5.10.26). Then

$$\varphi_i(t) = \xi + \int_\tau^t f(s, \varphi_i(s))ds, \quad i = 1, 2$$

for all $t \in (a, b)$, and

$$\varphi_1(t) - \varphi_2(t) = \int_\tau^t [f(s, \varphi_1(s)) - f(s, \varphi_2(s))]ds.$$

Let $r(t) = |\varphi_1(t) - \varphi_2(t)|$, and let $k > 0$ denote the Lipschitz constant for f. In the following we consider the case when $t \geq \tau$, and we leave the details of the proof for $t \leq \tau$ as an exercise. We have,

$$\begin{aligned} r(t) &\leq \int_\tau^t |f(s, \varphi_1(s)) - f(s, \varphi_2(s))|\, ds \leq \int_\tau^t k\, |\varphi_1(s) - \varphi_2(s)|\, ds \\ &= \int_\tau^t kr(s)ds; \end{aligned}$$

i.e.,

$$r(t) \leq \int_\tau^t kr(s)ds$$

for all $t \in [\tau, b)$. The conditions of Theorem 5.10.39 are clearly satisfied and we have: if $r(t) \leq \delta + \int_\tau^t kr(s)ds$, then $r(t) \leq \delta e^{\int_\tau^t kds}$. Since in the present case $\delta = 0$, it follows that

$$r(t) = 0 \text{ for all } t \in [\tau, b).$$

Therefore, $|\varphi_1(t) - \varphi_2(t)| = 0$ for all $t \in [\tau, b)$, and $\varphi_1(t) = \varphi_2(t)$ for all t in this interval. ■

Now suppose that in Eq. (5.10.25) f is continuous on some domain D of the (t, x) plane and assume that f is bounded on D; i.e., suppose there exists a constant $M > 0$ such that

$$\sup_{(t,x)\in D} |f(t, x)| \leq M.$$

Also, assume that $\tau \in (a, b)$, that $(\tau, \xi) \in D$, and that the initial-value problem (5.10.26) has a solution φ on a t interval (a, b) such that $(t, \varphi(t)) \in D$ for all $t \in (a, b)$. Then

$$\lim_{t\to a^+} \varphi(t) = \varphi(a^+) \quad \text{and} \quad \lim_{t\to b^-} \varphi(t) = \varphi(b^-)$$

exist. To prove this, let $t \in (a, b)$. Then

$$\varphi(t) = \xi + \int_\tau^t f(s, \varphi(s))ds.$$

If $a < t_1 < t_2 < b$, then

$$|\varphi(t_1) - \varphi(t_2)| \leq \int_{t_1}^{t_2} |f(s, \varphi(s))|\, ds \leq M\,|t_2 - t_1|.$$

Now let $t_1 \longrightarrow b$ and $t_2 \longrightarrow b$. Then $|t_1 - t_2| \longrightarrow 0$, and therefore $|\varphi(t_1) - \varphi(t_2)| \longrightarrow 0$. This limiting process yields thus a convergent Cauchy sequence; i.e., $\varphi(b^-)$ exists. The existence of $\varphi(a^+)$ is similarly established.

Next, let us assume that the points $(a, \varphi(a^+))$, $(b, \varphi(b^-))$ are in the domain D. We now show that the solution φ can be continued to the right of $t = b$. An identical procedure can be used to show that the solution φ can be continued to the left of $t = a$.

We define a function

$$\tilde{\varphi}(t) = \left\{ \begin{array}{ll} \varphi(t), & t \in (a, b) \\ \varphi(b^-), & t = b \end{array} \right\}.$$

Then

$$\tilde{\varphi}(t) = \xi + \int_\tau^t f(s, \tilde{\varphi}(s))ds$$

for all $t \in (a, b]$. Thus, the derivative of $\tilde{\varphi}(t)$ exists on the interval (a, b), and the left-hand derivative of $\tilde{\varphi}(t)$ at $t = b$ is given by

$$\tilde{\varphi}(b^-) = f(b, \tilde{\varphi}(b)).$$

Next, we consider the initial-value problem

$$\dot{x} = f(t, x)$$
$$x(b) = \varphi(b^-).$$

By Corollary 5.10.37, the differential equation $\dot{x} = f(t, x)$ has a solution ψ which passes through the point $(b, \varphi(b^-))$ and which exists on some interval $[b, b + \beta]$, $\beta > 0$. Now let

$$\hat{\varphi}(t) = \left\{ \begin{array}{ll} \tilde{\varphi}(t), & t \in (a, b] \\ \psi(t), & t \in [b, b + \beta] \end{array} \right\}.$$

To show that $\hat{\varphi}$ is a solution of the differential equation on the interval $(a, b + \beta]$, with $\hat{\varphi}(\tau) = \xi$, we must show that $\hat{\varphi}$ is continuous at $t = b$. Since

$$\varphi(b^-) = \xi + \int_\tau^b f(x, \hat{\varphi}(s))ds$$

and since

$$\hat{\varphi}(t) = \varphi(b^-) + \int_b^t f(s, \hat{\varphi}(s))ds,$$

we have

$$\hat{\varphi}(t) = \xi + \int_\tau^t f(s, \hat{\varphi}(s))ds$$

for all $t \in (a, b + \beta]$. The continuity of $\hat{\varphi}$ in the last equation implies the countinuity of $f(s, \hat{\varphi}(s))$. Differentiating the last equation, we have

$$\dot{\hat{\varphi}}(t) = f(t, \hat{\varphi}(t))$$

for all $t \in (a, b + \beta]$.

We call $\hat{\varphi}$ **a continuation** of the solution φ to the interval $(a, b + \beta]$. If f satisfies a Lipschitz condition on D with respect to x, then $\hat{\varphi}$ is unique, and we call $\hat{\varphi}$ **the continuation** of φ to the interval $(a, b + \beta]$.

We can repeat the above procedure of continuing solutions until the boundary of D is reached.

Now let the domain D be, in particular, a rectangle, as shown in Figure M. It is important to notice that, in general, we cannot continue solutions over the entire t interval T shown in this figure.

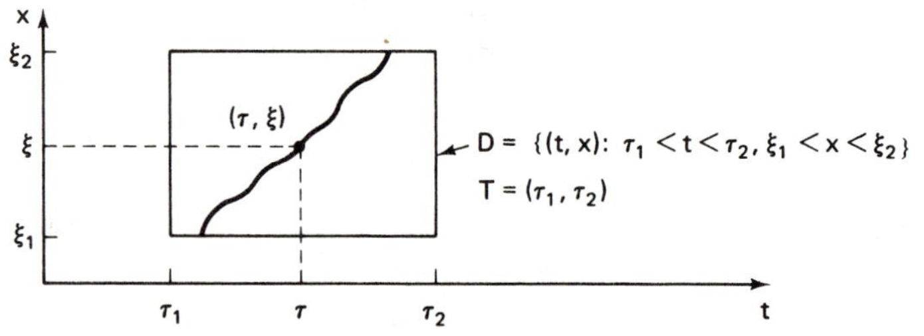

5.10.44. **Figure M.** Continuation of a solution to the boundary of domain D.

We summarize the above discussion in the following:

5.10.45. Theorem. In Eq. (5.10.25), let f be continuous and bound on a domain D of the (t, x) plane and let $(\tau, \xi) \in D$. Then all solutions of the initial-value problem (5.10.26) can be continued to the boundary of D.

We can readily extend Theorems 5.10.28, 5.10.33, Corollary 5.10.37, and Theorems 5.10.43 and 5.10.45 to initial-value problems characterized by systems of n first-order ordinary differential equations, as given in Definition 4.11.9 and Eq. 4.11.11. In doing so we replace $D \subset R^2$ by $D \subset R^{n+1}$, $x \in R$ by $\mathbf{x} \in R^n$, $f: D \longrightarrow R$ by $\mathbf{f}: D \longrightarrow R^n$, the absolute value $|x|$ by the quantity

$$|x| = \sum_{i=1}^{n} |x_i|, \tag{5.10.46}$$

and the metric $\rho(x, y) = |x - y|$ on R by the metric $\rho(\mathbf{x}, \mathbf{y}) = \sum_{i=1}^{n} |x_i - y_i|$ on R^n. (The reader can readily verify that the function given in Eq. (5.10.46) satisfies the axioms of a norm (see Theorem 4.9.31).) The definition of ϵ-approximate solution for the differential equation $\dot{\mathbf{x}} = \mathbf{f}(t, \mathbf{x})$ is identical to that given in Definition 5.10.27, save that scalars are replaced by vectors (e.g., the scalar function φ is replaced by the n-vector valued function $\boldsymbol{\varphi}$).

Also, the modifications involved in defining a Lipschitz condition for $\mathbf{f}(t, \mathbf{x})$ on $D \subset R^{n+1}$ are obvious.

5.10.47. Exercise. For the ordinary differential equation

$$\dot{\mathbf{x}} = \mathbf{f}(t, \mathbf{x}) \tag{5.10.48}$$

and for the initial-value problem

$$\dot{\mathbf{x}} = \mathbf{f}(t, \mathbf{x}), \qquad \mathbf{x}(\tau) = \boldsymbol{\xi} \tag{5.10.49}$$

characterized in Eq. (4.11.7) and Definition 4.11.9, respectively, state and prove results for existence, uniqueness, and continuation of solutions, which are analogous to Theorems 5.10.28, 5.10.33, Corollary 5.10.37, and Theorems 5.10.43 and 5.10.45.

In connection with Theorem 5.10.45 we noted that the solutions of initial-value problems described by non-linear ordinary differential equations can, in general, not be extended to the entire t interval T depicted in Figure M. We now show that in the case of initial-value problems characterized by linear ordinary differential equations it is possible to extend solutions to the entire interval T. First, we need some preliminary results.

Let

$$D = \{(t, \mathbf{x}) \colon a \leq t \leq b, \mathbf{x} \in R^n\} \tag{5.10.50}$$

where the function $|\cdot|$ is defined in Eq. (5.10.46). Consider the set of linear equations

$$\dot{x}_i = \sum_{j=1}^{n} a_{ij}(t) x_j \triangleq f_i(t, \mathbf{x}), \quad i = 1, \ldots, n \tag{5.10.51}$$

where the $a_{ij}(t)$, $i, j = 1, \ldots, n$, are assumed to be real and continuous functions defined on the interval $[a, b]$. We first show that $\mathbf{f}(t, \mathbf{x}) = [f_1(t, \mathbf{x}), \ldots, f_n(t, \mathbf{x})]^T$ satisfies a Lipschitz condition on D,

$$|\mathbf{f}(t, \mathbf{x}') - \mathbf{f}(t, \mathbf{x}'')| \leq k\,|\mathbf{x}' - \mathbf{x}''|$$

for all $(t, \mathbf{x}')$, $(t, \mathbf{x}'') \in D$, where $\mathbf{x}' = (x_1', \ldots, x_n')^T$, $\mathbf{x}'' = (x_1'', \ldots, x_n'')^T$, and $k = \max\limits_{1 \leq j \leq n} \sum_{i=1}^{n} |a_{ij}(t)|$. Indeed, we have

$$\begin{aligned}
|\mathbf{f}(t, \mathbf{x}') - \mathbf{f}(t, \mathbf{x}'')| &= \sum_{i=1}^{n} |f_i(t, \mathbf{x}') - f_i(t, \mathbf{x}'')| \\
&= \sum_{i=1}^{n} \left| \sum_{j=1}^{n} a_{ij}(t) x_j' - \sum_{j=1}^{n} a_{ij}(t) x_j'' \right| \\
&= \sum_{i=1}^{n} \left| \sum_{j=1}^{n} a_{ij}(t)(x_j' - x_j'') \right| \\
&\leq \sum_{i=1}^{n} \sum_{j=1}^{n} |a_{ij}(t)|\,|x_j' - x_j''| \\
&\leq k \sum_{j=1}^{n} |x_j' - x_j''| = k\,|\mathbf{x}' - \mathbf{x}''|.
\end{aligned}$$

Next, we prove the following:

5.10.52. Lemma. In Eq. (5.10.48), let $\mathbf{f}(t, \mathbf{x}) = (f_1(t, \mathbf{x}), \ldots, f_n(t, \mathbf{x}))^T$ be continuous on a domain $D \subset R^{n+1}$, and let $\mathbf{f}(t, \mathbf{x})$ satisfy a Lipschitz condition on D with respect to $\mathbf{x}$, with Lipschitz constant k. If $\boldsymbol{\varphi}_1$ and $\boldsymbol{\varphi}_2$ are unique solutions of the initial-value problem (5.10.49), with $\boldsymbol{\varphi}_1(\tau) = \boldsymbol{\xi}_1$, $\boldsymbol{\varphi}_2(\tau) = \boldsymbol{\xi}_2$ and with $(\tau, \boldsymbol{\xi}_1), (\tau, \boldsymbol{\xi}_2) \in D$, then

$$|\boldsymbol{\varphi}_1(t) - \boldsymbol{\varphi}_2(t)| \leq |\boldsymbol{\xi}_1 - \boldsymbol{\xi}_2| e^{k|t-\tau|} \tag{5.10.53}$$

for all $(t, \boldsymbol{\varphi}_1(t)), (t, \boldsymbol{\varphi}_2(t)) \in D$.

Proof. We assume that $t \geq \tau$, and we leave the details of the proof for $t \leq \tau$ as an exercise. We have

$$\boldsymbol{\varphi}_1(t) = \boldsymbol{\xi}_1 + \int_\tau^t \mathbf{f}(s, \boldsymbol{\varphi}_1(s))ds,$$

$$\boldsymbol{\varphi}_2(t) = \boldsymbol{\xi}_2 + \int_\tau^t \mathbf{f}(s, \boldsymbol{\varphi}_2(s))ds,$$

and

$$|\boldsymbol{\varphi}_1(t) - \boldsymbol{\varphi}_2(t)| \leq |\boldsymbol{\xi}_1 - \boldsymbol{\xi}_2| + k \int_\tau^t |\boldsymbol{\varphi}_1(s) - \boldsymbol{\varphi}_2(s)|\, ds. \tag{5.10.54}$$

Applying Theorem 5.10.39 to inequality (5.10.54), the desired inequality (5.10.53) results. ■

We are now in a position to prove the following important result for systems of linear ordinary differential equations.

5.10.55. Theorem. Let $D \subset R^{n+1}$ be given by Eq. (5.10.50), and let the real functions $a_{ij}(t)$, $i, j = 1, \ldots, n$, be continuous on the t interval $[a, b]$. Then there exists a *unique* solution to the initial-value problem

$$\left.\begin{aligned} \dot{x}_i &= \sum_{j=1}^n a_{ij}(t)x_j \triangleq f_i(t, \mathbf{x}), \quad i = 1, \ldots, n \\ x_i(\tau) &= \xi_i, \quad i = 1, \ldots, n \end{aligned}\right\} \tag{5.10.56}$$

with $(\tau, \xi_1, \ldots, \xi_n) \in D$. This solution can be extended to the *entire* interval $[a, b]$.

Proof. Since the vector $\mathbf{f}(t, \mathbf{x}) = (f_1(t, \mathbf{x}), \ldots, f_n(t, \mathbf{x}))^T$ is continuous on D, since $\mathbf{f}(t, \mathbf{x})$ satisfies a Lipschitz condition with respect to $\mathbf{x}$ on D, and since $(\tau, \boldsymbol{\xi}) \in D$ (where $\boldsymbol{\xi} = (\xi_1, \ldots, \xi_n)^T$), it follows from Theorem 5.10.43 (interpreted for systems of first-order ordinary differential equations) that the initial-value problem (5.10.56) has a unique solution $\boldsymbol{\psi}$ through the point

$(\tau, \boldsymbol{\xi})$ over some interval $[c, d] \subset [a, b]$. We must show that $\boldsymbol{\psi}$ can be continued to a unique solution $\boldsymbol{\varphi}$ over the entire interval $[a, b]$.

Let $\tilde{\boldsymbol{\psi}}$ be any solution of Eq. (5.10.56) through $(\tau, \boldsymbol{\xi})$ which exists on some subinterval of $[a, b]$. Applying Lemma 5.10.52 to $\boldsymbol{\varphi}_1 = \tilde{\boldsymbol{\psi}}$ and $\boldsymbol{\varphi}_2 = \mathbf{0}$, we have

$$|\tilde{\boldsymbol{\psi}}(t)| \leq |\boldsymbol{\xi}|\, e^{k(b-a)} \tag{5.10.57}$$

for all t in the domain of definition of $\tilde{\boldsymbol{\psi}}$. For purposes of contradiction, suppose that $\boldsymbol{\psi}$ does not have a continuation to $[a, b]$ and assume that $\boldsymbol{\psi}$ has a continuation $\tilde{\boldsymbol{\psi}}$ existing up to $t' < b$ and cannot be continued beyond t'. But inequality (5.10.57) implies that the path $(t, \tilde{\boldsymbol{\psi}}(t))$ remains inside a closed bounded subset of D. It follows from Theorem 5.10.45, interpreted for systems of first-order ordinary differential equations, that $\tilde{\boldsymbol{\psi}}$ may be continued beyond t'. We thus have arrived at a contradiction, which proves that a continuation $\boldsymbol{\varphi}$ of $\boldsymbol{\psi}$ exists on the entire interval $[a, b]$. This continuation is unique because $\mathbf{f}(t, \mathbf{x})$ satisfies a Lipschitz condition with respect to $\mathbf{x}$ on D. ■

5.10.58. Exercise. In Theorem 5.10.55, let $a_{ij}(t)$, $i, j = 1, \ldots, n$, be continuous on the open interval $(-\infty, \infty)$. Show that the initial-value problem (5.10.56) possesses unique solutions for every $(\tau, \xi) \in R^{n+1}$ which can be extended to the t interval $(-\infty, \infty)$.

5.10.59. Exercise. Let $D \subset R^{n+1}$ be given by Eq. (5.10.50), and let the real functions $a_{ij}(t)$, $v_i(t)$, $i, j = 1, \ldots, n$, be continuous on the t interval $[a, b]$. Show that there exists a unique solution to the initial-value problem

$$\dot{x}_i = \sum_{j=1}^{n} a_{ij}(t)x_j + v_i(t), \quad i = 1, \ldots, n,$$

$$x_i(\tau) = \xi_i, \quad i = 1, \ldots, n, \tag{5.10.60}$$

with $(\tau, \xi_1, \ldots, \xi_n) \in D$. Show that this solution can be extended to the entire interval $[a, b]$.

It is possible to relax the conditions on $v_i(t)$, $i = 1, \ldots, n$, in the above exercise considerably. For example, it can be shown that if $v_i(t)$ is piecewise continuous on $[a, b]$, then the assertions of Exercise 5.10.59 still hold.

We now address ourselves to the last item of the present section. Consider the initial-value problem (5.10.49) which we characterized in Definition 4.11.9. Assume that $\mathbf{f}(t, \mathbf{x})$ satisfies a Lipschitz condition on a domain $D \subset R^{n+1}$ and that $(\tau, \boldsymbol{\xi}) \in D$. Then the initial-value problem possesses a unique solution $\boldsymbol{\varphi}$ over some t interval containing τ. To indicate the dependence of $\boldsymbol{\varphi}$ on the initial point $(\tau, \boldsymbol{\xi})$, we write

$$\boldsymbol{\varphi}(t; \tau, \boldsymbol{\xi}),$$

where $\boldsymbol{\varphi}(\tau; \tau, \boldsymbol{\xi}) = \boldsymbol{\xi}$. We now ask: What are the effects of different initial conditions on the solution of Eq. (5.10.48)? Our next result provides the answer.

5.10.61. Theorem. In Eq. (5.10.49) let $\mathbf{f}(t, \mathbf{x})$ satisfy a Lipschitz condition with respect to $\mathbf{x}$ on $D \subset R^{n+1}$. Let $(\tau, \boldsymbol{\xi}) \in D$. Then the unique solution $\boldsymbol{\varphi}(t; \tau, \boldsymbol{\xi})$ of Eq. (5.10.49), existing on some bounded t interval containing τ, depends continuously on $\boldsymbol{\xi}$ on any such bounded interval. (This means if $\boldsymbol{\xi}_n \longrightarrow \boldsymbol{\xi}$, then $\boldsymbol{\varphi}(t; \tau, \boldsymbol{\xi}_n) \longrightarrow \boldsymbol{\varphi}(t; \tau, \boldsymbol{\xi})$.)

Proof. We have

$$\boldsymbol{\varphi}(t; \tau, \boldsymbol{\xi}_n) = \boldsymbol{\xi}_n + \int_\tau^t \mathbf{f}[s, \boldsymbol{\varphi}(s; \tau, \boldsymbol{\xi}_n)]ds$$

and

$$\boldsymbol{\varphi}(t; \tau, \boldsymbol{\xi}) = \boldsymbol{\xi} + \int_\tau^t \mathbf{f}[s, \boldsymbol{\varphi}(s; \tau, \boldsymbol{\xi})]ds.$$

It follows that for $t \geq \tau$ (the proof for $t \leq \tau$ is left as an exercise),

$$|\boldsymbol{\varphi}(t; \tau, \boldsymbol{\xi}_n) - \boldsymbol{\varphi}(t; \tau, \boldsymbol{\xi})| \leq |\boldsymbol{\xi}_n - \boldsymbol{\xi}| + \int_\tau^t |\mathbf{f}[s, \boldsymbol{\varphi}(s; \tau, \boldsymbol{\xi}_n)] - \mathbf{f}[s, \boldsymbol{\varphi}(s; \tau, \boldsymbol{\xi})]|\, ds$$

$$\leq |\boldsymbol{\xi}_n - \boldsymbol{\xi}| + k \int_\tau^t |\boldsymbol{\varphi}(s; \tau, \boldsymbol{\xi}_n) - \boldsymbol{\varphi}(s; \tau, \boldsymbol{\xi})|\, ds,$$

where k denotes a Lipschitz constant for $\mathbf{f}(t, \mathbf{x})$. Using Theorem 5.10.39, we obtain

$$|\boldsymbol{\varphi}(t; \tau, \boldsymbol{\xi}_n) - \boldsymbol{\varphi}(t; \tau, \boldsymbol{\xi})| \leq |\boldsymbol{\xi}_n - \boldsymbol{\xi}|\, e^{\int_\tau^t k\,ds} = |\boldsymbol{\xi}_n - \boldsymbol{\xi}|\, e^{k(t-\tau)}.$$

Thus if $\boldsymbol{\xi}_n \longrightarrow \boldsymbol{\xi}$, then $\boldsymbol{\varphi}(t; \tau, \boldsymbol{\xi}_n) \longrightarrow \boldsymbol{\varphi}(t; \tau, \boldsymbol{\xi})$. ■

It follows from the proof of the above theorem that the convergence is uniform with respect to t on any interval $[a, b]$ on which the solutions are defined.

5.10.62. Example. The initial-value problem

$$\left.\begin{aligned} \dot{x} &= 2x \\ x(\tau) &= \xi \end{aligned}\right\} \tag{5.10.63}$$

where $-\infty < \tau < \infty$, $-\infty < \xi < \infty$, has the unique solution

$$\varphi(t; \tau, \xi) = \xi e^{2(t-\tau)}, \quad -\infty < t < \infty,$$

which depends continuously on the initial value ξ. ■

Thus far, in the present section, we have concerned ourselves with problems characterized by *real ordinary differential equations*. It is an easy matter to verify that all the existence, uniqueness, continuation, and dependence (on initial conditions) results proved in the present section are also valid for initial-value problems described by *complex ordinary differential equations* such as those given, e.g., in Eq. (4.11.25). In this case, the norm of a complex vector $\mathbf{z} = (z_1, \ldots, z_n)^T$, $z_k = u_k + iv_k$, $k = 1, \ldots, n$, is given by

$$|\mathbf{z}| = \sum_{k=1}^{n} |z_k|,$$

where $|z_k| = (u_k^2 + v_k^2)^{1/2}$. The metric on C^n is in this case given by $\rho(\mathbf{z}_1, \mathbf{z}_2) = |\mathbf{z}_1 - \mathbf{z}_2|$.

5.11. REFERENCES AND NOTES

There are numerous excellent texts on metric spaces. Books which are especially readable include Copson [5.2], Gleason [5.3], Goldstein and Rosenbaum [5.4], Kantorovich and Akilov [5.5], Kolmogorov and Fomin [5.7], Naylor and Sell [5.8], and Royden [5.9]. Reference [5.8] includes some applications. The book by Kelley [5.6] is a standard reference on topology. An excellent reference on ordinary differential equations is the book by Coddington and Levinson [5.1].

REFERENCES

[5.1] E. A. Coddington and N. Levinson, *Theory of Ordinary Differential Equations*. New York: McGraw-Hill Book Company, Inc., 1955.

[5.2] E. T. Copson, *Metric Spaces*. Cambridge, England: Cambridge University Press, 1968.

[5.3] A. M. Gleason, *Fundamentals of Abstract Analysis*. Reading, Mass.: Addison-Wesley Publishing Co., Inc., 1966.

[5.4] M. E. Goldstein and B. M. Rosenbaum, "Introduction to Abstract Analysis," National Aeronautics and Space Administration, Report No. SP-203, Washington, D.C., 1969.

[5.5] L. V. Kantorovich and G. P. Akilov, *Functional Analysis in Normed Spaces*. New York: The Macmillan Company, 1964.

[5.6] J. Kelley, *General Topology*. Princeton, N.J.: D. Van Nostrand Company, Inc., 1955.

[5.7] A. N. Kolmogorov and S. V. Fomin, *Elements of the Theory of Functions and Functional Analysis*. Vol. 1. Albany, N.Y.: Graylock Press, 1957.

[5.8] A. W. NAYLOR and G. R. SELL, *Linear Operator Theory in Engineering and Science*. New York: Holt, Rinehart and Winston, 1971.
[5.9] H. L. ROYDEN, *Real Analysis*. New York: The Macmillan Company, 1965.
[5.10] A. E. TAYLOR, *General Theory of Functions and Integration*. New York: Blaisdell Publishing Company, 1965.

6

NORMED SPACES AND INNER PRODUCT SPACES

In Chapters 2–4 we concerned ourselves primarily with algebraic aspects of certain mathematical systems, while in Chapter 5 we addressed ourselves to topological properties of some mathematical systems. The stage is now set to combine topological and algebraic structures. In doing so, we arrive at linear topological spaces, namely normed linear spaces and inner product spaces, in general, and Banach spaces and Hilbert spaces, in particular. The properties of such spaces are the topic of the present chapter. In the next chapter we will study linear transformations defined on Banach and Hilbert spaces. The material of the present chapter and the next chapter constitutes part of a branch of mathematics called **functional analysis**.

Since normed linear spaces and inner product spaces are vector spaces as well as metric spaces, the results of Chapters 3 and 5 are applicable to the spaces considered in this chapter. Furthermore, since the Euclidean spaces considered in Chapter 4 are important examples of normed linear spaces and inner product spaces, the reader may find it useful to refer to Section 4.9 for proper motivation of the material to follow.

The present chapter consists of 16 sections. In the first 10 sections we consider some of the important general properties of normed linear spaces and Banach spaces. In sections 11 through 14 we examine some of the important general characteristics of inner product spaces and Hilbert spaces. (Inner product spaces are special types of normed linear spaces; Hilbert spaces are special cases of Banach spaces; Banach spaces are special kinds of normed linear spaces; and Hilbert spaces are special types of inner product spaces.) In section 15, we consider two applications. This chapter is concluded with a brief discussion of pertinent references in the last section.

6.1. NORMED LINEAR SPACES

Throughout this chapter, R denotes the field of real numbers, C denotes the field of complex numbers, F denotes either R or C, and X denotes a vector space over F.

6.1.1. Definition. Let $\|\cdot\|$ denote a mapping from X into R which satisfies the following properties for every $x, y \in X$ and every $\alpha \in F$:

(i) $\|x\| \geq 0$;
(ii) $\|x\| = 0$ if and only if $x = 0$;
(iii) $\|\alpha x\| = |\alpha| \cdot \|x\|$; and
(iv) $\|x + y\| \leq \|x\| + \|y\|$.

The function $\|\cdot\|$ is called a **norm** on X, the mathematical system consisting of $\|\cdot\|$ and X, $\{X; \|\cdot\|\}$, is called a **normed linear space**, and $\|x\|$ is called the **norm of** x. If $F = C$ we speak of a **complex normed linear space**, and if $F = R$ we speak of a **real normed linear space**.

Different norms defined on the same linear space X yield different normed linear spaces. If in a given discussion it is clear which particular norm is being used, we simply write X in place of $\{X; \|\cdot\|\}$ to denote the normed linear space under consideration. Properties (iii) and (iv) in Definition 6.1.1 are called the **homogeneity property** and the **triangle inequality** of a norm, respectively.

Let $\{X; \|\cdot\|\}$ be a normed linear space and let $x_i \in X$, $i = 1, \ldots, n$. Repeated use of the triangle inequality yields

$$\|x_1 + \ldots + x_n\| \leq \|x_1\| + \ldots + \|x_n\|.$$

The following result shows that every normed linear space has a metric associated with it, induced by the norm $\|\cdot\|$. Therefore, every normed linear space is also a metric space.

6.1.2. Theorem. Let $\{X; \|\cdot\|\}$ be a normed linear space, and let ρ be a real-valued function defined on $X \times X$ given by $\rho(x, y) = \|x - y\|$ for all $x, y \in X$. Then ρ is a metric on X and $\{X; \rho\}$ is a metric space.

6.1.3. Exercise. Prove Theorem 6.1.2.

This theorem tells us that all of the results in the previous chapter on metric spaces apply to normed linear spaces as well, provided we let $\rho(x, y) = \|x - y\|$. We will adopt the convention that when using the terminology of metric spaces (e.g., completeness, compactness, convergence, continuity, etc.) in a normed linear space $\{X; \|\cdot\|\}$, we mean with respect to the metric space $\{X; \rho\}$, where $\rho(x, y) = \|x - y\|$. Also, whenever we use metric space properties on F, i.e., on R or C, we mean with respect to the usual metric on R or C, respectively.

With the foregoing in mind, we now introduce the following important concept.

6.1.4. Definition. A complete normed linear space is called a **Banach space.**

Thus, $\{X; \{\|\cdot\|\}$ is a Banach space if and only if $\{X; \rho\}$ is a complete metric space, where $\rho(x, y) = \|x - y\|$.

6.1.5. Example. Let $X = R^n$, the space of n-tuples of real numbers, or let $X = C^n$, the space of n-tuples of complex numbers. From Example 3.1.10 we see that X is a vector space. For $x \in X$ given by $x = (\xi_1, \ldots, \xi_n)$, and for $p \in R$ such that $1 \leq p < \infty$, define

$$\|x\|_p = [|\xi_1|^p + \ldots + |\xi_n|^p]^{1/p}.$$

We can readily verify that $\|\cdot\|_p$ satisfies the axioms of a norm. Axioms (i), (ii), (iii) of Definition 6.1.1 follow trivially, while axiom (iv) is a direct consequence of Minkowski's inequality for finite sums (5.2.6). Letting $\rho_p(x, y) = \|x - y\|_p$, then $\{X; \rho_p\}$ is the metric space of Exercise 5.5.25. Since $\{X; \rho_p\}$ is complete, it follows that $\{R^n; \|\cdot\|_p\}$ and $\{C^n; \|\cdot\|_p\}$ are Banach spaces.

We may also define a norm on X by letting

$$\|x\|_\infty = \max_{1 \leq i \leq n} |\xi_i|.$$

It can readily be verified that $\{R^n; \|\cdot\|_\infty\}$ and $\{C^n; \|\cdot\|_\infty\}$ are also Banach spaces (see Exercise 5.5.25). ■

6.1.6. Example. Let $X = R^\infty$ (see Example 3.1.11) or $X = C^\infty$ (see Example 3.1.13), let $1 \leq p < \infty$, and as in Example 5.3.5, let

$$l_p = \left\{x \in X: \sum_{i=1}^{\infty} |\xi_i|^p < \infty\right\}.$$

Define

$$\|x\|_p = \left(\sum_{i=1}^{\infty} |\xi_i|^p\right)^{1/p}. \tag{6.1.7}$$

It is readily verified that $\|\cdot\|_p$ is a norm on the linear space l_p. Axioms (i), (ii), (iii) of Definition 6.1.1 follow trivially, while axiom (iv), the triangle inequality, follows from Minkowski's inequality for infinite sums (5.2.7). Invoking Example 5.5.26, it also follows that $\{l_p; \|\cdot\|_p\}$ is a Banach space. Henceforth, when we simply refer to the Banach space l_p, we assume that the norm on this space is given by Eq. (6.1.7).

Letting $p = \infty$ and

$$l_\infty = \{x \in X: \sup_i \{|\xi_i|\} < \infty\}$$

(refer to Example 5.3.8), and defining

$$\|x\|_\infty = \sup_i \{|\xi_i|\}, \tag{6.1.8}$$

it is readily verified that $\{l_\infty; \|\cdot\|_\infty\}$ is also a Banach space. When we simply refer to the Banach space l_∞, we have in mind the norm given in Eq. (6.1.8). ■

6.1.9. Example

(a) Let $\mathcal{C}[a, b]$ denote the linear space of real continuous functions on the interval $[a, b]$, as given in Example 3.1.19. For $x \in \mathcal{C}[a, b]$ define

$$\|x\|_p = \left[\int_a^b |x(t)|^p \, dt\right]^{1/p}, \quad 1 \leq p < \infty.$$

It is easily shown that $\{\mathcal{C}[a, b]; \|\cdot\|_p\}$ is a normed linear space. Axioms (i)–(iii) of Definition 6.1.1 follow trivially, while axiom (iv) follows from the Minkowski inequality for integrals (5.2.8). Let $\rho_p(x, y) = \|x - y\|_p$. Then $\{\mathcal{C}[a, b]; \rho_p\}$ is a metric space which is not complete (see Example 5.5.29 where we considered the special case $p = 2$). It follows that $\{\mathcal{C}[a, b]; \|\cdot\|_p\}$ is not a Banach space.

Next, define on the linear space $\mathcal{C}[a, b]$ the function $\|\cdot\|_\infty$ by

$$\|x\|_\infty = \sup_{t \in [a, b]} |x(t)|.$$

It is readily shown that $\{\mathcal{C}[a, b]; \|\cdot\|_\infty\}$ is a normed linear space. Let $\rho_\infty(x, y) = \|x - y\|_\infty$. In accordance with Example 5.5.28, $\{\mathcal{C}[a, b]; \rho_\infty\}$ is a complete metric space, and thus $\{\mathcal{C}[a, b]; \|\cdot\|_\infty\}$ is a Banach space.

The above discussion can be modified in an obvious way for the case where $\mathcal{C}[a, b]$ consists of complex-valued continuous functions defined on $[a, b]$. Here vector addition and multiplication of vectors by scalars are defined similarly as in Eqs. (3.1.20) and (3.1.21), respectively. Furthermore, it is easy to show that $\{\mathcal{C}[a, b]; \|\cdot\|_p\}$, $1 \leq p < \infty$, and $\{\mathcal{C}[a, b]; \|\cdot\|_\infty\}$ are normed linear spaces with norms defined similarly as above. Once more, the space $\{\mathcal{C}[a, b]; \|\cdot\|_p\}$, $1 \leq p < \infty$, is not a Banach space, while the space $\{\mathcal{C}[a, b]; \|\cdot\|_\infty\}$ is.

(b) The metric space $\{L_p[a, b]; \rho_p\}$ was defined in Example 5.5.31. It can be shown that $L_p[a, b]$ is a vector space over R. If we let

$$\|x\|_p = \left[\int_{[a, b]} |f|^p \, d\mu\right]^{1/p},$$

$p \geq 1$, for $f \in L_p[a, b]$, where the integral is the Lebesgue integral, then $\{L_p[a, b]; \|\cdot\|_p\}$ is a Banach space since $\{L_p[a, b]; \rho_p\}$ is complete, where $\rho_p(x, y) = \|x - y\|_p$. ■

6.1.10. Example. Let $\{X; \|\cdot\|_x\}$, $\{Y; \|\cdot\|_y\}$ be two normed linear spaces over F, and let $X \times Y$ denote the Cartesian product of X and Y. Defining vector addition on $X \times Y$ by

$$(x_1, y_1) + (x_2, y_2) = (x_1 + x_2, y_1 + y_2)$$

and multiplication of vectors by scalars as

$$\alpha(x, y) = (\alpha x, \alpha y),$$

we can readily show that $X \times Y$ is a linear space (see Eqs. (3.2.14), (3.2.15) and the related discussion). This space can be used to generate a normed linear space $\{X \times Y; \|\cdot\|\}$ by defining the norm $\|\cdot\|$ as

$$\|(x, y)\| = \|x\|_x + \|y\|_y.$$

Furthermore, if $\{X; \|\cdot\|_x\}$ and $\{Y; \|\cdot\|_y\}$ are Banach spaces, then it is easily shown that $\{X \times Y; \|\cdot\|\}$ is also a Banach space. ■

6.1.11. Exercise. Verify the assertions made in Examples 6.1.5 through 6.1.10.

We note that in a normed linear space $\{X; \|\cdot\|\}$ a sphere $S(x_0; r)$ with center $x_0 \in X$ and radius $r > 0$ is given by

$$S(x_0; r) = \{x \in X: \|x - x_0\| < r\}. \tag{6.1.12}$$

Referring to Theorem 5.4.27 and Exercise 5.4.31, recall that in a metric space the closure of a sphere (denoted by $\bar{S}(x_0; r)$) need not coincide with the closed sphere (denoted by $K(x_0; r)$). In a normed linear space we have the following result.

6.1.13. Theorem. Let X be a normed linear space, let $x_0 \in X$, and let $r > 0$. Let $\bar{S}(x_0; r)$ denote the closure of the open sphere $S(x_0; r)$ given by Eq. (6.1.12). Then $\bar{S}(x_0; r) = K(x_0; r)$, the closed sphere, where

$$K(x_0; r) = \{x \in X: \|x - x_0\| \leq r\}. \tag{6.1.14}$$

Proof. By Exercise 5.4.31 we know that $\bar{S}(x_0; r) \subset K(x_0; r)$. Thus, we need only show that $K(x_0; r) \subset \bar{S}(x_0; r)$. It is clearly sufficient to show that $\{x \in X: \|x - x_0\| = r\} \subset \bar{S}(x_0; r)$. To do so, let x be such that $\|x - x_0\| = r$, and let $0 < \epsilon < 1$. Let $y = \epsilon x_0 + (1 - \epsilon)x$. Then $y - x_0 = (1 - \epsilon)(x - x_0)$. Thus, $\|y - x_0\| = |1 - \epsilon| \cdot \|x - x_0\| < r$ and so $y \in S(x_0; r)$. Also, $y - x = \epsilon(x_0 - x)$. Therefore, $\|y - x\| = \epsilon \cdot r$. This means that $x \in \bar{S}(x_0; r)$, which completes the proof. ■

Thus, in a normed linear space we may call $\bar{S}(x_0; r)$ the closed sphere given by Eq. (6.1.14).

When regarded as a function from X into R, a norm has the following important property.

6.1.15. Theorem. Let $\{X; \|\cdot\|\}$ be a normed linear space. Then $\|\cdot\|$ is a continuous mapping of X into R.

Proof. We view $\|\cdot\|$ as a mapping from the metric space $\{X; \rho\}$, $\rho = \|x - y\|$, into the real numbers with the usual metric for R. Thus, for given $\epsilon > 0$, we wish to show that there is a $\delta > 0$ such that $\|x - y\| < \delta$ implies $|\|x\| - \|y\|| < \epsilon$. Now let $z = x - y$. Then $x = z + y$ and so $\|x\| \leq \|z\| + \|y\|$. This implies that $\|x\| - \|y\| \leq \|z\|$. Similarly, $y = x - z$, and so $\|y\| \leq \|x\| + \|-z\| = \|x\| + \|z\|$. Thus, $\|y\| - \|x\| \leq \|z\|$. It now follows that $|\|x\| - \|y\|| \leq \|z\| = \|x - y\|$. Letting $\delta = \epsilon$, the desired result follows. ■

In this chapter we will not always require that a particular normed linear space be a Banach space. Nonetheless, many important results of analysis require the completeness property. This is also true in applications. For example, in the solution of various types of equations (such as non-linear differential equations, integral equations, etc.) or in optimization problems or in non-linear feedback problems or in approximation theory, as well as many other areas of applications, we frequently obtain our desired solution in the form of a sequence generated by means of some iterative scheme. In such a sequence, each succeeding member is closer to the desired solution than its predecessor. Now even though the precise solution to which a sequence of this type may converge is unknown, it is usually imperative that the sequence converge to an element in that space which happens to be the setting of the particular problem in question.

6.2. LINEAR SUBSPACES

We now turn our attention briefly to linear subspaces of a normed linear space. We first recall Definition 3.2.1. A non-empty subset Y of a vector space X is called a **linear subspace** in X if (i) $x + y \in Y$ whenever x and y are in Y, and (ii) $\alpha x \in Y$ whenever $\alpha \in F$ and $x \in Y$. Next, consider a normed linear space $\{X; \|\cdot\|\}$, let Y be a linear subspace in X, and let $\|\cdot\|_1$ denote the restriction of $\|\cdot\|$ to Y; i.e.,

$$\|x\|_1 = \|x\| \text{ for all } x \in Y.$$

Then it is easy to show that $\{Y; \|\cdot\|_1\}$ is also a normed linear space. We call $\|\cdot\|_1$ the **norm induced by** $\|\cdot\|$ on Y and we say that $\{Y; \|\cdot\|_1\}$ is a **normed linear subspace** of $\{X; \|\cdot\|\}$, or simply a linear subspace of X. Since there is usually no room for confusion, we drop the subscript and simply denote this subspace by $\{Y; \|\cdot\|\}$. In fact, when it is clear which norm is being used, we usually refer to the normed linear spaces X and Y.

Our first result is an immediate consequence of Theorem 5.5.33.

6.2.1. Theorem. Let X be a Banach space, and let Y be a linear subspace of X. Then Y is a Banach space if and only if Y is closed.

In the following we give an example of a linear subspace of a Banach space which is not closed.

6.2.2. Example. Let X be the Banach space l_2 of Example 6.1.6, and let Y be the space of finitely non-zero sequences given in Example 3.1.14. It is easily shown that Y is a linear subspace of X. To show that Y is not closed, consider the sequence $\{y_n\}$ in Y defined by

$$\begin{aligned} y_1 &= (1, 0, 0, \ldots), \\ y_2 &= (1, 1/2, 0, 0, \ldots), \\ y_3 &= (1, 1/2, 1/4, 0, 0, \ldots), \\ &\ldots\ldots\ldots\ldots\ldots\ldots, \\ y_n &= (1, 1/2, \ldots, 1/2^n, 0, 0, \ldots). \end{aligned}$$

This sequence converges to the point $x = (1, 1/2, \ldots, 1/2^n, 1/2^{n+1}, \ldots) \in X$. Since $x \notin Y$, it follows from part (iii) of Theorem 5.5.8 that Y is not a closed subset of X. ■

Next, we prove:

6.2.3. Theorem. Let X be a Banach space, let Y be a linear subspace of X, and let $\bar{Y}$ denote the closure of Y. Then $\bar{Y}$ is a closed linear subspace of X.

Proof. Since $\bar{Y}$ is closed, we only have to show that $\bar{Y}$ is a linear subspace. Let $x, y \in \bar{Y}$, and let $\epsilon > 0$. Then there exist elements $x', y' \in Y$ such that $\|x - x'\| < \epsilon$ and $\|y - y'\| < \epsilon$. Hence, for arbitrary $\alpha, \beta \in F$, $\alpha x' + \beta y' \in Y$. Now $\|(\alpha x + \beta y) - (\alpha x' + \beta y')\| = \|\alpha(x - x') + \beta(y - y')\| \leq |\alpha| \cdot \|x - x'\| + |\beta| \cdot \|y - y'\| < (|\alpha| + |\beta|)\epsilon$. Since $\epsilon > 0$ is arbitrary, this implies that $\alpha x + \beta y$ is an adherent point of Y; i.e., $\alpha x + \beta y \in \bar{Y}$. This completes the proof of the theorem. ■

We conclude this section with the following useful result.

6.2.4. Theorem. Let X be a normed linear space, and let Y be a linear subspace of X. If Y is an open subset of X, then $Y = X$.

Proof. Let $x \in X$. We wish to show that $x \in Y$. Since $0 \in Y$, we may assume that $x \neq 0$. Since Y is open and $0 \in Y$, there is some $\epsilon > 0$ such that the sphere $S(0; \epsilon) \subset Y$. Let $z = \frac{\epsilon}{2||x||}x$. Then $||z|| < \epsilon$ and so $z \in Y$. Since Y is a linear subspace, it follows that $\frac{2||x||}{\epsilon}z = x \in Y$. ■

6.3. INFINITE SERIES

Having defined a norm on a linear space, we are in a position to consider the concept of infinite series in a meaningful way. Throughout this section we refer to a normed linear space $\{X; ||\cdot||\}$ simply as X.

6.3.1. Definition. Let $\{x_n\}$ be a sequence of elements in X. For each positive integer m, let

$$y_m = x_1 + \ldots + x_m.$$

We call $\{y_m\}$ the sequence of **partial sums** of $\{x_n\}$. If the sequence $\{y_m\}$ converges to a limit $y \in X$, we say the **infinite series**

$$x_1 + x_2 + \ldots + x_k + \ldots = \sum_{n=1}^{\infty} x_n$$

converges and we write

$$y = \sum_{n=1}^{\infty} x_n.$$

We say the **infinite series** $\sum_{n=1}^{\infty} x_n$ **diverges** if the sequence $\{y_m\}$ diverges.

The following result yields sufficient conditions for an infinite series to converge.

6.3.2. Theorem. Let X be a Banach space, and let $\{x_n\}$ be a sequence in X. If $\sum_{n=1}^{\infty} ||x_n|| < \infty$, then

(i) the infinite series $\sum_{n=1}^{\infty} x_n$ converges; and

(ii) $||\sum_{n=1}^{\infty} x_n|| \leq \sum_{n=1}^{\infty} ||x_n||$.

Proof. To prove the first part, let $y_m = x_1 + \ldots + x_m$. If $n > m$, then $y_n - y_m = x_{m+1} + \ldots + x_n$. Hence,

$$||y_n - y_m|| = ||\sum_{k=m+1}^{n} x_k|| \leq \sum_{k=m+1}^{n} ||x_k||.$$

Since $\sum_{n=1}^{\infty} ||x_n||$ is a convergent infinite series of real numbers, the sequence of partial sums $s_m = ||x_1|| + \ldots + ||x_m||$ is Cauchy. Hence, given $\epsilon > 0$, there is a positive integer N such that $n > m \geq N$ implies $|s_n - s_m| \leq \epsilon$. But $|s_n - s_m| \geq ||y_n - y_m||$, and so $\{y_m\}$ is a Cauchy sequence. Since X is complete, $\{y_m\}$ is convergent and conclusion (i) follows.

To prove the second part, let $y_m = x_1 + \ldots + x_m$, and let $y = \lim_m y_m = \sum_{n=1}^{\infty} x_n$. Then for each positive integer m we have $y = y - y_m + y_m$ and $||y|| \leq ||y - y_m|| + ||y_m|| \leq ||y - y_m|| + \sum_{i=1}^{m} ||x_i||$. Taking the limit as $m \longrightarrow \infty$, we have $||\sum_{i=1}^{\infty} x_i|| \leq \sum_{i=1}^{\infty} ||x_i||$. ■

6.4. CONVEX SETS

In the present section we consider the concepts of convexity and cones which arise naturally in many applications. *Throughout this section, X is a real normed linear space.*

Let x and y be two elements of X. We call the set $\overline{xy}$, defined by

$$\overline{xy} = \{z \in X: z = \alpha x + (1 - \alpha)y \text{ for all } \alpha \in R \text{ such that } 0 \leq \alpha \leq 1\},$$

the **line segment** joining x and y. Convex sets are now characterized as follows.

6.4.1. Definition. Let Y be a subset of X. Then Y is said to be **convex** if Y contains the line segment $\overline{xy}$ whenever x and y are two arbitrary points in Y. A convex set is called a **convex body** if it contains at least one interior point, i.e., if it completely contains some sphere.

In Figure A we depict a line segment $\overline{xy}$, a convex set, and a non-convex set in R^2.

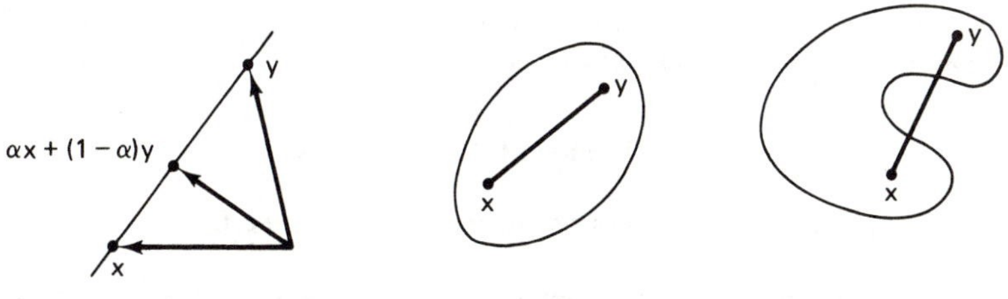

Line segment $\overline{xy}$ Convex set Non-convex set

6.4.2. Figure A

Note that an equivalent statement for Y to be convex is that if $x, y \in Y$ then $\alpha x + \beta y \in Y$ whenever α and β are positive constants such that $\alpha + \beta = 1$.

We cite a few examples.

6.4.3. Example. The empty set is convex. Also, a set consisting of one point is convex. In R^3, a cube and a sphere are convex bodies, while a plane and a line segment are convex sets but not convex bodies. Any linear subspace of X is a convex set. Also, any linear variety of X (see Definition 3.2.17) is a convex set. ■

6.4.4. Example. Let Y and Z be convex sets in X, let $\alpha, \beta \in R$, and let $\alpha Y = \{x \in X: x = \alpha y, y \in Y\}$. Then the set $\alpha Y + \beta Z$ is a convex set in X. ■

6.4.5. Exercise. Prove the assertions made in Examples 6.4.3 and 6.4.4.

6.4.6. Theorem. Let Y be a convex set in X, and let $\alpha, \beta \in R$ be positive scalars. Then $(\alpha + \beta)Y = \alpha Y + \beta Y$.

Proof. Regardless of convexity, if $x \in (\alpha + \beta)Y$, then $x = (\alpha + \beta)y = ay + \beta y \in \alpha Y + \beta Y$, and thus $(\alpha + \beta)Y \subset \alpha Y + \beta Y$. Now let Y be convex, and let $x = \alpha y + \beta z$, where $y, z \in Y$. Then

$$\frac{1}{\alpha + \beta}x = \frac{\alpha}{\alpha + \beta}y + \frac{\beta}{\alpha + \beta}z \in Y,$$

because

$$\frac{\alpha}{\alpha + \beta} + \frac{\beta}{\alpha + \beta} = \frac{\alpha + \beta}{\alpha + \beta} = 1.$$

Therefore, $x \in (\alpha + \beta)Y$ and thus $\alpha Y + \beta Y \subset (\alpha + \beta)Y$. This completes the proof. ■

We leave the proof of the next result as an exercise.

6.4.7. Theorem. Let $\mathfrak{C}$ be an arbitrary collection of convex sets. The intersection $\bigcap_{Y \in \mathfrak{C}} Y$ is also a convex set.

6.4.8. Exercise. Prove Theorem 6.4.7.

The preceding result gives rise to the following concept.

6.4.9. Definition. Let Y be any set in X. The **convex hull** of Y, also called the **convex cover** of Y, denoted by Y_c, is the intersection of all convex sets which contain Y.

We note that the convex hull of Y is the smallest convex set which contains Y. Examples of convex covers of sets in R^2 are depicted in Figure B.

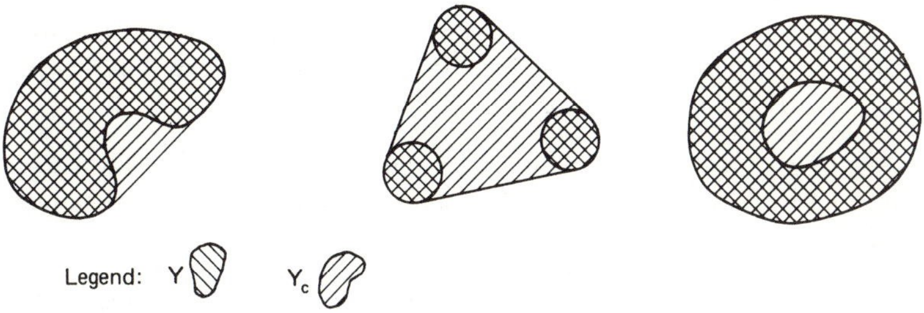

6.4.10. Figure B. Convex hulls.

6.4.11. Theorem. Let Y be any set in X. The convex hull of Y is the set of points expressible as $\alpha_1 y_1 + \alpha_2 y_2 + \ldots + \alpha_n y_n$, where $y_1, \ldots, y_n \in Y$, where $\alpha_i > 0$, $i = 1, \ldots, n$, where $\sum_{i=1}^{n} \alpha_i = 1$ and where n is not fixed.

Proof. If Z is the set of elements expressible as described above, then clearly Z is convex. Moreover, $Y \subset Z$, and hence $Y_c \subset Z$. To show that $Z \subset Y_c$, we show that Z is contained in every convex set which contains Y. We do so by induction on the number of elements of Y that appear in the representation of an element of Z. Let U be a convex set with $U \supset Y$. If $z = \alpha_1 z_1 \in Z$ for which $n = 1$, then $\alpha_1 = 1$ and $z \in U$. Now assume that an element of Z is in U if it is represented in terms of $n - 1$ elements of Y. Let $z = \alpha_1 z_1 + \ldots + \alpha_n z_n$ be in Z, let $\beta = \alpha_1 + \ldots + \alpha_{n-1}$, let $\beta_i = \alpha_i/\beta$, $i = 1, \ldots, n - 1$, and let $u = \beta_1 z_1 + \ldots + \beta_{n-1} z_{n-1}$. Then $u \in U$, by the induction hypothesis. But $z_n \in U$, $\alpha_n = 1 - \beta$, and $z = \beta u + (1 - \beta) z_n \in U$, since U is convex. This completes the induction, and thus $Z \subset U$ from which it follows that $Z \subset Y_c$. ■

6.4.12. Theorem. Let Y be a convex set in X. Then the closure of Y, $\bar{Y}$, is also a convex set.

6.4.13. Exercise. Prove Theorem 6.4.12.

Since the intersection of any number of closed sets is always closed, it follows from Theorem 6.4.7 that the intersection of an arbitrary number of closed convex sets is also a closed convex set.

We now consider some interesting aspects of norms in terms of convex sets.

6.4.14. Theorem. Any sphere in X is a convex set.

Proof. We consider without loss of generality the unit sphere,

$$Y = \{x \in X: ||x|| < 1\}.$$

If $x_0, y_0 \in Y$, then $||x_0|| < 1$ and $||y_0|| < 1$. Now if $\alpha \geq 0$ and $\beta \geq 0$, where $\alpha + \beta = 1$, then $||\alpha x_0 + \beta y_0|| \leq ||\alpha x_0|| + ||\beta y_0|| = \alpha||x_0|| + \beta||y_0|| < \alpha + \beta = 1$, and thus $\alpha x_0 + \beta y_0 \in Y$. ■

In view of Theorems 6.1.13, 6.4.12, and 6.4.14, it follows that a closed sphere $\bar{S}(x_0; r)$ is also convex. The following example, cast in R^2, is rather instructive.

6.4.15. Example. On R^2 we define the norm $|| \cdot ||_p$ of Example 6.1.5. A moment's reflection reveals that in case of $|| \cdot ||_2$, the unit sphere is a circle of radius 1; when the norm is $|| \cdot ||_\infty$, the unit sphere is a square with vertices (1, 1), (1, −1), (−1, 1), (−1, −1); if the norm is $|| \cdot ||_1$, the unit sphere is the square with vertices (0, 1), (1, 0), (−1, 0), (0, −1). If for the unit sphere corresponding to $|| \cdot ||_p$ we let p increase from 1 to ∞, then this sphere will deform in a continuous manner from the square corresponding to $|| \cdot ||_1$ to the square corresponding to $|| \cdot ||_\infty$. This is depicted in Figure C. We note that in all cases the unit sphere results in a convex set.

For the case of the real-valued function

$$||x|| = (|\xi_1|^p + |\xi_2|^p)^{1/p}, \quad 0 < p < 1 \tag{6.4.16}$$

the set determined by $||x|| \leq 1$ results in a set which is *not convex*. In particular, if $p = 2/3$, the set determined by $||x|| \leq 1$ yields the boundary and the interior of an asteroid, as shown in Figure C. The reason for the non-convexity of this set can be found in the fact that the function (6.4.16) does *not* represent a norm. In particular, it can be shown that (6.4.16) does not satisfy the triangle inequality. ■

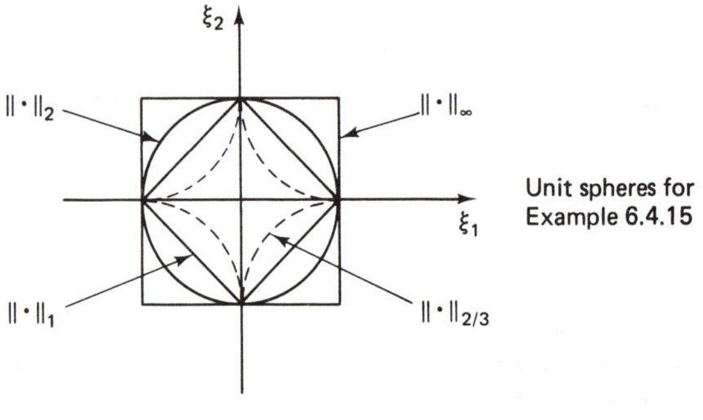

6.4.17. Figure C. Unit spheres for Example 6.4.15.

6.4.18. Exercise. Verify the assertions made in Example 6.4.15.

We conclude this section by introducing the notion of cone.

6.4.19. Definition. A set Y in X is called a **cone** with vertex at the origin if $y \in Y$ implies that $\alpha y \in Y$ for all $\alpha \geq 0$. If Y is a cone with vertex at the origin, then the set $x_0 + Y$, $x_0 \in X$, is called a **cone with vertex** x_0. A **convex cone** is a set which is both convex and a cone.

In Figure D examples of cones are shown.

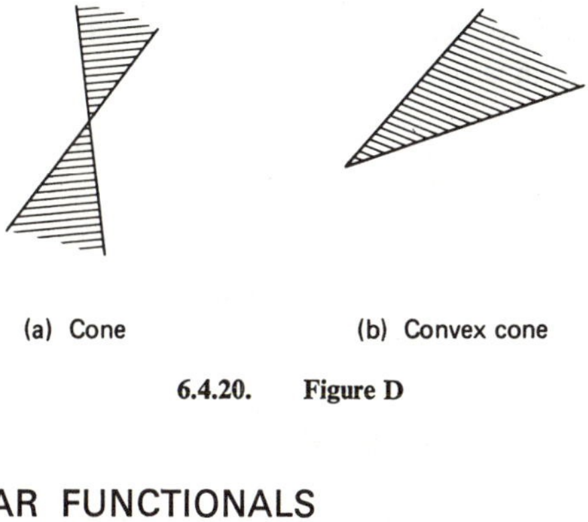

(a) Cone (b) Convex cone

6.4.20. Figure D

6.5. LINEAR FUNCTIONALS

Throughout this section X is a normed linear space.

We recall that a mapping, f, from X into F is called a **functional** on X (see Definition 3.5.1). If f is also linear, i.e., $f(\alpha x + \beta y) = \alpha f(x) + \beta f(y)$ for all $\alpha, \beta \in F$ and all $x, y \in X$, then f is called a **linear functional** (refer to Definition 3.5.1). Recall further that X^f, the set of all linear functionals on X, is a linear space over F (see Theorem 3.5.16). Let $f \in X^f$ and $x \in X$. In accordance with Eq. (3.5.10), we use the notation

$$f(x) = \langle x, f \rangle \tag{6.5.1}$$

to denote the value of f at x. Alternatively, we sometimes find it convenient to let $x' \in X^f$ denote a linear functional defined on X and write (see Eq. (3.5.11))

$$x'(x) = \langle x, x' \rangle. \tag{6.5.2}$$

Invoking Definition 5.7.1, we note that continuity of a functional at a point $x_0 \in X$ means, in the present context, that for every $\epsilon > 0$ there is a $\delta > 0$ such that $|f(x) - f(x_0)| < \epsilon$ whenever $||x - x_0|| < \delta$. Our first result shows that if a linear functional on X is continuous at one point of X then it is continuous at all points of X.

6.5.3. Theorem. If a linear functional f on X is continuous at some point $x_0 \in X$, then it is continuous for all $x \in X$.

Proof. If $\{y_n\}$ is a sequence in X such that $y_n \longrightarrow x_0$, then $f(y_n) \longrightarrow f(x_0)$, by Theorem 5.7.8. Now let $\{x_n\}$ be a sequence in X converging to $x \in X$. Then the sequence $\{y_n\}$ in X given by $y_n = x_n - x + x_0$ converges to x_0. By the linearity of f, we have

$$f(x_n) - f(x) = f(y_n) - f(x_0).$$

Since $|f(y_n) - f(x_0)| \longrightarrow 0$ as $y_n \longrightarrow x_0$, we have $|f(x_n) - f(x)| \longrightarrow 0$ as $x_n \longrightarrow x$, and therefore f is continuous at $x \in X$. Since x is arbitrary, the proof of the theorem is complete. ■

It is clear that if f is a linear functional and if $f(x) \neq 0$ for some $x \in X$, then the range of f is all of F; i.e., $\mathcal{R}(f) = F$.

For linear functionals we define boundedness as follows.

6.5.4. Definition. A linear functional f on X is said to be **bounded** if there exists a real constant $M \geq 0$ such that

$$|f(x)| \leq M\|x\|$$

for all $x \in X$. If f is not bounded, then it is said to be **unbounded**.

The following theorem shows that continuity and boundedness of linear functionals are equivalent.

6.5.5. Theorem. A linear functional f on a normed linear space X is bounded if and only if it is continuous.

Proof. Assume that f is bounded, and let M be such that $|f(x)| \leq M\|x\|$ for all $x \in X$. If $x_n \longrightarrow 0$, then $|f(x_n)| \leq M\|x_n\| \longrightarrow 0$. Hence, f is continuous at $x = 0$. From Theorem 6.5.3 it follows that f is continuous for all $x \in X$.

Conversely, assume that f is continuous at $x = 0$ and hence at any $x \in X$. There is a $\delta > 0$ such that $|f(x)| < 1$ whenever $\|x\| \leq \delta$. Now for any $x \neq 0$ we have $\|(\delta x)/\|x\|\| = \delta$, and thus

$$|f(x)| = \left|f\left(\frac{\delta x}{\|x\|} \frac{\|x\|}{\delta}\right)\right| = \left|f\left(\frac{\delta x}{\|x\|}\right)\right| \cdot \frac{\|x\|}{\delta} < \frac{\|x\|}{\delta}.$$

If we let $M = 1/\delta$, then $|f(x)| \leq M\|x\|$, and f is bounded. ■

We will see later, in Example 6.5.17, that there may exist linear functionals on a normed linear space which are unbounded. The class of linear functionals which are bounded has some interesting properties.

6.5.6. Theorem. Let X^f be the vector space of all linear functionals on X, and let X^* denote the family of all bounded linear functionals on X. Define the function $\|\cdot\|: X^* \longrightarrow R$ by

$$\|f\| = \sup_{x \neq 0} \frac{|f(x)|}{\|x\|} \text{ for } f \in X^*. \tag{6.5.7}$$

Then

(i) X^* is a linear subspace of X^f;
(ii) the function $\|\cdot\|$ defined in Eq. (6.5.7) is a norm on X^*; and
(iii) the normed space $\{X^*; \|\cdot\|\}$ is complete.

Proof. The proof of part (i) is straightforward and is left as an exercise.

To prove part (ii), note that if $f \neq 0$, then $\|f\| > 0$ and if $f = 0$, then $\|f\| = 0$. Also, since

$$\sup_{x \neq 0} \frac{|\alpha f(x)|}{\|x\|} = |\alpha| \sup_{x \neq 0} \frac{|f(x)|}{\|x\|},$$

it follows that $\|\alpha f\| = |\alpha| \|f\|$. Finally,

$$\|f_1 + f_2\| = \sup_{x \neq 0} \frac{|f_1(x) + f_2(x)|}{\|x\|} \leq \sup_{x \neq 0} \left\{\frac{|f_1(x)| + |f_2(x)|}{\|x\|}\right\}$$
$$\leq \sup_{x \neq 0} \frac{|f_1(x)|}{\|x\|} + \sup_{x \neq 0} \frac{|f_2(x)|}{\|x\|} = \|f_1\| + \|f_2\|.$$

Hence, $\|\cdot\|$ satisfies the axioms of a norm.

To prove part (iii), let $\{x'_n\} \in X^*$ be a Cauchy sequence. Then $\|x'_n - x'_m\| \longrightarrow 0$ as $m, n \longrightarrow \infty$. If we evaluate this sequence at any $x \in X$, then $\{x'_n(x)\}$ is a Cauchy sequence of scalars, because $|x'_n(x) - x'_m(x)| \leq \|x'_n - x'_m\| \|x\|$. This implies that for each $x \in X$ there is a scalar $x'(x)$ such that $x'_n(x) \longrightarrow x'(x)$. We observe that $x'(\alpha x + \beta y) = \lim_{n\to\infty} x'_n(\alpha x + \beta y) = \lim_{n\to\infty} [\alpha x'_n(x) + \beta x'_n(y)] = \alpha \lim_{n\to\infty} x'_n(x) + \beta \lim_{n\to\infty} x'_n(y) = \alpha x'(x) + \beta x'(y)$, i.e., $x'(\alpha x + \beta y) = \alpha x'(x) + \beta x'(y)$, and thus, x' is a linear functional. Next we show that x' is bounded. Since $\{x'_n\}$ is a Cauchy sequence, for $\epsilon > 0$ there is an M such that $|x'_n(x) - x'_m(x)| < \epsilon\|x\|$ for all $m, n > M$ and for all $x \in X$. But $x'_n(x) \longrightarrow x'(x)$, and hence $|x'(x) - x'_m(x)| < \epsilon\|x\|$ for all $m > M$. It now follows that

$$|x'(x)| = |x'(x) - x'_m(x) + x'_m(x)| \leq |x'(x) - x'_m(x)| + |x'_m(x)|$$
$$\leq \epsilon\|x\| + \|x'_m\| \|x\|,$$

and thus x' is a bounded linear functional. Finally, to show that $x'_m \longrightarrow x' \in X^*$, we note that $|x'(x) - x'_m(x)| < \epsilon\|x\|$ whenever $m > M$ from which we have $\|x' - x'_m\| < \epsilon$ whenever $m > M$. This proves the theorem. ■

6.5.8. Exercise. Prove part (i) of Theorem 6.5.6.

It is especially interesting to note that X^* is a Banach space whether X is or is not a Banach space. We are now in a position to make the following definition.

6.5.9. Definition. The set of all bounded linear functionals on a normed space X is called the **normed conjugate space** of X, or the **normed dual** of X, or simply the **dual** of X, and is denoted by X^*. For $f \in X^*$ we call $\|f\|$ defined by Eq. (6.5.7) the **norm** of f.

The next result states that the norm of a functional can be represented in various equivalent ways.

6.5.10. Theorem. Let f be a bounded linear functional on X, and let $\|f\|$ be the norm of f. Then

(i) $\|f\| = \inf_M \{M: |f(x)| \leq M\|x\| \text{ for all } x \in X\}$;

(ii) $\|f\| = \sup_{\|x\| \leq 1} \{|f(x)|\}$; and

(iii) $\|f\| = \sup_{\|x\| = 1} \{|f(x)|\}$.

6.5.11. Exercise. Prove Theorem 6.5.10.

Let us now consider the norms of some specific linear functionals.

6.5.12. Example. Consider the normed linear space $\{\mathcal{C}[a, b]; \|\cdot\|_\infty\}$. The mapping

$$f(x) = \int_a^b x(s)\, ds, \quad x \in \mathcal{C}[a, b]$$

is a linear functional on $\mathcal{C}[a, b]$ (cf. Example 3.5.2). The norm of this functional equals $(b - a)$, because

$$|f(x)| = \left| \int_a^b x(s)\, ds \right| \leq (b - a) \max_{a \leq s \leq b} |x(s)|. \quad \blacksquare$$

6.5.13. Example. Consider the space $\{\mathcal{C}[a, b]; \|\cdot\|_\infty\}$, let x_0 be a fixed element of $\mathcal{C}[a, b]$, and let x be any element of $\mathcal{C}[a, b]$. The mapping

$$f(x) = \int_a^b x(s)x_0(s)\, ds$$

is a linear functional on $\mathcal{C}[a, b]$ (cf. Example 3.5.2). This functional is bounded, because

$$|f(x)| = \left| \int_a^b x(s)x_0(s)\, ds \right| \leq \left(\int_a^b |x_0(s)|\, ds \right) \|x\|_\infty.$$

Since f is bounded and linear, it follows that it is continuous. We leave it to the reader to show that

$$\|f\| = \int_a^b |x_0(s)|\, ds. \quad \blacksquare$$

6.5.14. Example. Let $a = (\alpha_1, \ldots, \alpha_n)$ be a fixed element of F^n, and let $x = (\xi_1, \ldots, \xi_n)$ denote an arbitrary element of F^n. Then if

$$f(x) = \sum_{i=1}^{n} \alpha_i \xi_i,$$

it follows that f is a linear functional on F^n (cf. Example 3.5.6). Letting $\|x\| = (|\xi_1|^2 + \ldots + |\xi_n|^2)^{1/2}$, it follows from the Schwarz inequality (4.9.29) that

$$|f(x)| = \left|\sum_{i=1}^{n} \alpha_i \xi_i\right| \leq \|a\| \cdot \|x\|. \tag{6.5.15}$$

Thus, f is bounded and continuous. In order to determine the norm of f, we rewrite (6.5.15) as

$$\frac{|f(x)|}{\|x\|} \leq \sup_{x \neq 0} \frac{|f(x)|}{\|x\|} \leq \|a\|,$$

from which it follows that $\|f\| \leq \|a\|$. Next, by setting $x = a$, we have $|f(a)| = \|a\|^2$. Thus,

$$\frac{|f(a)|}{\|a\|} = \|a\|.$$

Therefore $\|f\| = \|a\|$. $\blacksquare$

6.5.16. Example. Analogous to the above example, let $a = (\alpha_1, \alpha_2, \ldots)$ be a fixed element of the Banach space l_2 (see Example 6.1.6), and let $x = (\xi_1, \xi_2, \ldots)$ be an arbitrary element of l_2. It follows that if

$$f(x) = \sum_{i=1}^{\infty} \alpha_i \xi_i,$$

then f is a linear functional on l_2. We can show that f is bounded by observing that

$$|f(x)| = \left|\sum_{i=1}^{\infty} \alpha_i \xi_i\right| \leq \sum_{i=1}^{\infty} |\alpha_i \xi_i| \leq \|a\| \cdot \|x\|,$$

which follows from Hölder's inequality for infinite sums (5.2.4). Thus, f is bounded and, hence, continuous. In a manner similar to that of Example 6.5.14, we can show that $\|f\| = \|a\|$. $\blacksquare$

We conclude this section with an example of an unbounded linear functional.

6.5.17. Example. Consider the space X of finitely non-zero sequences $x = (\xi_1, \xi_2, \ldots, \xi_n, 0, 0, \ldots)$ (cf. Example 3.1.14). Define $\|\cdot\|: X \to R$ as $\|x\| = \max_i |\xi_i|$. It is easy to show that $\{X; \|\cdot\|\}$ is a normed linear space. Furthermore, it is readily verified that the mapping

$$f(x) = \sum_{k=1}^{n} k\xi_k$$

is an *unbounded* linear functional on X. ■

6.5.18. Exercise. Verify the assertions made in Examples 6.5.12, 6.5.13, 6.5.14, 6.5.16, and 6.5.17.

6.6. FINITE-DIMENSIONAL SPACES

We now briefly turn our attention to finite-dimensional vector spaces. Throughout this section X denotes a normed linear space.

We recall that if $\{x_1, \ldots, x_n\}$ is a basis for a linear space X, then for each $x \in X$ there is a unique set of scalars $\{\xi_1, \ldots, \xi_n\}$ in F, called the coordinates of x with respect to this basis (see Definition 3.3.36). We now prove the following result.

6.6.1. Theorem. Let X be a finite-dimensional normed linear space, and let $\{x_1, \ldots, x_n\}$ be a basis for X. For each $x \in X$, let the coordinates of x with respect to this basis be denoted by $(\xi_1, \ldots, \xi_n) \in F^n$. For $i = 1, \ldots, n$, define the linear functionals $f_i: X \to F$ by $f_i(x) = \xi_i$. Then each f_i is a continuous linear functional.

Proof. The proof that f_i is linear is straightforward. To show that f_i is a bounded linear functional, we let

$$S = \{a = (\alpha_1, \ldots, \alpha_n) \in F^n: |\alpha_1| + |\alpha_2| + \ldots + |\alpha_n| = 1\}.$$

It is left as an exercise to show that S is a compact set in the metric space $\{F^n; \rho_1\}$ (see Example 5.3.1). Now let us define the function $g: S \to R$ by

$$g(a) = \|\alpha_1 x_1 + \ldots + \alpha_n x_n\|.$$

The reader can readily verify that g is a continuous function on S. Now let $m = \inf\{g(a): a \in S\}$. It follows from Theorem 5.7.15 that there is an $a_0 \in S$ such that $g(a_0) = m$. Note that $m \neq 0$ since $\{x_1, \ldots, x_n\}$ is a basis for X, and also $a_0 \neq 0$. Hence $m > 0$. It now follows that

$$\|\alpha_1 x_1 + \ldots + \alpha_n x_n\| \geq m$$

for every $a = (\alpha_1, \ldots, \alpha_n) \in S$. Since $|\alpha_1| + \ldots + |\alpha_n| = 1$ for $a \in S$, we see that

$$||\alpha_1 x_1 + \ldots + \alpha_n x_n|| \geq m(|\alpha_1| + \ldots + |\alpha_n|) \tag{6.6.2}$$

for all $a \in S$.

Next, for arbitrary $x \in X$ with coordinates $(\xi_1, \ldots, \xi_n) \in F^n$, we let $\beta = |\xi_1| + \ldots + |\xi_n|$. First, we suppose that $\beta > 0$. Then

$$\begin{aligned} ||x|| = ||\xi_1 x_1 + \ldots + \xi_n x_n|| &= \beta \left\| \frac{\xi_1}{\beta} x_1 + \ldots + \frac{\xi_n}{\beta} x_n \right\| \\ &\geq \beta m\left(\left| \frac{\xi_1}{\beta} \right| + \ldots + \left| \frac{\xi_n}{\beta} \right| \right) \\ &= m(|\xi_1| + \ldots + |\xi_n|), \end{aligned}$$

where inequality (6.6.2) has been used. Therefore, if $\beta \neq 0$, we have

$$(|\xi_1| + \ldots + |\xi_n|) \leq \frac{1}{m} ||x||. \tag{6.6.3}$$

Noting that inequality (6.6.3) is also true if $\beta = 0$, we conclude that this inequality is true for all $x \in X$. Since $|f_i(x)| = |\xi_i| \leq |\xi_1| + \ldots + |\xi_n|$, $i = 1, \ldots, n$, we see that $|f_i(x)| \leq (1/m)||x||$ for any $x \in X$. Hence, f_i is a bounded linear functional and, consequently, it is continuous. ■

6.6.4. Exercise. Prove that the set S and the function g have the properties asserted in the proof of Theorem 6.6.1.

The preceding theorem allows us to prove the following important result.

6.6.5. Theorem. Let X be a finite-dimensional normed linear space. Then X is complete.

Proof. Let $\{x_1, \ldots, x_n\}$ be a basis for X, let $\{y_k\}$ be a Cauchy sequence in X, and for each k let the coordinates of y_k with respect to $\{x_1, \ldots, x_n\}$ be given by $(\eta_{k1}, \ldots, \eta_{kn})$. It follows from Theorem 6.6.1 that there is a constant M such that $|\eta_{kj} - \eta_{ij}| \leq M||y_k - y_i||$ for $j = 1, \ldots, n$ and all $i, k = 1, 2, \ldots$. Hence, each sequence $\{\eta_{kj}\}$ is a Cauchy sequence in F, i.e., in R or C, and is therefore convergent. Let $\eta_{0j} = \lim_k \eta_{kj}$ for $j = 1, \ldots, n$. If we let

$$y_0 = \eta_{01} x_1 + \ldots + \eta_{0n} x_n,$$

it follows that $\{y_k\}$ converges to y_0. This proves that X is complete. ■

The next result follows from Theorems 6.6.5 and 6.2.1.

6.6.6. Theorem. Let X be a normed linear space, and let Y be a finite-dimensional linear subspace of X. Then (i) Y is complete, and (ii) Y is closed.

6.6.7. Exercise. Prove Theorem 6.6.6.

Our next result is an immediate consequence of Theorem 6.6.1.

6.6.8. Theorem. Let X be a finite-dimensional normed linear space, and let f be a linear functional on X. Then f is continuous.

6.6.9. Exercise. Prove Theorem 6.6.8.

We recall from Definition 5.6.30 and Theorem 5.6.31 that a subset Y of a metric space X is relatively compact if every sequence of elements in Y contains a subsequence which converges to an element in X. This property can be useful in characterizing finite-dimensional subspaces in an arbitrary normed linear space as we shall see in the next theorem. Note also that in view of Definition 5.1.19 a subset Y in a normed linear space X is bounded if and only if there is a $\lambda > 0$ such that $\|y\| \leq \lambda$ for all $y \in Y$.

6.6.10. Theorem. Let X be a normed linear space, and let Y be a linear subspace of X. Then Y is finite dimensional if and only if every bounded subset of Y is relatively compact.

Proof. (Necessity) Assume that Y is finite dimensional, and let $\{x_1, \ldots, x_n\}$ be a basis for Y. Then for any $y \in Y$ there is a unique set $\{\eta_1, \ldots, \eta_n\}$ such that $y = \eta_1 x_1 + \ldots + \eta_n x_n$. Let A be a bounded subset of Y, and let $\{y_k\}$ be a sequence in A. Then we can write $y_k = \eta_{1k}x_1 + \ldots + \eta_{nk}x_n$ for $k = 1, 2, \ldots$. There exists a $\lambda > 0$ such that $\|y_k\| \leq \lambda$ for all k. Consider $|\eta_{1k}| + \ldots + |\eta_{nk}|$. We wish to show that this sum is bounded. Suppose that it is not. Then for each positive integer m, we can find a y_{k_m} such that $|\eta_{1k_m}| + \ldots + |\eta_{nk_m}| \triangleq \gamma_m \geq m$. Now let $y'_{k_m} = (1/\gamma_m)y_{k_m}$. It follows that

$$\|y'_{k_m}\| = \frac{1}{\gamma_m}\|y_{k_m}\| \leq \frac{\lambda}{\gamma_m} \leq \frac{\lambda}{m}.$$

Thus, $y'_{k_m} \longrightarrow 0$ as $m \longrightarrow \infty$. On the other hand,

$$y'_{k_m} = \eta'_{1k_m}x_1 + \ldots + \eta'_{nk_m}x_n$$

where $\eta'_{ik_m} = \eta_{ik_m}/\gamma_m$ for $i = 1, \ldots, n$. Since $|\eta'_{1k_m}| + \ldots + |\eta'_{nk_m}| = 1$, the coordinates $\{\eta'_{1k_m}, \ldots, \eta'_{nk_m}\}$ form a bounded sequence in F^n and as such contain a convergent subsequence. Let $\{\eta'_{10}, \ldots, \eta'_{n0}\}$ be the limit of such a convergent subsequence whose indices we denote by k_{mj}. If we let $y'_0 = \eta'_{10}x_1 + \ldots + \eta'_{n0}x_n$, then we have

$$\|y'_0 - y'_{k_{mj}}\| \leq |\eta'_{10} - \eta'_{1k_{mj}}| \cdot \|x_1\| + \ldots + |\eta'_{n0} - \eta'_{nk_{mj}}| \cdot \|x_n\| \longrightarrow 0 \text{ as } mj \longrightarrow \infty.$$

Thus, $y'_{k_{mj}} \longrightarrow y'_0$. Since $y'_{k_{mj}} \longrightarrow 0$, it follows that $y'_0 = 0$. But this is impossible because $\{x_1, \ldots, x_n\}$ is a linearly independent set. We conclude that the sum

$|\eta_{1k}| + \ldots + |\eta_{nk}|$ is bounded. Consequently, there is a subsequence $\{\eta_{1kj}, \ldots, \eta_{nkj}\}$ which is convergent in F^n. Let $\{\eta_{10}, \ldots, \eta_{n0}\}$ be the limit of the convergent subsequence, and let $y_0 = \eta_{10}x_1 + \ldots + \eta_{n0}x_n$. Then $y_{kj} \to y_0$. Thus, $\{y_k\}$ contains a convergent subsequence, and this proves that A is relatively compact.

(Sufficiency) Assume that every bounded subset of Y is relatively compact. Let $x_1 \in Y$ be such that $||x_1|| = 1$, and let $V_1 = V(\{x_1\})$ be the linear subspace generated by $\{x_1\}$ (see Definition 3.3.6). If $V_1 = Y$, then we are done. If $V_1 \neq Y$, let $y_2 \in Y$ be such that $y_2 \notin V_1$. Let $d = \inf_\alpha ||y_2 - \alpha x_1||$. Since V_1 is closed by Theorem 6.6.6, we must have $d > 0$; otherwise $y_2 \in \bar{V}_1$. For every $\eta > 0$ there is an $x_0 \in V_1$ such that $d \leq ||y_2 - x_0|| < d + \eta$. Now let $x_2 = (y_2 - x_0)/||y_2 - x_0||$. Then $x_2 \notin V_1$, $||x_2|| = 1$, and

$$||x_2 - x|| = \left|\left| \frac{y_2 - x_0}{||y_2 - x_0||} - x \right|\right| = \frac{1}{||y_2 - x_0||}||y_2 - x'||$$
$$> \frac{||y_2 - x'||}{d + \eta} \geq \frac{d}{d + n} = 1 - \frac{\eta}{d + \eta},$$

where $x' = x_0 + ||y_2 - x_0||x \in V_1$ for all $x \in V_1$. Since η is arbitrary, we can choose η so that $||x_2 - x_1|| \geq \frac{1}{2}$.

Now let V_2 be the linear subspace generated by $\{x_1, x_2\}$. If $V_1 = Y$, we are done. If not, we can proceed in the manner used above to select an $x_3 \notin V_2$, $||x_3|| = 1$, $||x_1 - x_3|| \geq \frac{1}{2}$, and $||x_2 - x_3|| \geq \frac{1}{2}$. If we continue this process, then we either have $V(\{x_1, \ldots, x_n\}) = Y$ for some n, or or else we obtain an infinite sequence $\{x_n\}$ such that $||x_n|| = 1$ and $||x_n - x_m|| \geq \frac{1}{2}$ for all $m \neq n$. The second alternative is impossible, since $\{x_n\}$ is a bounded sequence and as such must contain a convergent subsequence. This completes the proof. ■

6.7. GEOMETRIC ASPECTS OF LINEAR FUNCTIONALS

Throughout this section X denotes a real normed linear space. Before giving geometric interpretations of linear functionals we introduce the notions of maximal subspace and hyperplane.

6.7.1. Definition. A linear subspace Y of linear space X is called **maximal** if it is not all of X and if there exists no linear subspace Z of X such that $Y \neq Z$, $Z \neq X$ and $Y \subset Z$.

Recall that if Y is a linear subspace of X and if $z \in Y$, then we call the set $Z = z + Y$ a linear variety (see Definition 3.2.17). In this case we also say that Z is a **translation** of Y.

6.7.2. Definition. A **hyperplane** Y in a linear space X is a maximal linear variety resulting from the translation of a maximal linear subspace.

If a hyperplane Y contains the origin, then it is simply a maximal linear subspace and all hyperplanes Z obtained by translating Y are said to be **parallel** to Y.

The following theorem provides us with an important characterization of hyperplanes in terms of linear functionals.

6.7.3. Theorem. If $f \neq 0$ is a linear functional on X and if α is any fixed scalar, then the set $Y = \{x: f(x) = \alpha\}$ is a hyperplane. It contains the origin 0 if and only if $\alpha = 0$. Conversely, if Y is a hyperplane in a linear space X, then there is a linear functional f on X and a fixed scalar α such that $Y = \{x: f(x) = \alpha\}$.

Proof. Consider the first part. Since $f \neq 0$ there is an x_1 such that $f(x_1) = \beta \neq 0$. If $x_0 = (\alpha/\beta)x_1$, then $f(x_0) = (\alpha/\beta)f(x_1) = \alpha$ and thus $x_0 \in Y$. Let $Y_0 = -x_0 + Y$. It is readily verified that $Y_0 = \{x: f(x) = 0\}$ and that Y_0 is a linear subspace, so that Y is a linear variety. Since $Y_0 \neq X$, we can write every element of X as the sum of an element of Y_0 and a multiple of y, where $y \in X - Y_0$. If $x \in X$, if y is any element in $X - Y_0$ such that $f(y) \neq 0$, and if

$$z = x - \frac{f(x)}{f(y)}y,$$

then $f(z) = 0$, and thus x has the required form. Now assume that Y_1 is a linear subspace of X for which $Y_0 \subset Y_1$ and $Y_1 \neq Y_0$. We can choose $y \in Y_1 - Y_0$, and the above argument shows that $X \subset Y_1$ so that $Y_1 = X$. This shows that Y_0 is maximal and that Y is a hyperplane.

The assertion that Y contains 0 if and only if $\alpha = 0$ follows readily.

Consider now the last part of the thorem. If Y is a hyperplane in X, then Y is the translation of a linear subspace Z in X; i.e., $Y = x_0 + Z$, with x_0 fixed. If $x_0 \notin Z$, and if $V(Y + x_0)$ denotes the linear subspace generated by the set $Y + x_0$, then $V(Y + x_0) = X$. If for $x = \alpha x_0 + z$, $z \in Z$, we define $f(x) = \alpha$, then $Y = \{x: f(x) = 1\}$. On the other hand, if $x_0 \in Z$, then we take $x_1 \notin Z$, $X = V(Z + x_1)$, $Y = Z$, and define for $x = \alpha x_1 + z$, $f(x) = \alpha$. Then $Y = \{x: f(x) = 0\}$. This concludes the proof of the theorem. ■

In the proof of the above theorem we established also the following result:

6.7.4. Theorem. Let $f \neq 0$ be a linear functional on the linear space X, and let $Z = \{x: f(x) = 0\}$. If $x_0 \in X - Z$, then every $x \in X$ can be expressed as

$$x = \frac{f(x)}{f(x_0)}x_0 + z, \; z \in Z.$$

The next result shows that it is possible to establish a unique correspondence between hyperplanes and linear functionals. This result follows readily from Theorem 6.7.3.

6.7.5. Theorem. Let Y be a hyperplane in a linear space X. If Y does not contain the origin, there is a unique linear functional f on X such that $Y = \{x : f(x) = 1\}$.

6.7.6. Exercise. Prove Theorem 6.7.5.

6.7.7. Theorem. Let Y be a maximal linear subspace in a Banach space X. Then either $Y = \bar{Y}$ or $\bar{Y} = X$; i.e., either Y is closed or else Y is dense in X.

Proof. Since Y is a linear subspace, $\bar{Y}$ is a linear subspace of X by Theorem 6.2.3. Now $Y \subset \bar{Y}$. Hence, if $Y \neq \bar{Y}$ we must have $\bar{Y} = X$, since Y is a maximal linear subspace. ■

In the next result we will show that Y is closed if and only if the functional f associated with Y is bounded (i.e., continuous). Thus, corresponding to any hyperplane in a normed linear space there is a functional that is bounded whenever the hyperplane is closed and vice versa.

6.7.8. Theorem. Let f be a non-zero linear functional on X, and let $Y = \{x : f(x) = \alpha\}$ be a hyperplane in X. Then Y is closed for every α if and only if f is bounded.

Proof. Since f is bounded, it is continuous. If $\{x_n\}$ is a sequence in Y which converges to $x \in X$, then $f(x_n) \longrightarrow f(x) = \alpha$, so that $x \in Y$, and thus Y is closed.

Conversely, let $Z = \{x : f(x) = 0\}$ be closed. In view of Theorem 6.7.4, there exists an $x_0 \in X - Z$ such that $X = [x_0 + Z]$. Now let $\{x_n\}$ be a sequence in X such that $x_n \longrightarrow x \in X$. Then it is possible to express each x_n and x as $x_n = c_n x_0 + z_n$ and $x = cx_0 + z$, where $z_n, z \in Z$. Let $d = \inf_{z \in Z} \|x_0 - z\|$. Since Z is closed, $d > 0$. Now

$$\begin{aligned} \|x - x_n\| &= \|(c - c_n)x_0 - (z - z_n)\| \\ &\geq \inf_{(z - z_n) \in Z} \|(c - c_n)x_0 - (z - z_n)\| = |c - c_n| d. \end{aligned}$$

Thus $c_n \longrightarrow c$. Moreover, since $f(x_n) = c_n f(x_0) + f(z_n) = c_n f(x_0) \longrightarrow cf(x_0) = f(x)$, it follows that f is continuous on X, and hence bounded. ■

We now introduce the concept of a half-space.

6.7.9. Definition. Let f be a non-zero linear functional on X, and let $\alpha \in R$. Let Y be the hyperplane given by $Y = \{x : f(x) = \alpha\}$. Let $Y_1, Y_2, Y_3,$

and Y_4 be subsets of X defined by $Y_1 = \{x: f(x) < \alpha\}$, $Y_2 = \{x: f(x) \leq \alpha\}$, $Y_3 = \{x: f(x) > \alpha\}$, and $Y_4 = \{x: f(x) \geq \alpha\}$. Then each of the sets Y_1, Y_2, Y_3, and Y_4 is called a **half space** determined by Y. In addition, let Z_1 and Z_2 be subsets of X. We say that Y **separates** Z_1 and Z_2 if either (i) $Z_1 \subset Y_2$ and $Z_2 \subset Y_4$ or (ii) $Z_1 \subset Y_4$ and $Z_2 \subset Y_2$.

6.7.10. Exercise. Show that each of the sets Y_1, Y_2, Y_3, Y_4 in the preceding definition is convex. Also, show that if in the above definition f is continuous, then Y_1 and Y_3 are open sets in X, and Y_2 and Y_4 are closed sets in X.

In order to demonstrate some of the notions introduced, we conclude this section with the following example.

6.7.11. Example. Let $X = R^2$, and let $x = (\xi_1, \xi_2) \in X$. Let $y = (\eta_1, \eta_2)$ be any fixed vector in X, and define the linear functional f on X as

$$f(x) = \eta_1\xi_1 + \eta_2\xi_2.$$

The set

$$Y_0 = \{x \in R^2: f(x) = \eta_1\xi_1 + \eta_2\xi_2 = 0\}$$

is a line through the origin of R^2 which is normal to the vector y. If $x_1 \notin Y_0$, the hyperplane

$$Y = x_1 + Y_0 = \{x_1 + u: u \in Y_0\}$$

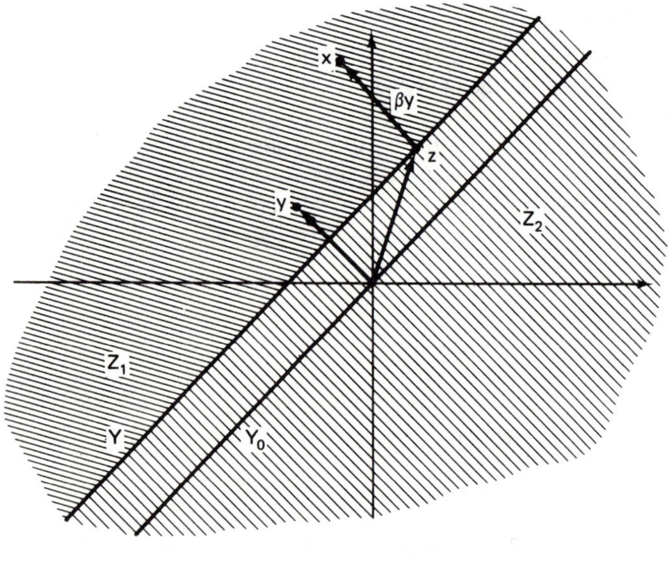

6.7.12. Figure E. Half spaces.

is a linear variety which is parallel to Y_0. The hyperplane Y divides R^2 into two open half-spaces Z_1 and Z_2 as depicted in Figure E. It should be noted that $x \in X$ can now be written as $x = z + \beta y$, $z \in Y$, where $x \in Z_1$ if $\beta > 0$ and $x \in Z_2$ if $\beta < 0$. ∎

6.8. EXTENSION OF LINEAR FUNCTIONALS

In this section we state and prove the Hahn-Banach theorem. This result is very important in analysis and has important implications in applications. We would like to point out that the present form of this theorem is not the most general version of the Hahn-Banach theorem.

Throughout this section X will denote a real normed linear space.

6.8.1. Definition. Let Y be a linear subspace of X, let Z be a proper linear subspace of Y, let f be a bounded linear functional defined on Z, and let $\tilde{f}$ be a bounded linear functional defined on Y. If $\tilde{f}(x) = f(x)$ whenever $x \in Z$, then $\tilde{f}$ is called an **extension** of f from Z to Y. If the spaces X, Y, Z are normed and if $\|f\|_Z = \|\tilde{f}\|_Y$, then $\tilde{f}$ is called a **norm preserving** extension of f.

We now prove the following version of the **Hahn-Banach theorem.**

6.8.2. Theorem. Every bounded linear functional f defined on a linear subspace Y of a real normed linear space X can be extended to the entire space X with preservation of norm. Specifically, one can find a bounded linear functional $\tilde{f}$ such that

(i) $\tilde{f}(x) = f(x)$ for every $x \in Y$; and
(ii) $\|\tilde{f}\|_X = \|f\|_Y$.

Proof. Although this theorem is true for X not separable, we shall give the proof only for the case where X is separable (see Definition 5.4.33 for separability). We assume that Y is a proper linear subspace of X, for otherwise there is nothing to prove. Let $x_1 \in X$ but $x_1 \notin Y$, and let us define the subset

$$Y_1 = \{x \in X: x = \alpha x_1 + y, \alpha \in R, y \in Y\}.$$

It is straightforward to verify that Y_1 is a linear subspace of X, and furthermore that for each $x \in Y_1$ there is a unique $\alpha \in R$ and a unique $y \in Y$ such that $x = \alpha x_1 + y$. If an extension $\tilde{f}$ of f from Y to Y_1 exists, then it has the form

$$\tilde{f}(x) = \alpha \tilde{f}(x_1) + f(y),$$

and if we let $c = -\tilde{f}(x_1)$, then $\tilde{f}(x) = f(y) - c\alpha$. From this it is clear that the extension is specified by prescribing the constant αc. In order that the norm of the functional not be increased when it is continued from Y to Y_1, we must find a c such that the inequality

$$|f(y) - \alpha c| \leq ||f|| \, ||y + \alpha x_1||$$

holds for all $y \in Y$.

If $y \in Y$, then $y/\alpha \in Y$ and the above inequality can be written as

$$|f(\alpha z) - \alpha c| \leq ||f|| \, ||\alpha z + \alpha x_1||$$

or

$$|f(z) - c| \leq ||f|| \, ||z + x_1||.$$

This inequality can be rewritten as

$$-||f|| \, ||z + x_1|| \leq f(z) - c \leq ||f|| \, ||z + x_1||$$

or, equivalently, as

$$f(z) - ||f|| \, ||z + x_1|| \leq c \leq f(z) + ||f|| \, ||z + x_1|| \tag{6.8.3}$$

for all $z \in Y$. We now must show that such a number c does indeed always exist. To do this, it suffices to show that for any $y_1, y_2 \in Y$, we have

$$c_1 \triangleq f(y_1) - ||f|| \, ||y_1 + x_1|| \leq f(y_2) + ||f|| \, ||y_2 + x_1|| \triangleq c_2. \tag{6.8.4}$$

But this inequality follows directly from

$$\begin{aligned} f(y_1) - f(y_2) &\leq ||f|| \, ||y_1 - y_2|| = ||f|| \, ||y_1 + x_1 - x_1 - y_2|| \\ &\leq ||f|| \, ||y_1 + x_1|| + ||f|| \, ||y_2 + x_1||. \end{aligned}$$

In view of (6.8.3) and (6.8.4) it follows that $c_1 \leq c \leq c_2$. If we now let

$$\tilde{f}(x) = f(y) - \alpha c, \; x \in Y_1,$$

we have $||\tilde{f}|| = ||f||$, and $\tilde{f}$ is an extension of f from Y to Y_1.

Next, since X is separable it contains a denumerable everywhere dense set $\{x_1, x_2, \ldots, x_n, \ldots\}$. From this set of vectors we select, one at a time, a linearly independent subset of vectors $\{y_1, y_2, \ldots, y_n, \ldots\}$ which belongs to $X - Y$. The set $\{y_1, y_2, \ldots, y_n, \ldots\}$ together with the linear subspace Y generates a subspace W dense in X.

Following the above procedure, we now extend the functional f to a functional $\tilde{f}$ on the subspace W by extending f from Y to Y_1, then to Y_2, etc., where

$$Y_1 = \{x : x = \alpha y_1 + y; \; y \in Y, \alpha \in R\}$$

and

$$Y_2 = \{x : x = \alpha y_2 + y; \; y \in Y_1, \alpha \in R\}, \quad \text{etc.}$$

Finally, we extend $\tilde{f}$ from the dense subspace W to the space X. At the remaining points of X the functional is defined by continuity. If $x \in X$, then there exists a sequence $\{w_n\}$ of vectors in W converging to x. By continuity, if $\lim_{n\to\infty} w_n = x$, then $\tilde{f}(x) = \lim_{n\to\infty} \tilde{f}(w_n)$. The inequality $|\tilde{f}(x)| \leq ||f|| \, ||x||$ follows from

$$|\tilde{f}(x)| = \lim_{n\to\infty} |\tilde{f}(w_n)| \leq \lim_{n\to\infty} ||f|| \, ||w_n|| = ||f|| \, ||x||.$$

This completes the proof of the theorem. ∎

The next result is a direct consequence of Theorem 6.8.2.

6.8.5. Corollary. Let $x_0 \in X$, $x_0 \neq 0$. Then there exists a bounded non-zero linear functional f defined on all of X such that $f(x_0) = ||x_0||$ and $||f|| = 1$.

Proof. Let Y be the linear subspace of X given by $Y = \{y \in X: y = \alpha x_0, \alpha \in R\}$. For $y \in Y$, define $f_0(y) = \alpha||x_0||$, where $y = \alpha x_0$. Then $||y|| = |\alpha| \cdot ||x_0||$, and so $\dfrac{|f_0(y)|}{||y||} = 1$ for all $y \in Y$. This implies that $||f_0|| = 1$. The proof now follows from Theorem 6.8.2. ∎

The next result is also a consequence of the Hahn-Banach theorem.

6.8.6. Corollary. Let $x_0 \in X$, $x_0 \neq 0$, and let $\gamma > 0$. Then there exists a bounded non-zero linear functional f defined on all of X such that $||f|| = \gamma$ and $f(x_0) = ||f|| \cdot ||x_0||$.

The above corollary guarantees the existence of non-trivial bounded linear functionals.

6.8.7. Exercise. Prove Corollary 6.8.6.

In the next example a geometric interpretation of Corollary 6.8.5 is given.

6.8.8. Example. Let $x_0 \in X$, $x_0 \neq 0$, and let f be a linear functional defined on X such that $f(x_0) = ||x_0||$ and $||f|| = 1$. Let K be the closed sphere given by $K = \{x \in X: ||x|| \leq ||x_0||\}$. Now if $x \in K$, then $f(x) \leq |f(x)| \leq ||f|| \cdot ||x|| \leq ||x_0||$, and so x belongs to the half-space $\{x \in X: f(x) \leq ||x_0||\}$. Thus, the hyperplane $\{x \in X: f(x) = ||x_0||\}$ is tangent to the closed sphere (as illustrated in Figure F). ∎

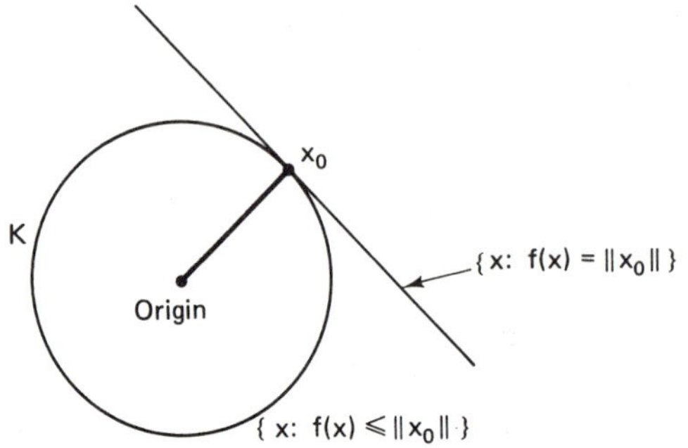

6.8.9. Figure F. Illustration of Corollary 6.8.5.

In closing this section, we mention two of the more important consequences of the Hahn-Banach theorem with significant practical implications. One of these states that given a convex set Y in X containing an interior point and given a fixed point not in the interior of Y, there is a hyperplane separating the fixed point and the convex set Y. The second of these asserts that if Y_1 and Y_2 are convex sets in X, if Y_1 has interior points, and if Y_2 contains no interior point of Y_1, then there is a closed hyperplane which separates Y_1 and Y_2.

6.9. DUAL SPACE AND SECOND DUAL SPACE

In this section we briefly reconsider dual space X^* (see Definition 6.5.9), and we introduce the dual space of X^*, called the **second dual space**. Throughout this section X is a real normed linear space, and X^f is the algebraic conjugate of X.

We begin by determining the dual spaces of some common normed linear spaces.

6.9.1. Example. Let $X = R^n$, let $x = (\xi_1, \ldots, \xi_n)$ denote an arbitrary element of R^n, let $a = (\alpha_1, \ldots, \alpha_n)$ be some fixed element in R^n, let $\|x\| = \sqrt{\xi_1^2 + \ldots + \xi_n^2}$, and recall from Example 6.5.14 that the functional $f(x) = \alpha_1\xi_1 + \ldots + \alpha_n\xi_n$ is a bounded linear functional on X and $\|f\| = \|a\|$. If we define a set of basis vectors in R^n as $e_1 = (1, 0, \ldots, 0), \ldots, e_n = (0, \ldots, 0, 1)$, then $x \in R^n$ may be expressed as $x = \sum_{i=1}^{n} \xi_i e_i$. If we let $\alpha_i = f(e_i)$, where f is any bounded linear functional on R^n, then

$$f(x) = \sum_{i=1}^{n} f(\xi_i e_i) = \sum_{i=1}^{n} \xi_i f(e_i) = \sum_{i=1}^{n} \alpha_i \xi_i.$$

Thus, the dual space X^* of $R^n = X$ is itself the space R^n in the sense that the elements of X^* consist of all functionals of the form $f(x) = \sum_{i=1}^{n} \alpha_i \xi_i$. Furthermore, the norm on X^* is $\|f\| = (\sum_{i=1}^{n} \alpha_i^2)^{1/2}$. ■

6.9.2. Exercise. Let $X = R^n$, where the norm of $x = (\xi_1, \ldots, \xi_n) \in X$ is given by $\|x\| = \max_{1 \le i \le n} |\xi_i|$ (see Example 6.1.5). Show that if $f \in X^*$, then there is an $a = (\alpha_1, \ldots, \alpha_n) \in R^n$ such that $f(x) = \alpha_1\xi_1 + \ldots + \alpha_n\xi_n$, so that $X^* = R^n$, and show that the norm on X^* is given by $\|f\| = \sum_{i=1}^{n} |\alpha_i|$.

6.9.3. Exercise. Let $X = R^n$, and define the norm of $x = (\xi_1, \ldots, \xi_n) \in X$ by $\|x\| = (|\xi_1|^p + \ldots + |\xi_n|^p)^{1/p}$, where $1 < p < \infty$ (see Example 6.1.5). Show that if $f \in X^*$ then there is an $a = (\alpha_1, \ldots, \alpha_n) \in R^n$ such that $f(x) = \alpha_1\xi_1 + \ldots + \alpha_n\xi_n$, i.e., $X^* = R^n$, and show that the norm on X^* is given by $\|f\| = (|\alpha_1|^q + \ldots + |\alpha_n|^q)^{1/q}$, where q is such that $\frac{1}{p} + \frac{1}{q} = 1$.

6.9.4. Exercise. Let X be the space l_p, $1 \le p < \infty$, defined in Example 6.1.6 and let $\frac{1}{p} + \frac{1}{q} = 1$. If $p = 1$, we take $q = \infty$. Show that the dual space of l_p is l_q. Specifically, show that every bounded linear functional on l_p is uniquely representable as

$$f(x) = \sum_{i=1}^{\infty} \alpha_i \xi_i,$$

where $a = (\alpha_1, \ldots, \alpha_k, \ldots)$ is an element of l_q. Also, show that every element a of l_q defines an element of $(l_p)^*$ in the same way and that

$$\|f\| = \begin{Bmatrix} \left(\sum_{i=1}^{\infty} |\alpha_i|^q\right)^{1/q} & \text{if } 1 < p < \infty \\ \sup_i |\alpha_i| & \text{if } p = 1. \end{Bmatrix}.$$

Since X^* is a normed linear space (see Theorem 6.5.6) it is possible to form the dual space of X^*, which we will denote by X^{**} and which will be referred to as the **second dual space** of X. As before, we will use the notation x'' for elements of X^{**} and we will write

$$x''(x') = \langle x', x'' \rangle,$$

where $x' \in X^*$. If X^f denotes the algebraic conjugate of X, then the reader can readily show that even though $X^* \subset X^f$ and $X^{**} \subset (X^*)^f$, in general, X^{**} is not a linear subspace of X^{ff}.

Let us define a mapping J of X into X^{**} by the relation

$$\langle x', Jx \rangle = \langle x, x' \rangle, \; x \in X, \; x' \in X^* \tag{6.9.5}$$

or, equivalently, by

$$Jx = x'', \quad x''(x') = x'(x). \tag{6.9.6}$$

We call this mapping J the **canonical mapping** of X into X^{**}. The functional x'' defined on X^* in this way is linear, because

$$\begin{aligned} x''(\alpha x'_1 + \beta x'_2) = \langle x, \alpha x'_1 + \beta x'_2 \rangle &= \alpha\langle x, x'_1\rangle + \beta\langle x, x'_2\rangle \\ &= \alpha x''(x'_1) + \beta x''(x'_2), \end{aligned}$$

and thus $x'' \in (X^*)^f$. Since

$$|x''(x')| = |x'(x)| = |\langle x, x'\rangle| \le \|x\|\|x'\|,$$

it follows that $\|x''\| \le \|x\|$ and thus $x'' \in X^{**}$. We can actually show that $\|x''\| = \|x\|$. This is obvious for $x = 0$. If $x \neq 0$, then in view of Corollary 6.8.6 there exists a non-zero $x' \in X^*$ such that $\langle x, x'\rangle = \|x\|\|x'\|$, and thus $\|x''\| = \|x\|$. From this it follows that the norm of every $x \in X$ can be defined in two ways: as the norm of an element in X and as the norm of a linear functional on X^*, i.e., as the norm of an element in X^{**}. We summarize this discussion in the following result:

6.9.7. Theorem. X is isometric to some linear subspace in X^{**}.

If we agree not to distinguish between isometric spaces, then Theorem 6.9.7 can simply be stated as $X \subset X^{**}$.

6.9.8. Definition. A normed linear space is said to be **reflexive** if the canonical mapping (6.9.6), $J: X \longrightarrow X^{**}$, is onto. If we again agree not to distinguish between isometric spaces, we write in this case $X^{**} = X$. If $X \neq X^{**}$, then X is said to be **irreflexive**.

6.9.9. Example. The space R_p^n, $1 \le p \le \infty$ is reflexive. ■

6.9.10. Example. The spaces l_p, $1 < p < \infty$, are reflexive. ■

6.9.11. Example. The space l_1 is irreflexive. ■

6.9.12. Exercise. Prove the assertions made in Examples 6.9.9 through 6.9.11.

6.10. WEAK CONVERGENCE

Having introduced the normed dual space, we are now in a position to consider the notion of weak convergence, a concept which arises frequently in analysis and which plays an important role in certain applications. *Throughout this section X denotes a normed linear space and X^* is the dual space of X.*

6.10.1. Definition. A sequence $\{x_n\}$ of elements in X is said to **converge weakly** to the element $x \in X$ if for every $x' \in X^*$, $\langle x_n, x' \rangle \longrightarrow \langle x, x' \rangle$. In this case we write $x_n \longrightarrow x$ weakly. If a sequence $\{x_n\}$ converges to $x \in X$, i.e., if $\|x_n - x\| \longrightarrow 0$ as $n \longrightarrow \infty$, then we call this convergence **strong convergence** or **convergence in norm** to distinguish it from weak convergence.

6.10.2. Theorem. Let $\{x_n\}$ be a sequence in X which converges in norm to $x \in X$. Then $\{x_n\}$ converges weakly to x.

Proof. Assume that $\|x_n - x\| \longrightarrow 0$ as $n \longrightarrow \infty$. Then for any $x' \in X^*$ we have

$$|\langle x_n, x' \rangle - \langle x, x' \rangle| \leq \|x'\| \|x_n - x\| \longrightarrow 0 \text{ as } n \longrightarrow \infty,$$

and thus $x_n \longrightarrow x$ weakly. ■

Thus, strong convergence implies weak convergence. However, the converse is not true, in general, as the following example shows.

6.10.3. Example. Consider in l_2 the sequence of vectors $x_1 = (1, 0, \ldots, 0, \ldots)$, $x_2 = (0, 1, 0, \ldots, 0, \ldots)$, $x_3 = (0, 0, 1, \ldots, 0, \ldots)$, ... To show that $\{x_n\}$ converges weakly we note that every $x' \in l_2 = X^*$ can be represented as the scalar product with some fixed vector $y = (\eta_1, \eta_2, \ldots, \eta_n, \ldots)$; i.e., if $x = (\xi_1, \xi_2, \ldots, \xi_n, \ldots)$, then

$$\langle x, x' \rangle = \sum_{i=1}^{\infty} \xi_i \eta_i$$

(see Exercise 6.9.4). For the case of the sequence $\{x_n\}$ we now have

$$\langle x_n, x' \rangle = \eta_n,$$

and since $\eta_n \longrightarrow 0$ as $n \longrightarrow \infty$ for every $y \in l_2$, it follows that $\langle x_n, x' \rangle \longrightarrow 0$ as $n \longrightarrow \infty$ for every $x' \in l_2$. Thus, $\{x_n\}$ converges to 0 weakly. However, $x_n \not\longrightarrow 0$ strongly, because $\|x_n\| = 1$. ■

We leave the proof of the next result as an exercise to the reader.

6.10.4. Theorem. If X is finite dimensional, weak and strong convergence are equivalent.

6.10.5. Exercise. Prove Theorem 6.10.4.

Analogous to the concept of weak convergence of elements of a normed linear space X we can introduce the notion of weak convergence of elements of X^*.

6.10.6. Definition. A sequence of functionals $\{x'_n\}$ in X^* **converges weakstar** (i.e., weak*) to the linear functional $x' \in X^*$ if for every $x \in X$ we have $\langle x, x'_n \rangle \longrightarrow \langle x, x' \rangle$. We say that $x'_n \longrightarrow x'$ weak*.

Since strong convergence in X^* implies weak convergence in X^*, it follows that if a sequence of linear functionals $\{x'_n\}$ in X^* converges to the linear functional $x' \in X^*$, then $x'_n \longrightarrow x'$ weak*.

Let us consider an example.

6.10.7. Example. Let $[a, b]$ be an interval on the real line containing the origin, i.e., $a < 0 < b$, and let $\{\mathcal{C}[a, b]; \|\cdot\|_\infty\}$ be the Banach space of real-valued continuous functions as defined in Example 6.1.9. Let $\{\varphi_n\}$ be a sequence of functions in $\mathcal{C}[a, b]$ satisfying the following conditions for $n = 1, 2, \ldots$:

(i) $\varphi_n(t) \geq 0$ for all $t \in [a, b]$;

(ii) $\varphi_n(t) = 0$ if $|t| > 1/n$ and $t \in [a, b]$; and

(iii) $\int_a^b \varphi_n(t)\, dt = 1.$

For each $n = 1, 2, \ldots$, we can define a continuous linear functional x'_n on X (see Example 6.5.13) by

$$\langle x, x'_n \rangle = \int_a^b x(t)\varphi_n(t)\, dt$$

where $x \in \mathcal{C}[a, b]$. Now let x'_0 be defined on $\mathcal{C}[a, b]$ by

$$\langle x, x'_0 \rangle = x(0)$$

for all $x \in \mathcal{C}[a, b]$. It is clear that $x'_0 \in X^*$. We now show that $x'_n \longrightarrow x'_0$ weak*. By the mean value theorem from the calculus, there is a t_n such that $-1/n \leq t_n \leq 1/n$ and

$$\int_{-1/n}^{1/n} \varphi_n(t)x(t)\, dt = x(t_n) \int_{-1/n}^{1/n} \varphi_n(t)\, dt = x(t_n)$$

for each $n = 1, 2, \ldots$, and $x \in \mathcal{C}[a, b]$. Thus, $\langle x, x'_n \rangle \longrightarrow x(0)$ for every $x \in \mathcal{C}[a, b]$; i.e., $x'_n \longrightarrow x_0$ weak*. We see that the sequence of functions $\{\varphi_n\}$ does not approach a limit in $\mathcal{C}[a, b]$. In particular, there is no $\varphi_0 \in \mathcal{C}[a, b]$ such that $x(0) = \int_a^b x(t)\varphi_0(t)\, dt$. Frequently, in applications, it is convenient to say the sequence $\{\varphi_n\}$ converges to the so-called "δ function" which has this property. We see that the sequence $\{\varphi_n\}$ converges to the δ function in the sense of weak* convergence. ■

6.10.8. Theorem. Let X be a separable normed linear space. Every bounded sequence of linear functionals in X^* contains a weakly convergent subsequence.

Proof. Since X is separable, we can choose a denumerable everywhere dense set $\{x_1, x_2, \ldots, x_n, \ldots\}$ in X. Now let $\{x'_n\}$ be a sequence in X^*. Since this sequence is bounded in norm, the sequence $\{\langle x_1, x'_n \rangle\}$ is a bounded sequence in either R or C. It now follows that we can select from $\{x'_n\}$ a subsequence

$\{x'_{0_n}\}$ such that the sequence $\{\langle x_1, x'_{0_n}\rangle\}$ converges. Again, from the subsequence $\{x'_{0_n}\}$ we can select another subsequence $\{x'_{1_n}\}$ such that the sequence $\{\langle x_2, x'_{1_n}\rangle\}$ converges. Continuing this procedure, we obtain the sequences

$$\begin{array}{l} x'_{0_1}, x'_{0_2}, \ldots, x'_{0_n}, \ldots \\ x'_{1_1}, x'_{1_2}, \ldots, x'_{1_n}, \ldots \\ x'_{2_1}, x'_{2_2}, \ldots, x'_{2_n}, \ldots \\ \ldots\ldots\ldots\ldots\ldots\ldots \end{array}$$

By taking the diagonal of the above array, we obtain the subsequence of linear functionals $x'_{0_1}, x'_{1_2}, x'_{2_3}, \ldots$. For this subsequence, the sequence $x'_{0_1}(x_n), x'_{1_2}(x_n), x'_{2_3}(x_n), \ldots$ converges for all n. But then $x'_{0_1}(x), x'_{1_2}(x), x'_{2_3}(x), \ldots$ converges for all $x \in X$. This completes the proof of the theorem. ■

The concepts of weak convergence and weak* convergence give rise to various generalizations, some of which we briefly mention.

Let X be a normed linear space and, let X^* be its normed dual. We call a set $Y \subset X^*$ **weak* compact** if every infinite sequence from Y contains a weak* convergent subsequence. We say that a functional defined on X, which in general may be non-linear, is **weakly continuous** at a point $x_0 \in X$ if for every $\epsilon > 0$ there is a $\delta > 0$ and a finite collection $\{x'_1, x'_2, \ldots, x'_n\}$ in X^*, such that $|f(x) - f(x_0)| < \epsilon$ for all x such that $|\langle x, x'_i\rangle| < \delta$ for $i = 1, 2, \ldots, n$. We can define weak* continuity of a functional similarly by interchanging the roles of X and X^*.

It can be shown that if X is a real normed linear space and X^* is its normed dual, then any closed sphere in X^* is weak* compact.

The reader can readily show that if f is a weakly continuous functional, then $x_n \longrightarrow x$ weakly implies that $f(x_n) \longrightarrow f(x)$.

6.11. INNER PRODUCT SPACES

We recall (see Definition 3.6.19 and the discussion following this definition) that if X is a complex linear space, a function defined on $X \times X$ into C, which we denote by (x, y) for $x, y \in X$, is called an **inner product** if

(i) $(x, x) > 0$ for all $x \neq 0$ and $(x, x) = 0$ if $x = 0$;

(ii) $(x, y) = \overline{(y, x)}$ for all $x, y \in X$;

(iii) $(\alpha x + \beta y, z) = \alpha(x, z) + \beta(y, z)$ for all $x, y, z \in X$ and for all $\alpha, \beta \in C$; and

(iv) $(x, \alpha y + \beta z) = \bar{\alpha}(x, y) + \bar{\beta}(x, z)$ for all $x, y, z \in X$ and for all $\alpha, \beta \in C$.

In the case of real linear spaces, the preceding characterization of an inner product is identical, except we omit complex conjugates in (ii) and (iv). We call a complex (real) linear space X on which an inner product, $(\cdot, \cdot)$, is defined a complex (real) **inner product space** which we denote by $\{X; (\cdot, \cdot)\}$ (see Definition 3.6.20). If the particular inner product being used in a given discussion is understood, we simply write X to denote the inner product space. In accordance with our discussion following Definition 3.6.20, recall also that different inner products defined on the same linear space yield different inner product spaces. Finally, refer also to the discussion following Definition 3.6.20 for the characterization of an (inner product) subspace.

We have already extensively studied finite-dimensional real inner product spaces, i.e., Euclidean vector spaces, in Sections 4.9 and 4.10. Our subsequent presentation will be in a more general setting, where X need not be finite dimensional and where X may be a complex vector space. In fact, *unless otherwise stated,* $\{X; (\cdot, \cdot)\}$ *will denote in this section an arbitrary complex inner product space.* Since the proofs of several of the following theorems are nearly identical to corresponding ones in Sections 4.9 and 4.10, we will leave such proofs as exercises.

One of our first objectives will be to show that every inner product space $\{X; (\cdot, \cdot)\}$ has a norm associated with it which is induced by its inner product $(\cdot, \cdot)$. We find it convenient to consider first the **Schwarz inequality**, given in the following theorem.

6.11.1. Theorem. For any $x \in X$, let us define the function $\|\cdot\|: X \to R$ by $\|x\| = (x, x)^{1/2}$. Then for all $x, y \in X$,

$$|(x, y)| \leq \|x\| \cdot \|y\|. \tag{6.11.2}$$

6.11.3. Exercise. Prove Theorem 6.11.1 (see Theorem 4.9.28).

Using the above results, we can now readily show that the function $\|\cdot\|$ defined by $\|x\| = (x, x)^{1/2}$ is a norm.

6.11.4. Theorem. Let X be an inner product space. Then the function $\|\cdot\|: X \to R$ defined by

$$\|x\| = (x, x)^{1/2} \tag{6.11.5}$$

is a norm; i.e., for every $x, y \in X$ and for every $\alpha \in C$, we have

(i) $\|x\| \geq 0$;
(ii) $\|x\| = 0$ if and only if $x = 0$;
(iii) $\|\alpha x\| = |\alpha| \|x\|$; and
(iv) $\|x + y\| \leq \|x\| + \|y\|$.

6.11.6. Exercise. Prove Theorem 6.11.4 (see Theorem 4.9.31).

Theorem 6.11.4 allows us to view every inner product space as a normed linear space, provided that we use Eq. (6.11.5) to define the norm on X. Moreover, in view of Theorem 6.1.2, we may view every inner product space as a metric space, provided that we define the metric by $\rho(x, y) = ||x - y||$. *Subsequently, we adopt the convention that when using the properties and terminology of a normed linear space in connection with an inner product space we mean the norm induced by the inner product, as given in Eq. (6.11.5).*

We are now in a position to make the following important definition.

6.11.7. Definition. A complete inner product space is called a **Hilbert space**.

Thus, every Hilbert space is also a Banach space (and also a complete metric space). Some authors insist that Hilbert spaces be infinite dimensional. We shall not follow that practice. An arbitrary inner product space (not necessarily complete) is sometimes also called a **pre-Hilbert space**.

6.11.8. Example. Let X be a finite-dimensional (real or complex) inner product space. It follows from Theorem 6.6.5 that X is a Hilbert space. ■

6.11.9. Example. Let l_2 be the (complex) linear space defined in Example 6.1.6. Let $x = (\xi_1, \xi_2, \ldots) \in l_2$, $y = (\eta_1, \eta_2, \ldots) \in l_2$, and define (x, y): $l_2 \times l_2 \longrightarrow C$ as

$$(x, y) = \sum_{i=1}^{\infty} \xi_i \bar{\eta}_i.$$

It can readily be shown that $(\cdot, \cdot)$ is an inner product on X. Since l_2 is complete relative to the norm induced by this inner product (see Example 6.1.6), it follows that l_2 is a Hilbert space. ■

6.11.10. Example

(a) Let $X = \mathcal{C}[a, b]$ denote the linear space of complex-valued continuous functions defined on $[a, b]$ (see Example 6.1.9). For $x, y \in \mathcal{C}[a, b]$ define

$$(x, y) = \int_a^b x(t)\bar{y}(t)\, dt.$$

It is readily verified that this space is a pre-Hilbert space. In view of Example 6.1.9 this space is not complete relative to the norm $||x|| = (x, x)^{1/2}$, and hence it is not a Hilbert space.

(b) We extend the space of real-valued functions, $L_p[a, b]$, defined in Example 5.5.31 for the case $p = 2$, to complex-valued functions to be the set of all functions $f: [a, b] \longrightarrow C$ such that $f = u + iv$ for $u, v \in L_2[a, b]$. Denoting this space also by $L_2[a, b]$, we define

$$(f, g) = \int_{[a, b]} f \bar{g}\, d\mu,$$

for $f, g \in L_2[a, b]$, where integration is in the Lebegue sense. The space $\{L_2[a, b]; (\cdot, \cdot)\}$ is a Hilbert space. ■

In the next example we consider the Cartesian product of Hilbert spaces.

6.11.11. Example. Let $\{X_i\}$, $i = 1, \ldots, n$, denote a finite collection of Hilbert spaces over C, and let $X = X_1 \times \ldots \times X_n$. If $x \in X$, then $x = (x_1, \ldots, x_n)$ with $x_i \in X_i$. Defining vector addition and multiplication of vectors by scalars in the usual manner (see Eqs. (3.2.14), (3.2.15), and the related discussion, and see Example 6.1.10) it follows that X is a linear space. If $x, y \in X$ and if $(x_i, y_i)_i$ denotes the inner product of x_i and y_i on X_i, then it is easy to show that

$$(x, y) = \sum_{i=1}^{n} (x_i, y_i)_i$$

defines an inner product on X. The norm induced on X by this inner product is

$$||x|| = (x, x)^{1/2} = \left(\sum_{i=1}^{n} ||x_i||_i^2\right)^{1/2}$$

where $||x_i||_i = (x_i, x_i)_i^{1/2}$. It is readily verified that X is complete, and thus X is a Hilbert space. ■

6.11.12. Exercise. Verify the assertions made in Example 6.11.11.

In Theorem 6.1.15 we saw that in a normed linear space $\{X; ||\cdot||\}$, the norm $||\cdot||$ is a continuous mapping of X into R. Our next result establishes the continuity of an inner product. In the following, $x_n \longrightarrow x$ implies convergence with respect to the norm induced by the inner product $(\cdot, \cdot)$ on X.

6.11.13. Theorem. Let $\{x_n\}$ be a sequence in X such that $x_n \longrightarrow x$, where $x \in X$, and let $\{y_n\}$ be a sequence in X. Then

(i) $(z, x_n) \longrightarrow (z, x)$ for all $z \in X$;
(ii) $(x_n, z) \longrightarrow (x, z)$ for all $z \in X$;
(iii) $||x_n|| \longrightarrow ||x||$; and
(iv) if $\sum_{n=1}^{\infty} y_n$ is convergent in X, then $(\sum_{n=1}^{\infty} y_n, z) = \sum_{n=1}^{\infty} (y_n, z)$ for all $z \in X$.

6.11.14. Exercise. Prove Theorem 6.11.13.

Next, let us recall that two vectors $x, y \in X$ are said to be **orthogonal** if $(x, y) = 0$ (see Definition 3.6.22). In this case we write $x \perp y$. If $Y \subset X$

and $x \in X$ is such that $x \perp y$ for all $y \in Y$, then we write $x \perp Y$. Also, if $Z \subset X$ and $Y \subset X$ and if $z \perp Y$ for all $z \in Z$, then we write $Y \perp Z$. Furthermore, observe that $x \perp x$ implies that $x = 0$. Finally, the notion of inner product allows us to consider the concepts of alignment and colinearity of vectors.

6.11.15. Definition. Let X be an inner product space. The vectors $x, y \in X$ are said to be **colinear** if $(x, y) = \pm||x|| \cdot ||y||$ and **aligned** if $(x, y) = ||x|| \cdot ||y||$.

Our next result is proved by straightforward computation.

6.11.16. Theorem. For all $x, y \in X$ we have

(i) $||x + y||^2 + ||x - y||^2 = 2||x||^2 + 2||y||^2$; and
(ii) if $x \perp y$, then $||x + y||^2 = ||x||^2 + ||y||^2$.

6.11.17. Exercise. Prove Theorem 6.11.16.

Parts (i) and (ii) of Theorem 6.11.16 are referred to as the **parallelogram law** and the **Pythagorean theorem**, respectively (refer to Theorems 4.9.33 and 4.9.38).

6.11.18. Definition. Let $\{x_\alpha : \alpha \in I\}$ be an indexed set of elements in X, where I is an arbitrary index set (i.e., I is not necessarily the integers). Then $\{x_\alpha : \alpha \in I\}$ is said to be an **orthogonal set of vectors** if $x_\alpha \perp x_\beta$ for all $\alpha, \beta \in I$ such that $\alpha \neq \beta$. A vector $x \in X$ is called a **unit vector** if $||x|| = 1$. An orthogonal set of vectors is called an **orthonormal set** if every element of the set is a unit vector. Finally, if $\{x_i\}$ is a sequence of elements in X, we define an **orthogonal sequence** and an **orthonormal sequence** in an obvious manner.

Using an inductive process we can generalize part (ii) of Theorem 6.11.16 as follows.

6.11.19. Theorem. Let $\{x_1, \ldots, x_n\}$ be a finite orthogonal set in X. Then

$$\left\|\sum_{j=1}^{n} x_j\right\|^2 = \sum_{j=1}^{n} ||x_j||^2.$$

We note that if $x \neq 0$ and if $y = x/||x||$, then $||y|| = 1$. Hence, it is possible to convert every orthogonal set of vectors into an orthonormal set.

Let us now consider a specific example.

6.11.20. Example. Let X denote the space of continuous complex-valued functions on the interval $[0, 1]$. In accordance with Example 6.11.10, we define an inner product on X by

$$(f, g) = \int_0^1 f(t)\overline{g(t)}\, dt. \tag{6.11.21}$$

We now show that the set of vectors defined by

$$f_n(t) = e^{2\pi n t i}, \quad n = 0, \pm 1, \pm 2, \ldots, i = \sqrt{-1} \tag{6.11.22}$$

is an orthonormal set in X. Substituting Eq. (6.11.22) into Eq. (6.11.21), we obtain

$$(f_n, f_m) = \int_0^1 f_n(t)\overline{f_m(t)}\, dt = \int_0^1 e^{2\pi(n-m)it}\, dt$$

$$= \frac{e^{2\pi(n-m)i} - 1}{2\pi(n-m)i}.$$

Since $e^{2\pi k i} = \cos 2\pi k + i \sin 2\pi k$, we have

$$(f_n, f_m) = 0, \quad m \neq n;$$

i.e., if $m \neq n$, then $f_n \perp f_m$. On the other hand,

$$(f_n, f_n) = \int_0^1 e^{2\pi(n-n)it}\, dt = 1;$$

i.e., if $n = m$, then $(f_n, f_n) = 1$ and $\|f_n\| = 1$. ■

The next result arises often in applications.

6.11.23. Theorem. If $\{x_1, \ldots, x_n\}$ is a finite orthonormal set in X, then

(i) $\sum_{i=1}^{n} |(x, x_i)|^2 \leq \|x\|^2$ for all $x \in X$; and (6.11.24)

(ii) $(x - \sum_{i=1}^{n} (x, x_i)x_i) \perp x_j$ for any $j = 1, \ldots, n$.

6.11.25. Exercise. Prove Theorem 6.11.23 (see Theorem 4.9.58).

On passing to the limit as $n \longrightarrow \infty$ in (6.11.24), we obtain the following result.

6.11.26. Theorem. If $\{x_i\}$ is any countable orthonormal set in X, then

$$\sum_{i=1}^{\infty} |(x, x_i)|^2 \leq \|x\|^2 \tag{6.11.27}$$

for every $x \in X$.

The relationship (6.11.27) is known as the **Bessel inequality**. The scalars $\alpha_i = (x, x_i)$ are called the **Fourier coefficients** of x with respect to the orthonormal set $\{x_i\}$.

The next result is a generalization of Theorem 4.9.17.

6.11.28. Theorem. In an inner product space X we have $(x, y) = 0$ for all $x \in X$ if and only if $y = 0$.

6.11.29. Exercise. Prove Theorem 6.11.28.

From our discussion thus far it should be clear that not every normed linear space can be made into an inner product space. The following theorem gives us sufficient conditions for which a normed linear space is also an inner product space.

6.11.30. Theorem. Let X be a normed linear space. If for all $x, y \in X$,

$$||x + y||^2 + ||x - y||^2 = 2(||x||^2 + ||y||^2), \tag{6.11.31}$$

then it is possible to define an inner product on X by

$$(x, y) = \tfrac{1}{4}\{||x + y||^2 - ||x - y||^2 + i||x + iy||^2 - i||x - iy||^2\} \tag{6.11.32}$$

for all $x, y \in X$, where $i = \sqrt{-1}$.

6.11.33. Exercise. Prove Theorem 6.11.30.

6.11.34. Corollary. If X is a real normed linear space whose norm satisfies Eq. (6.11.31) for all $x, y \in X$, then it is possible to define an inner product on X by

$$(x, y) = \tfrac{1}{4}\{||x + y||^2 - ||x - y||^2\}$$

for all $x, y \in X$.

6.11.35. Exercise. Prove Corollary 6.11.34.

In view of part (i) of Theorem 6.11.16 and in view of Theorem 6.11.30, condition (6.11.31) is both necessary and sufficient that a normed linear space be also an inner product space. Furthermore, it can also be shown that Eq. (6.11.32) uniquely defines the inner product on a normed linear space.

We conclude this section with the following exercise.

6.11.36. Exercise. Let l_p, $1 \leq p \leq \infty$, be the normed linear space defined in Example 6.1.6. Show that l_p is an inner product space if and only if $p = 2$.

6.12. ORTHOGONAL COMPLEMENTS

In this section we establish some interesting structural properties of Hilbert spaces. Specifically, we will show that any vector x of a Hilbert space X can uniquely be represented as the sum of two vectors y and z, where y

is in a subspace Y of X and z is orthogonal to Y. This is known as the *projection theorem*. In proving this theorem we employ the so-called "classical projection theorem," a result of great importance in its own right. This theorem extends the following familiar result to the case of (infinite-dimensional) Hilbert spaces: in the three-dimensional Euclidean space the shortest distance between a point and a plane is along a vector through the point and perpendicular to the plane. Both the classical projection theorem and the projection theorem are of great importance in applications.

Throughout this section, $\{X; (\cdot, \cdot)\}$ *is a complex inner product space.*

6.12.1. Definition. Let Y be a non-void subset of X. The set of all vectors orthogonal to Y, denoted by $Y^\perp$, is called the **orthogonal complement** of Y. The orthogonal complement of $Y^\perp$ is denoted by $(Y^\perp)^\perp \triangleq Y^{\perp\perp}$, the orthogonal complement of $Y^{\perp\perp}$ is denoted by $(Y^{\perp\perp})^\perp \triangleq Y^{\perp\perp\perp}$, etc.

6.12.2. Example. Let X be the space E^3 depicted in Figure G, and let Y be the x_1-axis. Then $Y^\perp$ is the x_2x_3-plane, $Y^{\perp\perp}$ is the x_1-axis, $Y^{\perp\perp\perp}$ is again the x_2x_3-plane, etc. Thus, *in the present case*, $Y^{\perp\perp} = Y$, $Y^{\perp\perp\perp} = Y^\perp$, $Y^{\perp\perp\perp\perp} = Y^{\perp\perp}$, $Y^{\perp\perp\perp\perp\perp} = Y^{\perp\perp\perp} = Y^\perp$, etc. ■

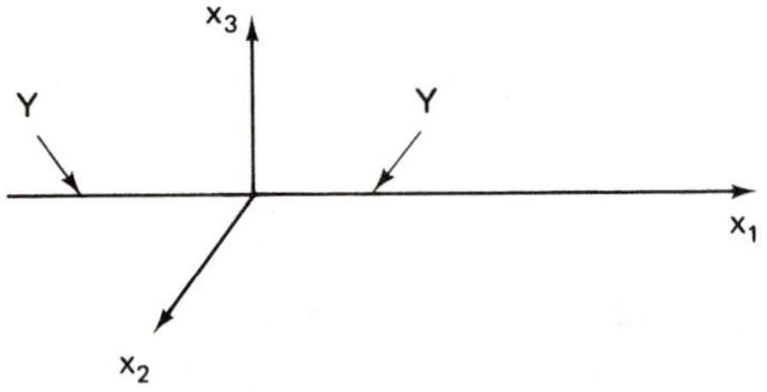

6.12.3 Figure G

We now state and prove several properties of the orthogonal complement. The proof of the first result is left as an exercise.

6.12.4. Theorem. In an inner product space X, $\{0\}^\perp = X$ and $X^\perp = \{0\}$.

6.12.5. Exercise. Prove Theorem 6.12.4.

6.12.6. Theorem. Let Y be a non-void subset of X. Then $Y^\perp$ is a closed linear subspace of X.

Proof. If $x, y \in Y^\perp$, then $(x, z) = 0$ and $(y, z) = 0$ for all $z \in Y$. Hence, $(\alpha x + \beta y, z) = \alpha(x, z) + \beta(y, z) = 0$, and thus $(\alpha x + \beta y) \perp z$ for all $z \in Y$, or $(\alpha x + \beta y) \in Y^\perp$. Therefore, $Y^\perp$ is a linear subspace of X.

To show that $Y^\perp$ is closed, assume that x_0 is a point of accumulation of $Y^\perp$. Then there is a sequence $\{x_n\}$ from $Y^\perp$ such that $\|x_n - x_0\| \longrightarrow 0$ as $n \longrightarrow \infty$. By Theorem 6.11.13 we have $0 = (x_n, z) \longrightarrow (x_0, z)$ as $n \longrightarrow \infty$ for all $z \in Y$. Therefore $x_0 \in Y^\perp$ and $Y^\perp$ is closed. ■

Before considering the next result we require the following concept.

6.12.7. Definition. Let Y be a non-void subset of X, and let $V(Y)$ be the linear subspace generated by Y (see Definition 3.3.6). Let $\overline{V(Y)}$ denote the closure of $V(Y)$. We call $\overline{V(Y)}$ the **closed linear subspace generated by** Y.

Note that in view of Theorem 6.2.3, $\overline{V(Y)}$ is indeed a linear subspace of X.

6.12.8. Theorem. Let Y and Z be non-void subsets of X. Then

(i) either $Y \cap Y^\perp = \varnothing$ or $Y \cap Y^\perp = \{0\}$;
(ii) $Y \subset Y^{\perp\perp}$;
(iii) if $Y \subset Z$, then $Z^\perp \subset Y^\perp$;
(iv) $Y^\perp = Y^{\perp\perp\perp}$; and
(v) $Y^{\perp\perp}$ is the smallest closed linear subspace of X which contains Y; i.e., $Y^{\perp\perp} = \overline{V(Y)}$.

Proof. To prove part (i), assume that $Y \cap Y^\perp \neq \varnothing$, and let $x \in Y \cap Y^\perp$. Then $x \in Y$ and $x \in Y^\perp$ and so $(x, x) = 0$. This implies that $x = 0$.

The proof of part (ii) is left as an exercise.

To prove part (iii), let $y \in Z^\perp$. Then $y \perp x$ for all $x \in Z$. Since $Z \supset Y$, it follows that $y \perp x$ for all $x \in Y$. Thus, $y \in Y^\perp$ whenever $y \in Z^\perp$ and $Y^\perp \supset Z^\perp$.

To prove part (iv) we note that, by part (ii) of this theorem, $Y^\perp \subset Y^{\perp\perp\perp}$. On the other hand, since $Y \subset Y^{\perp\perp}$, by part (iii) of this theorem, $Y^\perp \supset Y^{\perp\perp\perp}$. Thus, $Y^\perp = Y^{\perp\perp\perp}$.

The proof of part (v) is also left as an exercise. ■

6.12.9. Exercise. Prove parts (ii) and (v) of Theorem 6.12.8.

In view of part (iv) of the above theorem, we can write $Y^\perp = Y^{\perp\perp\perp} = Y^{\perp\perp\perp\perp\perp} = \ldots$, and $Y^{\perp\perp} = Y^{\perp\perp\perp\perp} = Y^{\perp\perp\perp\perp\perp\perp} = \ldots$.

Before giving the classical projection theorem, we state and prove the following preliminary result.

6.12.10. Theorem. Let Y be a linear subspace of X, and let x be an arbitrary vector in X. Let

$$\delta = \inf \{\|y - x\| : y \in Y\}.$$

If there exists a $y_0 \in Y$ such that $||y_0 - x|| = \delta$, then y_0 is unique, and moreover $y_0 \in Y$ is the unique element in Y such that $||y_0 - x|| = \delta$ if and only if $(x - y_0) \perp Y$.

Proof. Let us first show that if $||y_0 - x|| = \delta$, then $(x - y_0) \perp Y$. In doing so we assume to the contrary that there is a $y \in Y$ not orthogonal to $x - y_0$. We also assume, without loss of generality, that y is a unit vector and that $(x - y_0, y) = \alpha \neq 0$. Defining a vector $z \in Y$ as $z = y_0 + \alpha y$, we have

$$\begin{aligned}||x - z||^2 &= ||x - y_0 - \alpha y||^2 = (x - y_0 - \alpha y, x - y_0 - \alpha y)\\&= (x - y_0, x - y_0) - (x - y_0, \alpha y) - (\alpha y, x - y_0) + (\alpha y, \alpha y)\\&= ||x - y_0||^2 - \bar{\alpha}\alpha - \alpha\bar{\alpha} + \alpha\bar{\alpha}||y||^2\\&= ||x - y_0||^2 - |\alpha|^2 < ||x - y_0||^2;\end{aligned}$$

i.e., $||x - z|| < ||x - y_0||$. From this it follows that if $x - y_0$ is not orthogonal to every $y \in Y$, then $||y_0 - x|| \neq \delta$. This completes the first part of the proof.

Next, assume that $(x - y_0) \perp Y$. We must show that y_0 is a unique vector such that $||x - y|| > ||x - y_0||$ for all $y \neq y_0$. For any $y \in Y$ we have, in view of part (ii) of Theorem 6.11.16,

$$||x - y||^2 = ||x - y_0 + y_0 - y||^2 = ||x - y_0||^2 + ||y_0 - y||^2.$$

From this it follows that $||x - y|| > ||x - y_0||$ for all $y \neq y_0$. This completes the proof of the theorem. ■

In Figure H the meaning of Theorem 6.12.10 is illustrated pictorially for a subset Y of E^3.

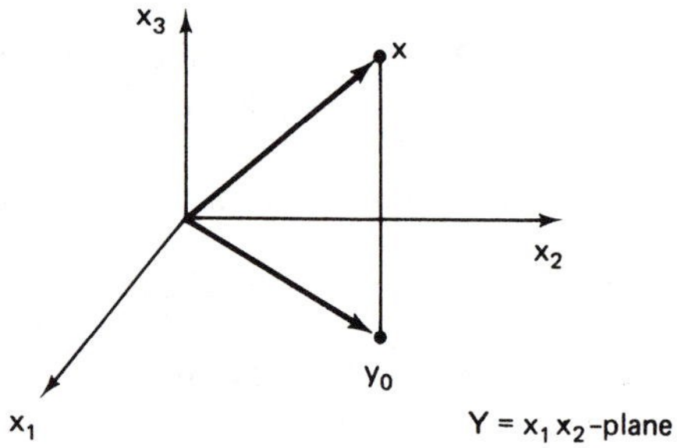

6.12.11. Figure H

The preceding theorem does not ensure the existence of the vector y_0. However, if we require in Theorem 6.12.10 that Y be a *closed* linear subspace in a Hilbert space X, then the existence of the unique vector y_0 is guaranteed.

This important result, which we will prove below, is called the **classical projection theorem**.

6.12.12. Theorem. Let X be a Hilbert space, and let Y be a closed linear subspace of X. Let x be an arbitrary vector in X, and let

$$\delta = \inf \{\|y - x\| : y \in Y\}.$$

Then there exists a unique vector $y_0 \in Y$ such that $\|y_0 - x\| = \delta$. Moreover, $y_0 \in Y$ is the unique vector such that $\|y_0 - x\| = \inf \{\|y - x\|: y \in Y\}$ if and only if the vector $(x - y_0) \perp Y$.

Proof. In view of Theorem 6.12.10 we only have to establish the existence of a vector $y_0 \in Y$ such that $\|x - y_0\| = \delta$. Assume that $x \notin Y$ (if $x \in Y$, then $x = y_0$ and we are done). Since δ is the infimum of $\|y - x\|$ for all $y \in Y$, there is a sequence $\{y_n\}$ in Y such that $\|x - y_n\| \longrightarrow \delta$ as $n \longrightarrow \infty$. We now show that $\{y_n\}$ is a Cauchy sequence. By part (i) of Theorem 6.11.16 we have

$$\|(y_m - x) + (x - y_n)\|^2 + \|(y_m - x) - (x - y_n)\|^2$$
$$= 2\|y_m - x\|^2 + 2\|x - y_n\|^2.$$

This equation yields, after some straightforward manipulations, the relation

$$\|y_m - y_n\|^2 = 2\|y_m - x\|^2 + 2\|x - y_n\|^2 - 4\left\|x - \frac{(y_m + y_n)}{2}\right\|^2.$$

Since Y is a linear subspace, it follows that for each $y_m, y_n \in Y$ we have $(y_m + y_n)/2 \in Y$. Thus, $\|x - (y_m + y_n)/2\| \geq \delta$ and

$$\|y_m - y_n\|^2 \leq 2\|y_m - x\|^2 + 2\|x - y_n\|^2 - 4\delta^2.$$

Also, since $\|y_m - x\|^2 \longrightarrow \delta^2$ as $m \longrightarrow \infty$, it follows that $\|y_m - y_n\|^2 \longrightarrow 0$ as $m, n \longrightarrow \infty$. Hence, $\{y_n\}$ is a Cauchy sequence. Since Y is a closed linear subspace of a Hilbert space, it is itself a Hilbert space and as such $\{y_n\}$ has a limit $y_0 \in Y$. Finally, by the continuity of the norm (see Theorem 6.1.15), it follows that $\lim_{n \to \infty} \|x - y_n\| = \|x - y_0\| = \delta$. This proves the theorem. ■

The next result is a consequence of the preceding theorem.

6.12.13. Theorem. If Y and Z are closed linear subspaces of a Hilbert space X, if $Y \subset Z$, and if $Y \neq Z$, then there exists a non-zero vector in Z, say z, such that $z \perp Y$.

Proof. Let x be any vector in Z which is not in Y (there is one such vector by hypothesis). If we define δ as above, i.e., $\delta = \inf \{\|y - x\|: y \in Y\}$, then there exists by Theorem 6.12.12 a vector $y_0 \in Y$ such that $\|x - y_0\| = \delta$. Now let $z = y_0 - x$. Then $z \perp Y$ by Theorem 6.12.12. ■

From part (ii) of Theorem 6.12.8 we have, in general, $Y^{\perp\perp} \supset Y$. Under certain conditions equality holds.

6.12.14. Theorem. Let Y be a linear subspace of a Hilbert space X. Then $\bar{Y} = Y^{\perp\perp}$.

Proof. From part (ii) of Theorem 6.12.8 we have $Y \subset Y^{\perp\perp}$. Since $Y^{\perp\perp}$ is closed by Theorem 6.12.6, it follows that $\bar{Y} \subset Y^{\perp\perp}$. For purposes of contradiction, let us now assume that $\bar{Y} \neq Y^{\perp\perp}$. Then Theorem 6.12.13 establishes the existence of a vector $z \in Y^{\perp\perp}$ such that $z \neq 0$ and such that $z \perp \bar{Y}$. Thus, $z \in \bar{Y}^{\perp}$. Since $Y \subset \bar{Y}$, it follows that $z \in Y^{\perp}$. Therefore, we have $z \in Y^{\perp} \cap Y^{\perp\perp}$ and $z \neq 0$, which is a contradiction to part (i) of Theorem 6.12.8. Hence, we must have $\bar{Y} = Y^{\perp\perp}$. ■

We note that if, in particular, Y is a closed linear subspace of X, then $Y = Y^{\perp\perp}$.

In connection with the next result, recall the definition of the sum of two subsets of X (see Definition 3.2.8).

6.12.15. Theorem. If Y and Z are closed linear subspaces of a Hilbert space X, and if $Y \perp Z$, then $Y + Z$ is a closed linear subspace of X.

Proof. In view of Theorem 3.2.10, $Y + Z$ is a linear subspace of X. To show that $Y + Z$ is closed, it suffices to show that if u is a point of accumulation for $Y + Z$, then $u = y + z$ for some $y \in Y$ and for some $z \in Z$.

Let u be a point of accumulation of $Y + Z$. Then there is a sequence of vectors $\{u_n\}$ in $Y + Z$ with $||u_n - u|| \longrightarrow 0$ as $n \longrightarrow \infty$. In this sequence we have for each n, $u_n = y_n + z_n$ with $y_n \in Y$ and $z_n \in Z$. Suppose now that $\{u_n\}$ converges to a vector $u \in X$. By the Pythagorean theorem (see Theorem 6.11.16) we have

$$||u_n - u_m||^2 = ||y_n - y_m + z_n - z_m||^2 = ||y_n - y_m||^2 + ||z_n - z_m||^2.$$

But $||u_n - u_m|| \longrightarrow 0$ as $m, n \longrightarrow \infty$, because u_n having a limit is a Cauchy sequence. Therefore, $||y_n - y_m||^2 \longrightarrow 0$ and $||z_n - z_m||^2 \longrightarrow 0$ as $m, n \longrightarrow \infty$. But this implies that the sequences $\{y_n\}$, $\{z_n\}$ are also Cauchy sequences. Since Y and Z are closed, these sequences have limits $y \in Y$ and $z \in Z$, respectively. Finally, we note that

$$||u_n - (y + z)|| = ||y_n - y + z_n - z|| \leq ||y_n - y|| + ||z_n - z|| \longrightarrow 0$$

as $n \longrightarrow \infty$. Therefore, since z_n cannot approach two distinct limits, we have $u = y + z$. This completes the proof. ■

Before proceeding to the next result, we recall from Definition 3.2.13 that a linear space X is the direct sum of two linear subspaces Y and Z if for every $x \in X$ there is a unique $y \in Y$ and a unique $z \in Z$ such that $x =$

$y + z$. We write, in this case, $X = Y \oplus Z$. The following result is known as the **projection theorem**.

6.12.16. Theorem. If Y is a closed linear subspace of a Hilbert space X, then $X = Y \oplus Y^{\perp}$.

Proof. Let $Z = Y + Y^{\perp}$. By hypothesis, Y is a closed linear subspace and so is $Y^{\perp}$ in view of Theorem 6.12.6. From the previous result it now follows that Z is also a closed linear subspace. Next, we show that $Z = X$. Since $Y \subset Z$ and $Y^{\perp} \subset Z$ it follows from part (iii) of Theorem 6.12.8 that $Z^{\perp} \subset Y^{\perp}$ and also that $Z^{\perp} \subset Y^{\perp\perp}$, so that $Z^{\perp} \subset Y^{\perp} \cap Y^{\perp\perp}$. But from part (i) of Theorem 6.12.8 we have $Y^{\perp} \cap Y^{\perp\perp} = \{0\}$. Therefore, the zero vector is the only element in both $Y^{\perp}$ and $Y^{\perp\perp}$, and thus $Z^{\perp} = \{0\}$. Since Z is a closed linear subspace we have from Theorems 6.12.4 and 6.12.14,

$$Z = Z^{\perp\perp} = (Z^{\perp})^{\perp} = \{0\}^{\perp} = X.$$

We have thus shown that we can represent every $x \in X$ as the sum $x = y + z$, where $y \in Y$ and $z \in Y^{\perp}$. To show that this representation is unique we consider $x = y_1 + z_1$ and $x = y_2 + z_2$, where $y_1, y_2 \in Y$ and $z_1, z_2 \in Y^{\perp}$. Then $(x - x) = 0 = y_1 - y_2 + z_1 - z_2$ or $y_1 - y_2 = z_2 - z_1$. Now clearly $(y_1 - y_2) \in Y$ and $(z_2 - z_1) \in Y^{\perp}$. Since $y_1 - y_2 = z_2 - z_1$ we also have $(y_1 - y_2) \in Y^{\perp}$ and $(z_2 - z_1) \in Y$. From this it follows that $y_1 - y_2 = z_2 - z_1 = 0$; i.e., $y_1 = y_2$ and $z_1 = z_2$. Therefore, x is unique. ■

The above theorem allows us to write any vector x of a Hilbert space X as the sum of two vectors y and z; i.e., $x = y + z$, where y is in a closed linear subspace Y of X and z is in $Y^{\perp}$. It is this theorem which gave rise to the expression **orthogonal complement**.

If X is a Hilbert space and if Y is a closed linear subspace of X and if $x = y + z$, where $y \in Y$ and $z \in Y^{\perp}$, then we define the mapping P as

$$Px = y.$$

We call the function P the **projection** of x onto Y. Note that $P(Px) \triangleq P^2x = Py = y$; i.e., $P^2 = P$. We will examine the properties of projections in greater detail in the next chapter. (Refer also to Definition 3.7.1 and Theorem 3.7.4.)

6.13. FOURIER SERIES

In the previous section we examined some of the structural properties of Hilbert spaces. Presently, we will concern ourselves with the representation of elements in Hilbert space. We will see that the vectors of a Hilbert space can under certain conditions be represented as a linear combination of a finite or infinite number of vectors from an orthonormal set. In this connection we will touch upon the concept of basis in Hilbert space. The property which makes all this possible is, of course, the inner product.

Much of the material in this section is concerned with an abstract approach to the topic of Fourier series. Since the reader is probably already familiar with certain facets of Fourier analysis, he or she is now in a position to recognize the power and the beauty of the abstract approach.

Throughout this section $\{X; (\cdot, \cdot)\}$ *is a complex inner product space*, and convergence of an infinite series is to be understood in the sense of Definition 6.3.1.

We now consider the representation of a vector y of a finite-dimensional linear subspace Y in an inner product space.

6.13.1. Theorem. Let X be an inner product space, let $\{y_1, \ldots, y_n\}$ be a finite orthonormal set in X, and let Y be the linear subspace of X generated by $\{y_1, \ldots, y_n\}$. Then the vectors $\{y_1, \ldots, y_n\}$ form a basis for Y and, moreover, in the representation of a vector $y \in Y$ by the sum

$$y = \alpha_1 y_1 + \ldots + \alpha_n y_n,$$

the coefficients α_i are specified by

$$\alpha_i = (y, y_i), \quad i = 1, \ldots, n.$$

6.13.2. Exercise. Prove Theorem 6.13.1. (Refer to Theorems 4.9.44 and 4.9.51.)

We now generalize the preceding result.

6.13.3. Theorem. Let X be a Hilbert space and let $\{x_i\}$ be a countably infinite orthonormal sequence in X. A series $\sum_{i=1}^{\infty} \alpha_i x_i$ is convergent to an element $x \in X$, i.e.,

$$x = \sum_{i=1}^{\infty} \alpha_i x_i,$$

if and only if $\sum_{i=1}^{\infty} |\alpha_i|^2 < \infty$. In this case we have the relation

$$\alpha_i = (x, x_i), \quad i = 1, 2, \ldots.$$

Proof. Assume that $\sum_{i=1}^{\infty} |\alpha_i|^2 < \infty$, and let $s_n = \sum_{i=1}^{n} \alpha_i x_i$. If $n > m$, then

$$\|s_n - s_m\|^2 = \left\| \sum_{i=m+1}^{n} \alpha_i x_i \right\|^2 = \left(\sum_{i=m+1}^{n} \alpha_i x_i, \sum_{j=m+1}^{n} \alpha_j x_j \right)$$
$$= \sum_{i=m+1}^{n} |\alpha_i|^2 \longrightarrow 0$$

as $n, m \longrightarrow \infty$. Therefore, $\{s_n\}$ is a Cauchy sequence and as such it has a limit, say x, in the Hilbert space X. Thus $\lim_{n \to \infty} s_n = x$.

Conversely, if $\{s_n\}$ converges then it is a Cauchy sequence and $\|s_n - s_m\|^2 = \sum_{i=m+1}^{n} |\alpha_i|^2 \longrightarrow 0$ as $n, m \longrightarrow \infty$. From this it follows that $\sum_{i=m+1}^{n} |\alpha_i|^2 \longrightarrow 0$ and $\sum_{i=m+1}^{\infty} |\alpha_i|^2 < \infty$.

Now assume that $\sum_{i=1}^{\infty} |\alpha_i|^2 < \infty$, and let $x = \lim_{n \to \infty} s_n$. We must show that $\alpha_i = (x, x_i)$. From Theorem 6.13.1 we have $\alpha_i = (s_n, x_i)$, $i = 1, \ldots, n$. But $s_n \longrightarrow x$, and hence by the continuity of the inner product we have $(s_n, x_i) \longrightarrow (x, x_i)$ as $n \longrightarrow \infty$. Therefore, $\alpha_i = (x, x_i)$, which completes the proof. ■

In the next result we use the concept of closed linear subspace generated by a set (see Definition 6.12.7).

6.13.4. Theorem. Let $\{x_i\}$ be an orthonormal sequence in a Hilbert space X, and let Y be the closed linear subspace generated by $\{x_i\}$. Corresponding to each $x \in X$ the series

$$\sum_{i=1}^{\infty} (x, x_i)x_i \tag{6.13.5}$$

converges to an element $\tilde{x} \in Y$. Moreover, $(x - \tilde{x}) \perp Y$.

6.13.6. Exercise. Prove Theorem 6.13.4. (Hint: Utilize Theorems 6.11.26, 6.13.3, and the continuity of the inner product.)

A more general version of Theorem 6.13.4 can be established by replacing the orthonormal sequence $\{x_i\}$ by an arbitrary orthonormal set Z.

In view of Theorem 6.13.4 any element x of a Hilbert space X can unambiguously be represented by a series of the form 6.13.5 provided that the closed linear subspace Y generated by the orthonormal sequence $\{x_i\}$ is equal to the space X. The scalars (x, x_i) in 6.13.5 are called **Fourier coefficients** of x with respect to the $\{x_i\}$.

6.13.7. Definition. Let X be a Hilbert space. An orthonormal set Y in X is said to be **complete** if there exists no orthonormal set of which Y is a proper subset.

The next result enables us to characterize complete orthonormal sets.

6.13.8. Theorem. Let X be a Hilbert space, and let Y be an orthonormal set in X. Then the following statements are equivalent:

(i) Y is complete;
(ii) if $(x, y) = 0$ for all $y \in Y$, then $x = 0$; and
(iii) $\overline{V(Y)} = X$.

6.13.9. Exercise. Prove Theorem 6.13.8 for the case where Y is an orthonormal sequence $\{x_i\}$.

As a specific example of a complete orthonormal set, we consider the set of elements $e_1 = (1, 0, \ldots, 0, \ldots)$, $e_2 = (0, 1, 0, \ldots, 0, \ldots)$, $e_3 = (0, 0, 1, 0, \ldots, 0, \ldots), \ldots$ in the Hilbert space l_2 (see Example 6.11.9). It is readily verified that $Y = \{e_i\}$ is an orthonormal set in l_2. Now let $x = (\xi_1, \xi_2, \ldots, \xi_n, \ldots) \in l_2$, and corresponding to x let $x_k = \sum_{i=1}^{k} \xi_i e_i$. Then $\|x - x_k\|^2 = \sum_{i=k+1}^{\infty} |\xi_i|^2$, and thus $\lim_{k\to\infty} \|x - x_k\| = 0$. Hence, $\overline{V(Y)} = l_2$ and Y is complete by the preceding theorem.

Many of the subsequent results involving countable orthonormal sets may be shown to hold for uncountable orthonormal sets as well (refer to Definition 1.2.48). The proofs of these generalized results usually require a postulate known as **Zorn's lemma**. (Consult the references cited at the end of this chapter for a discussion of this lemma.) Although the proofs of such generalized results are not particularly difficult, they do involve an added level of abstraction which we do not wish to pursue in this book. In connection with generalized results of this type, it is also necessary to use the notion of *cardinal number* of a set, introduced at the end of Section 1.2.

The next result is known as **Parseval's formula** (refer also to Corollary 4.9.49).

6.13.10. Theorem. Let X be a Hilbert space and let the sequence $\{x_i\}$ be orthonormal in X. Then

$$\|x\|^2 = \sum_{i=1}^{\infty} |(x, x_i)|^2 \tag{6.13.11}$$

for every $x \in X$ if and only if the sequence $\{x_i\}$ is complete.

Proof. Assume to the contrary that the sequence $\{x_i\}$ is not complete. Then there exists some $z \neq 0$ such that $(z, x_i) = 0$ for all i. Thus, there exists a $z \in X$ such that $\|z\|^2 \neq \sum_{i=1}^{\infty} |(z, x_i)|^2$. This proves the first part.

Now assume that the sequence $\{x_i\}$ is complete. In view of Theorems 6.13.4 and 6.13.8 we have

$$x = \sum_{i=1}^{\infty} (x, x_i) x_i = \sum_{i=1}^{\infty} \alpha_i x_i.$$

Since $\{x_i\}$ is orthonormal we obtain

$$||x||^2 = \left(\sum_{i=1}^{\infty} \alpha_i x_i, \sum_{j=1}^{\infty} \alpha_j x_j\right) = \sum_{i=1}^{\infty} \sum_{j=1}^{\infty} \alpha_i \bar{\alpha}_j (x_i, x_j) = \sum_{i=1}^{\infty} |\alpha_i|^2$$
$$= \sum_{i=1}^{\infty} |(x, x_i)|^2.$$

This completes the proof. ■

A more general version of Theorem 6.13.10 can be established by replacing the orthonormal sequence by an orthonormal set.

The next result, known as the **Gram-Schmidt procedure**, allows us to construct orthonormal sets in inner-product spaces (compare with Theorem 4.9.55).

6.13.12. Theorem. Let X be an inner-product space. Let $\{x_i\}$ be a finite or a countably infinite sequence of linearly independent vectors. Then there exists an orthonormal sequence $\{y_i\}$ having the same cardinal number as the sequence $\{x_i\}$ and generating the same linear subspace as $\{x_i\}$.

Proof. Since $x_1 \neq 0$, let us define y_1 as

$$y_1 = \frac{x_1}{||x_1||}.$$

It is clear that y_1 and x_1 generate the same linear subspace. Next, let

$$z_2 = x_2 - (x_2, y_1)y_1.$$

Since

$$(z_2, y_1) = (x_2 - (x_2, y_1)y_1, y_1) = (x_2, y_1) - (x_2, y_1)(y_1, y_1)$$
$$= (x_2, y_1) - (x_2, y_1) = 0,$$

it follows that $z_2 \perp y_1$. We now let $y_2 = z_2/||z_2||$. Note that $z_2 \neq 0$, because x_2 and y_1 are linearly independent. Also, y_1 and y_2 generate the same linear subspace as x_1 and x_2, because y_2 is a linear combination of y_1 and y_2.

Proceeding in the fashion described above we define $z_1, z_2, \ldots$ and $y_1, y_2, \ldots$ recursively as

$$z_n = x_n - \sum_{i=1}^{n-1} (x_n, y_i)y_i$$

and

$$y_n = \frac{z_n}{||z_n||}.$$

As before, we can readily verify that $z_n \perp y_i$ for all $i < n$, that $z_n \neq 0$, and that the $\{y_i\}$, $i = 1, \ldots, n$, generate the same linear subspace as the $\{x_i\}$, $i = 1, \ldots, n$. If the set $\{x_i\}$ is finite, the process terminates. Otherwise it is continued indefinitely by induction.

The sequence $\{y_i\}$ thus constructed can be put into a one-to-one correspondence with the sequence $\{x_i\}$. Therefore, these sequences have the same cardinal number. ■

The following result can be established by use of Zorn's lemma.

6.13.13. Theorem. Let X be an inner product space containing a non-zero element. Then X contains a complete orthonormal set. If Y is any orthonormal set in X, then there is a complete orthonormal set containing Y as a subset.

Indeed, it is also possible to prove the following result: if in an inner product space Y_1 and Y_2 are two complete orthonormal sets, then Y_1 and Y_2 have the same cardinal number, so that a one-to-one mapping of set Y_1 onto set Y_2 can be established. This result, along with Theorem 6.13.13, allows us to conclude that with each Hilbert space X there is associated in a natural way a cardinal number η. This, in turn, enables us to consider η as the **dimension of a Hilbert space** X. For the case of finite-dimensional spaces this concept and the usual definition of dimension coincide. However, in general, these two notions are not to be viewed as one and the same concept.

Next, recall that in Chapter 5 we defined a metric space X to be separable if there is a countable subset everywhere dense in X (see Definition 5.4.33). Since normed linear spaces and inner product spaces are also metric spaces, we speak also of separable Banach spaces and separable Hilbert spaces. In the case of Hilbert spaces, we can characterize separability in the following equivalent way.

6.13.14. Theorem. A Hilbert space X is separable if and only if it contains a complete orthonormal sequence.

6.13.15. Exercise. Prove Theorem 6.13.14.

Since in a separable Hilbert space X with a complete orthonormal sequence $\{x_i\}$ one can represent every $x \in X$ as

$$x = \sum_{i=1}^{\infty} (x, x_i)x_i,$$

we refer to a complete orthonormal sequence $\{x_i\}$ in a separable Hilbert space X as a **basis** for X. Caution should be taken here not to confuse this concept with the definition of basis introduced in Chapter 3. (See Definitions 3.3.6 and 3.3.22.) In that case we defined each x in a vector space to have a representation as a finite linear combination of vectors x_i. Indeed, the concept of Hamel basis (see Definition 3.3.22), which is a purely algebraic concept, is of very little value in spaces which are not finite dimensional. In such spaces, orthonormal basis as defined above is much more useful.

We conclude this section with the following result.

6.13.16. Theorem. Let Y be an orthonormal set in a separable Hilbert space X. Then Y is either a finite set or a countably infinite set.

6.13.17. Exercise. Prove Theorem 6.13.16.

6.14. THE RIESZ REPRESENTATION THEOREM

In this section we state and prove an important result known as the **Riesz representation theorem**. A direct consequence of this theorem is that the dual space X^* of a Hilbert space X is itself a Hilbert space. Throughout this section, $\{X; (\cdot, \cdot)\}$ is a Hilbert space.

We begin by first noting that for a fixed $y \in X$,

$$f(x) = (x, y) \tag{6.14.1}$$

is a linear functional in x. By means of (6.14.1) distinct vectors $y \in X$ are associated with distinct functionals. From the Schwarz inequality we have

$$|(x, y)| \leq ||x|| ||y||.$$

Hence, $||f|| \leq ||y||$ and f is bounded (i.e., $f \in X^*$). From this it follows that if X is a Hilbert space, then bounded linear functionals are determined by the elements of X itself. In the next theorem we show that every element y of X determines a unique bounded linear functional f (i.e., a unique element of X^*) of the form (6.14.1) and that $||f|| = ||y||$. From this we conclude that the dual space X^* of the Hilbert space X is itself a Hilbert space. (Compare the following with Theorem 4.9.63.)

6.14.2. Theorem. (Riesz) Let f be a bounded linear functional on X. Then there is a unique $y \in X$ such that $f(x) = (x, y)$ for all $x \in X$. Moreover, $||f|| = ||y||$, and every y determines a unique element of the dual space X^* in this way.

Proof. For fixed $y \in X$, define the linear functional f on X by Eq. (6.14.1). From the Schwarz inequality we have $|f(x)| = |(x, y)| \leq ||y|| ||x||$ so that f is a bounded linear functional and $||f|| \leq ||y||$. Letting $x = y$ we have $|f(y)| = |(y, y)| = ||y|| ||y||$, from which it follows that $||f|| = ||y||$.

Next, let f be a bounded linear functional defined on the Hilbert space X. Let Z be the set of all vectors $z \in X$ such that $f(z) = 0$. By Theorem 3.4.19, Z is a linear subspace of X. Now let $\{z_n\}$ be a sequence of vectors in Z, and let $x_0 \in X$ be a point of accumulation of $\{z_n\}$. In view of the continuity of f we now have $0 = f(z_n) \longrightarrow f(x_0)$ as $n \longrightarrow \infty$. Thus, $x_0 \in Z$ and Z is closed.

If $Z = X$, then for all $x \in X$ we have $f(x) = 0$, and the equality $f(x) = (x, y) = 0$ for all $x \in X$ holds if and only if $y = 0$.

Now consider the case $Z \subset X$, $X \neq Z$. From above, Z is a closed linear subspace of X. We can therefore utilize Theorem 6.12.16 to represent X by the direct sum

$$X = Z \oplus Z^{\perp}.$$

Since $Z \subset X$ and $Z \neq X$, there exists in view of Theorem 6.12.13 a non-zero vector $u \in X$ such that $u \perp Z$; i.e., $u \in Z^{\perp}$. Also, since $u \neq 0$ and since $u \in Z^{\perp}$, it follows from part (i) of Theorem 6.12.8 that $u \notin Z$, and hence $f(u) \neq 0$. Since $Z^{\perp}$ is a linear subspace of X, we may assume without loss of generality that $f(u) = 1$. We now show that u is a scalar multiple of our desired vector y in Eq. (6.14.1).

For any fixed $x \in X$ we can write

$$f(x - f(x)u) = f(x) - f(x)f(u) = f(x) - f(x) = 0,$$

and thus $(x - f(x)u) \in Z$. From before, we have $u \perp Z$ and hence $(x - f(x)u, u) = 0$, or $(x, u) = f(x)\|u\|^2$, or $f(x) = (x, u/\|u\|^2)$. Letting $y = u/\|u\|^2$ yields now the desired form

$$f(x) = (x, y).$$

To show that the vector y is unique we assume that $f(x) = (x, y')$ and $f(x) = (x, y'')$ for all $x \in X$. Then $(x, y') - (x, y'') = 0$, or $(x, y' - y'') = 0$, or $(y' - y'', x) = 0$ for all $x \in X$. It now follows from Theorem 6.11.28 that $y' = y''$. This completes the proof of the theorem. ■

6.14.3. Exercise. Show that every Hilbert space X is reflexive (refer to Definition 6.9.8).

6.14.4. Exercise. Two normed linear spaces over the same field are said to be **congruent** if they are isomorphic (see Definition 3.4.76) and isometric (see Definition 5.9.16). Let X be a Hilbert space. Show that X is congruent to X^*.

6.15. SOME APPLICATIONS

We now consider two applications to some of the material of the present chapter. This section consists of three parts. In the first of these we consider the problem of approximating elements in a Hilbert space by elements in a finite-dimensional subspace. In the second part we briefly consider random variables, while in the third part we concern ourselves with the estimation of random variables.

A. Approximation of Elements in Hilbert Space (Normal Equations)

In many applications it is necessary to approximate functions by simpler ones. This problem can often be implemented by approximating elements from an appropriate Hilbert space by elements belonging to a suitable linear subspace. In other words, we need to consider the problem of approximating a vector x in a Hilbert space X by a vector y_0 in a linear subspace Y of X.

Let $y_i \in X$ for $i = 1, \ldots, n$, and let $Y = V(\{y_i\})$ denote the linear subspace of X generated by $\{y_1, \ldots, y_n\}$. Since Y is finite dimensional, it is closed. Now for any fixed $x \in X$ we wish to find that element of Y which minimizes $\|x - y\|$ for all $y \in Y$. If $y_0 \in Y$ is that element, then we say that y_0 **approximates** x. We call $(x - y_0)$ the **error vector** and $\|x - y_0\|$ the **error.** Since any vector in Y can be expressed as a linear combination $y = \alpha_1 y_1 + \ldots + \alpha_n y_n$, our problem is reduced to finding the set of α_i, $i = 1, \ldots, n$, for which the error $\|x - \alpha_1 y_1 - \ldots - \alpha_n y_n\|$ is minimized. But in view of the classical projection theorem (Theorem 6.12.12), $y_0 \in Y$ which minimizes the error is unique and, moreover, $(x - y_0) \perp y_i$, $i = 1, \ldots, n$. From this we obtain the n simultaneous linear equations

$$\mathbf{G}^T(y_1, \ldots, y_n)\begin{bmatrix} \alpha_1 \\ \cdot \\ \cdot \\ \cdot \\ \alpha_n \end{bmatrix} = \begin{bmatrix} (x, y_1) \\ \cdot \\ \cdot \\ \cdot \\ (x, y_n) \end{bmatrix}, \tag{6.15.1}$$

where in Eq. (6.15.1) $\mathbf{G}^T(y_1, \ldots, y_n)$ is the transpose of the matrix

$$\mathbf{G}(y_1, \ldots, y_n) = \begin{bmatrix} (y_1, y_1) & \cdots & (y_1, y_n) \\ (y_2, y_1) & \cdots & (y_2, y_n) \\ \cdot & & \cdot \\ \cdot & & \cdot \\ \cdot & & \cdot \\ (y_n, y_1) & \cdots & (y_n, y_n) \end{bmatrix}. \tag{6.15.2}$$

The matrix (6.15.2) is called the **Gram matrix** of $y_1, \ldots, y_n$. The determinant of (6.15.2) is called the **Gram determinant** and is denoted by $\Delta(y_1, \ldots, y_n)$. The equations (6.15.1) are called the **normal equations**. It is clear that in a real Hilbert space $\mathbf{G}(y_1, \ldots, y_n) = \mathbf{G}^T(y_1, \ldots, y_n)$, and that in a complex Hilbert space $\mathbf{G}(y_1, \ldots, y_n) = \overline{\mathbf{G}^T(y_1, \ldots, y_n)}$.

In order to approximate $x \in X$ by $y_0 \in Y$ we only need to solve Eq. (6.15.1) for the α_i, $i = 1, \ldots, n$. The next result gives conditions under which Eq. (6.15.1) possesses a unique solution for the α_i.

6.15.3. Theorem. A set of elements $\{y_1, \ldots, y_n\}$ of a Hilbert space X is linearly independent if and only if the Gram determinant $\Delta(y_1, \ldots, y_n) \neq 0$.

Proof. We prove this result by proving the equivalent statement $\Delta(y_1, \ldots, y_n) = 0$ if and only if the vectors $\{y_1, \ldots, y_n\}$ are linearly dependent.

Assume that $\{y_1, \ldots, y_n\}$ is a set of linearly dependent vectors in X. Then there exists a set of scalars $\{\alpha_1, \ldots, \alpha_n\}$, not all zero, such that

$$\alpha_1 y_1 + \ldots + \alpha_n y_n = 0. \tag{6.15.4}$$

Taking the inner product of Eq. (6.15.4) with the vectors $\{y_1, \ldots, y_n\}$ yields the n linear equations

$$\begin{array}{c} \alpha_1(y_1, y_1) + \cdots + \alpha_n(y_1, y_n) = 0 \\ \vdots \qquad\qquad \vdots \\ \alpha_1(y_n, y_1) + \cdots + \alpha_n(y_n, y_n) = 0 \end{array} \tag{6.15.5}$$

Taking the $\{\alpha_1, \ldots, \alpha_n\}$ as unknowns, we see that for a non-trivial solution $(\alpha_1, \ldots, \alpha_n)$ to exist we must have $\Delta(y_1, \ldots, y_n) = 0$.

Conversely, assume that $\Delta(y_1, \ldots, y_n) = 0$. Then a non-trivial solution $(\alpha_1, \ldots, \alpha_n)$ exists for Eq. (6.15.5). After rewriting Eq. (6.15.5) as

$$\Big(y_i, \sum_{j=1}^{n} \alpha_j y_j\Big) = 0, \quad i = 1, \ldots, n,$$

we obtain

$$\Big(\sum_{i=1}^{n} \alpha_i y_i, \sum_{j=1}^{n} \alpha_j y_j\Big) = \Big\|\sum_{i=1}^{n} \alpha_i y_i\Big\|^2 = 0,$$

which implies that $\sum_{i=1}^{n} \alpha_i y_i = 0$. Therefore, the set $\{y_1, \ldots, y_n\}$ is linearly dependent. This completes the proof. ■

The next result establishes an expression for the error $\|x - y_0\|$. The proof of this result follows directly from the classical projection theorem.

6.15.6. Theorem. Let X be a Hilbert space, let $x \in X$, let $\{y_1, \ldots, y_n\}$ be a set of linearly independent vectors in X, let Y be the linear subspace of X generated by $\{y_1, \ldots, y_n\}$, and let $y_0 \in Y$ be such that

$$\|x - y_0\| = \min_{y \in Y} \|x - y\| = \min \|x - \alpha_1 y_1 - \ldots - \alpha_n y_n\|.$$

Then

$$\|x - y_0\| = \frac{\Delta(y_1, \ldots, y_n, x)}{\Delta(y_1, \ldots, y_n)},$$

where

$$\Delta(y_1, \ldots, y_n, x) = \det \begin{bmatrix} (y_1, y_1) & \cdots & (y_1, y_n) & (y_1, x) \\ (y_2, y_1) & \cdots & (y_2, y_n) & (y_2, x) \\ \vdots & & \vdots & \vdots \\ (y_n, y_1) & \cdots & (y_n, y_n) & (y_n, x) \\ (x, y_1) & \cdots & (x, y_n) & (x, x) \end{bmatrix}$$

6.15.7. Exercise. Prove Theorem 6.15.6.

B. Random Variables

A rigorous development of the theory of probability is based on measure and integration theory. Since knowledge of this theory by the reader has not been assumed, a brief discussion of some essential concepts will now be given.

We begin by introducing some terminology. If Ω is a non-void set, a family of subsets, $\mathfrak{F}$, of Ω is called a **σ-algebra** (or a **σ-field**) if (i) for all $E, F \in \Omega$ we have $E \cup F \in \Omega$ and $E - F \in \Omega$, (ii) for any countable sequence of sets $\{E_n\}$ in $\mathfrak{F}$ we have $\bigcup_{n=1}^{\infty} E_n \in \mathfrak{F}$, and (iii) $\Omega \in \mathfrak{F}$. It readily follows that a σ-algebra is a family of subsets of Ω which is closed under all countable set operations.

A function $P: \mathfrak{F} \longrightarrow R$, where $\mathfrak{F}$ is a σ-algebra, is called a **probability measure** if (i) $0 \leq P(E) \leq 1$ for all $E \in \mathfrak{F}$, (ii) $P(\varnothing) = 0$ and $P(\Omega) = 1$, and (iii) for any countable collection of sets $\{E_n\}$ in $\mathfrak{F}$ such that $E_i \cap E_j = \varnothing$ if $i \neq j$, we have $P(\bigcup_{n=1}^{\infty} E_n) = \sum_{n=1}^{\infty} P(E_n)$.

A **probability space** is a triple $\{\Omega, \mathfrak{F}, P\}$, where Ω is a non-void set, $\mathfrak{F}$ is a σ-algebra of subsets of Ω, and P is a probability measure on $\mathfrak{F}$. We call elements $\omega \in \Omega$ **outcomes** (usually thought of as occurring at random), and we call elements $E \in \mathfrak{F}$ **events.**

A function $X: \Omega \longrightarrow R$ is called a **random variable** if $\{\omega: X(\omega) \leq x\} \in \mathfrak{F}$ for all $x \in R$. The set $\{\omega: X(\omega) \leq x\}$ is usually written in shorter form as $\{X \leq x\}$. If X is a random variable, then the function $F_X: R \longrightarrow R$ defined by $F_X(x) = P\{X < x\}$ for $x \in R$ is called the **distribution function** of X. If X_i, $i = 1, \ldots, n$ are random variables, we define the **random vector X** as $\mathbf{X} = (X_1, \ldots, X_n)^T$. Also, for $\mathbf{x} = (x_1, \ldots, x_n)^T \in R^n$, the **event** $\{X_1 < x_1, \ldots, X_n < x_n\}$ is defined to be $\{\omega: X_1(\omega) < x_1\} \cap \{\omega: X_2(\omega) < x_2\} \cap \ldots \cap \{\omega: X_n(\omega) < x_n\}$. Furthermore, for a random vector $\mathbf{X}$, the function $F_{\mathbf{X}}: R^n \longrightarrow R$, defined by $F_{\mathbf{X}}(\mathbf{x}) = P\{X_1 < x_1, \ldots, X_n < x_n\}$, is called the **distribution function** of $\mathbf{X}$.

If X is a random variable and g is a function, $g: R \longrightarrow R$, such that the Stieltjes integral $\int_{-\infty}^{\infty} g(x)dF_X$ exists, then the **expected value** of $g(X)$ is defined to be $E\{g(X)\} = \int_{-\infty}^{\infty} g(x)dF_X(x)$. Similarly, if $\mathbf{X}$ is a random vector and if g is a function, $g: R^n \longrightarrow R$ such that $\int_{R^n} g(\mathbf{x})dF_{\mathbf{X}}(\mathbf{x})$ exists, then the **expected value** of $g(\mathbf{X})$ is defined to be $E\{g(\mathbf{X})\} = \int_{R^n} g(\mathbf{x})dF_{\mathbf{X}}(\mathbf{x})$. Some of the expected values of primary interest are $E(X)$, the **expected value** of X, $E(X^2)$, the **second moment** of X, and $E\{[X - E(X)]^2\}$, the **variance** of X.

If we let $\mathfrak{L}_2$ denote the family of random variables defined on a probability space $\{\Omega, \mathfrak{F}, P\}$ such that $E(X^2) < \infty$, then this space is a vector space over R with the usual definition of addition and multiplication by a scalar. We say two random variables, X_1 and X_2, are **equal almost surely** if $P\{\omega: X_1(\omega) \neq X_2(\omega)\} = 0$. If we let L_2 denote the family of equivalence classes of all random variables which are almost surely equal (as in Example 5.5.31), then $\{L_2; (\ ,\ )\}$ is a real Hilbert space where the inner product is defined by

$$(X, Y) = E(XY) \text{ for } X, Y \in L_2.$$

Throughout the remainder of this section, we let $\{\Omega, \mathfrak{F}, P\}$ denote our underlying probability space, and we assume that all random variables belong to the Hilbert space L_2 with inner product $(X, Y) = E(XY)$.

C. *Estimation of Random Variables*

The special class of estimation problems which we consider may be formulated as follows: given a set of random variables $\{Y_1, \ldots, Y_m\}$, find the best estimate of another random variable, X. The sense in which an estimate is "best" will be defined shortly. Here we view the set $\{Y_1, \ldots, Y_m\}$ to be **observations** and the random variable X as the unknown.

For any mapping $f: R^m \longrightarrow R$ such that $f(Y_1, \ldots, Y_m) \in L_2$ for all observations $\{Y_1, \ldots, Y_m\}$, we call $\hat{X} = f(Y_1, \ldots, Y_m)$ an **estimate** of X. If f is linear, we call $\hat{X}$ a **linear estimate**.

Next, let f be linear; i.e., let f be a linear functional on R^m. Then there is a vector $\mathbf{a}^T = (\alpha_1, \ldots, \alpha_m) \in R^m$ such that $f(\mathbf{y}) = \mathbf{a}^T\mathbf{y}$ for all $\mathbf{y}^T = (\eta_1, \ldots, \eta_m) \in R^m$. Now a linear estimate, $\hat{X} = \alpha_1 Y_1 + \ldots + \alpha_m Y_m$, is called the **best linear estimate** of X, given $\{Y_1, \ldots, Y_m\}$, if $E\{[X - \alpha_1 Y_1 - \ldots - \alpha_m Y_m]^2\}$ is minimum with respect to $\mathbf{a} \in R^m$.

The classical projection theorem (see Theorem 6.12.12) tells us that the best linear estimate of X is the projection of X onto the linear vector space

$V(\{Y_1, \ldots, Y_m\})$. Furthermore, Eq. (6.15.1) gives us the explicit form for α_i, $i = 1, \ldots, m$. We are now in a position to summarize the above discussion in the following theorem, which is usually called the **orthogonality principle**.

6.15.8. Theorem. Let $X, Y_1, \ldots, Y_m$ belong to L_2. Then $\hat{X} = \alpha_1 Y_1 + \ldots + \alpha_m Y_m$ is the best linear estimate of X if and only if $\{\alpha_1, \ldots, \alpha_m\}$ are such that $E\{[X - \hat{X}]Y_i\} = 0$ for $i = 1, \ldots, m$.

We also have the following result.

6.15.9. Corollary. Let $X, Y_1, \ldots, Y_m$ belong to L_2. Let $\mathbf{G} = [\gamma_{ij}]$, where $\gamma_{ij} = E\{Y_i Y_j\}$, $i, j = 1, \ldots, m$, and let $\mathbf{b}^T = (\beta_1, \ldots, \beta_m) \in R^m$, where $\beta_i = E\{XY_i\}$ for $i = 1, \ldots, m$. If $\mathbf{G}$ is non-singular, then $\hat{X} = \alpha_1 Y_1 + \ldots + \alpha_m Y_m$ is the best linear estimate of X if and only if $\mathbf{a}^T = \mathbf{b}^T \mathbf{G}^{-1}$.

6.15.10. Exercise. Prove Theorem 6.15.8 and Corollary 6.15.9.

Let us now consider a specific case.

6.15.11. Example. Let $X, V_1, \ldots, V_m$ be random variables in L_2 such that $E\{X\} = E\{V_i\} = E\{XV_i\} = 0$ for $i = 1, \ldots, m$, and let $\mathbf{R} = [\rho_{ij}]$ be non-singular where $\rho_{ij} = E[V_i V_j]$ for $i, j = 1, \ldots, m$. Suppose that the measurements $\{Y_1, \ldots, Y_m\}$ of X are given by $Y_i = X + V_i$ for $i = 1, \ldots, m$. Then we have $E\{Y_i Y_j\} = E\{[X + V_i][X + V_j]\} = \sigma_x^2 + \rho_{ij}$ for $i, j = 1, \ldots, m$, where $\sigma_x^2 \triangleq E\{X^2\}$. Also, $E\{XY_i\} = E\{X(X + V_i)\} = \sigma_x^2$ for $i = 1, \ldots, m$. Thus, $\mathbf{G} = [\gamma_{ij}]$, where $\gamma_{ij} = \sigma_x^2 + \rho_{ij}$ for $i, j = 1, \ldots, m$, $\mathbf{b}^T = (\beta_1, \ldots, \beta_m)$, where $\beta_i = \sigma_x^2$ for $i = 1, \ldots, m$, and $\mathbf{a}^T = \mathbf{b}^T \mathbf{G}^{-1}$. ■

6.15.12. Exercise. In the preceding example, show that if $\rho_{ij} = \sigma_v^2 \delta_{ij}$ for $i, j = 1, \ldots, m$, where δ_{ij} is the Kronecker delta, then

$$\alpha_i = \frac{\sigma_x^2}{m\sigma_x^2 + \sigma_v^2} \text{ for } i = 1, \ldots, m.$$

The next result provides us with a useful means for finding the best linear estimate of a random variable X, given a set of random variables $\{Y_1, \ldots, Y_k\}$, if we already have the best linear estimate, given $\{Y_1, \ldots, Y_{k-1}\}$.

6.15.13. Theorem. Let $k \geq 2$, and let $Y_1, \ldots, Y_k$ be random variables in L_2. Let $\mathcal{Y}_j = V(\{Y_1, \ldots, Y_j\})$, the linear vector space generated by the random variables $\{Y_1, \ldots, Y_j\}$, for $1 \leq j \leq k$. Let $\hat{Y}_k(k-1)$ denote the best linear estimate of Y_k, given $\{Y_1, \ldots, Y_{k-1}\}$, and let $\tilde{Y}_k(k-1) = Y_k - \hat{Y}_k(k-1)$. Then $\mathcal{Y}_k = \mathcal{Y}_{k-1} \oplus V(\{\tilde{Y}_k(k-1)\})$.

Proof. By the classical projection theorem (see Theorem 6.12.12), $\tilde{Y}_k(k-1) \perp \mathcal{Y}_{k-1}$. Now for arbitrary $Z \in \mathcal{Y}_k$, we must have $Z = \zeta_1 Y_1 + \ldots + \zeta_{k-1}Y_{k-1} + \zeta_k Y_k$ for some $(\zeta_1, \ldots, \zeta_k)$. We can rewrite this as $Z = Z_1 + Z_2$, where $Z_1 = \zeta_1 Y_1 + \ldots + \zeta_{k-1}Y_{k-1} + \zeta_k \hat{Y}_k(k-1)$ and $Z_2 = \zeta_k \tilde{Y}_k(k-1)$. Since $Z_1 \in \mathcal{Y}_{k-1}$ and $Z_2 \perp \mathcal{Y}_{k-1}$, it follows from Theorem 6.12.12 that Z_1 and Z_2 are unique. Since $Z_1 \in \mathcal{Y}_{k-1}$ and $Z_2 \in V(\{\tilde{Y}_k(k-1)\})$, the theorem is proved. ■

We can extend the problem of estimation of (scalar) random variables to random vectors. Let $X_1, \ldots, X_n$ be random variables in L_2, and let $\mathbf{X} = (X_1, \ldots, X_n)^T$ be a random vector. Let $Y_1, \ldots, Y_m$ be random variables in L_2. We call $\hat{\mathbf{X}} = (\hat{X}_1, \ldots, \hat{X}_n)^T$ the best linear estimate of $\mathbf{X}$, given $\{Y_1, \ldots, Y_m\}$, if $\hat{X}_i$ is the best linear estimate of X_i, given $\{Y_1, \ldots, Y_m\}$ for $i = 1, \ldots, n$. Clearly, the orthogonality principle must hold for each X_i; i.e., we must have $E\{(X_i - \hat{X}_i)Y_j\} = 0$ for $i = 1, \ldots, n$ and $j = 1, \ldots, m$. In this case $\hat{\mathbf{X}}$ can be expressed as $\hat{\mathbf{X}} = \mathbf{AY}$, where $\mathbf{A}$ is an $(n \times m)$ matrix of real numbers and $\mathbf{Y} = (Y_1, \ldots, Y_m)^T$. Corollary 6.15.9 assumes now the following matrix form.

6.15.14. Theorem. Let $X_1, \ldots, X_n, Y_1, \ldots, Y_m$ be random variables in L_2. Let $\mathbf{G} = [\gamma_{ij}]$, where $\gamma_{ij} = E\{Y_i Y_j\}$ for $i, j = 1, \ldots, m$, and let $\mathbf{B} = [\beta_{ij}]$, where $\beta_{ij} = E\{X_i Y_j\}$ for $i = 1, \ldots, n$. If $\mathbf{G}$ is non-singular, then $\hat{\mathbf{X}} = \mathbf{AY}$ is the best linear estimate of $\mathbf{X}$, given $\mathbf{Y}$, if and only if $\mathbf{A} = \mathbf{BG}^{-1}$.

6.15.15. Exercise. Prove Theorem 6.15.14.

We note that $\mathbf{B}$ and $\mathbf{G}$ in the above theorem can be written in an alternate way. That is, we can say that

$$\hat{\mathbf{X}} = E\{\mathbf{XY}^T\}[E\{\mathbf{YY}^T\}]^{-1}\mathbf{Y} \tag{6.15.16}$$

is the best linear estimate of $\mathbf{X}$. By the expected value of a matrix of random variables, we mean the expected value of each element of the matrix.

In the remainder of this section we apply the preceding development to dynamic systems.

Let $J = \{1, 2, \ldots\}$ denote the set of positive integers. We use the notation $\{\mathbf{X}(k)\}$ to denote a sequence of random vectors; i.e., $\mathbf{X}(k)$ is a random vector for each $k \in J$. Let $\{\mathbf{U}(k)\}$ be a sequence of random vectors, $\mathbf{U}(k) = [U_1(k), \ldots, U_p(k)]^T$, with the properties

$$E\{\mathbf{U}(k)\} = \mathbf{0} \tag{6.15.17}$$

and

$$E\{\mathbf{U}(k)\mathbf{U}^T(j)\} = \mathbf{Q}(k)\delta_{jk} \tag{6.15.18}$$

for all $j, k \in J$, where $\mathbf{Q}(k)$ is a symmetric positive definite $(p \times p)$ matrix for all $k \in J$. Next, let $\{\mathbf{V}(k)\}$ be a sequence of random vectors, $\mathbf{V}(k) =$

$[V_1(k), \ldots, V_m(k)]^T$, with the properties

$$E\{\mathbf{V}(k)\} = \mathbf{0} \tag{6.15.19}$$

and

$$E\{\mathbf{V}(k)\mathbf{V}^T(j)\} = \mathbf{R}(k)\delta_{jk} \tag{6.15.20}$$

for all $j, k \in J$, where $\mathbf{R}(k)$ is a symmetric positive definite $(m \times m)$ matrix for all $k \in J$.

Now let $\mathbf{X}(1)$ be a random vector, $\mathbf{X}(1) = [X_1(1), \ldots, X_n(1)]^T$, with the properties

$$E\{\mathbf{X}(1)\} = \mathbf{0} \tag{6.15.21}$$

and

$$E\{\mathbf{X}(1)\mathbf{X}^T(1)\} = \mathbf{P}(1), \tag{6.15.22}$$

where $\mathbf{P}(1)$ is an $(n \times n)$ symmetric positive definite matrix. We assume further that the relationships among the random vectors are such that

$$E\{\mathbf{U}(k)\mathbf{V}^T(j)\} = \mathbf{0}, \tag{6.15.23}$$

$$E\{\mathbf{X}(1)\mathbf{U}^T(k)\} = \mathbf{0}, \tag{6.15.24}$$

and

$$E\{\mathbf{X}(1)\mathbf{V}^T(k)\} = \mathbf{0} \tag{6.15.25}$$

for all $k, j \in J$.

Next, let $\mathbf{A}(k)$ be a real $(n \times n)$ matrix for each $k \in J$, let $\mathbf{B}(k)$ be a real $(n \times p)$ matrix for each $k \in J$, and let $\mathbf{C}(k)$ be a real $(m \times n)$ matrix for each $k \in J$. We let $\{\mathbf{X}(k)\}$ and $\{\mathbf{Y}(k)\}$ be the sequences of random vectors generated by the difference equations

$$\mathbf{X}(k+1) = \mathbf{A}(k)\mathbf{X}(k) + \mathbf{B}(k)\mathbf{U}(k) \tag{6.15.26}$$

and

$$\mathbf{Y}(k) = \mathbf{C}(k)\mathbf{X}(k) + \mathbf{V}(k) \tag{6.15.27}$$

for $k = 1, 2, \ldots$.

We are now in a position to consider the following **estimation problem**: given the set of observations, $\{\mathbf{Y}(1), \ldots, \mathbf{Y}(k)\}$, find the best linear estimate of the random vector $\mathbf{X}(k)$. We could view the observed random variables as a single random vector, say $\mathcal{Y}^T = [\mathbf{Y}^T(1), \mathbf{Y}^T(2), \ldots, \mathbf{Y}^T(k)]$, and apply Theorem 6.15.14; however, it turns out that a rather elegant and significant algorithm exists for this problem, due to R. E. Kalman, which we consider next.

In the following, we adopt some additional convenient notation. For each $k, j \in J$, we let $\hat{\mathbf{X}}(j \mid k)$ denote the best linear estimate of $\mathbf{X}(j)$, given $\{\mathbf{Y}(1), \ldots, \mathbf{Y}(k)\}$. This notation is valid for $j < k$ and $j \geq k$; however, we shall limit our attention to the situation where $j \geq k$. In the present context, a recursive algorithm means that $\hat{\mathbf{X}}(k+1 \mid k+1)$ is a function only of $\hat{\mathbf{X}}(k \mid k)$ and $\mathbf{Y}(k+1)$. The following theorem, which is the last result of this section, provides the desired algorithm explicitly.

6.15.28. Theorem (Kalman). Given the foregoing assumptions for the dynamic system described by Eqs. (6.15.26) and (6.15.27), the best linear estimate of $\mathbf{X}(k)$, given $\{\mathbf{Y}(1), \ldots, \mathbf{Y}(k)\}$, is provided by the following set of difference equations:

$$\hat{\mathbf{X}}(k \mid k) = \hat{\mathbf{X}}(k \mid k-1) + \mathbf{K}(k)[\mathbf{Y}(k) - \mathbf{C}(k)\hat{\mathbf{X}}(k \mid k-1)], \quad (6.15.29)$$

and

$$\hat{\mathbf{X}}(k+1 \mid k) = \mathbf{A}(k)\hat{\mathbf{X}}(k \mid k), \quad (6.15.30)$$

where

$$\mathbf{K}(k) = \mathbf{P}(k \mid k-1)\mathbf{C}^T(k)[\mathbf{C}(k)\mathbf{P}(k \mid k-1)\mathbf{C}^T(k) + \mathbf{R}(k)]^{-1}, \quad (6.15.31)$$

$$\mathbf{P}(k \mid k) = [\mathbf{I} - \mathbf{K}(k)\mathbf{C}(k)]\mathbf{P}(k \mid k-1), \quad (6.15.32)$$

and

$$\mathbf{P}(k+1 \mid k) = \mathbf{A}(k)\mathbf{P}(k \mid k)\mathbf{A}^T(k) + \mathbf{B}(k)\mathbf{Q}(k)\mathbf{B}^T(k) \quad (6.15.33)$$

for $k = 1, 2, \ldots$, with initial conditions

$$\hat{\mathbf{X}}(1 \mid 0) = \mathbf{0}$$

and

$$\mathbf{P}(1 \mid 0) = \mathbf{P}(1).$$

Proof. Assume that $\hat{\mathbf{X}}(k \mid k-1)$ is known for $k \in J$. We may interpret $\hat{\mathbf{X}}(1 \mid 0)$ as the best linear estimate of $\mathbf{X}(1)$, given no observations. We wish to find $\hat{\mathbf{X}}(k \mid k)$ and $\hat{\mathbf{X}}(k+1 \mid k)$. It follows from Theorem 6.15.13 (extended to the case of random vectors) that there is a matrix $\mathbf{K}(k)$ such that $\hat{\mathbf{X}}(k \mid k) = \hat{\mathbf{X}}(k \mid k-1) + \mathbf{K}(k)\tilde{\mathbf{Y}}(k \mid k-1)$, where $\tilde{\mathbf{Y}}(k \mid k-1) = \mathbf{Y}(k) - \hat{\mathbf{Y}}(k \mid k-1)$, and $\hat{\mathbf{Y}}(k \mid k-1)$ is the best linear estimate of $\mathbf{Y}(k)$, given $\{\mathbf{Y}(1), \ldots, \mathbf{Y}(k-1)\}$. It follows immediately from Eqs. (6.15.23) and (6.15.27) and the orthogonality principle that $\hat{\mathbf{Y}}(k \mid k-1) = \mathbf{C}(k)\hat{\mathbf{X}}(k \mid k-1)$. Thus, we have shown that Eq. (6.15.29) must be true. In order to determine $\mathbf{K}(k)$, let $\tilde{\mathbf{X}}(k \mid k-1) = \mathbf{X}(k) - \hat{\mathbf{X}}(k \mid k-1)$. Then it follows from Eqs. (6.15.26) and (6.15.29) that

$$\tilde{\mathbf{X}}(k \mid k) = \tilde{\mathbf{X}}(k \mid k-1) - \mathbf{K}(k)[\mathbf{C}(k)\tilde{\mathbf{X}}(k \mid k-1) + \mathbf{V}(k)].$$

To satisfy the orthogonality principle, we must have $E\{\tilde{\mathbf{X}}(k \mid k)\mathbf{Y}^T(j)\} = \mathbf{0}$ for $j = 1, \ldots, k$. We see that this is satisfied for any $\mathbf{K}(k)$ for $j = 1, \ldots, k-1$. In order to satisfy $E\{\tilde{\mathbf{X}}(k \mid k)\mathbf{Y}^T(k)\} = 0$, $\mathbf{K}(k)$ must satisfy

$$\mathbf{0} = E\{\tilde{\mathbf{X}}(k \mid k-1)\mathbf{Y}^T(k)\} - \mathbf{K}(k)[\mathbf{C}(k)E\{\tilde{\mathbf{X}}(k \mid k-1)\mathbf{Y}^T(k)\} + E\{\mathbf{V}(k)\mathbf{Y}^T(k)\}]. \quad (6.15.34)$$

Let us first consider the term

$$E\{\tilde{\mathbf{X}}(k \mid k-1)\mathbf{Y}^T(k)\} = E\{\tilde{\mathbf{X}}(k \mid k-1)\mathbf{X}^T(k)\mathbf{C}^T(k) + \tilde{\mathbf{X}}(k \mid k-1)\mathbf{V}^T(k)\}. \quad (6.15.35)$$

We observe that $\mathbf{X}(k)$, the solution to the difference equation (6.15.26) at (time) k, is a linear combination of $\mathbf{X}(1)$ and $\mathbf{U}(1), \ldots, \mathbf{U}(k-1)$. In view of Eqs. (6.15.23) and (6.15.25) it follows that $E\{\mathbf{X}(j)\mathbf{V}^T(k)\} = \mathbf{0}$ for all $k, j \in J$. Hence, $E\{\tilde{\mathbf{X}}(k \mid k-1)\mathbf{V}^T(k)\} = \mathbf{0}$, since $\tilde{\mathbf{X}}(k \mid k-1)$ is a linear combination of $\mathbf{X}(k)$ and $\mathbf{Y}(1), \ldots, \mathbf{Y}(k-1)$.

Next, we consider the term

$$\begin{aligned} E\{\tilde{\mathbf{X}}(k \mid k-1)\mathbf{X}^T(k)\} &= E\{\tilde{\mathbf{X}}(k \mid k-1)[\mathbf{X}^T(k) \\ &\qquad - \hat{\mathbf{X}}^T(k \mid k-1) + \hat{\mathbf{X}}^T(k \mid k-1)]\} \\ &= E\{\tilde{\mathbf{X}}(k \mid k-1)[\tilde{\mathbf{X}}^T(k \mid k-1) + \hat{\mathbf{X}}^T(k \mid k-1)]\} \\ &= \mathbf{P}(k \mid k-1) \end{aligned} \tag{6.15.36}$$

where

$$\mathbf{P}(k \mid k-1) \triangleq E\{\tilde{\mathbf{X}}(k \mid k-1)\tilde{\mathbf{X}}^T(k \mid k-1)\}$$

and $E\{\tilde{\mathbf{X}}(k \mid k-1)\hat{\mathbf{X}}^T(k \mid k-1)\} = \mathbf{0}$, since $\hat{\mathbf{X}}(k \mid k-1)$ is a linear combination of $\{\mathbf{Y}(1), \ldots, \mathbf{Y}(k-1)\}$.

Now consider

$$E\{\mathbf{V}(k)\mathbf{Y}^T(k)\} = E\{\mathbf{V}(k)[\mathbf{X}^T(k)\mathbf{C}^T(k) + \mathbf{V}^T(k)]\} = \mathbf{R}(k). \tag{6.15.37}$$

Using Eqs. (6.15.35), (6.15.36), and (6.15.37), Eq. (6.15.34) becomes

$$\mathbf{0} = \mathbf{P}(k \mid k-1)\mathbf{C}^T(k) - \mathbf{K}(k)[\mathbf{C}(k)\mathbf{P}(k \mid k-1)\mathbf{C}^T(k) + \mathbf{R}(k)]. \tag{6.15.38}$$

Solving for $\mathbf{K}(k)$, we obtain Eq. (6.15.31).

To obtain Eq. (6.15.32), let $\tilde{\mathbf{X}}(k \mid k) = \hat{\mathbf{X}}(k) - \tilde{\mathbf{X}}(k \mid k)$ and $\mathbf{P}(k \mid k) = E\{\tilde{\mathbf{X}}(k \mid k)\tilde{\mathbf{X}}^T(k \mid k)\}$. In view of Eqs. (6.15.27) and (6.15.29) we have

$$\begin{aligned} \tilde{\mathbf{X}}(k \mid k) &= \tilde{\mathbf{X}}(k \mid k-1) - \mathbf{K}(k)[\mathbf{C}(k)\tilde{\mathbf{X}}(k \mid k-1) + \mathbf{V}(k)] \\ &= [\mathbf{I} - \mathbf{K}(k)\mathbf{C}(k)]\tilde{\mathbf{X}}(k \mid k-1) - \mathbf{K}(k)\mathbf{V}(k). \end{aligned}$$

From this it follows that

$$\begin{aligned} \mathbf{P}(k \mid k) &= [\mathbf{I} - \mathbf{K}(k)\mathbf{C}(k)]\mathbf{P}(k \mid k-1) \\ &\qquad - [\mathbf{I} - \mathbf{K}(k)\mathbf{C}(k)]\mathbf{P}(k \mid k-1)\mathbf{C}^T(k)\mathbf{K}^T(k) + \mathbf{K}(k)\mathbf{R}(k)\mathbf{K}^T(k) \\ &= [\mathbf{I} - \mathbf{K}(k)\mathbf{C}(k)]\mathbf{P}(k \mid k-1) \\ &\qquad - \{\mathbf{P}(k \mid k-1)\mathbf{C}^T(k) - \mathbf{K}(k)[\mathbf{C}(k)\mathbf{P}(k \mid k-1)\mathbf{C}^T(k) + \mathbf{R}(k)]\} \\ &\qquad \times \mathbf{K}^T(k). \end{aligned}$$

Using Eq. (6.15.38), it follows that Eq. (6.15.32) must be true.

To show that $\hat{\mathbf{X}}(k+1 \mid k)$ is given by Eq. (6.15.30), we simply show that the orthogonality principle is satisfied. That is,

$$\begin{aligned} &E\{[\mathbf{X}(k+1) - \mathbf{A}(k)\hat{\mathbf{X}}(k \mid k)]\mathbf{Y}^T(j)\} \\ &\qquad = E\{\mathbf{A}(k)[\mathbf{X}(k) - \hat{\mathbf{X}}(k \mid k)]\mathbf{Y}^T(j)\} + E\{\mathbf{B}(k)\mathbf{U}(k)\mathbf{Y}^T(j)\} = \mathbf{0} \end{aligned}$$

for $j = 1, \ldots, k$.

Finally, to verify Eq. (6.15.33), we have from Eqs. (6.15.26) and (6.15.30)

$$\tilde{\mathbf{X}}(k+1\,|\,k) = \mathbf{A}(k)\tilde{\mathbf{X}}(k\,|\,k) + \mathbf{B}(k)\mathbf{U}(k).$$

From this, Eq. (6.15.33) follows immediately. We note that $\hat{\mathbf{X}}(1\,|\,0) = \mathbf{0}$ and $\mathbf{P}(1\,|\,0) = \mathbf{P}(1)$. This completes the proof. ■

6.16. NOTES AND REFERENCES

The material of the present chapter as well as that of the next chapter constitutes part of what usually goes under the heading of functional analysis. Thus, these two chapters should be viewed as a whole rather than two separate parts.

There are numerous excellent sources dealing with Hilbert and Banach spaces. We cite a representative sample of these which the reader should consult for further study. References [6.6]–[6.8], [6.10], and [6.12] are at an introductory or intermediate level, whereas references [6.2]–[6.4] and [6.13] are at a more advanced level. The books by Dunford and Schwartz and by Hille and Phillips are standard and encyclopedic references on functional analysis; the text by Yosida constitutes a concise treatment of this subject, while the monograph by Halmos contains a compact exposition on Hilbert space. The book by Taylor is a standard reference on functional analysis at the intermediate level. The texts by Kantorovich and Akilov, by Kolmogorov and Fomin, and by Liusternik and Sobolev are very readable presentations of this subject. The book by Naylor and Sell, which presents a very nice introduction to functional analysis, includes some interesting examples. For references with applications of functional analysis to specific areas, including those in Section 6.15, see, e.g., Byron and Fuller [6.1], Kalman et al. [6.5], Luenberger [6.9], and Porter [6.11].

REFERENCES

[6.1] F. W. Byron and R. W. Fuller, *Mathematics of Classical and Quantum Physics*. Vols. I, II. Reading, Mass.: Addison-Wesley Publishing Co., Inc., 1969 and 1970.*

[6.2] N. Dunford and J. Schwartz, *Linear Operators*. Parts I and II. New York: Interscience Publishers, 1958 and 1964.

[6.3] P. R. Halmos, *Introduction to Hilbert Space*. New York: Chelsea Publishing Company, 1957.

[6.4] E. Hille and R. S. Phillips, *Functional Analysis and Semi-Groups*. Providence, R.I.: American Mathematical Society, 1957.

[6.5] R. E. Kalman, P. L. Falb, and M. A. Arbib, *Topics in Mathematical System Theory*. New York: McGraw-Hill Book Company, 1969.

*Reprinted in one volume by Dover Publications, Inc., New York, 1992.

[6.6] L. V. KANTOROVICH and G. P. AKILOV, *Functional Analysis in Normed Spaces*. New York: The Macmillan Company, 1964.
[6.7] A. N. KOLMOGOROV and S. V. FOMIN, *Elements of the Theory of Functions and Functional Analysis*. Vols. I, II. Albany, N.Y.: Graylock Press, 1957 and 1961.
[6.8] L. A. LIUSTERNIK and V. J. SOBOLEV, *Elements of Functional Analysis*. New York: Frederick Ungar Publishing Company, 1961.
[6.9] D. G. LUENBERGER, *Optimization by Vector Space Methods*. New York: John Wiley & Sons, Inc., 1969.
[6.10] A. W. NAYLOR and G. R. SELL, *Linear Operator Theory*. New York: Holt, Rinehart and Winston, 1971.
[6.11] W. A. PORTER, *Modern Foundations of Systems Engineering*. New York: The Macmillan Company, 1966.
[6.12] A. E. TAYLOR, *Introduction to Functional Analysis*. New York: John Wiley & Sons, Inc., 1958.
[6.13] K. YOSIDA, *Functional Analysis*. Berlin: Springer-Verlag, 1965.

7

LINEAR OPERATORS

In the present chapter we concern ourselves with linear operators defined on Banach and Hilbert spaces and we study some of the important properties of such operators. We also consider selected applications in this chapter.

This chapter consists of ten parts. Throughout, we consider primarily bounded linear operators, which we introduce in the first section. In the second section we look at inverses of linear transformations, in section three we introduce conjugate and adjoint operators, and in section four we study hermitian operators. In the fifth section we present additional special linear transformations, including normal operators, projections, unitary operators, and isometric operators. The spectrum of an operator is considered in the sixth, while completely continuous operators are introduced in the seventh section. In the eighth section we present one of the main results of the present chapter, the spectral theorem for completely continuous normal operators. Finally, in section nine we study differentiation of operators (which need not be linear) defined on Banach and Hilbert spaces.

Section ten, which consists of three subsections, is devoted to selected topics in applications. Items touched upon include applications to integral equations, an example from optimal control, and minimization of functionals (method of steepest descent). The chapter is concluded with a brief discussion of pertinent references in the eleventh section.

7.1. BOUNDED LINEAR TRANSFORMATIONS

Throughout this section X and Y denote vector spaces over the same field F, where F is either R (the real numbers) or C (the complex numbers).

We begin by pointing to several concepts considered previously. Recall from Chapter 1 that a *transformation* or *operator* T is a mapping of a subset $\mathfrak{D}(T)$ of X into Y. Unless specified to the contrary, we will assume that $X = \mathfrak{D}(T)$. Since a transformation is a mapping we distinguish, as in Chapter 1, between operators which are *onto* or *surjective*, *one-to-one* or *injective*, and *one-to-one and onto* or *bijective*. If T is a transformation of X into Y we write $T: X \longrightarrow Y$. If $x \in X$ we call $y = T(x)$ the *image of x in Y under T*, and if $V \subset X$ we define the *image of set V in Y under T* as the set

$$T(V) = \{y \in Y: y = T(v), v \in V \subset X\}.$$

On the other hand, if $W \subset Y$, then the *inverse image of set W under T* is the set

$$T^{-1}(W) = \{x \in X: y = T(x) \in W \subset Y\}.$$

We define the *range of T*, denoted $\mathfrak{R}(T)$, by

$$\mathfrak{R}(T) = \{y \in Y: y = T(x), x \in X\};$$

i.e., $\mathfrak{R}(T) = T(X)$. Recall that if a transformation T of X into Y is injective, then the inverse of T, denoted T^{-1}, exists (see Definition 1.2.9). Thus, if $y = T(x)$ and if T is injective, then $x = T^{-1}(y)$.

In Definition 3.4.1 we defined a *linear operator* (or a *linear transformation*) as a mapping of X into Y having the property that

(i) $T(x + y) = T(x) + T(y)$ for all $x, y \in X$; and

(ii) $T(\alpha x) = \alpha T(x)$ for all $\alpha \in F$ and all $x \in X$.

As in Chapter 3, we denote the class of all linear transformations from X into Y by $L(X, Y)$. Also, in the case of linear transformations we write Tx in place of $T(x)$.

Of great importance are bounded linear operators, which turn out to be also continuous. We have the following definition.

7.1.1. Definition. Let X and Y be normed linear spaces. A linear operator $T: X \longrightarrow Y$ is said to be **bounded** if there is a real number $\gamma \geq 0$ such that

$$\|Tx\|_Y \leq \gamma \|x\|_X$$

for all $x \in X$.

The notation $\|x\|_X$ indicates that the norm on X is used, while the notation $\|Tx\|_Y$ indicates that the norm on Y is employed. However, since the norms of the various spaces are usually understood, it is customary to drop the subscripts and simply write $\|x\|$ and $\|Tx\|$.

Our first result allows us to characterize a bounded linear operator in an equivalent way.

7.1.2. Theorem. Let $T \in L(X, Y)$. Then T is bounded if and only if T maps the unit sphere into a bounded subset of Y.

7.1.3. Exercise. Prove Theorem 7.1.2.

In Chapter 5 we introduced continuous functions (see Definition 5.7.1). The definition of continuity of an operator in the setting of normed linear spaces can now be rephrased as follows.

7.1.4. Definition. An operator $T: X \longrightarrow Y$ (not necessarily linear) is said to be **continuous** at a point $x_0 \in X$ if for every $\epsilon > 0$ there is a $\delta > 0$ such that

$$\|T(x) - T(x_0)\| < \epsilon$$

whenever $\|x - x_0\| < \delta$.

The reader can readily prove the next result.

7.1.5. Theorem. Let $T \in L(X, Y)$. If T is continuous at a single point $x_0 \in X$, then it is continuous at all $x \in X$.

7.1.6. Exercise. Prove Theorem 7.1.5.

In this chapter we will mainly concern ourselves with bounded linear operators. Our next result shows that in the case of linear operators boundedness and continuity are equivalent.

7.1.7. Theorem. Let $T \in L(X, Y)$. Then T is continuous if and only if it is bounded.

Proof. Assume that T is bounded, and let γ be such that $\|Tx\| \leq \gamma\|x\|$ for all $x \in X$. Now consider a sequence $\{x_n\}$ in X such that $x_n \longrightarrow 0$ as $n \longrightarrow \infty$. Then $\|Tx_n\| \leq \gamma\|x_n\| \longrightarrow 0$ as $n \longrightarrow \infty$, and hence T is continuous at the point $0 \in X$. From Theorem 7.1.5 it follows that T is continuous at all points $x \in X$.

Conversely, assume that T is continuous at $x = 0$, and hence at all $x \in X$. Since $T0 = 0$ we can find a $\delta > 0$ such that $\|Tx\| < 1$ whenever $\|x\| \leq \delta$. For any $x \neq 0$ we have $\|(\delta x)/\|x\|\| = \delta$, and hence

$$\|Tx\| = \left\|T\left(\frac{\delta x}{\|x\|} \cdot \frac{\|x\|}{\delta}\right)\right\| = \left(\frac{\|x\|}{\delta}\right)\left\|T\left(\frac{\delta x}{\|x\|}\right)\right\| \leq \frac{\|x\|}{\delta}.$$

If we let $\gamma = 1/\delta$, then $\|Tx\| \leq \gamma\|x\|$, and T is bounded. ■

Now let $S, T \in L(X, Y)$. In Eq. (3.4.42) we defined the *sum of linear operators* $(S + T)$ by

$$(S + T)x = Sx + Tx, \; x \in X,$$

and in Eq. (3.4.43) we defined *multiplication of T by a scalar $\alpha \in F$* as

$$(\alpha T)x = \alpha(Tx), \; x \in X, \alpha \in F.$$

We also recall (see Eq. (3.4.44)) that the *zero transformation*, 0, of X into Y is defined by $0x = 0$ for all $x \in X$ and that the negative of a transformation T, denoted by $-T$, is defined by $(-T)x = -Tx$ for all $x \in X$ (see Eq. 3.4.45)). Furthermore, the identity transformation $I \in L(X, X)$ is defined by $Ix = x$ for all $x \in X$ (see Eq. (3.4.56)). Referring to Theorem 3.4.47, we recall that $L(X, Y)$ is a linear space over F.

Next, let X, Y, Z be vector spaces over F, and let $S \in L(Y, Z)$ and $T \in L(X, Y)$. The *product* of S and T, denoted by ST, was defined in Eq. (3.4.50) as the mapping of X into Z such that

$$(ST)x = S(Tx), \; x \in X.$$

It can readily be shown that $ST \in L(X, Z)$. Furthermore, if $X = Y = Z$, then $L(X, X)$ is an associative algebra with identity I (see Theorem 3.4.59). Note however that the algebra $L(X, X)$ is, in general, not commutative because, in general,

$$ST \neq TS.$$

In the following, we will use the notation $B(X, Y)$ to denote the set of all bounded linear transformations from X into Y; i.e.,

$$B(X, Y) \triangleq \{T \in L(X, Y): T \text{ is bounded}\}. \tag{7.1.8}$$

The reader should have no difficulty in proving the next theorem.

7.1.9. Theorem. The space $B(X, Y)$ is a linear space over F.

7.1.10. Exercise. Prove Theorem 7.1.9.

Next, we wish to define a norm on $B(X, Y)$.

7.1.11. Definition. Let $T \in B(X, Y)$. The **norm** of T, denoted $\|T\|$, is defined by

$$\|T\| = \inf\{\gamma: \|Tx\| \leq \gamma\|x\| \text{ for all } x \in X\}. \tag{7.1.12}$$

Note that $\|T\|$ is finite and that

$$\|Tx\| \leq \|T\| \cdot \|x\|$$

for all $x \in X$. In proving that the function $\|\cdot\|: B(X, Y) \longrightarrow R$ satisfies all the axioms of a norm (see Definition 6.1.1), we need the following result.

7.1.13. Theorem. Let $T \in B(X, Y)$. Then $\|T\|$ can equivalently be expressed in any one of the following forms:

(i) $\|T\| = \inf_{\gamma} \{\gamma: \|Tx\| \leq \gamma \|x\| \text{ for all } x \in X\}$;

(ii) $\|T\| = \sup_{x \neq 0} \{\|Tx\|/\|x\|: x \in X\}$;

(iii) $\|T\| = \sup_{\|x\| \leq 1} \{\|Tx\|: x \in X\}$; and

(iv) $\|T\| = \sup_{\|x\| = 1} \{\|Tx\|: x \in X\}$.

7.1.14. Exercise. Prove Theorem 7.1.13.

We now show that the function $\|\cdot\|$ defined in Eq. (7.1.12) satisfies all the axioms of a norm.

7.1.15. Theorem. The linear space $B(X, Y)$ is a normed linear space (with norm defined by Eq. (7.1.12)); i.e.,

(i) for every $T \in B(X, Y)$, $\|T\| \geq 0$, and $\|T\| = 0$ if and only if $T = 0$;

(ii) $\|S + T\| \leq \|S\| + \|T\|$ for every $S, T \in B(X, Y)$; and

(iii) $\|\alpha T\| = |\alpha| \|T\|$ for every $T \in B(X, Y)$ and for every $\alpha \in F$.

Proof. The proof of part (i) is obvious. To verify (ii) we note that

$$\|(S + T)x\| = \|Sx + Tx\| \leq \|Sx\| + \|Tx\| \leq (\|S\| + \|T\|)\|x\|.$$

If $x = 0$, then we are finished. If $x \neq 0$, then

$$\|S + T\| = \sup_{x \neq 0} \frac{\|(S + T)x\|}{\|x\|} \leq \|S\| + \|T\| \text{ for all } x \in X, x \neq 0.$$

We leave the proof of part (iii), which is similar, as an exercise. ■

For the space $B(X, X)$ we have the following results.

7.1.16. Theorem. If $S, T \in B(X, X)$, then $ST \in B(X, X)$ and

$$\|ST\| \leq \|S\| \cdot \|T\|.$$

Proof. For each $x \in X$ we have

$$\|(ST)x\| = \|S(Tx)\| \leq \|S\| \cdot \|Tx\| \leq \|S\| \cdot \|T\| \cdot \|x\|,$$

which shows that $ST \in B(X, X)$. If $x \neq 0$, then

$$\|ST\| = \sup_{x \neq 0} \frac{\|(ST)x\|}{\|x\|} \leq \|S\| \cdot \|T\|,$$

completing the proof. ■

7.1.17. Theorem. Let I denote the identity operator on X. Then $I \in B(X, X)$, and $\|I\| = 1$.

7.1.18. Exercise. Prove Theorem 7.1.17.

We now consider some specific cases.

7.1.19. Example. Let $X = l_2$, the Banach space of Example 6.1.6. For $x = (\xi_1, \xi_2, \ldots) \in X$, let us define $T: X \longrightarrow X$ by

$$Tx = (0, \xi_2, \xi_3, \ldots).$$

The reader can readily verify that T is a linear operator which is neither injective nor surjective. We see that

$$||Tx||^2 = \sum_{i=2}^{\infty} |\xi_i|^2 \leq \sum_{i=1}^{\infty} |\xi_i|^2 = ||x||^2.$$

Thus, T is a bounded linear operator. To compute $||T||$ we observe that $||Tx|| \leq ||x||$, which implies that $||T|| \leq 1$. Choosing, in particular, $x = (0, 1, 0, \ldots) \in X$, we have $||Tx|| = ||x|| = 1$ and

$$1 = ||Tx|| \leq ||T|| \cdot ||x|| = ||T||.$$

Thus, it must be that $||T|| = 1$. ■

7.1.20. Example. Let $X = \mathfrak{C}[a, b]$, and let $||\cdot||_\infty$ be the norm on $\mathfrak{C}[a, b]$ defined in Example 6.1.9. Let $k: [a, b] \times [a, b] \longrightarrow R$ be a real-valued function, continuous on the square $a \leq s \leq b$, $a \leq t \leq b$. Define the operator $T: X \longrightarrow X$ by

$$[Tx](s) = \int_a^b k(s, t)x(t)\, dt$$

for $x \in X$. Then $T \in L(X, X)$ (see Example 3.4.6). Then

$$\begin{aligned} ||Tx|| &= \sup_{a \leq s \leq b} \left| \int_a^b k(s, t)x(t)\, dt \right| \\ &\leq \left[\sup_{a \leq s \leq b} \int_a^b |k(s, t)|\, dt \right] \cdot [\sup_{a \leq t \leq b} |x(t)|] \\ &= \gamma_0 \cdot ||x||. \end{aligned}$$

This shows that $T \in B(X, Y)$ and that $||T|| \leq \gamma_0$. It can, in fact, be shown that $||T|| = \gamma_0$. ■

For norms of linear operators on finite-dimensional spaces, we have the following important result.

7.1.21. Theorem. Let $T \in L(X, Y)$. If X is finite dimensional, then T is continuous.

Proof. Let $\{x_1, \ldots, x_n\}$ be a basis for X. For each $x \in X$, there is a unique set of scalars $\{\xi_1, \ldots, \xi_n\}$ such that $x = \xi_1 x_1 + \ldots + \xi_n x_n$. If we define the linear functionals $f_i: X \longrightarrow F$ by $f_i(x) = \xi_i$, $i = 1, \ldots, n$, then by Theorem

6.6.1 we know that each f_i is a continuous linear functional. Thus, there exists a set of real numbers $\{\gamma_1, \ldots, \gamma_n\}$ such that $|f_i(x)| \leq \gamma_i \|x\|$ for $i = 1, \ldots, n$. Now

$$Tx = \xi_1 Tx_1 + \ldots + \xi_n Tx_n.$$

If we let $\beta = \max_i \|Tx_i\|$ and $\gamma_0 = \max_i \gamma_i$, then it follows that $\|Tx\| \leq n\beta\gamma_0\|x\|$. Thus, T is bounded and hence continuous. ■

Next, we concern ourselves with various norms of linear transformations on the finite dimensional space R^n.

7.1.22. Example. Let $X = R^n$, and let $\{u_1, \ldots, u_n\}$ be the natural basis for R^n (see Example 4.1.15). For any $A \in L(X, X)$ there is an $n \times n$ matrix, say $\mathbf{A} = [a_{ij}]$ (see Definition 4.2.7), which represents A with respect to $\{u_1, \ldots, u_n\}$. Thus, if $Ax = y$, where $x = (\xi_1, \ldots, \xi_n) \in X$ and $y = (\eta_1, \ldots, \eta_n) \in X$, we may represent this transformation by $\mathbf{y} = \mathbf{Ax}$ (see Eq. (4.2.17)). In Example 6.1.5 we defined several norms on R^n, namely

$$\|x\|_p = [|\xi_1|^p + \ldots + |\xi_n|^p]^{1/p}, \quad 1 \leq p < \infty$$

and

$$\|x\|_\infty = \max_i \{|\xi_i|\}.$$

It turns out that different norms on R^n give rise to different norms of transformation A. (In this case we speak of the **norm of** A **induced** by the norm defined on R^n.) In the present example we derive expressions for the norm of A in terms of the elements of matrix $\mathbf{A}$ when the norm on R^n is given by $\|\cdot\|_1$, $\|\cdot\|_2$, and $\|\cdot\|_\infty$.

(i) Let $p = 1$; i.e., $\|x\| = |\xi_1| + \ldots + |\xi_n|$. Then $\|A\| = \max_j \sum_{i=1}^n |a_{ij}|$. To prove this, we see that

$$\begin{aligned}\|Ax\| &= \sum_{i=1}^n \left| \sum_{j=1}^n a_{ij}\xi_j \right| \leq \sum_{i=1}^n \sum_{j=1}^n |a_{ij}\xi_j| \\ &= \sum_{j=1}^n |\xi_j| \sum_{i=1}^n |a_{ij}| \leq \sum_{j=1}^n |\xi_j| \cdot \left\{\max_{1 \leq j \leq n} \sum_{i=1}^n |a_{ij}|\right\} \\ &= \left\{\max_{1 \leq j \leq n} \sum_{i=1}^n |a_{ij}|\right\} \cdot \|x\|.\end{aligned}$$

Let j_0 be such that $\sum_{i=1}^n |a_{ij_0}| = \max_{1 \leq j \leq n} \sum_{i=1}^n |a_{ij}| \triangleq \gamma_0$. Then $\|A\| \leq \gamma_0$. To show that equality must hold, let $x_0 = (\xi_1, \ldots, \xi_n) \in R^n$ be given by $\xi_{j_0} = 1$, and $\xi_i = 0$ if $i \neq j_0$. Then

$$\|Ax_0\| = \sum_{i=1}^n |a_{ij_0}| \quad \text{and} \quad \|x_0\| = 1.$$

From this it follows that $\|A\| \geq \gamma_0$, and so we conclude that $\|A\| = \gamma_0$.

(ii) Let $p = 2$; i.e., $||x|| = (|\xi_1|^2 + \ldots + |\xi_n|^2)^{1/2}$. Let $\mathbf{A}^T$ denote the transpose of $\mathbf{A}$ (see Eq. (4.2.9)), and let $\{\lambda_1, \ldots, \lambda_k\}$ be the distinct eigenvalues of the matrix $\mathbf{A}^T\mathbf{A}$ (see Definition 4.5.6). Let $\lambda_0 = \max_j \lambda_j$. Then $||A|| = \sqrt{\lambda_0}$. To prove this we note first that by Theorem 4.10.28 the eigenvalues of $\mathbf{A}^T\mathbf{A}$ are all real. We show first that they are, in fact, non-negative.

Let $\{\mathbf{x}_1, \ldots, \mathbf{x}_k\}$ be eigenvectors of $\mathbf{A}^T\mathbf{A}$ corresponding to the eigenvalues $\{\lambda_1, \ldots, \lambda_k\}$, respectively. Then for each $i = 1, \ldots, k$ we have $\mathbf{A}^T\mathbf{A}\mathbf{x}_i = \lambda_i\mathbf{x}_i$. Thus, $\mathbf{x}_i^T\mathbf{A}^T\mathbf{A}\mathbf{x}_i = \lambda_i\mathbf{x}_i^T\mathbf{x}_i$. From this it follows that

$$\lambda_i = \frac{\mathbf{x}_i^T\mathbf{A}^T\mathbf{A}\mathbf{x}_i}{\mathbf{x}_i^T\mathbf{x}_i} \geq 0.$$

For arbitrary $x \in X$ it follows from Theorem 4.10.44 that $x = x_1 + \ldots + x_k$, where $A^TAx_i = \lambda_i x_i$, $i = 1, \ldots, k$. Hence, $A^TAx = \lambda_1 x_1 + \ldots + \lambda_k x_k$. By Theorem 4.9.41 we have $||Ax||^2 = \mathbf{x}^T\mathbf{A}^T\mathbf{A}\mathbf{x}$. Thus,

$$||Ax||^2 = \mathbf{x}^T\mathbf{A}^T\mathbf{A}\mathbf{x} = \sum_{i=1}^{k} \lambda_i||\mathbf{x}_i||^2 \leq \lambda_0 \sum_{i=1}^{k} ||\mathbf{x}_i||^2 = \lambda_0||x||^2,$$

from which it follows that $||A|| \leq \sqrt{\lambda_0}$. If we let x be an eigenvector corresponding to λ_0, then we must have $||Ax||^2 = \lambda_0||x||^2$, and so equality is achieved. Thus, $||A|| = \sqrt{\lambda_0}$.

(iii) Let $||x|| = \max_i \{|\xi_i|\}$. Then $||A|| = \max_i \left(\sum_{j=1}^{n} |a_{ij}|\right)$. The proof of this part is left as an exercise. ∎

7.1.23. Exercise. Prove part (iii) of Example 7.1.22.

Next, we prove the following important result concerning the completeness of $B(X, Y)$

7.1.24. Theorem. If Y is complete, then the normed linear space $B(X, Y)$ is also complete.

Proof. Let $\{T_n\}$ be a Cauchy sequence in the normed linear space $B(X, Y)$. Choose N such that for a given $\epsilon > 0$, $||T_m - T_n|| < \epsilon$ whenever $m \geq N$ and $n \geq N$. Since the T_n are bounded we have for each $x \in X$,

$$||T_m x - T_n x|| \leq ||T_m - T_n||||x|| < \epsilon||x||$$

whenever $m, n \geq N$. From this it follows that $\{T_n x\}$ is a Cauchy sequence in Y. But Y is complete, by hypothesis. Therefore, $T_n x$ has a limit in Y which depends on $x \in X$. Let us denote this limit by Tx; i.e., $\lim_{n\to\infty} T_n x = Tx$. To show that T is linear we note that

$$T(x + y) = \lim_{n\to\infty} T_n(x + y) = \lim_{n\to\infty} T_n x + \lim_{n\to\infty} T_n y = Tx + Ty$$

and

$$T(\alpha x) = \lim_{n\to\infty} T_n(\alpha x) = \alpha \lim_{n\to\infty} T_n x = \alpha Tx.$$

Thus, T is a linear operator of X into Y. We show next that T is bounded and hence continuous. Since every Cauchy sequence in a normed linear space is bounded, it follows that the sequence $\{T_n\}$ is bounded, and thus $\|T_n\| \leq M$ for all n, where M is some constant. We have

$$\|Tx\| = \|\lim_{n\to\infty} T_n x\| = \lim_{n\to\infty} \|T_n x\| \leq \sup (\|T_n\| \|x\|)$$
$$= (\sup \|T_n\|)\|x\|.$$

This proves that T is bounded and therefore continuous, and $T \in B(X, Y)$. Finally, we must show that $T_n \longrightarrow T$ as $n \longrightarrow \infty$ in the norm of $B(X, Y)$. From before, we have $\|T_m x - T_n x\| < \epsilon\|x\|$ whenever $m, n > N$. If we let $n \longrightarrow \infty$, then $\|T_m x - Tx\| < \epsilon\|x\|$ for every $x \in X$ provided that $m \geq N$. This implies that $\|T_m - T\| < \epsilon$ whenever $m \geq N$. But $T_m \longrightarrow T$ as $m \longrightarrow \infty$ with respect to the norm defined on $B(X, Y)$. Therefore, $B(X, Y)$ is complete and the theorem is proved. ■

In Definition 3.4.16 we defined the null space of $T \in L(X, Y)$ as

$$\mathfrak{N}(T) = \{x \in X : Tx = 0\}. \tag{7.1.25}$$

We then showed that the range space $\mathfrak{R}(T)$ is a linear subspace of Y and that $\mathfrak{N}(T)$ is a linear subspace of X. For the case of bounded linear transformations we have the following result.

7.1.26. Theorem. Let $T \in B(X, Y)$. Then $\mathfrak{N}(T)$ is a closed linear subspace of X.

Proof. $\mathfrak{N}(T)$ is a linear subspace of X by Theorem 3.4.19. That it is closed follows from part (ii) of Theorem 5.7.9, since $\mathfrak{N}(T) = T^{-1}(\{0\})$ and since $\{0\}$ is a closed subset of Y. ■

We conclude this section with the following useful result for continuous linear transformations.

7.1.27. Theorem. Let $T \in L(X, Y)$. Then T is continuous if and only if

$$T\left(\sum_{i=1}^{\infty} \alpha_i x_i\right) = \sum_{i=1}^{\infty} \alpha_i T x_i$$

for every convergent series $\sum_{i=1}^{\infty} \alpha_i x_i$ in X.

The proof of this theorem follows readily from Theorem 5.7.8. We leave the details as an exercise.

7.1.28. Exercise. Prove Theorem 7.1.27.

7.2. INVERSES

Throughout this section X and Y denote vector spaces over the same field F where F is either R (the real numbers) or C (the complex numbers).

We recall that a linear operator $T: X \longrightarrow Y$ has an inverse, T^{-1}, if it is injective, and if this is so, then T^{-1} is a linear operator from $\mathcal{R}(T)$ onto X (see Theorem 3.4.32). We have the following result concerning the continuity of T^{-1}.

7.2.1. Theorem. Let $T \in L(X, Y)$. Then T^{-1} exists, and $T^{-1} \in B(\mathcal{R}(T), X)$ if and only if there is an $\alpha > 0$ such that $\|Tx\| \geq \alpha\|x\|$ for all $x \in X$. If this is so, $\|T^{-1}\| \leq 1/\alpha$.

Proof. Assume that there is a constant $\alpha > 0$ such that $\alpha\|x\| \leq \|Tx\|$ for all $x \in X$. Then $Tx = 0$ implies $x = 0$, and T^{-1} exists by Theorem 3.4.32. For $y \in \mathcal{R}(T)$ there is an $x \in X$ such that $y = Tx$ and $T^{-1}y = x$. Thus,

$$\alpha\|x\| = \alpha\|T^{-1}y\| \leq \|Tx\| = \|y\|,$$

or

$$\|T^{-1}y\| \leq \frac{1}{\alpha}\|y\|.$$

Hence, T^{-1} is bounded and $\|T^{-1}\| \leq 1/\alpha$.

Conversely, assume that T^{-1} exists and is bounded. Then, for $x \in X$ there is a $y \in \mathcal{R}(T)$ such that $y = Tx$, and also $x = T^{-1}y$. Since T^{-1} is bounded we have

$$\|x\| = \|T^{-1}y\| \leq \|T^{-1}\|\|y\| = \|T^{-1}\|\|Tx\|,$$

or

$$\|Tx\| \geq \frac{1}{\|T^{-1}\|}\|x\|. \quad \blacksquare$$

The next result, called the **Neumann expansion theorem**, gives us important information concerning the existence of the inverse of a certain class of bounded linear transformations.

7.2.2. Theorem. Let X be a Banach space, let $T \in B(X, X)$, let $I \in B(X, X)$ denote the identity operator, and let $\|T\| < 1$. Then the range of $(I - T)$ is X, the inverse of $(I - T)$ exists and is bounded and satisfies the inequality

$$\|(I - T)^{-1}\| \leq \frac{1}{1 - \|T\|}. \tag{7.2.3}$$

Furthermore, the series $\sum_{n=0}^{\infty} T^n$ in $B(X, X)$ converges uniformly to $(I - T)^{-1}$ with respect to the norm of $B(X, X)$; i.e.,

$$(I - T)^{-1} = I + T + T^2 + \ldots + T^n + \ldots. \tag{7.2.4}$$

Proof. Since $\|T\| < 1$, it follows that the series $\sum_{n=0}^{\infty} \|T\|^n$ converges. In view of Theorem 7.1.16 we have $\|T^n\| \leq \|T\|^n$, and hence the series $\sum_{n=0}^{\infty} T^n$ converges in the space $B(X, X)$, because this space is complete in view of Theorem 7.1.24. If we set

$$S = \sum_{n=0}^{\infty} T^n,$$

then

$$ST = TS = \sum_{n=0}^{\infty} T^{n+1},$$

and

$$(I - T)S = S(I - T) = I.$$

It now follows from Theorem 3.4.65 that $(I - T)^{-1}$ exists and is equal to S. Furthermore, $S \in B(X, X)$. The inequality (7.2.3) follows now readily and is left as an exercise. ■

7.2.5. Exercise. Prove inequality (7.2.3).

The next result, which is of great significance, is known as the **Banach inverse theorem.**

7.2.6. Theorem. Let X and Y be Banach spaces, and let $T \in B(X, Y)$. If T is bijective then T^{-1} is bounded.

Proof. The proof of this theorem is rather lengthy and requires two preliminary results which we state and prove separately.

7.2.7. Proposition. If A is any subset of X such that $\bar{A} = X$ ($\bar{A}$ denotes the closure of A), then any $x \in X$ such that $x \neq 0$ can be written in the form

$$x = x_1 + x_2 + \ldots + x_n + \ldots,$$

where $x_n \in A$ and $\|x_n\| \leq 3\|x\|/2^n$, $n = 1, 2, \ldots$.

Proof. The sequence $\{x_k\}$ is constructed as follows. Let $x_1 \in A$ be such that $\|x - x_1\| \leq \frac{1}{2}\|x\|$. This can certainly be done since $\bar{A} = X$. Now choose $x_2 \in A$ such that $\|x - x_1 - x_2\| \leq \frac{1}{4}\|x\|$. We continue in this manner and obtain

$$\|x - x_1 - \ldots - x_n\| \leq \frac{1}{2^n}\|x\|.$$

We can always choose such an $x_n \in A$, because $x - x_1 - \ldots - x_{n-1} \in X$ and $\bar{A} = X$. By construction of $\{x_n\}$, $\left\| x - \sum_{k=1}^{n} x_k \right\| \longrightarrow 0$ as $n \longrightarrow \infty$. Hence,

$x = \sum_{k=1}^{\infty} x_k$. We now compute $\|x_n\|$. First, we see that

$$\|x_1\| = \|x_1 - x + x\| \leq \|x_1 - x\| + \|x\| \leq \tfrac{3}{2}\|x\|,$$
$$\|x_2\| = \|x_2 + x_1 - x - x_1 + x\| \leq \|x - x_1 - x_2\| + \|x - x_1\| \leq \tfrac{3}{4}\|x\|,$$

and, in general,

$$\begin{aligned}\|x_n\| &= \|x_n + x_{n-1} + \ldots + x_1 - x + x - x_1 - \ldots - x_{n-1}\| \\ &\leq \|x - x_1 - \ldots - x_n\| + \|x - x_1 - \ldots - x_{n-1}\| \\ &\leq \frac{3}{2^n}\|x\|,\end{aligned}$$

which proves the proposition. ■

7.2.8. Proposition. If $\{A_n\}$ is any countable collection of subsets of X such that $X = \bigcup_{n=1}^{\infty} A_n$, then there is a sphere $S(x_0; \epsilon) \subset X$ and a set A_n such that $S(x_0; \epsilon) \subset \bar{A}_n$.

Proof. The proof is by contradiction. Without loss of generality, assume that

$$A_1 \subset A_2 \subset A_3 \subset \ldots.$$

For purposes of contradiction assume that for every $x \in X$ and every n there is an $\epsilon_n > 0$ such that $S(x; \epsilon_n) \cap A_n = \varnothing$. Now let $x_1 \in X$ and $\epsilon_1 > 0$ be such that $S(x_1; \epsilon_1) \cap A_1 = \varnothing$. Let $x_2 \in X$ and $\epsilon_2 > 0$ be such that $S(x_2; \epsilon_2) \subset S(x_1; \epsilon_1)$ and $S(x_2; \epsilon_2) \cap A_2 = \varnothing$. We see that it is possible to construct a sequence of closed nested spheres, $\{K_n\}$, (see Definition 5.5.34) in such a fashion that the diameter of these spheres, diam (K_n), converges to zero. In view of part (ii) of Theorem 5.5.35, $\bigcap_{k=1}^{\infty} K_n \neq \varnothing$. Let $x \in \bigcap_{k=1}^{\infty} K_n$. Then $x \notin A_n$ for all n. But this contradicts the fact that $X = \bigcup_{n=1}^{\infty} A_n$. This completes the proof of the proposition. ■

Proof of Theorem 7.2.6. Let

$$A_k = \{y \in Y: \|T^{-1}y\| \leq k\|y\|\}, \quad k = 1, 2, \ldots.$$

Clearly, $Y = \bigcup_{k=1}^{\infty} A_k$. By Proposition 7.2.8 there is a sphere $S(y_0; \epsilon) \subset Y$ and a set A_n such that $S(y_0; \epsilon) \subset \bar{A}_n$. We may assume that $y_0 \in A_n$. Let β be such that $0 < \beta < \epsilon$, and let us define the sets B and B_0 by

$$B = \{y \in S(y_0; \epsilon): \beta < \|y - y_0\|\}$$

and

$$B_0 = \{y \in Y: y = z - y_0, z \in B\}.$$

We now show that there is an A_K such that $B_0 \subset \bar{A}_K$. Let $y \in B \cap A_n$. Then $y - y_0 \in B_0$. We then have

$$\begin{aligned} \|T^{-1}(y - y_0)\| &\leq \|T^{-1}y\| + \|T^{-1}y_0\| \\ &\leq n[\|y\| + \|y_0\|] \\ &\leq n[\|y - y_0\| + 2\|y_0\|] \\ &= n\|y - y_0\|\left[1 + \frac{2\|y_0\|}{\|y - y_0\|}\right] \\ &\leq n\|y - y_0\|\left[1 + \frac{2\|y_0\|}{\beta}\right]. \end{aligned}$$

Now let K be a positive integer such that

$$n\left[1 + \frac{2\|y_0\|}{\beta}\right] \leq K.$$

It then follows that $y - y_0 \in A_K$. It follows readily that $B_0 \subset \bar{A}_K$.

Now let y be an arbitrary element in Y. It is always possible to choose a real number λ such that $\lambda y \in B_0$. Thus, there is a sequence $\{y_i\}$ such that $y_i \in A_K$ for all i and $\lim y_i = \lambda y$. This means that the sequence $\left\{\frac{1}{\lambda} y_i\right\}$ converges to y. We observe from the definition of A_K that if $y_i \in A_K$, then $\frac{1}{\lambda} y_i \in A_K$ for any real number λ. Hence, we have shown that $Y \subset \bar{A}_K$.

Finally, for arbitrary $y \in Y$ we can write, by Proposition 7.2.7,

$$y = y_1 + y_2 + \ldots + y_n + \ldots,$$

where $\|y_n\| < 3\|y\|/2^n$. Let $x_k = T^{-1}y_k$, $k = 1, \ldots,$ and consider the infinite series $\sum_{k=1}^{\infty} x_k$. This series converges, since

$$\|x_k\| = \|T^{-1}y_k\| \leq K\|y_k\| < \frac{3K\|y\|}{2^k}$$

so that

$$\|x\| \leq \sum_{k=1}^{\infty} \|x_k\| \leq 3K\|y\| \sum_{k=1}^{\infty} \frac{1}{2^k} = 3K\|y\|.$$

Since T is continuous and since $\sum_{k=1}^{\infty} x_k$ converges, it follows that $Tx = T\left(\sum_{k=1}^{\infty} x_k\right) = \sum_{k=1}^{\infty} Tx_k = \sum_{k=1}^{\infty} y_k = y$. Hence, $x = T^{-1}y$. Therefore, $\|x\| = \|T^{-1}y\| \leq 3K\|y\|$. This implies that T^{-1} is bounded, which was to be proved. ■

Utilizing the principle of contraction mappings (see Theorem 5.8.5), we now establish results related to inverses which are important in applications. In the setting of normed linear spaces we can restate the definition of a *contraction mapping* as being a function $T: X \longrightarrow X$ (T is not necessarily linear) such that

$$\|T(x) - T(y)\| \leq \alpha\|x - y\|$$

for all $x, y \in X$, with $0 < \alpha < 1$. The *principle of contraction mappings* asserts that if T is a contraction mapping, then the equation

$$T(x) = x$$

has one and only one solution $x \in X$.

We now state and prove the following result.

7.2.9. Theorem. Let X be a Banach space, let $T \in B(X, X)$, let $\lambda \in F$, and let $\lambda \neq 0$.

(i) If $|\lambda| > \|T\|$, then $Tx = \lambda x$ has a unique solution, namely $x = 0$;

(ii) if $|\lambda| > \|T\|$, then $(T - \lambda I)^{-1}$ exists and is continuous on X;

(iii) if $|\lambda| > \|T\|$, then for a given $y \in X$ there is one and only one vector $x \in X$ such that $(T - \lambda I)x = y$, and

$$x = -\left[\frac{y}{\lambda} + \frac{Ty}{\lambda^2} + \ldots\right]; \text{ and}$$

(iv) if $\|I - T\| < 1$, then T^{-1} exists and is continuous on X.

Proof.

(i) For any $x, y \in X$, we have

$$\|\lambda^{-1}Tx - \lambda^{-1}Ty\| = |\lambda^{-1}|\|T(x - y)\| \leq |\lambda^{-1}|\|T\|\|x - y\|.$$

Thus, if $\|T\| < |\lambda|$, then $\lambda^{-1}T$ is a contraction mapping. In view of the principle of contraction mappings there is a unique $x \in X$ with $\lambda^{-1}Tx = x$, or $Tx = \lambda x$. The unique solution has to be $x = 0$, because $T0 = 0$.

(ii) Let $L = \frac{1}{\lambda}T$. Then $\|L\| = \frac{1}{|\lambda|}\|T\| < 1$. It now follows from Theorem 7.2.2 that $(L - I)^{-1}$ exists and is continuous on X. Thus, $(\lambda L - \lambda I)^{-1} = (T - \lambda I)^{-1}$ exists and is continuous on X. This completes the proof of part (ii). The proofs of the remaining parts are left as an exercise. ■

7.2.10. Exercise. Prove parts (iii) and (iv) of Theorem 7.2.9.

7.3. CONJUGATE AND ADJOINT OPERATORS

Associated with every bounded linear operator defined on a normed linear space is a transformation called its **conjugate**, and associated with every bounded linear operator defined on an inner product space is a transformation called its **adjoint**. These operators, which we consider in this section, are of utmost importance in analysis as well as in applications.

Throughout this section X and Y are normed linear spaces over F, where F is either R (the real numbers) or C (the complex numbers). In some cases we may further assume that X and Y are inner product spaces, and in other instances we may require that X and/or Y be complete.

Let X^f and Y^f denote the algebraic conjugate of X and Y, respectively (refer to Definition 3.5.18). Utilizing the notation of Section 3.5, we write $x' \in X^f$ and $y' \in Y^f$ to denote elements of these spaces. If $T \in L(X, Y)$, we defined the transpose of T, T^T, to be a mapping from Y^f to X^f determined by the equation

$$\langle x, T^T y' \rangle = \langle Tx, y' \rangle \text{ for all } x \in X, y' \in Y^f$$

(see Definition 3.5.27), and we showed that $T^T \in L(Y^f, X^f)$.

Now let us assume that $T\colon X \longrightarrow Y$ is a bounded linear operator on X into Y. Let X^* and Y^* denote the normed conjugate spaces of X and Y, respectively (refer to Definition 6.5.9). If $y' \in Y^*$, then $y'(y) = \langle y, y' \rangle$ is defined for every $y \in Y$ and, in particular, it is defined for every $y = Tx$, $x \in X$. The quantity $\langle Tx, y' \rangle = y'(Tx)$ is a scalar for each $x \in X$. Writing $x'(x) = \langle Tx, y' \rangle = y'(Tx)$, we have defined a functional x' on X. Since y' is a linear transformation (it is a bounded linear functional) and since T is a linear transformation (it is a bounded linear operator), it follows readily that x' is a linear functional. Also, since T is bounded, we have

$$|x'(x)| = |y'(Tx)| = |\langle Tx, y' \rangle| \leq \|y'\| \|Tx\| \leq \|y'\| \|T\| \|x\|,$$

and therefore x' is a bounded linear functional and $x' \in X^*$. We have thus assigned to each functional $y' \in Y^*$ a functional $x' \in X^*$; i.e., we have established a linear operator which maps Y^* into X^*. This operator is called the *conjugate operator* of the operator T and is denoted by T'. We now have

$$x' = T'y'.$$

The definition of $T'\colon Y^* \longrightarrow X^*$ is usually expressed by the relation

$$\langle x, T'y' \rangle = \langle Tx, y' \rangle, x \in X, y' \in Y^*.$$

Utilizing operator notation rather than bracket notation, the definition of the conjugate operator T' satisfies the equation

$$x'(x) = y'(Tx) = (T'y')(x), x \in X,$$

and we may therefore write

$$y'T = T'y',$$

where $y'T$ denotes the functional on X consisting of the operators T and y', and $T'y'$ is the functional obtained by operating on y' by T'.

The reader can readily show that T' is unique and linear. If $Y^* = Y^f$, which is the case if Y is finite dimensional, then the conjugate T' and the transpose T^T are identical concepts. However, since, in general, Y^* is a proper subspace of Y^f, T^T is an extension of T' or, conversely, T' is a restriction of T^T to the space Y^*.

We summarize the above discussion in the following definition and Figure A.

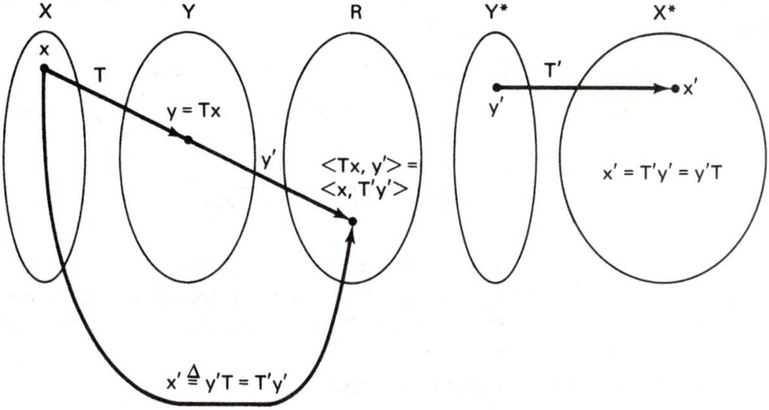

7.3.1. Figure A

7.3.2. Definition. Let T be a bounded linear operator on X into Y. The **conjugate operator** of T, $T'\colon Y^* \longrightarrow X^*$ is defined by the formula

$$\langle x, T'y'\rangle = \langle Tx, y'\rangle, x \in X, y' \in Y^*.$$

7.3.3. Exercise. Show that the conjugate operator T' is unique and linear.

Before exploring the properties of conjugate operators, we introduce another important operator which is closely related to the conjugate operator, the so-called "adjoint operator." In this case we focus our attention on Hilbert spaces.

Let X and Y denote Hilbert spaces, and let the symbol $(,)$ denote the inner product on both X and Y. If T is a bounded linear transformation on X into Y, then in view of the above discussion there is a unique bounded linear operator from Y^* into X^*, called the **conjugate** of T. But in view of Theorem 6.14.2, the dual spaces X^*, Y^* may be identified with X and Y, respectively, because X and Y are Hilbert spaces. This gives rise to a new type of bounded linear operator from Y into X, called the **adjoint** of T, which we consider in place of T'.

Let $y_0 \in Y$ be fixed, and let $x'(x) = (x, x') = (Tx, y_0)$, where $T \in B(X, Y)$ and $x' \in X^*$. By Theorem 6.14.2 there is a unique $x_0 \in X$ such that

$x'(x) = (x, x_0)$. Writing $x_0 = T^*y_0$ we define in this way a transformation of Y into X. We call this transformation the adjoint of T. Dropping the subscript zero, we characterize the adjoint of T by the formula

$$(Tx, y) = (x, T^*y), x \in X, y \in Y.$$

We will now show that $T^*: Y \longrightarrow X$ is linear, unique, and bounded. To prove linearity, let $x \in X$, $y_1, y_2 \in Y$, let $\alpha, \beta \in F$, and note that

$$\begin{aligned}(x, T^*(\alpha y_1 + \beta y_2)) &= (Tx, \alpha y_1 + \beta y_2) = \bar{\alpha}(Tx, y_1) + \bar{\beta}(Tx, y_2) \\ &= \bar{\alpha}(x, T^*y_1) + \bar{\beta}(x, T^*y_2) = (x, \alpha T^*y_1 + \beta T^*y_2).\end{aligned}$$

From this it follows that

$$T^*(\alpha y_1 + \beta y_2) = \alpha T^*y_1 + \beta T^*y_2,$$

and therefore T^* is linear.

To show that T^* is unique we note that if $(x, T^*y) = (x, S^*y)$, then $(x, T^*y) - (x, S^*y) = 0$ implies $(x, (T^* - S^*)y) = 0$ for all $x \in X$. From this it follows that $(T^* - S^*)y \perp x$ for all $x \in X$, and thus $(T^* - S^*)y = 0$ for all $y \in Y$. Therefore, $T^* = S^*$.

To verify that T^* is bounded we observe that

$$\begin{aligned}\|T^*x\|^2 = |(T^*x, T^*x)| = |(T(T^*x), x)| &\leq \|T(T^*x)\|\|x\| \\ &\leq \|T\|\|T^*x\|\|x\|,\end{aligned}$$

and thus

$$\|T^*x\| \leq \|T\|\|x\|.$$

From this it follows that T^* is bounded and furthermore $\|T^*\| \leq \|T\|$.

We now give the following formal definition.

7.3.4. Definition. Let X and Y be Hilbert spaces, and let T be a bounded linear operator on X into Y. The **adjoint operator** $T^*: Y \longrightarrow X$ is defined by the formula

$$(Tx, y) = (x, T^*y), x \in X, y \in Y.$$

Summarizing the above discussion we have the following result.

7.3.5. Theorem. The adjoint operator T^* given in Definition 7.3.4 is linear, unique, and bounded.

The reader is cautioned that many authors use the terms *conjugate operator* and *adjoint operator* interchangeably. Also, the symbol T^* is used by many authors to denote both adjoint and conjugate operators.

Some of the important properties of conjugate operators are summarized in the following result.

7.3.6. Theorem. Conjugate transformations have the following properties:

(i) $\|T'\| = \|T\|$;
(ii) $I' = I$, where I is the identity operator on a normed linear space X;
(iii) $0' = 0$, where 0 is the zero operator on a normed linear space X;
(iv) $(S + T)' = S' + T'$, where $S, T \in B(X, Y)$ and where X, Y are normed linear spaces;
(v) $(\alpha T)' = \alpha T'$, where $T \in B(X, Y), \alpha \in F$, and X, Y are normed linear spaces;
(vi) $(ST)' = T'S'$, where $T \in B(X, Y)$, $S \in B(Y, Z)$, and X, Y, Z are normed linear spaces; and
(vii) if T^{-1} exists and if $T^{-1} \in B(Y, X)$, then $(T')^{-1}$ exists, and moreover $(T')^{-1} = (T^{-1})'$.

Proof. To prove part (i) we note that

$$|\langle x, T'y'\rangle| = |\langle Tx, y'\rangle| \leq \|y'\|\|Tx\| \leq \|y'\|\|T\|\|x\|.$$

From this it follows that $\|T'y'\| \leq \|T\|\|y'\|$, and therefore

$$\|T'\| \leq \|T\|.$$

Next, let $x_0 \in X, x_0 \neq 0$. In view of the Hahn-Banach theorem (see Corollary 6.8.5) there is a $y'_0 \in Y^*$, $\|y'_0\| = 1$, such that $\langle Tx_0, y'_0\rangle = \|Tx_0\|$. Therefore,

$$\|Tx_0\| = |\langle x_0, T'y'_0\rangle| \leq \|T'y'_0\|\|x_0\| \leq \|T'\|\|x_0\|,$$

from which it follows that

$$\|T\| \leq \|T'\|.$$

Therefore, $\|T\| = \|T'\|$.

The proofs of properties (ii)–(vi) are straightforward. To prove (iv), for example, we note that

$$\begin{aligned}\langle x, (S + T)'y'\rangle &= \langle (S + T)x, y'\rangle = \langle Sx + Tx, y'\rangle \\ &= \langle Sx, y'\rangle + \langle Tx, y'\rangle = \langle x, S'y'\rangle + \langle x, T'y'\rangle \\ &= \langle x, S'y' + T'y'\rangle = \langle x, (S' + T')y'\rangle.\end{aligned}$$

From this it follows that $(S + T)' = S' + T'$.

To prove part (vii) assume that $T \in B(X, Y)$ has a bounded inverse T^{-1}: $Y \longrightarrow X$. To show that T': $Y^* \longrightarrow X^*$ has an inverse we must show that it is injective. Let $y'_1, y'_2 \in Y^*$ be such that $y'_1 \neq y'_2$. Then

$$\langle x, T'y'_1\rangle - \langle x, T'y'_2\rangle = \langle Tx, y'_1 - y'_2\rangle \neq 0$$

for some $x \in X$. From this it follows that $T'y'_1 \neq T'y'_2$, and T' is one-to-one. We can, in fact, show that T' is onto. We note that for any $x' \in X^*$ and any

$x \in X$, $Tx = y$, and we have

$$\langle x, x' \rangle = \langle T^{-1}y, x' \rangle = \langle y, (T^{-1})'x' \rangle = \langle Tx, (T^{-1})'x' \rangle$$
$$= \langle x, T'(T^{-1})'x' \rangle.$$

From this it follows that

$$x' = T'(T^{-1})'x'.$$

This shows that $x' \in \mathcal{R}(T')$ and that $(T')^{-1} = (T^{-1})'$. ■

7.3.7. Exercise. Prove parts (ii), (iii), (v), and (vi) of Theorem 7.3.6.

In the next theorem some of the important properties of adjoint operators are summarized.

7.3.8. Theorem. Let X, Y, and Z be Hilbert spaces, and let I and 0 denote the identity and zero transformation on X, respectively. Then

(i) $||T^*|| = ||T||$, where $T \in B(X, Y)$;
(ii) $I^* = I$;
(iii) $0^* = 0$;
(iv) $(S + T)^* = S^* + T^*$, where $S, T \in B(X, Y)$;
(v) $(\alpha T)^* = \bar{\alpha} T^*$, where $T \in B(X, Y)$ and $\alpha \in F$;
(vi) $(ST)^* = T^*S^*$, where $T \in B(X, Y)$, $S \in B(Y, Z)$;
(vii) if $T^{-1} \in B(Y, X)$ exists, then $(T^*)^{-1} \in B(X, Y)$ exists, and moreover $(T^*)^{-1} = (T^{-1})^*$;
(viii) if for $T \in B(X, Y)$ we define $(T^*)^* = T^{**}$, then $T^{**} = T$; and
(ix) $||T^*T|| = ||T||^2$, where $T \in B(X, Y)$.

Proof. To prove part (i) we note that

$$||T^*x||^2 = |(T^*x, T^*x)| = |(T(T^*x), x)| \leq ||T(T^*x)|| \, ||x||$$
$$\leq ||T|| \, ||T^*x|| \, ||x||,$$

or

$$||T^*x|| \leq ||T|| \, ||x||.$$

From the last inequality it follows that $||T^*|| \leq ||T||$. Reversing the roles of T and T^* we obtain

$$||Tx||^2 = |(Tx, Tx)| = |(T^*(Tx), x)| \leq ||T^*(Tx)|| \, ||x|| \leq ||T^*|| \, ||Tx|| \, ||x||,$$

or

$$||Tx|| \leq ||T^*|| \, ||x||.$$

From this it follows that $||T|| \leq ||T^*||$, and therefore $||T|| = ||T^*||$.

The proofs of properties (ii)–(viii) are trivial. To prove part (ix), we first note that

$$\|T^*T\| \leq \|T^*\|\|T\| = \|T\|\|T\| = \|T\|^2.$$

On the other hand,

$$\|Tx\|^2 = (Tx, Tx) = (T^*Tx, x) \leq \|T^*Tx\|\|x\| \leq \|T^*T\|\|x\|\|x\|.$$

Taking the square root on both sides of the above inequality we obtain

$$\|Tx\| \leq \sqrt{\|T^*T\|}\,\|x\|,$$

and thus $\|T\| \leq \sqrt{\|T^*T\|}$, or $\|T\|^2 \leq \|T^*T\|$. Hence, $\|T^*T\| = \|T\|^2$. ■

7.3.9. Exercise. Prove parts (ii)–(viii) of Theorem 7.3.8.

From the above discussion it is obvious that adjoint operators are distinct from conjugate operators even though many of their properties appear to be identical, especially for the case of real spaces. We now cite a few examples to illustrate some of the concepts considered above.

7.3.10. Example. Let $X = C^n$ be the Hilbert space with inner product defined in Example 3.6.24, and let $A \in L(X, X)$ be represented (with respect to the natural basis for X) by the $n \times n$ matrix $\mathbf{A} = [a_{ij}]$. The transformation $y = Ax$ can be written in the form

$$y_i = \sum_{j=1}^{n} a_{ij}x_j, \quad i = 1, 2, \ldots, n,$$

where y_i is the ith component of the vector $y \in X$. Let A^* denote the adjoint of A on the Hilbert space X, and let A^* be represented by the $n \times n$ matrix $[a_{ij}^*]$. Now if $u = (u_1, \ldots, u_n) \in X$, then

$$(Ax, u) = (y, u) = \sum_{i=1}^{n} y_i\bar{u}_i = \sum_{i=1}^{n} \bar{u}_i\Big(\sum_{j=1}^{n} a_{ij}x_j\Big),$$

and

$$(x, A^*u) = \sum_{i=1}^{n} x_i\overline{\Big(\sum_{j=1}^{n} a_{ij}^*u_j\Big)}.$$

In order that $(Ax, u) = (x, A^*u)$ we must have $a_{ij}^* = \bar{a}_{ji}$; i.e., the matrix of A^* is the transpose of the conjugate of the matrix of A. ■

7.3.11. Example. Let $X = Y = L_2[a, b]$, $a < b$ (see Example 6.11.10), and define the *Fredholm operator* T by

$$y(t) = (Tx)(t) = \int_a^b k(s, t)x(s)ds, \; t \in [a, b],$$

where it is assumed that the kernel function $k(s, t)$ is well enough behaved so that

$$\int_a^b \int_a^b |k(s, t)|^2 \, dtds < \infty.$$

Now if $u \in L_2[a, b]$, then

$$(Tx, u) = (y, u) = \int_a^b y(t)\overline{u(t)}\,dt = \int_a^b \overline{u(t)}\left(\int_a^b k(s, t)x(s)ds\right)dt$$
$$= \int_a^b x(s)\left(\int_a^b k(s, t)\overline{u(t)}dt\right)ds.$$

From this it follows that the adjoint T^* of T maps u into the function

$$z(t) = (T^*u)(t) = \int_a^b \overline{k(t, s)}u(s)ds;$$

i.e., the adjoint of T is obtained by interchanging the roles of s and t in the kernel and by utilizing the complex conjugate of k. ■

7.3.12. Exercise. Let $X = Y = l_2$ (see Example 6.1.6) and define T: $l_2 \longrightarrow l_2$ by

$$T(\xi_1, \xi_2, \ldots, \xi_n, \ldots) = (0, \xi_1, \xi_2, \ldots, \xi_n, \ldots) = y,$$

for all $x = (\xi_1, \xi_2, \ldots, \xi_n, \ldots) \in l_2$. Show that $T^*: l_2 \longrightarrow l_2$ is the operator defined by

$$T^*(\eta_1, \eta_2, \ldots, \eta_n, \ldots) = (\eta_2, \eta_3, \ldots, \eta_n, \ldots)$$

for all $y = (\eta_1, \eta_2, \ldots, \eta_n, \ldots) \in l_2$.

Recalling the definition of orthogonal complement (refer to Definition 6.12.1), we have the following important results for bounded linear operators on Hilbert spaces.

7.3.13. Theorem. Let T be a bounded linear operator on a Hilbert space X into a Hilbert space Y. Then,

(i) $\{\overline{\mathfrak{R}(T)}\}^\perp = \mathfrak{N}(T^*)$;
(ii) $\overline{\mathfrak{R}(T)} = \mathfrak{N}(T^*)^\perp$;
(iii) $\mathfrak{N}(T) = \{\overline{\mathfrak{R}(T^*)}\}^\perp$;
(iv) $\overline{\mathfrak{R}(T^*)} = \mathfrak{N}(T)^\perp$;
(v) $\mathfrak{N}(T^*) = \mathfrak{N}(TT^*)$; and
(vi) $\overline{\mathfrak{R}(T)} = \overline{\mathfrak{R}(TT^*)}$.

Proof. We prove (i) and (v) and leave the proofs of (ii)–(iv) and (vi) as an exercise.

To prove (i), we first show that $\mathfrak{R}(T)^\perp = \mathfrak{N}(T^*)$. Let $y \in \mathfrak{R}(T)^\perp$. Then $(y, Tx) = 0$ for all $x \in X$, and hence $(T^*y, x) = 0$ for all $x \in X$. This can be true only if $T^*y = 0$; i.e., $y \in \mathfrak{N}(T^*)$. On the other hand, if $y \in \mathfrak{N}(T^*)$, then $(T^*y, x) = 0$ for all $x \in X$. Thus, $(y, Tx) = 0$ for every $x \in X$, which implies that $y \in \mathfrak{R}(T)^\perp$. Now $\mathfrak{R}(T)$ need not be closed. However, by Theorem 6.12.14 $\overline{\mathfrak{R}(T)} = \mathfrak{R}(T)^{\perp\perp}$. Therefore, $\{\overline{\mathfrak{R}(T)}\}^\perp = \mathfrak{R}(T)^{\perp\perp\perp} = \mathfrak{R}(T)^\perp = \mathfrak{N}(T^*)$.

To prove (v), let $y \in \mathfrak{N}(T^*)$. Then $T^*y = 0$ and $TT^*y = 0$. This implies that $\mathfrak{N}(T^*) \subset \mathfrak{N}(TT^*)$. Next, let $y \in \mathfrak{N}(TT^*)$. Then $TT^*y = 0$ and $(y, TT^*y) = 0$. This implies that $(T^*y, T^*y) = 0$ so that $T^*y = 0$. Therefore, $y \in \mathfrak{N}(T^*)$ and $\mathfrak{N}(TT^*) \subset \mathfrak{N}(T^*)$, completing the proof of part (v). ∎

7.3.14. Exercise. Prove parts (ii)–(iv) and (vi) of Theorem 7.3.13.

We conclude this section with the following results.

7.3.15. Theorem. Let $T \in B(X, X)$, where X is a Hilbert space, and let M and N be subsets of X. Define $T(M)$ as

$$T(M) = \{y: y = Tx, x \in M\}.$$

If $T(M) \subset N$, then $T^*(N^\perp) \subset M^\perp$.

Proof. Let $z \perp N$. Then for $x \in M$ we have $(Tx, z) = 0 = (x, T^*z)$. Therefore, $T^*z \perp x$ for all $x \in M$ and $T^*z \in M^\perp$. ∎

7.3.16. Theorem. Let $T \in B(X, X)$, where X is a Hilbert space, and let M and N be closed linear subspaces of X. Then $T(M) \subset N$ if and only if $T^*(N^\perp) \subset M^\perp$.

Proof. If $T(M) \subset N$, then by Theorem 7.3.15 $T^*(N^\perp) \subset M^\perp$. Conversely, if $T^*(N^\perp) \subset M^\perp$, then by Theorem 7.3.15, $T^{**}(M^{\perp\perp}) \subset N^{\perp\perp}$. But $T^{**} = T$ and if M and N are closed linear subspaces, then $M^{\perp\perp} = M$ and $N^{\perp\perp} = N$. Therefore, $T(M) \subset N$. ∎

7.4. HERMITIAN OPERATORS

Throughout this section X denotes a complex Hilbert space. We shall be primarily concerned with operators $T \in B(X, X)$. By T^* we shall always mean the adjoint of T.

For our first result, recall the definition of bilinear functional (Definition 3.6.4).

7.4.1. Theorem. Let $T \in B(X, X)$ and define the function $\varphi: X \times X \longrightarrow C$ by $\varphi(x, y) = (Tx, y)$ for all $x, y \in X$. Then φ is a bilinear functional.

7.4.2. Exercise. Prove Theorem 7.4.1.

Of central importance in this section is the following class of operators.

7.4.3. Definition. A bounded linear transformation $T \in B(X, X)$ is said to be **hermitian** if $T = T^*$.

Some authors call such transformations **self-adjoint operators** (see Definition 4.10.20).

The next two results allow us to characterize a hermitian operator in an equivalent manner. The first of these involves symmetric bilinear forms (see Definition 3.6.10).

7.4.4. Theorem. Let $T \in B(X, X)$. Then T is hermitian if and only if the bilinear transformation $\varphi(x, y) = (Tx, y)$ is symmetric.

Proof. If $T^* = T$, then $\varphi(x, y) = (Tx, y) = (x, T^*y) = (x, Ty) = \overline{(Ty, x)} = \overline{\varphi(y, x)}$, and therefore φ is symmetric.

Conversely, assume that $\varphi(x, y) = \overline{\varphi(y, x)}$. Then $\overline{\varphi(y, x)} = \overline{(Ty, x)} = (x, Ty) = \varphi(x, y) = (Tx, y) = (x, T^*y)$; i.e., $(x, Ty) = (x, T^*y)$ for all $x, y \in X$. From this it follows that

$$((T^* - T)x, y) = 0,$$

and thus $(T^* - T)x \perp y$ for all $x \in X$. This implies that $T^*x = Tx$ for all $x \in X$ or $T^* = T$. ■

7.4.5. Theorem. Let $T \in B(X, X)$. Then T is hermitian if and only if (Tx, x) is real for every $x \in X$.

Proof. If T is hermitian, then $(Tx, y) = \overline{(Ty, x)}$. Setting $x = y$, we obtain $(Tx, x) = \overline{(Tx, x)}$, which implies that (Tx, x) is real.

Conversely, suppose $(x, Tx) = \overline{(x, Tx)}$ for all $x \in X$. Then $(x, Tx) = (Tx, x)$. Now consider (x, Ty) for arbitrary $x, y \in X$. It is easily verified that

$$(x + y, T(x + y)) - (x - y, T(x - y)) + i(x + iy, T(x + iy)) - i(x - iy, T(x - iy)) = 4(x, Ty) \tag{7.4.6}$$

where $i = \sqrt{-1}$. Also,

$$(T(x + y), x + y) - (T(x - y), x - y) + i(T(x + iy), x + iy)) - i(T(x - iy), x - iy) = 4(Tx, y). \tag{7.4.7}$$

Since the left-hand sides of Eqs. (7.4.6) and (7.4.7) are equal, it follows that $(x, Ty) = (Tx, y)$ for all $x, y \in X$, and hence $T = T^*$. ■

The norm of a hermitian operator can be found as follows.

7.4.8. Theorem. Let $T \in B(X, X)$ be a hermitian operator. Then the norm of T can be expressed in the following equivalent ways:

(i) $\|T\| = \sup \{|(Tx, x)| : \|x\| = 1\}$; and
(ii) $\|T\| = \sup \{|(Tx, y)| : \|x\| = \|y\| = 1\}$.

7.4.9. Exercise. Prove Theorem 7.4.8.

In the next theorem, some of the more important properties of hermitian operators are given.

7.4.10. Theorem. Let $S, T \in B(X, X)$ be hermitian operators, and let α be a real scalar. Then

(i) $(S + T)$ is a hermitian operator;
(ii) αT is a hermitian operator;
(iii) if T is bijective, then T^{-1} is hermitian; and
(iv) ST is hermitian if and only if $ST = TS$.

7.4.11. Exercise. Prove Theorem 7.4.10.

Since in the case of hermitian operators (Tx, x) is real for all $x \in X$, the following definition concerning definiteness applies (recall Definition 3.6.10).

7.4.12. Definition. Let $T \in B(X, X)$ be a hermitian operator. Then T is said to be **positive** if $(Tx, x) \geq 0$ for all $x \in X$. In this case we write $T \geq 0$. If $(Tx, x) > 0$ for all $x \neq 0$, we say that T is **strictly positive**.

7.4.13. Definition. Let $S, T \in B(X, X)$ be hermitian operators. If the hermitian operator $T + (-S) = T - S \geq 0$, then we write $T \geq S$.

7.4.14. Theorem. Let $S, T, U \in B(X, X)$ be hermitian operators, and let α be a real scalar. Then,

(i) if $S \geq 0, T \geq 0$, then $(S + T) \geq 0$;
(ii) if $\alpha > 0, T \geq 0$, then $\alpha T \geq 0$;
(iii) if $S \leq T, T \leq U$, then $S \leq U$; and
(iv) for any $V \in B(X, X)$, if $T \geq 0$, then $V^*TV \geq 0$. In particular, $V^*V \geq 0$.

Proof. The proofs of parts (i)–(iii) are obvious. For example, if $S \geq 0$, $T \geq 0$, then $(Sx, x) + (Tx, x) = (Sx + Tx, x) = ((S + T)x, x) \geq 0$ and $(S + T) \geq 0$.

To prove part (iv) we note that $(V^*TVx, x) = (TVx, Vx) \geq 0$, since $Vx = y$ is a vector in X and $(Ty, y) \geq 0$ for all $y \in X$. If we consider, in particular, $T = I = I^*$, then $V^*V \geq 0$. ■

The proof of the next result follows by direct verification of the formulas involved.

7.4.15. Theorem. Let $A \in B(X, X)$, and let

$$U = \frac{1}{2}[A + A^*] \quad \text{and} \quad V = \frac{1}{2i}[A - A^*],$$

where $i = \sqrt{-1}$. Then

(i) U and V are hermitian operators; and

(ii) if $A = C + iD$, where C and D are hermitian, then $C = U$ and $D = V$.

7.4.16. Exercise. Prove Theorem 7.4.15.

Let us now consider some specific cases.

7.4.17. Example. Let $X = C^n$ with inner product given in Example 3.6.24. Let $A \in B(X, X)$, and let $\{e_1, \ldots, e_n\}$ be any orthonormal basis for X. As we saw in Example 7.3.10, if A is represented by the matrix $\mathbf{A}$, then A^* is represented by $\mathbf{A}^* = \bar{\mathbf{A}}^T$. In this case A is hermitian if and only if $\mathbf{A} = \bar{\mathbf{A}}^T$. ■

7.4.18. Example. Let $X = L_2[a, b]$ (see Example 6.11.10), and define $T \in B(X, X)$ by

$$y = Tx = tx(t).$$

Then for any $z \in X$ we have

$$(Tx, z) = \int_a^b tx(t)\overline{z(t)}\, dt = \int_a^b x(t)\overline{tz(t)}\, dt$$
$$= (x, Tz) = (T^*x, z).$$

Thus, $T = T^*$ and T is hermitian. ■

7.4.19. Exercise. Let $X = L_2[a, b]$, and define $T: X \longrightarrow X$ by

$$z = Tx = \int_a^t x(s)\, ds.$$

Show that $T^* \neq T$ and therefore T is not hermitian.

7.4.20. Exercise. Let $X = L_2[a, b]$ and consider the Fredholm operator given in Example 7.3.11; i.e.,

$$y(t) = (Tx)(t) = \int_a^b k(s, t)x(s)ds, \; t \in [a, b].$$

Show that $T = T^*$ if and only if $k(t, s) = \overline{k(s, t)}$.

We conclude this section with the following result, which we will subsequently require.

7.4.21. Theorem. Let X be a Hilbert space, let $T \in B(X, X)$ be a hermitian operator, and let $\lambda \in R$. Then there exists a real number $\gamma > 0$ such that $\gamma \|x\| \leq \|(T - \lambda I)x\|$ for all $x \in X$ if and only if $(T - \lambda I)$ is bijective and $(T - \lambda I)^{-1} \in B(X, X)$, in which case $\|(T - \lambda I)^{-1}\| \leq 1/\gamma$.

Proof. Let $T_\lambda = T - \lambda I$. It follows from Theorem 7.4.10 that T_λ is also hermitian.

To prove sufficiency, let $T_\lambda^{-1} \in B(X, X)$. It follows that for all $y \in X$, $\|T_\lambda^{-1}y\| \leq \|T_\lambda^{-1}\| \cdot \|y\|$. Letting $y = T_\lambda x$ and $\gamma = \|T_\lambda^{-1}\|^{-1}$, we have $\|T_\lambda x\| \geq \gamma\|x\|$ for all $x \in X$.

To prove necessity, let $\gamma > 0$ be such that $\gamma\|x\| \leq \|T_\lambda x\|$ for all $x \in X$. We see that $T_\lambda x = 0$ implies $x = 0$; i.e., $\mathfrak{N}(T_\lambda) = \{0\}$, and so T_λ is injective. We next show that $\overline{\mathfrak{R}(T_\lambda)} = X$. It follows from Theorem 6.12.16 that $X = \overline{\mathfrak{R}(T_\lambda)} \oplus \overline{\mathfrak{R}(T_\lambda)}^\perp$. From Theorem 7.3.13, we have $\overline{\mathfrak{R}(T_\lambda)}^\perp = \mathfrak{N}(T_\lambda^*)$. Since T_λ is hermitian, $\mathfrak{N}(T_\lambda^*) = \mathfrak{N}(T_\lambda) = \{0\}$. Hence, $\overline{\mathfrak{R}(T_\lambda)} = X$. We next show that $\mathfrak{R}(T_\lambda) = \overline{\mathfrak{R}(T_\lambda)}$, i.e. the range of T_λ is closed. Let $\{y_n\}$ be a sequence in $\mathfrak{R}(T_\lambda)$ such that $y_n \longrightarrow y$. Then there is a sequence $\{x_n\}$ in X such that $T_\lambda x_n = y_n$. For any positive integers m, n, $\gamma\|x_m - x_n\| \leq \|T_\lambda x_m - T_\lambda x_n\| = \|y_m - y_n\|$. Since $\{y_n\}$ is Cauchy, $\{x_n\}$ must also be Cauchy. Let $x_n \longrightarrow x$. Then $y_n = T_\lambda x_n \longrightarrow T_\lambda x = y$. Thus, $y \in \mathfrak{R}(T_\lambda)$ and so $\mathfrak{R}(T_\lambda)$ is closed. This proves that T_λ is bijective. Finally, $\gamma\|T_\lambda^{-1}y\| \leq \|y\|$ for all $y \in X$ implies $T_\lambda^{-1} \in B(X, X)$ and $\|T_\lambda^{-1}\| \leq 1/\gamma$. This completes the proof of the theorem. ■

7.5. OTHER LINEAR OPERATORS: NORMAL OPERATORS, PROJECTIONS, UNITARY OPERATORS, AND ISOMETRIC OPERATORS

In this section we consider additional important types of linear operators. Throughout this section X is a complex Hilbert space, T^* denotes the adjoint of $T \in B(X, X)$, and $I \in B(X, X)$ denotes the identity operator.

7.5.1. Definition. An operator $T \in B(X, X)$ is said to be a **normal operator** if $T^*T = TT^*$.

7.5.2. Definition. An operator $T \in B(X, X)$ is said to be an **isometric operator** if $T^*T = I$.

7.5.3. Definition. An operator $T \in B(X, X)$ is said to be an **unitary operator** if $T^*T = TT^* = I$.

Our first result is for normal operators.

7.5.4. Theorem. Let $T \in B(X, X)$. Let $U, V \in B(X, X)$ be hermitian operators such that $T = U + iV$. Then T is normal if and only if $UV = VU$.

7.5.5. Exercise. Prove Theorem 7.5.4. Recall that U and V are unique by Theorem 7.4.15.

For the next result, recall that a linear subspace Y of X is invariant under a linear transformation T if $T(Y) \subset Y$ (see Definition 3.7.9). Also, recall that a closed linear subspace Y of a Hilbert space X is itself a Hilbert space with inner product induced by the inner product on X (see Theorem 6.2.1).

7.5.6. Theorem. Let $T \in B(X, X)$ be a normal operator, and let Y be a closed linear subspace of X which is invariant under T. Let T_1 be the restriction of T to Y. Then $T_1 \in B(Y, Y)$ and T_1 is normal.

7.5.7. Exercise. Prove Theorem 7.5.6.

For isometric operators we have the following result.

7.5.8. Theorem. Let $T \in B(X, X)$. Then the following are equivalent:

(i) T is isometric;

(ii) $(Tx, Ty) = (x, y)$ for all $x, y \in X$; and

(iii) $\|Tx - Ty\| = \|x - y\|$ for all $x, y \in X$.

Proof. If T is isometric, then $(x, y) = (Ix, y) = (T^*Tx, y) = (Tx, Ty)$ for all $x, y \in X$.

Next, assume that $(Tx, Ty) = (x, y)$. Then $\|Tx - Ty\|^2 = \|T(x - y)\|^2 = (T(x - y), T(x - y)) = ((x - y), (x - y)) = \|x - y\|^2$; i.e., $\|Tx - Ty\| = \|x - y\|$.

Finally, assume that $\|Tx - Ty\| = \|x - y\|$. Then $(T^*Tx, x) = (Tx, Tx) = \|Tx\|^2 = \|x\|^2 = (x, x)$; i.e., $(T^*Tx, x) = (x, x)$ for all $x \in X$. But this implies that $T^*T = I$; i.e., T is isometric. ■

From Theorem 7.5.8 there follows the following corollary.

7.5.9. Corollary. If $T \in B(X, X)$ is an isometric operator, then $\|Tx\| = \|x\|$ for all $x \in X$ and $\|T\| = 1$.

For unitary operators we have the following result.

7.5.10. Theorem. Let $T \in B(X, X)$. Then the following are equivalent:

(i) T is unitary;

(ii) T^* is unitary;

(iii) T and T^* are isometric;

(iv) T is isometric and T^* is injective;
(v) T is isometric and surjective; and
(vi) T is bijective and $T^{-1} = T^*$.

7.5.11. Exercise. Prove Theorem 7.5.10.

Before considering projections, let us briefly return to Section 3.7. Recall that if (a linear space) X is the direct sum of two linear subspaces X_1 and X_2, i.e., $X = X_1 \oplus X_2$, then for each $x \in X$ there exist unique $x_1 \in X_1$ and $x_2 \in X_2$ such that $x = x_1 + x_2$. We call a mapping $P: X \longrightarrow X$ defined by $Px = x_1$ the projection on X_1 along X_2. Recall that $P \in L(X, X)$, $\mathcal{R}(P) = X_1$, and $\mathfrak{N}(P) = X_2$. Furthermore, recall that if $P \in L(X, X)$ is such that $P^2 = P$, then P is said to be idempotent and this condition is both necessary and sufficient for P to be a projection on $\mathcal{R}(P)$ along $\mathfrak{N}(P)$ (see Theorem 3.7.4). Now if X is a Hilbert space and if $X_1 = Y$ is a closed linear subspace of X, then $X_2 = Y^\perp$ and $X = Y \oplus Y^\perp$ (see Theorem 6.12.16). If for this particular case P is the projection on Y along $Y^\perp$, then P is an orthogonal projection (see Definition 3.7.16). In this case we shall simply call P the orthogonal projection on Y.

7.5.12. Theorem. Let Y be a closed linear subspace of X such that $Y \neq \{0\}$ and $Y \neq X$. Let P be the orthogonal projection onto Y. Then

(i) $P \in B(X, X)$;
(ii) $\|P\| = 1$; and
(iii) $P^* = P$.

Proof. We know that $P \in L(X, X)$. To show that P is bounded let $x = x_1 + x_2$, where $x_1 \in Y$ and $x_2 \in Y^\perp$. Then $\|Px\| = \|x_1\| \leq \|x\|$. Hence, P is bounded and $\|P\| \leq 1$. If $x_2 = 0$, then $\|Px\| = \|x\|$ and so $\|P\| = 1$.

To prove (iii), let $x, y \in X$ be given by $x = x_1 + x_2$ and $y = y_1 + y_2$, respectively, where $x_1, y_1 \in Y$ and $x_2, y_2 \in Y^\perp$. Then $(x, Py) = (x_1 + x_2, y_1) = (x_1, y_1)$ and $(Px, y) = (x_1, y_1 + y_2) = (x_1, y_1)$. Thus, $(x, Py) = (Px, y)$ for all $x, y \in X$.

This implies that $P = P^*$. ■

From the above theorem it follows that an orthogonal projection is a hermitian operator.

7.5.13. Theorem. Let Y be a closed linear subspace of X, and let P be the orthogonal projection onto Y. If

$$Y_1 = \{x \in X: Px = x\}$$

and if Y_2 is the range of P, then $Y = Y_1 = Y_2$.

Proof. Since $Y_2 = Y$, since $Y \subset Y_1$, and since $Y_1 \subset Y_2$, it follows that $Y = Y_1 = Y_2$. ■

7.5.14. Theorem. Let $P \in L(X, X)$. If P is idempotent and hermitian, then

$$Y = \{x \in X: Px = x\}$$

is a closed linear subspace of X and P is the orthogonal projection onto Y.

Proof. Since P is a linear operator we have

$$P(\alpha x + \beta y) = \alpha Px + \beta Py.$$

If $x, y \in Y$, then $Px = x$ and $Py = y$, and it follows that

$$P(\alpha x + \beta y) = \alpha x + \beta y.$$

Therefore, $(\alpha x + \beta y) \in Y$ and Y is a linear subspace of X. We must show that Y is a closed linear subspace. First, however, we show that P is bounded and therefore continuous. Since

$$||Pz||^2 = (Pz, Pz) = (P^*Pz, z) = (P^2z, z) = (Pz, z) \leq ||Pz|| \, ||z||,$$

we have $||Pz|| \leq ||z||$ and $||P|| = 1$.

To show that Y is a closed linear subspace of X let x_0 be a point of accumulation of the space Y. Then there is a sequence of vectors $\{x_n\}$ in Y such that $\lim_n ||x_n - x_0|| = 0$. Since $x_n \in Y$, we can put $Px_n = x_n$ and we have $||Px_n - x_0|| \longrightarrow 0$ as $n \longrightarrow \infty$. Since P is bounded, it is continuous and thus we also have $||Px_n - Px_0|| \longrightarrow 0$ as $n \longrightarrow \infty$, and hence $x_0 \in Y$.

Finally, we must show that P is an orthogonal projection. Let $x \in Y$, and let $y \in Y^\perp$. Then $(Py, x) = (y, Px) = (y, x) = 0$, since $x \perp y$. Therefore, $Py \perp x$ and $Py \in Y^\perp$. But $P(Py) = Py$, since $P^2 = P$ and thus $Py \in Y$. Therefore, it follows that $Py = 0$, because $Py \in Y$ and $Py \in Y^\perp$. Now let $z = x + y \in X$, where $x \in Y$ and $y \in Y^\perp$. Then $Pz = Px + Py = x + 0 = x$. Hence, P is an orthogonal projection onto Y. ■

The next result is a direct consequence of Theorem 7.5.14.

7.5.15. Corollary. Let Y be a closed linear subspace of X, and let P be the orthogonal projection onto Y. Then $P(Y^\perp) = \{0\}$.

7.5.16. Exercise. Prove Corollary 7.5.15.

The next result yields the representation of an orthogonal projection onto a finite-dimensional subspace of X.

7.5.17. Theorem. Let $\{x_1, \ldots, x_n\}$ be a finite orthonormal set in X, and let Y be the linear subspace of X generated by $\{x_1, \ldots, x_n\}$. Then the orthogonal projection of X onto Y is given by

$$Px = \sum_{i=1}^{n} (x, x_i)x_i \text{ for all } x \in X.$$

Proof. We first note that Y is a closed linear subspace of X by Theorem 6.6.6. We now show that P is a projection by proving that $P^2 = P$. For any $j = 1, \ldots, n$ we have

$$Px_j = \sum_{i=1}^{n} (x_j, x_i)x_i = x_j. \tag{7.5.18}$$

Hence, for any $x \in X$ we have

$$\begin{aligned} P^2x &= P(Px) = P\Big(\sum_{i=1}^{n} (x, x_i)x_i\Big) = \sum_{i=1}^{n} (x, x_i)Px_i \\ &= \sum_{i=1}^{n} (x, x_i)x_i = Px. \end{aligned}$$

Next, we show that $\mathcal{R}(P) = Y$. It is clear that $\mathcal{R}(P) \subset Y$. To show that $Y \subset \mathcal{R}(P)$, let $y \in Y$. Then

$$y = \eta_1 x_1 + \ldots + \eta_n x_n$$

for some $\{\eta_1, \ldots, \eta_n\}$. It follows from Eq. (7.5.18) that $Py = y$ and so $y \in \mathcal{R}(P)$.

Finally, to show that P is an orthogonal projection, we must show that $\mathcal{R}(P) \perp \mathfrak{N}(P)$. To do so, let $x \in \mathfrak{N}(P)$ and let $y \in \mathcal{R}(P)$. Then

$$\begin{aligned} (x, y) &= (x, Py) = \Big(x, \sum_{i=1}^{n} (y, x_i)x_i\Big) = \sum_{i=1}^{n} \overline{(y, x_i)}(x, x_i) \\ &= \sum_{i=1}^{n} (x, x_i)(x_i, y) = \Big(\sum_{i=1}^{n} (x, x_i)x_i, y\Big) = (Px, y) \\ &= (0, y) = 0. \end{aligned}$$

This completes the proof. ■

Referring to Definition 3.7.12 we recall that if Y and Z are linear subspaces of (a linear space) X such that $X = Y \oplus Z$, and if $T \in L(X, X)$ is such that both Y and Z are invariant under T, then T is said to be reduced by Y and Z. When X is a Hilbert space, we make the following definition.

7.5.19. Definition. Let Y be a closed linear subspace of X, and let $T \in L(X, X)$. Then Y is said to **reduce** T if Y and $Y^\perp$ are invariant under T.

Note that in view of Theorem 6.12.16, Definitions 3.7.12 and 7.5.19 are consistent.

The proof of the next theorem is straightforward.

7.5.20. Theorem. Let Y be a closed linear subspace of X, and let $T \in B(X, X)$. Then

(i) Y is invariant under T if and only if $Y^\perp$ is invariant under T^*; and
(ii) Y reduces T if and only if Y is invariant under T and T^*.

7.5.21. Exercise. Prove Theorem 7.5.20.

7.5.22. Theorem. Let Y be a closed linear subspace of X, let P be the orthogonal projection onto Y, let $T \in B(X, X)$, and let I denote the identity operator on X. Then

(i) Y is invariant under T if and only if $TP = PTP$;
(ii) Y reduces T if and only if $TP = PT$; and
(iii) $(I - P)$ is the orthogonal projection onto $Y^\perp$.

Proof. To prove (i), assume that $TP = PTP$. Then for any $x \in Y$ we have $Tx = T(Px) = P(TPx) \in Y$, since P applied to any vector of X is in Y.

Conversely, if Y is invariant under T, then for any vector $x \in X$ we have $T(Px) \in Y$, because $Px \in Y$. Thus, $P(TPx) = TPx$ for every $x \in X$.

To prove (ii), assume that $PT = TP$. Then $PTP = P^2T = PT = TP$. Therefore, $PTP = TP$, and it follows from (i) that Y is invariant under T.

To prove that Y reduces T we must show that Y is invariant under T^*. Since P is hermitian we have $T^*P = (PT)^* = (TP)^* = P^*T^* = PT^*$; i.e., $T^*P = PT^*$. But above we showed that $PTP = TP$. Applying this to T^* we obtain $T^*P = PT^*P$. In view of (i), Y is now invariant under T^*. Therefore, the closed linear space reduces the linear operator T.

Conversely, assume that Y reduces T. By part (i), $TP = PTP$ and $T^*P = PT^*P$. Thus, $PT = (T^*P)^* = (PT^*P)^* = PTP = TP$; i.e., $TP = PT$.

To prove (iii) we first show that $(I - P)$ is hermitian. We note that $(I - P)^* = I^* - P^* = I - P$. Next, we show that $(I - P)$ is idempotent. We observe that $(I - P)^2 = (I - 2P + P^2) = (I - 2P + P) = (I - P)$.

Finally, we note that $(I - P)x = x$ if and only if $Px = 0$, which implies that $x \in Y^\perp$. Thus,

$$Y^\perp = \{x \in X: (I - P)x = x\}.$$

It follows from Theorem 7.5.14 that $(I - P)$ is a projection onto $Y^\perp$. ■

The next result follows immediately from part (iii) of the preceding theorem.

7.5.23. Theorem. Let Y be a closed linear subspace of X, and let P be the orthogonal projection on Y. If $\|Px\| = \|x\|$, then $Px = x$, and consequently $x \in Y$.

7.5.24. Exercise. Prove Theorem 7.5.23.

We leave the proof of the following result as an exercise.

7.5.25. Theorem. Let Y and Z be closed linear subspaces of X, and let P and Q be the orthogonal projections on Y and Z, respectively. Let 0 denote the zero transformation in $B(X, X)$. The following are equivalent:

(i) $Y \perp Z$;
(ii) $PQ = 0$;
(iii) $QP = 0$;
(iv) $P(Z) = \{0\}$; and
(v) $Q(Y) = \{0\}$.

7.5.26. Exercise. Prove Theorem 7.5.25.

For the product of two orthogonal projections we have the following result.

7.5.27. Theorem. Let Y_1 and Y_2 be closed linear subspaces of X, and let P_1 and P_2 be the orthogonal projections onto Y_1 and Y_2, respectively. The product transformation P_1P_2 is an orthogonal projection if and only if P_1 commutes with P_2. In this case the range of P_1P_2 is $Y_1 \cap Y_2$.

Proof. Assume that $P_1P_2 = P_2P_1$. Then $(P_1P_2)^* = P_2^*P_1^* = P_2P_1 = P_1P_2$; i.e., if $P_1P_2 = P_2P_1$ then $(P_1P_2)^* = (P_1P_2)$. Also, $(P_1P_2)^2 = P_1P_2P_1P_2 = P_1P_1P_2P_2 = P_1P_2$; i.e., if $P_1P_2 = P_2P_1$, then P_1P_2 is idempotent. Therefore, P_1P_2 is an orthogonal projection.

Conversely, assume that P_1P_2 is an orthogonal projection. Then $(P_1P_2)^* = P_2^*P_1^* = P_2P_1$ and also $(P_1P_2)^* = P_1P_2$. Hence, $P_1P_2 = P_2P_1$.

Finally, we must show that the range of P_1P_2 is equal to $Y_1 \cap Y_2$. Assume that $x \in \mathcal{R}(P_1P_2)$. Then $P_1P_2x = x$, because P_1P_2 is an orthogonal projection. Also, $P_1P_2x = P_1(P_2x) \in Y_1$, because any vector operated on by P_1 is in Y_1. Similarly, $P_2P_1x = P_2(P_1x) \in Y_2$. Now, by hypothesis, $P_1P_2 = P_2P_1$, and therefore $P_1P_2x = P_2P_1x = x \in Y_1 \cap Y_2$. Thus, whenever $x \in \mathcal{R}(P_1P_2)$, then $x \in Y_1 \cap Y_2$. This implies that $\mathcal{R}(P_1P_2) \subset Y_1 \cap Y_2$. To show that $\mathcal{R}(P_1P_2) \supset Y_1 \cap Y_2$, assume that $x \in Y_1 \cap Y_2$. Then $P_1P_2x = P_1(P_2x) = P_1x = x \in \mathcal{R}(P_1P_2)$. Thus, $Y_1 \cap Y_2 \subset \mathcal{R}(P_1P_2)$. Therefore, $\mathcal{R}(P_1P_2) = Y_1 \cap Y_2$. ■

7.5.28. Theorem. Let Y and Z be closed linear subspaces of X, and let P and Q be the orthogonal projections onto Y and Z, respectively. The following are equivalent:

(i) $P \leq Q$;
(ii) $\|Px\| \leq \|Qx\|$ for all $x \in X$;
(iii) $Y \subset Z$;
(iv) $QP = P$; and
(v) $PQ = P$.

Proof. Assume that $P \leq Q$. Since P and Q are orthogonal projections, they are hermitian. For a hermitian operator, $P \geq 0$ means $(Px, x) \geq 0$ for all $x \in X$. If $P \leq Q$, then $(Px, x) \leq (Qx, x)$ for all $x \in X$ or $(P^2x, x) \leq (Q^2x, x)$ or $(Px, Px) \leq (Qx, Qx)$ or $||Px||^2 \leq ||Qx||^2$, and hence $||Px|| \leq ||Qx||$ for all $x \in X$.

Next, assume that $||Px|| \leq ||Qx||$ for all $x \in X$. If $x \in Y$, then $Px = x$ and

$$(x, x) = (Px, Px) = ||Px||^2 \leq ||Qx||^2 \leq ||Q|| ||x||^2 = ||x||^2 = (x, x),$$

and therefore $||Qx|| = ||x||$. From Theorem 7.5.23 it now follows that $Qx = x$, and hence $x \in Z$. Thus, whenever $x \in Y$ then $x \in Z$ and $Z \supset Y$.

Now assume that $Z \supset Y$ and let $y = Px$, where x is any vector in X. Then $QPx = Qy = y = Px$ for all $x \in X$ and $QP = P$.

Suppose now that $QP = P$. Then $(QP)^* = P^*$, or $P^*Q^* = PQ = P^* = P$; i.e., $PQ = P$.

Finally, assume that $PQ = P$. For any $x \in X$ we have $(Px, x) = ||Px||^2 = ||PQx||^2 \leq ||P||^2 ||Qx||^2 = ||Qx||^2 = (Qx, Qx) = (Q^2x, x) = (Qx, x)$; i.e., $(Px, x) \leq (Qx, x)$ from which we have $P \leq Q$. ■

We leave the proof of the next result as an exercise.

7.5.29. Theorem. Let Y_1 and Y_2 be closed linear subspaces of X, and let P_1 and P_2 be the orthogonal projections onto Y_1 and Y_2, respectively. The difference transformation $P = P_1 - P_2$ is an orthogonal projection if and only if $P_2 \leq P_1$. The range of P is $Y_1 \cap Y_2^\perp$.

7.5.30. Exercise. Prove Theorem 7.5.29.

We close this section by considering some specific cases.

7.5.31. Example. Let R denote the transformation from E^2 into E^2 given in Example 4.10.48. That transformation is represented by the matrix

$$\mathbf{R}_\theta = \begin{bmatrix} \cos\theta & -\sin\theta \\ \sin\theta & \cos\theta \end{bmatrix}$$

with respect to an orthonormal basis $\{e_1, e_2\}$. By direct computation we obtain

$$\mathbf{R}_\theta^* = \begin{bmatrix} \cos\theta & \sin\theta \\ -\sin\theta & \cos\theta \end{bmatrix}.$$

It readily follows that $R^*R = RR^* = I$. Therefore, R is a linear transformation which is isometric, unitary, and normal. ■

7.5.32. Exercise. Let $X = L_2[0, \infty)$ and define the **truneation operator** P_T by $y = P_T x$, where

$$y(t) = \begin{cases} x(t) & \text{for all } 0 \leq t \leq T \\ 0 & \text{for all } t > T \end{cases}$$

Show that P_T is an orthogonal projection with range

$$\mathfrak{R}(P_T) = \{x \in X: x(t) = 0 \text{ for } t > T\},$$

and null space

$$\mathfrak{N}(P_T) = \{x \in X: x(t) = 0 \text{ for all } t \leq T\}.$$

Additional examples of different types of operators are considered in Section 7.10.

7.6. THE SPECTRUM OF AN OPERATOR

In Chapter 4 we introduced and discussed eigenvalues and eigenvectors of linear transformations defined on finite-dimensional vector spaces. In the present section we continue this discussion in the setting of infinite-dimensional spaces.

Unless otherwise stated, X will denote a complex Banach space and I will denote the identity operator on X. However, in our first definition, X may be an arbitrary vector space over a field F.

7.6.1. Definition. Let $T \in L(X, X)$. A scalar $\lambda \in F$ is called an **eigenvalue** of T if there exists an $x \in X$ such that $x \neq 0$ and such that $Tx = \lambda x$. Any vector $x \neq 0$ satisfying the equation $Tx = \lambda x$ is called an **eigenvector** of T corresponding to the eigenvalue λ.

7.6.2. Definition. Let X be a complex Banach space and let $T: X \longrightarrow X$. The set of all $\lambda \in F = C$ such that

(i) $\mathfrak{R}(T - \lambda I)$ is dense in X;
(ii) $(T - \lambda I)^{-1}$ exists; and
(iii) $(T - \lambda I)^{-1}$ is continuous (i.e., bounded)

is called the **resolvent set** of T and is denoted by $\rho(T)$. The complement of $\rho(T)$ is called the **spectrum** of T and is denoted by $\sigma(T)$.

The preceding definitions require some comments. First, note that if λ is an eigenvalue of T, there is an $x \neq 0$ such that $(T - \lambda I)x = 0$. From Theorem 3.4.32 this is true if and only if $(T - \lambda I)$ does not have an inverse. Hence, if λ is an eigenvalue of T, then $\lambda \in \sigma(T)$. Note, however, that there are other ways that a complex number λ may fail to be in $\rho(T)$. These possibilities are enumerated in the following definition.

7.6.3. Definition. The set of all eigenvalues of T is called the **point spectrum** of T. The set of all λ such that $(T - \lambda I)^{-1}$ exists but $\mathcal{R}(T - \lambda I)$ is not dense in X is called the **residual spectrum** of T. The set of all λ such that $(T - \lambda I)^{-1}$ exists and such that $\mathcal{R}(T - \lambda I)$ is dense in X but $(T - \lambda I)^{-1}$ is not continuous is called the **continuous spectrum**. We denote these sets by $P\sigma(T)$, $R\sigma(T)$, and $C\sigma(T)$, respectively.

Clearly, $\sigma(T) = P\sigma(T) \cup C\sigma(T) \cup R\sigma(T)$. Furthermore, when X is finite dimensional, then $\sigma(T) = P\sigma(T)$. We summarize the preceding definition in the following table.

	$(T - \lambda I)^{-1}$ exists *and* $(T - \lambda I)^{-1}$ is continuous	$(T - \lambda I)^{-1}$ exists *but* $(T - \lambda I)^{-1}$ is not continuous	$(T - \lambda I)^{-1}$ does not exist
$\overline{\mathcal{R}(T - \lambda I)} = X$	$\lambda \in \rho(T)$	$\lambda \in C\sigma(T)$	$\lambda \in P\sigma(T)$
$\overline{\mathcal{R}(T - \lambda I)} \neq X$	$\lambda \in R\sigma(T)$	$\lambda \in R\sigma(T)$	$\lambda \in P\sigma(T)$

7.6.4. Table A. Characterization of the resolvent set and the spectrum of an operator

7.6.5. Example. Let $X = l_2$ be the Hilbert space of Example 6.11.9, let $x = (\xi_1, \xi_2, \ldots) \in X$, and define $T \in B(X, X)$ by

$$Tx = (\xi_1, \tfrac{1}{2}\xi_2, \tfrac{1}{3}\xi_3, \ldots).$$

For each $\lambda \in C$ we want to determine (a) whether $(T - \lambda I)^{-1}$ exists; (b) if so, whether $(T - \lambda I)^{-1}$ is continuous; and (c) whether $\overline{\mathcal{R}(T - \lambda I)} = X$.

First we consider the point spectrum of T. If $Tx = \lambda x$ then $\left(\frac{1}{k} - \lambda\right)\xi_k = 0$, $k = 1, 2, \ldots$. This holds for non-trivial x if and only if $\lambda = 1/k$ for some k. Hence,

$$P\sigma(T) = \left\{\frac{1}{k} : k = 1, 2, \ldots\right\}.$$

Next, assume that $\lambda \notin P\sigma(T)$, so that $(T - \lambda I)^{-1}$ exists, and let us investigate the continuity of $(T - \lambda I)^{-1}$. We see that if $y = (\eta_1, \eta_2, \ldots) \in \mathcal{R}(T - \lambda I)$, then $(T - \lambda I)^{-1}y = x$ is given by

$$\xi_k = \frac{\eta_k}{\frac{1}{k} - \lambda} = \frac{k\eta_k}{1 - \lambda k}.$$

Now if $\lambda = 0$, then $\|(T - \lambda I)^{-1}y\|^2 = \sum_{k=1}^{\infty} k^2\eta_k^2$ and $(T - \lambda I)^{-1}$ is not bounded and hence not continuous. On the other hand, if $\lambda \neq 0$, then $(T - \lambda I)^{-1}$ is continuous since $|\xi_k| \leq \gamma|\eta_k|$ for all k, where

$$\gamma = \sup_k \frac{1}{\left|\lambda - \frac{1}{k}\right|} < \infty, \quad \lambda \notin P\sigma(T).$$

Next, let us examine $\mathfrak{R}(T - \lambda I)$. It is clear that if $y = (\eta_1, \eta_2, \ldots, \eta_n, 0, 0, \ldots)$, then there is an $x \in X$ such that $(T - \lambda I)x = y$ for any λ. This implies that $\mathfrak{R}(T - \lambda I)$ is dense in X; i.e., $\overline{\mathfrak{R}(T - \lambda I)} = X$. Note, however, that if $\lambda = 0$, then $\mathfrak{R}(T - \lambda I) \neq X$. That is, $\mathfrak{R}(T) \neq X$, but $\overline{\mathfrak{R}(T)} = X$. For example, $y = (1, 1/2, 1/3, \ldots) \notin \mathfrak{R}(T)$; however, $(1, 1/2, \ldots, 1/k, 0, 0, \ldots) \in \mathfrak{R}(T)$ for any k, and hence $y \in \overline{\mathfrak{R}(T)}$.

Summarizing, we have

$$P\sigma(T) = \left\{\lambda : \lambda = \frac{1}{k}, k = 1, 2, \ldots\right\},$$

$$C\sigma(T) = \{\lambda : \lambda = 0\},$$

$$R\sigma(T) = \varnothing,$$

and

$$\rho(T) = [P\sigma(T) \cup C\sigma(T)]^{\sim}. \quad \blacksquare$$

7.6.6. Exercise. Let $X = l_2$, the Hilbert space of Example 6.11.9, let $x = (\xi_1, \xi_2, \xi_3, \ldots)$, and define the **right shift operator** $T_r: X \longrightarrow X$ and the **left shift operator** $T_l: X \longrightarrow X$ by

$$y = T_r x = (0, \xi_1, \xi_2, \ldots)$$

and

$$y = T_l x = (\xi_2, \xi_3, \xi_4, \ldots),$$

respectively. Show that

$$\rho(T_r) = \rho(T_l) = \{\lambda \in C : |\lambda| > 1\},$$

$$C\sigma(T_r) = C\sigma(T_l) = \{\lambda \in C : |\lambda| = 1\},$$

$$R\sigma(T_r) = P\sigma(T_l) = \{\lambda \in C : |\lambda| < 1\},$$

$$P\sigma(T_r) = R\sigma(T_r) = \varnothing.$$

We now examine some of the properties of the resolvent set and the spectrum.

7.6.7. Theorem. Let $T \in B(X, X)$. If $|\lambda| > \|T\|$, then $\lambda \in \rho(T)$ or, equivalently, if $\lambda \in \sigma(T)$, then $|\lambda| \leq \|T\|$.

7.6.8. Exercise. Prove Theorem 7.6.7 (use Theorem 7.2.2).

7.6.9. Theorem. Let $T \in B(X, X)$. Then $\rho(T)$ is open and $\sigma(T)$ is closed.

Proof. Since $\sigma(T)$ is the complement of $\rho(T)$, it is closed if and only if $\rho(T)$ is open. Let $\lambda_0 \in \rho(T)$. Then $(T - \lambda_0 I)$ has a continuous inverse. For arbitrary λ we now have

$$\begin{aligned}
&||I - (T - \lambda_0 I)^{-1}(T - \lambda I)|| \\
&\qquad = ||(T - \lambda_0 I)^{-1}(T - \lambda_0 I) - (T - \lambda_0 I)^{-1}(T - \lambda I)|| \\
&\qquad = ||(T - \lambda_0 I)^{-1}[(T - \lambda_0 I) - (T - \lambda I)]|| \\
&\qquad = ||(\lambda - \lambda_0)(T - \lambda_0 I)^{-1}|| \\
&\qquad = |\lambda - \lambda_0| ||(T - \lambda_0 I)^{-1}||.
\end{aligned}$$

Now for $|\lambda - \lambda_0|$ sufficiently small, we have

$$||I - (T - \lambda_0 I)^{-1}(T - \lambda I)|| = |\lambda - \lambda_0| ||(T - \lambda_0)^{-1}|| < 1.$$

Now in Theorem 7.2.2 we showed that if $T \in B(X, X)$, then T has a continuous inverse if $||I - T|| < 1$. In our case it now follows that $(T - \lambda_0 I)^{-1}(T - \lambda I)$ has a continuous inverse, and therefore $(T - \lambda I)$ has a continuous inverse whenever $|\lambda - \lambda_0|$ is sufficiently small. This implies that $\lambda \in \rho(T)$ and $\rho(T)$ is open. Hence, $\sigma(T)$ is closed. ■

For normal, hermitian, and isometric operators we have the following result.

7.6.10. Theorem. Let X be a Hilbert space, let $T \in B(X, X)$, let λ be an eigenvalue of T, and let $Tx = \lambda x$. Then

(i) if T is hermitian then λ is real;
(ii) if T is isometric, then $|\lambda| = 1$;
(iii) if T is normal, then $\bar{\lambda}$ is an eigenvalue of T^* and $T^*x = \bar{\lambda}x$; and
(iv) if T is normal, if μ is an eigenvalue of T such that $\mu \neq \lambda$, and if $Ty = \mu y$, then $x \perp y$.

Proof. Without loss of generality, assume that x is a unit vector.

To prove (i) note that $\lambda = \lambda||x||^2 = \lambda(x, x) = (\lambda x, x) = (Tx, x)$, which is real by Theorem 7.4.5. Therefore, $(Tx, x) = (x, Tx) = \overline{(Tx, x)} = \bar{\lambda}$; i.e., $\lambda = \bar{\lambda}$ and λ is real.

To verify (ii), note that if T is isometric, then $||Tx|| = ||x|| = 1$, by Corollary 7.5.9. Since $Tx = \lambda x$ it follows that $||\lambda x|| = 1$ or $|\lambda| ||x|| = 1$, and hence $|\lambda| = 1$.

To prove (iii), assume that T is normal; i.e., $T^*T = TT^*$. Then

$$\begin{aligned}(T - \lambda I)(T - \lambda I)^* &= (T - \lambda I)(T^* - \bar{\lambda} I)\\ &= (T - \lambda I)T^* - (T - \lambda I)\bar{\lambda} I\\ &= TT^* - \lambda T^* - \bar{\lambda} T + \lambda\bar{\lambda} I\\ &= T^*T - \bar{\lambda} T - \lambda T^* + \lambda\bar{\lambda} I\\ &= (T^* - \bar{\lambda} I)(T - \lambda I) = (T - \lambda I)^*(T - \lambda I);\end{aligned}$$

i.e.,

$$(T - \lambda I)(T - \lambda I)^* = (T - \lambda I)^*(T - \lambda I),$$

and $(T - \lambda I)$ is normal. Also, we can readily verify that $\|(T - \lambda I)x\| = \|(T - \lambda I)^*x\|$. Since $(T - \lambda I)x = 0$, it follows that $(T - \lambda I)^*x = 0$, or $(T^* - \bar{\lambda} I)x = 0$, or $T^*x = \bar{\lambda} x$. Therefore, $\bar{\lambda}$ is an eigenvalue of T^* with eigenvector x.

To prove the last part assume that $\lambda \neq \mu$ and that T is normal. Then

$$\begin{aligned}(\lambda - \mu)(x, y) &= \lambda(x, y) - \mu(x, y) = (\lambda x, y) - (x, \bar{\mu} y)\\ &= (Tx, y) - (x, T^*y) = (Tx, y) - (Tx, y) = 0;\end{aligned}$$

i.e., $(\lambda - \mu)(x, y) = 0$. Since $\lambda \neq \mu$ we have $x \perp y$. ■

The next two results indicate what happens to the spectrum of an operator T when it is subjected to various elementary transformations.

7.6.11. Theorem. Let $T \in B(X, X)$, and let $p(T)$ denote a polynomial in T. Then

$$\sigma(p(T)) = p(\sigma(T)) = \{p(\lambda): \lambda \in \sigma(T)\}.$$

7.6.12. Exercise. Prove Theorem 7.6.11.

7.6.13. Theorem. Let $T \in B(X, X)$ be a bijective mapping. Then

$$\sigma(T^{-1}) = [\sigma(T)]^{-1} \triangleq \left\{\frac{1}{\lambda}: \lambda \in \sigma(T)\right\}.$$

Proof. Since T^{-1} exists, $0 \notin \sigma(T)$ and so the definition of $[\sigma(T)]^{-1}$ makes sense. Now for any $\lambda \neq 0$, consider the identity

$$\left(T^{-1} - \frac{1}{\lambda} I\right) = (\lambda I - T)\frac{1}{\lambda} T^{-1}.$$

It follows that if $\lambda \notin \sigma(T)$, then $\left(T^{-1} - \frac{1}{\lambda} I\right)$ has a continuous inverse; i.e., $\lambda \notin \sigma(T)$ implies that $\frac{1}{\lambda} \notin \sigma(T^{-1})$. In other words, $\sigma(T^{-1}) \subset [\sigma(T)]^{-1}$. To prove that $[\sigma(T)]^{-1} \subset \sigma(T^{-1})$ we proceed similarly, interchanging the roles of T and T^{-1}. ■

We now introduce the concept of the approximate point spectrum of an operator.

7.6.14. Definition. Let $T \in B(X, X)$. Then $\lambda \in C$ is said to belong to the **approximate point spectrum** of T if for every $\epsilon > 0$ there exists a non-zero vector $x \in X$ such that $\|Tx - \lambda x\| < \epsilon\|x\|$. We denote the approximate point spectrum by $\pi(T)$. If $\lambda \in \pi(T)$, then λ is called an **approximate eigenvalue** of T.

Clearly, $P\sigma(T) \subset \pi(T)$. Other properties of $\pi(T)$ are as follows.

7.6.15. Theorem. Let X be a Hilbert space, and let $T \in B(X, X)$. Then $\pi(T) \subset \sigma(T)$.

Proof. Assume that $\lambda \notin \sigma(T)$. Then $(T - \lambda I)$ has a continuous inverse, and for any $x \in X$ we have

$$\|x\| = \|(T - \lambda I)^{-1}(T - \lambda I)x\| \leq \|(T - \lambda I)^{-1}\| \|(T - \lambda I)x\|.$$

Now let $\epsilon = 1/\|(T - \lambda I)^{-1}\|$. Then we have, from above, $\|Tx - \lambda x\| \geq \epsilon\|x\|$ for every $x \in X$ and $\lambda \notin \pi(T)$. Therefore, $\sigma(T) \supset \pi(T)$. ■

We leave the proof of the next result as an exercise.

7.6.16. Theorem. Let X be a Hilbert space, and let $T \in B(X, X)$ be a normal operator. Then $\pi(T) = \sigma(T)$.

7.6.17. Exercise. Prove Theorem 7.6.16.

We can use the approximate point spectrum to establish some of the properties of the spectrum of hermitian operators.

7.6.18. Theorem. Let X be a Hilbert space, and let $T \in B(X, X)$ be hermitian. Then

(i) $\sigma(T)$ is a subset of the real line;
(ii) $\|T\| = \sup \{|\lambda| : \lambda \in \sigma(T)\}$; and
(iii) $\sigma(T)$ is not empty and either $+\|T\|$ or $-\|T\|$ belongs to $\sigma(T)$.

Proof. To prove (i), note that if T is hermitian it is normal and $\sigma(T) = \pi(T)$. Let $\lambda \in \pi(T)$, and assume that $\lambda \neq 0$ is complex. Then for any $x \neq 0$ we have

$$\begin{aligned} 0 < |\lambda - \bar{\lambda}| \|x\|^2 &= |\lambda - \bar{\lambda}|(x, x) = |((T - \lambda I)x, x) - ((T - \bar{\lambda} I)x, x)| \\ &\leq |((T - \lambda I)x, x)| + |((T - \bar{\lambda} I)x, x)| \leq \|(T - \lambda I)x\| \|x\| \\ &\qquad + \|(T - \bar{\lambda} I)x\| \|x\| \\ &= 2\|(T - \lambda I)x\| \|x\|; \end{aligned}$$

i.e.,

$$0 < |\lambda - \bar{\lambda}| \|x\| \leq 2\|(T - \lambda I)x\|$$

for all $x \in X$. But this implies that $\lambda \notin \pi(T)$, contrary to the original assumption. Hence, it must follow that $\lambda = \bar{\lambda}$, which implies that λ is real.

To prove (ii), first note that $||T|| \geq \sup \{|\lambda|: \lambda \in \sigma(T)\}$ for any $T \in B(X, X)$ (see Theorem 7.6.7). To show that equality holds, if T is hermitian, we first must show that $||T||^2 \in \pi(T^2) = \sigma(T^2)$. For all real λ and all $x \in X$ we can write

$$||T^2x - \lambda^2 x||^2 = (T^2x - \lambda^2 x, T^2x - \lambda^2 x)$$
$$= (T^2x, T^2x) - (T^2x, \lambda^2 x) - (\lambda^2 x, T^2x) + \lambda^4(x, x).$$

Since $(T^2x, x) = (Tx, T^*x) = (Tx, Tx)$, we now have

$$(T^2x - \lambda^2 x, T^2x - \lambda^2 x) = (T^2x, T^2x) - 2\lambda^2(Tx, Tx) + \lambda^4(x, x),$$

or

$$||T^2x - \lambda^2 x||^2 = ||T^2x||^2 - 2\lambda^2||Tx||^2 + \lambda^4||x||^2. \qquad (7.6.19)$$

Now let $\{x_n\}$ be a sequence of unit vectors such that $||Tx_n|| \longrightarrow ||T||$. If $\lambda = ||T||$, then we have, from Eq. (7.6.19),

$$||T^2x_n - \lambda^2 x_n||^2 = ||T^2x_n||^2 - 2\lambda^2||Tx_n||^2 + \lambda^4$$
$$\leq (||T|| \, ||Tx_n||)^2 - 2\lambda^2||Tx_n||^2 + \lambda^4 = \lambda^2||Tx_n||^2 - 2\lambda^2||Tx_n||^2 + \lambda^4$$
$$= \lambda^4 - \lambda^2||Tx_n||^2 \longrightarrow 0 \text{ as } n \longrightarrow \infty;$$

i.e., $||T^2x_n - \lambda^2 x_n|| \longrightarrow 0$ as $n \longrightarrow \infty$, and thus $\lambda^2 \in \pi(T^2) = \sigma(T^2)$.

Using Theorems 7.6.11 and 7.6.15 and the fact that $\lambda^2 \in \pi(T^2)$, it now follows that

$$||T|| = \sup \{|\lambda|: \lambda \in \sigma(T)\}.$$

The proof of (iii) is left as an exercise. ■

7.6.20. Exercise. Prove part (iii) of Theorem 7.6.18.

7.6.21. Theorem. Let X be a Hilbert space, and let $T \in B(X, X)$. Then $\pi(T)$ is closed.

7.6.22. Exercise. Prove Theorem 7.6.21.

In the following we let $T \in B(X, X)$, $\lambda \in C$, and we let $\mathfrak{N}_\lambda(T)$ be the null space of $T - \lambda I$; i.e.,

$$\mathfrak{N}_\lambda(T) = \{x \in X: (T - \lambda I)x = 0\} = \mathfrak{N}(T - \lambda I). \qquad (7.6.23)$$

It follows from Theorem 7.1.26 that $\mathfrak{N}_\lambda(T)$ is a closed linear subspace of X. For the next result, recall Definition 3.7.9 for the meaning of an invariant subspace.

7.6.24. Theorem. Let X be a Hilbert space, let $\lambda \in C$, and let $S, T \in B(X, X)$. If $ST = TS$, then $\mathfrak{N}_\lambda(T)$ is invariant under S.

Proof. Let $x \in \mathfrak{N}_\lambda(T)$. We want to show that $Sx \in \mathfrak{N}_\lambda(T)$; i.e., $TSx = \lambda Sx$. Since $x \in \mathfrak{N}_\lambda(T)$, we have $Tx = \lambda x$. Thus, $STx = \lambda Sx$. Since $ST = TS$, we have $TSx = \lambda Sx$. ∎

7.6.25. Corollary. $\mathfrak{N}_\lambda(T)$ is invariant under T.

Proof. Since $TT = TT$, the result follows from Theorem 7.6.24. ∎

For the next result, recall Definition 7.5.19.

7.6.26. Theorem. Let X be a Hilbert space, let $\lambda \in C$, and let $T \in B(X, X)$. If T is normal, then

(i) $\mathfrak{N}_\lambda(T) = \mathfrak{N}_{\bar{\lambda}}(T^*)$;
(ii) $\mathfrak{N}_\lambda(T) \perp \mathfrak{N}_\mu(T)$ if $\lambda \neq \mu$; and
(iii) $\mathfrak{N}_\lambda(T)$ reduces T.

Proof. The proofs of parts (i) and (ii) are left as an exercise.

To prove (iii), we see that $\mathfrak{N}_\lambda(T)$ is invariant under T from Corollary 7.6.25. To prove that $\mathfrak{N}_\lambda(T)^\perp$ is invariant under T, let $y \in \mathfrak{N}_\lambda(T)^\perp$. We want to show that $(x, Ty) = 0$ for all $x \in \mathfrak{N}_\lambda(T)$. If $x \in \mathfrak{N}_\lambda(T)$, we have $Tx = \lambda x$, and so, by part (i), $T^*x = \bar{\lambda}x$. Now $(x, Ty) = (T^*x, y) = (\bar{\lambda}x, y) = \bar{\lambda}(x, y) = 0$. This implies that $Ty \in \mathfrak{N}_\lambda(T)^\perp$, and so $\mathfrak{N}_\lambda(T)^\perp$ is invariant under T. This completes the proof of part (iii). ∎

7.6.27. Exercise. Prove parts (i) and (ii) of Theorem 7.6.26.

Before considering the last result of this section, we make the following definition.

7.6.28. Definition. A family of closed linear subspaces in a Hilbert space X is said to be **total** if the only vector $y \in X$ orthogonal to each member of the family is $y = 0$.

7.6.29. Theorem. Let X be a Hilbert space and let $S, T \in B(X, X)$. If the family of closed linear subspaces of X given by $\{\mathfrak{N}_\lambda(T): \lambda \in C\}$ is total, then $TS = ST$ if and only if $\mathfrak{N}_\lambda(T)$ is invariant under S for all $\lambda \in C$.

Proof. The necessity follows from Theorem 7.6.24. To prove sufficiency, assume that $\mathfrak{N}_\lambda(T)$ is invariant under S for all $\lambda \in C$. Let $\mathfrak{N}$ denote the null space of $TS - ST$; i.e., $\mathfrak{N} = \mathfrak{N}(TS - ST)$. If $x \in \mathfrak{N}_\lambda(T)$, then $Sx \in \mathfrak{N}_\lambda(T)$ by hypothesis. Hence, $TSx = T(Sx) = \lambda(Sx) = S(\lambda x) = S(Tx) = STx$ for all $x \in \mathfrak{N}_\lambda(T)$. Thus, $(TS - ST)x = 0$ for any $x \in \mathfrak{N}_\lambda(T)$, and so $\mathfrak{N}_\lambda(T) \subset \mathfrak{N}$. If there is a vector $y \perp \mathfrak{N}$, then it follows that $y \perp \mathfrak{N}_\lambda(T)$ for all $\lambda \in C$. By hypothesis, the family $\{\mathfrak{N}_\lambda(T): \lambda \in C\}$ is total, and thus $y = 0$. It follows that $\mathfrak{N}^\perp = \{0\}$ and $\mathfrak{N}^{\perp\perp} = \{0\}^\perp$ and $\mathfrak{N}^{\perp\perp} = \mathfrak{N}$, because $\mathfrak{N}$ is a closed linear subspace of X. Therefore, $\mathfrak{N} = X$; i.e., $(TS - ST)x = 0$ for all $x \in X$. Hence, $TS = ST$. ∎

7.7. COMPLETELY CONTINUOUS OPERATORS

Throughout this section X is a normed linear space over the field of complex numbers C.

Recall that a set $Y \subset X$ is bounded if there is a constant k such that for all $x \in Y$ we have $\|x\| \leq k$. Also, recall that a set Y is relatively compact if each sequence $\{x_n\}$ of elements chosen from Y contains a convergent subsequence (see Definition 5.6.30 and Theorem 5.6.31). When Y contains only a finite number of elements then any sequence constructed from Y must include some elements infinitely many times, and thus Y contains a convergent subsequence. From this it follows that any set containing a finite number of elements is relatively compact. Every relatively compact set is contained in a compact set and hence is bounded. For the finite-dimensional case it is also true that every bounded set is relatively compact (e.g., in R^n the Bolzano-Weierstrass theorem guarantees this). However, in the infinite-dimensional case it does not follow that every bounded set is also relatively compact.

In analysis and in applications linear operators which transform bounded sets into relatively compact sets are of great importance. Such operators are called **completely continuous operators** or **compact operators**. We give the following formal definition.

7.7.1. Definition. Let X and Y be normed linear spaces, and let T be a linear transformation with domain X and range in Y. Then T is said to be **completely continuous** or **compact** if for each bounded sequence $\{x_n\}$ in X, the sequence $\{Tx_n\}$ contains a subsequence converging to some element of $y \in Y$.

We have the following equivalent characterization of a completely continuous operator.

7.7.2. Theorem. Let X and Y be normed linear spaces, and let $T \in B(X, Y)$. Then T is completely continuous if and only if the sequence $\{Tx_n\}$ contains a subsequence convergent to some $y \in Y$ for all sequences $\{x_n\}$ such that $\|x_n\| \leq 1$ for all n.

7.7.3. Exercise. Prove Theorem 7.7.2.

Clearly, if an operator T is completely continuous, then it is continuous. On the other hand, the fact that T may be continuous does not ensure that it is completely continuous.

We now cite some examples.

7.7.4. Example. Let $T: X \longrightarrow X$ be the zero operator; i.e., $Tx = 0$ for all $x \in X$. Then T is clearly completely continuous. ■

7.7.5. Example. Let $X = \mathcal{C}[a, b]$, and let $\|\cdot\|_\infty$ be the norm on $\mathcal{C}[a, b]$ as defined in Example 6.1.9. Let $k: [a, b] \times [a, b] \longrightarrow R$ be a real-valued function continuous on the square $a \leq s \leq b, a \leq t \leq b$. Defining $T: X \longrightarrow X$ by

$$[Tx](s) = \int_a^b k(s, t)x(t)dt$$

for all $x \in X$, we saw in Example 7.1.20 that T is a bounded linear operator. We now show that T is completely continuous.

Let $\{x_n\}$ be a bounded sequence in X; i.e., there is a $K > 0$ such that $\|x_n\|_\infty \leq K$ for all n. It readily follows that if $y_n = Tx_n$, then $\|y_n\| \leq \gamma_0 \|x_n\|$, where $\gamma_0 = \sup_{a \leq s \leq b} \int_a^b |k(s, t)|\, dt$ (see Example 7.1.20). We now show that $\{y_n\}$ is an equicontinuous set of functions on $[a, b]$ (see Definition 5.8.11). Let $\epsilon > 0$. Then, because of the uniform continuity of k on $[a, b] \times [a, b]$, there is a $\delta > 0$ such that $|k(s_1, t) - k(s_2, t)| < \dfrac{\epsilon}{K(b - a)}$ if $|s_1 - s_2| < \delta$ for every $t \in [a, b]$. Thus

$$|y_n(s_1) - y_n(s_2)| \leq \int_a^b |k(s_1, t) - k(s_2, t)||x(t)|\, dt < \epsilon$$

for all n and all s_1, s_2 such that $|s_1 - s_2| < \delta$. This implies the set $\{y_n\}$ is equicontinuous, and so by the Arzela-Ascoli theorem (Theorem 5.8.12), the set $\{y_n\}$ is relatively compact in $\mathcal{C}[a, b]$; i.e., it has a convergent subsequence. This implies that T is completely continuous.

It can be shown that if $X = L_2[a, b]$ and if T is the Fredholm operator defined in Example 7.3.11, then T is also a completely continuous operator. ■

The next result provides us with an example of a continuous linear transformation which is not completely continuous.

7.7.6. Theorem. Let $I \in B(X, X)$ denote the identity operator on X. Then I is completely continuous if and only if X is finite dimensional.

Proof. The proof is an immediate consequence of Theorem 6.6.10. ■

We now consider some of the general properties of completely continuous operators.

7.7.7. Theorem. Let X and Y be normed linear spaces, let $S, T \in B(X, Y)$ be completely continuous operators, and let $\alpha, \beta \in C$. Then the operator $(\alpha S + \beta T)$ is completely continuous.

Proof. Given a sequence $\{x_n\}$ with $\|x_n\| \leq 1$, there is a subsequence $\{x_{n_k}\}$ such that the sequence $\{Sx_{n_k}\}$ has a limit u; i.e., $Sx_{n_k} \longrightarrow u$. From the sequence $\{x_{n_k}\}$ we pick another subsequence $\{x_{n_{k_j}}\}$ such that $Tx_{n_{k_j}} \longrightarrow v$. Then

$$(\alpha S + \beta T)x_{n_{k_j}} = \alpha S x_{n_{k_j}} + \beta T x_{n_{k_j}} \longrightarrow \alpha u + \beta v$$

as $n_k, n_{k_j} \longrightarrow \infty$. ■

We leave the proofs of the next results as an exercise.

7.7.8. Theorem. Let $T \in B(X, X)$ be completely continuous. Let Y be a closed linear subspace of X which is invariant under T. Let T_1 be the restriction of T to Y. Then $T_1 \in B(Y, Y)$ and T_1 is completely continuous.

7.7.9. Exercise. Prove Theorem 7.7.8.

7.7.10. Theorem. Let $T \in B(X, X)$ be a completely continuous operator, and let $S \in B(X, X)$ be any bounded linear operator. Then ST and TS are completely continuous.

7.7.11. Exercise. Prove Theorem 7.7.10.

7.7.12. Corollary. Let X and Y be normed linear spaces, and let $T \in B(X, Y)$ and $S \in B(Y, X)$. If T is completely continuous, then ST is completely continuous.

7.7.13. Exercise. Prove Corollary 7.7.12.

7.7.14. Example. A consequence of the above corollary is that if $T \in B(X, X)$ is completely continuous and X is infinite dimensional, then T cannot be a bijective mapping of X onto X. For, suppose T were bijective. Then we would have $T^{-1}T = I$. By the Banach inverse theorem (see Theorem 7.2.6) T^{-1} would then be continuous, and by the preceding theorem the identity mapping would be completely continuous. However, according to Theorem 7.7.6, this is possible only when X is finite dimensional.

Pursuing this example further, let $X = \mathcal{C}[a, b]$ with $\|\cdot\|_\infty$ as defined in Example 6.1.9. Let $T: X \longrightarrow X$ be defined by $Tx(t) = \int_a^t x(\tau)d\tau$ for $a \leq t \leq b$ and $x \in X$. It is easily shown that T is a completely continuous operator on X. It is, however, not bijective since $\mathcal{R}(T)$ is the family of all functions which are continuously differentiable in X, and thus $\mathcal{R}(T)$ is clearly a proper subset of X. The operator T is injective, since $Tx = 0$ implies $x = 0$. The inverse T^{-1} is given by $T^{-1}y(t) = dy(t)/dt$ for $y \in \mathcal{R}(T)$ and $a \leq t \leq b$. We saw in Example 5.7.4 that T^{-1} is not continuous. ■

In our next result we require the following definition.

7.7.15. Definition. Let X and Y be normed linear spaces, and let $T \in B(X, Y)$. The operator T is said to be **finite dimensional** if $T(X)$ is finite dimensional; i.e., the range of T is finite dimensional.

7.7.16. Theorem. Let X and Y be normed linear spaces, and let $T \in B(X, Y)$. If T is a finite-dimensional operator, then it is a completely continuous operator.

Proof. Let $\{x_n\}$ be a sequence in X such that $\|x_n\| \leq 1$ for all n. Then $\{Tx_n\}$ is a bounded sequence in $T(X)$. It follows from Theorem 6.6.10 that the set $\{Tx_n\}$ is relatively compact, and as such this set has a convergent subsequence in $T(X)$. It follows from Theorem 7.7.2 that T is completely continuous. ■

The proof of the next result utilizes what is called the **diagonalization process.**

7.7.17. Theorem. Let X and Y be Banach spaces, and let $\{T_n\}$ be a sequence of completely continuous operators mapping X into Y. If the sequence $\{T_n\}$ converges in norm to an operator T, then T is completely continuous.

Proof. Let $\{x_n\}$ be an arbitrary sequence in X with $\|x_n\| \leq 1$. We must show that the sequence $\{Tx_n\}$ contains a convergent subsequence.

By assumption, T_1 is a completely continuous operator, and thus we can select a convergent subsequence from the sequence $\{T_1 x_n\}$. Let

$$x_{11}, x_{21}, x_{31}, \ldots, x_{n1}, \ldots$$

denote the inverse images of the members of this convergent subsequence. Next, let us apply T_2 to each member of the above subsequence. Since T_2 is completely continuous, we can again select a convergent subsequence from the sequence $\{T_2 x_{n1}\}$. The inverse images of the terms of this sequence are

$$x_{12}, x_{22}, x_{32}, \ldots, x_{n2}, \ldots.$$

Continuing this process we can generate the array

$$\begin{array}{l} x_{11}, x_{21}, x_{31}, x_{41}, x_{51}, \ldots \\ x_{12}, x_{22}, x_{32}, x_{42}, x_{52}, \ldots \\ x_{13}, x_{23}, x_{33}, x_{43}, x_{53}, \ldots \\ \ldots\ldots\ldots\ldots\ldots\ldots\ldots\ldots \end{array}$$

Using this array, let us now form the diagonal sequence

$$x_{11}, x_{22}, x_{33}, \ldots.$$

Now each of the operators $T_1, T_2, T_3, \ldots, T_n, \ldots$ transforms this sequence into a convergent sequence. To show that T is completely continuous we must show that T also transforms this sequence into a convergent sequence. Now

$$\|Tx_{nn} - Tx_{mm}\| = \|Tx_{nn} - T_k x_{nn} + T_k x_{nn} - T_k x_{mm} + T_k x_{mm} - Tx_{mm}\|$$
$$\leq \|Tx_{nn} - T_k x_{nn}\| + \|T_k x_{nn} - T_k x_{mm}\| + \|T_k x_{mm} - Tx_{mm}\|$$
$$\leq \|T - T_k\|(\|x_{nn}\| + \|x_{mm}\|) + \|T_k x_{nn} - T_k x_{mm}\|;$$

i.e.,

$$\|Tx_{nn} - Tx_{mm}\| \leq \|T - T_k\|(\|x_{nn}\| + \|x_{mm}\|) + \|T_k x_{nn} - T_k x_{mm}\|.$$

Since the sequence $\{T_k x_{nn}\}$ converges, we can choose $m, n > N$ such that $\|T_k x_{nn} - T_k x_{mm}\| < \epsilon/2$, and also we can choose k so that $\|T - T_k\| < \epsilon/4$. We now have

$$\|Tx_{nn} - Tx_{mm}\| < \epsilon$$

whenever $m, n > N$ and $\{Tx_{nn}\}$ is a Cauchy sequence. Since Y is a complete space it follows that this sequence converges in Y and by Theorem 7.7.2 the desired result follows. ■

Theorem 7.7.7 implies that the family of completely continuous operators forms a linear subspace of $B(X, Y)$. The preceding theorem states that if Y is complete, then this linear subspace is closed.

7.7.18. Theorem. Let X be a Hilbert space, and let $T \in B(X, X)$. Then

(i) T is completely continuous if and only if T^*T is completely continuous; and

(ii) T is completely continuous if and only if T^* is completely continuous.

Proof. We prove (i) and leave the proof of (ii) as an exercise. Assume that T is completely continuous. It then follows from Theorem 7.7.10 that T^*T is completely continuous.

Conversely, assume that T^*T is completely continuous, and let $\{x_n\}$ be a sequence in X such that $\|x_n\| \leq 1$. It follows that there is a subsequence $\{x_{n_k}\}$ such that $T^*Tx_{n_k} \longrightarrow x \in X$ as $n_k \longrightarrow \infty$. Now

$$\|Tx_{n_j} - Tx_{n_k}\|^2 = \|T(x_{n_j} - x_{n_k})\|^2 = (T(x_{n_j} - x_{n_k}), T(x_{n_j} - x_{n_k}))$$
$$= (T^*T(x_{n_j} - x_{n_k}), (x_{n_j} - x_{n_k})) \leq \|T^*T(x_{n_j} - x_{n_k})\| \cdot \|x_{n_j} - x_{n_k}\|$$
$$\leq 2\|T^*Tx_{n_j} - T^*Tx_{n_k}\| \longrightarrow 0$$

as $n_j, n_k \longrightarrow \infty$. Thus, $\{Tx_{n_j}\}$ is a Cauchy sequence and so it is convergent. It follows from Theorem 7.7.2 that T is completely continuous. ■

7.7.19. Exercise. Prove part (ii) of Theorem 7.7.18.

In the remainder of this section we turn our attention to the properties of eigenvalues of completely continuous operators.

7.7.20. Theorem. Let X be a Hilbert space, let $T \in B(X, X)$, and let $\lambda \in C$. If T is completely continuous and if $\lambda \neq 0$, then

$$\mathfrak{N}_\lambda(T) = \{x: Tx = \lambda x\}$$

is finite dimensional.

Proof. The proof is by contradiction. Assume that $\mathfrak{N}_\lambda(T)$ is not finite dimensional. Then there is an orthonormal infinite sequence $x_1, x_2, \ldots, x_n, \ldots$ in $\mathfrak{N}_\lambda(T)$, and

$$\|Tx_n - Tx_m\|^2 = \|\lambda x_n - \lambda x_m\|^2 = |\lambda|^2 \|x_n - x_m\|^2 = 2|\lambda|^2;$$

i.e., $\|Tx_n - Tx_m\| = \sqrt{2}\,|\lambda| \neq 0$ for all $m \neq n$. Therefore, no subsequence of $\{Tx_n\}$ can be a Cauchy sequence, and hence no subsequence of $\{Tx_n\}$ can converge. This completes the proof. ■

In the next result $\pi(T)$ denotes the approximate point spectrum of T.

7.7.21. Theorem. Let X be a Hilbert space, let $T \in B(X, X)$, and let $\lambda \in C$. If T is completely continuous, if $\lambda \neq 0$, and if $\lambda \in \pi(T)$, then λ is an eigenvalue.

Proof. For each positive integer n there is an $x_n \in X$ such that $\|Tx_n - \lambda x_n\| < \frac{1}{n}\|x_n\|$ for $\lambda \in \pi(T)$. We may assume that $\|x_n\| = 1$. Since T is completely continuous, there is a subsequence of $\{x_n\}$, say $\{x_{n_k}\}$ such that $\{Tx_{n_k}\}$ is convergent. Let $\lim_{n_k} Tx_{n_k} = y \in X$. It now follows that $\|y - \lambda x_{n_k}\| \longrightarrow 0$ as $n_k \longrightarrow \infty$; i.e., $\lambda x_{n_k} \longrightarrow y$. Now $\|y\| \neq 0$, because $\|y\| = \lim_{n_k} \|\lambda x_{n_k}\| = |\lambda| \lim_{n_k} \|x_{n_k}\| = |\lambda| \neq 0$. By the continuity of T we now have

$$Ty = T(\lim_{n_k} \lambda x_{n_k}) = \lim_{n_k} T(\lambda x_{n_k}) = \lambda \lim_{n_k} Tx_{n_k} = \lambda y.$$

Hence, $Ty = \lambda y$, $y \neq 0$. Thus, λ is an eigenvalue of T and y is the corresponding eigenvector. ■

The proof of the next result is an immediate consequence of Theorems 7.6.16 and 7.7.21.

7.7.22. Theorem. Let X be a Hilbert space, and let $T \in B(X, X)$ be completely continuous and normal. If $\lambda \in \sigma(T)$ and $\lambda \neq 0$, then λ is an eigenvalue of T.

7.7.23. Exercise. Prove Theorem 7.7.22.

The above theorem states that, with the possible exception of $\lambda = 0$, the spectrum of a completely continuous normal operator consists entirely of eigenvalues; i.e., if $\lambda \neq 0$, either $\lambda \in P\sigma(T)$ or $\lambda \in \rho(T)$.

7.7.24. Theorem. Let X be a Hilbert space, and let $T \in B(X, X)$. If T is completely continuous and hermitian, then T has an eigenvalue, λ, with $|\lambda| = \|T\|$.

Proof. The proof follows directly from part (iii) of Theorem 7.6.18 and Theorem 7.7.22. ■

7.7.25. Theorem. Let X be a Hilbert space, and let $T \in B(X, X)$. If T is normal and completely continuous, then T has at least one eigenvalue.

Proof. If $T = 0$, then $\lambda = 0$ clearly satisfies the conclusion of the theorem. So let us assume that $T \neq 0$. Also, if $T = T^*$, the conclusion of the theorem follows from Theorem 7.7.24. So let us assume that $T \neq T^*$. Let $U = \frac{1}{2}(T + T^*)$ and $V = \frac{1}{2i}(T - T^*)$. It follows from Theorem 7.4.15 that U and V are hermitian. Furthermore, by Theorem 7.5.4 we have $UV = VU$. From Theorems 7.7.7 and 7.7.18, U and V are completely continuous. By assumption, $V \neq 0$. By the preceding theorem, V has a non-zero eigenvalue which we shall call β. It follows from Theorem 7.1.26 that $\mathfrak{N}_\beta(V) = \mathfrak{N}(V - \beta I) \triangleq N$ is a closed linear subspace of X. Since $UV = VU$, Theorem 7.6.24 implies that N is invariant under U. Now let U_1 be the restriction of U to the linear subspace N. It follows that U_1 is completely continuous by Theorem 7.7.8. It is readily verified that U_1 is a hermitian operator on the inner product subspace N (see Eq. (3.6.21)). Hence, U_1 is completely continuous and hermitian. This implies that there is an $\alpha \in C$ and an $x \in N$ such that $x \neq 0$ and $U_1 x = \alpha x$. This means $Ux = \alpha x$. Now since $x \in N$, we must have $Vx = \beta x$. It follows that $\lambda = \alpha + i\beta$ is an eigenvalue of T with corresponding eigenvector x, since $Tx = [U + iV]x = \alpha x + i\beta x = (\alpha + i\beta)x = \lambda x$. This completes the proof. ■

We now state and prove the last result of this section.

7.7.26. Theorem. Let X be a Hilbert space, and let $T \in B(X, X)$. If T is normal and completely continuous, then T has an eigenvalue λ such that $|\lambda| = \|T\|$.

Proof. Let $S = T^*T$. Then S is hermitian and completely continuous by Theorem 7.7.18. Also, $S \geq 0$ because $(Sx, x) = (T^*Tx, x) = (Tx, Tx) = \|Tx\|^2 \geq 0$. This last condition implies that S has no negative eigenvalues. Specifically, if λ is an eigenvalue of S, then there is an $x \neq 0$ in X such that $Sx = \lambda x$. Now

$$0 \leq (Sx, x) \leq (\lambda x, x) = \lambda(x, x) = \lambda \|x\|^2,$$

and since $\|x\| \neq 0$, we have $\lambda \geq 0$. By Theorem 7.7.24, S has an eigenvalue, μ, where $\pm\mu = \|S\| = \|T^*T\| = \|T\|^2$. Now let $N \triangleq \mathfrak{N}(S - \mu I) = \mathfrak{N}_\mu(S)$, and note that N contains a non-zero vector. Since T is normal, $TS = T(T^*T)$

$= (T^*T)T = ST$. Similarly, we have $T^*S = ST^*$. By Theorem 7.6.24, N is invariant under T and under T^*. By Theorem 7.5.6 this means T remains normal when its domain of definition is restricted to N. By Theorem 7.7.25, there is a $\lambda \in C$ and a vector $x \neq 0$ in N such that $Tx = \lambda x$, and thus $T^*x = \bar{\lambda}x$. Now since $Sx = T^*Tx = T^*(\lambda x) = \lambda T^*x = \lambda\bar{\lambda}x = |\lambda|^2 x$ for this $x \neq 0$, and since $Tx = \mu x$ for all $x \in N$, it follows that $|\lambda|^2 = \mu = \|S\| = \|T^*T\| = \|T\|^2$. Therefore, $|\lambda| = \|T\|$ and λ is an eigenvalue of T. ■

7.8. THE SPECTRAL THEOREM FOR COMPLETELY CONTINUOUS NORMAL OPERATORS

The main result of this section is referred to as the **spectral theorem** (for completely continuous operators). Some of the direct consequences of this theorem provide an insight into the geometric properties of normal operators. Results such as the spectral theorem play a central role in applications. In Section 7.10 we will apply this theorem to integral equations.

Throughout this section, X is a complex Hilbert space.

We require some preliminary results.

7.8.1. Theorem. Let $T \in B(X, X)$ be completely continuous and normal. For each $\epsilon > 0$, let A_ϵ be the annulus in the complex plane defined by

$$A_\epsilon = \{\lambda \in C: \epsilon \leq |\lambda| \leq \|T\|\}.$$

Then the number of eigenvalues of T contained in A_ϵ is finite.

Proof. To the contrary, let us assume that for some $\epsilon > 0$ the annulus A_ϵ contains an infinite number of eigenvalues. By the Bolzano-Weierstrass theorem, there is a point of accumulation λ_0 of the eigenvalues in the annulus A_ϵ. Let $\{\lambda_n\}$ be a sequence of distinct eigenvalues such that $\lambda_n \longrightarrow \lambda_0$ as $n \longrightarrow \infty$, and let $Tx_n = \lambda_n x_n$, $\|x_n\| = 1$. Since T is a completely continuous operator, there is a subsequence $\{x_{n_k}\}$ of $\{x_n\}$ for which the sequence $\{Tx_{n_k}\}$ converges to an element $u \in X$; i.e., $Tx_{n_k} \longrightarrow u$ as $n_k \longrightarrow \infty$. Thus, since $Tx_{n_k} = \lambda_{n_k}x_{n_k}$, we have $\lambda_{n_k}x_{n_k} \longrightarrow u$. But $1/\lambda_{n_k} \longrightarrow 1/\lambda_0$ because $\lambda_n \neq 0$. Therefore $x_{n_k} \longrightarrow (1/\lambda_0)u$. But the x_{n_k} are distinct eigenvectors corresponding to distinct eigenvalues. By part (iv) of Theorem 7.6.10 $\{x_{n_k}\}$ is an orthonormal sequence and $x_{n_k} \longrightarrow (1/\lambda_0)u$. But $\|x_{n_k} - x_{n_j}\|^2 = 2$, and thus $\{x_{n_k}\}$ cannot be a Cauchy sequence. Yet, it is convergent by assumption; i.e., we have arrived at a contradiction. Therefore, our initial assumption is false and the theorem is proved. ■

Our next result is a direct consequence of the preceding theorem.

7.8.2. Theorem. Let $T \in B(X, X)$ be completely continuous and normal. Then the number of eigenvalues of T is at most denumerable. If the set of eigenvalues is denumerable, then we have a point of accumulation at zero and only at zero (in the complex plane). The non-zero eigenvalues can be ordered so that

$$\|T\| = |\lambda_1| \geq |\lambda_2| \geq |\lambda_3| \geq \ldots \geq |\lambda_n| \geq \ldots.$$

7.8.3. Exercise. Prove Theorem 7.8.2.

The next result is known as the **spectral theorem**. Here we let $\lambda_0 = 0$, and we let $\{\lambda_1, \lambda_2, \ldots\}$ be the non-zero eigenvalues of a completely continuous operator $T \in B(X, X)$. Note that λ_0 may or may not be an eigenvalue of T. If λ_0 is an eigenvalue, then $\mathfrak{N}(T)$ need not be finite dimensional. However, by Theorem 7.7.20, $\mathfrak{N}(T - \lambda_i I)$ is finite dimensional for $i = 1, 2, \ldots$.

7.8.4. Theorem. Let $T \in B(X, X)$ be completely continuous and normal, let $\lambda_0 = 0$, and let $\{\lambda_1, \lambda_2, \ldots\}$ be the non-zero distinct eigenvalues of T (this collection may be finite). Let $\mathfrak{N}_i = \mathfrak{N}(T - \lambda_i I)$ for $i = 0, 1, 2, \ldots$. Then the family of closed linear subspaces $\{\mathfrak{N}_i\}_{i=0}^{\infty}$ of X is total.

Proof. The fact that each $\mathfrak{N}_i$ is a closed linear subspace of X follows from Theorem 7.1.26. Now let $Y = \bigcup_n \mathfrak{N}_n$, and let $N = Y^{\perp}$. We wish to show that $N = \{0\}$. By Theorem 6.12.6, N is a closed linear subspace of X. We will show first that Y is invariant under T^*. Let $x \in Y$. Then $x \in \mathfrak{N}_n$ for some n and $Tx = \lambda_n x$. Now $\lambda_n(T^*x) = T^*(\lambda_n x) = T^*Tx = T(T^*x)$; i.e., $T(T^*x) = \lambda_n(T^*x)$ and so $T^*x \in \mathfrak{N}_n$, which implies $T^*x \in Y$. Therefore, Y is invariant under T^*. From Theorem 7.3.15 it follows that $Y^{\perp}$ is invariant under T. Hence, N is an invariant closed linear subspace under T. It follows from Theorems 7.7.8 and 7.5.6 that if T_1 is the restriction of T to N, then $T_1 \in B(N, N)$ and T_1 is completely continuous and normal. Now let us suppose that $N \neq \{0\}$. By Theorem 7.7.25 there is a non-zero $x \in N$ and a $\lambda \in C$ such that $T_1 x = \lambda x$. But if this is so, λ is an eigenvalue of T and it follows that $x \in \mathfrak{N}_n$ for some n. Hence, $x \in N \cap Y$, which is impossible unless $x = 0$. This completes the proof. ■

In proving an alternate form of the spectral theorem, we require the following result.

7.8.5. Theorem. Let $\{N_k\}$ be a sequence of orthogonal closed linear subspaces of X; i.e., $N_k \perp N_j$ for all $j \neq k$. Then the following statements are equivalent:

(i) $\{N_k\}$ is a total family;
(ii) X is the smallest closed linear subspace which contains every N_k; and

(iii) for every $x \in X$ there is a unique sequence $\{x_k\}$ such that

(a) $x_k \in N_k$ for every k,

(b) $\sum_{k=1}^{\infty} x_k = x$; and

(c) $\sum_{k=1}^{\infty} \|x_k\|^2 < \infty$.

Proof. We first prove the equivalence of statements (i) and (ii). Let $Y = \bigcup_n N_n$. Then $Y \subset Y^{\perp\perp}$ by Theorem 6.12.8. Furthermore, $Y^{\perp\perp}$ is the smallest closed linear subspace which contains Y by Theorem 6.12.8. Now suppose $\{N_k\}$ is a total family. Then $Y^{\perp} = \{0\}$. Hence, $Y^{\perp\perp} = X$ and so X is the smallest closed linear subspace which contains every N_k.

On the other hand, suppose X is the smallest closed linear subspace which contains every N_k. Then $X = Y^{\perp\perp}$ and $Y^{\perp\perp\perp} = \{0\}$. But $Y^{\perp\perp\perp} = Y^{\perp}$. Thus, $Y^{\perp} = \{0\}$, and so $\{N_k\}$ is a total family.

We now prove the equivalence of statements (i) and (iii). Let $\{N_k\}$ be a total family, and let $x \in X$. For every $k = 1, 2, \ldots$, there is an $x_k \in N_k$ and a $y_k \in N_k^{\perp}$ such that $x = x_k + y_k$. If $x_k = 0$, then $(x, x_k) = 0$. If $x_k \neq 0$, then $(x, x_k/\|x_k\|) = (x_k + y_k, x_k/\|x_k\|) = \|x_k\|$. Thus, it follows from Bessel's inequality that

$$\sum_{k=1}^{\infty} \left|\left(x, \frac{x_k}{\|x_k\|}\right)\right|^2 \leq \|x\|^2 < \infty.$$

Hence, $\sum_{k=1}^{\infty} \|x_k\|^2 < \infty$. Next, let $x_0 = \sum_{k=1}^{\infty} x_k$. Then $x_0 \in X$. For fixed j, let $y \in N_j$. Then $(x - x_0, y) = (x_j + y_j - x_0, y) = (x_j, y) + (y_j, y) - (x_0, y) = (x_j, y) - \left(\sum_{k=1}^{\infty} x_k, y\right) = (x_j, y) - \sum_{k=1}^{\infty} (x_k, y) = (x_j, y) - (x_j, y) = 0$. Thus, $(x - x_0)$ is orthogonal to every element of N_j for every j. Since $\{N_k\}$ is a total family, we have $x = x_0$. To prove uniqueness, suppose that $x = \sum_{k=1}^{\infty} x_k = \sum_{k=1}^{\infty} x_k'$, where $x_k, x_k' \in N_k$ for all k. Then $\sum_{k=1}^{\infty} (x_k - x_k') = 0$. Since $x_k - x_k' \in N_k$ we have $(x_k - x_k') \perp (x_j - x_j')$ for $j \neq k$, and so $\left\|\sum_{k=1}^{\infty} (x_k - x_k')\right\|^2 = \sum_{k=1}^{\infty} \|x_k - x_k'\|^2 = 0$. Thus, $\|x_k - x_k'\| = 0$ for all k, and x_k is unique for each k.

To prove that (iii) implies (i), assume that $x \in N_k^{\perp}$ for every k. By hypothesis, $x = \sum_{k=1}^{\infty} x_k$, where $x_k \in N_k$ for all k. Hence, for any j we have

$$0 = (x, x_j) = \left(\sum_{k=1}^{\infty} x_k, x_j\right) = \sum_{k=1}^{\infty} (x_k, x_j) = (x_j, x_j) = \|x_j\|^2,$$

and $x_j = 0$ for all j. This means $x = 0$, and so $\{N_k\}$ is a total family. This completes the proof. ■

In Definition 3.2.13 we introduced the direct sum of a finite number of linear subspaces. The preceding theorem permits us to extend this definition in a meaningful way to a countable number of linear subspaces.

7.8.6. Definition. Let $\{Y_k\}$ be a sequence of mutually orthogonal closed linear subspaces of X, and let $\overline{V(\{Y_k\})}$ be the closed linear subspace generated by $\{Y_k\}$. If every $x \in \overline{V(\{Y_k\})}$ is uniquely representable as $x = \sum_{k=1}^{\infty} x_k$, where $x_k \in Y_k$ for every k, then we say $\overline{V(\{Y_k\})}$ is the **direct sum** of $\{Y_k\}$. In this case we write

$$\overline{V(\{Y_k\})} = Y_1 \oplus Y_2 \oplus \ldots \oplus Y_n \oplus \ldots .$$

We are now in a position to present another version of the **spectral theorem.**

7.8.7. Theorem. Let $T \in B(X, X)$ be completely continuous and normal, let $\lambda_0 = 0$, and let $\{\lambda_1, \lambda_2, \ldots, \lambda_n, \ldots\}$ be the non-zero distinct eigenvalues of T. Let $\mathfrak{N}_i = \mathfrak{N}(T - \lambda_i I)$ for $i = 0, 1, 2, \ldots$, and let P_i be the projection on $\mathfrak{N}_i$ along $\mathfrak{N}_i^{\perp}$. Then

(i) P_i is an orthogonal projection for each i;

(ii) $P_i P_j = 0$ for all i, j such that $i \neq j$;

(iii) $\sum_{j=0}^{\infty} P_j = I$; and

(iv) $T = \sum_{j=1}^{\infty} \lambda_j P_j$.

Proof. The proof of each part follows readily from results already obtained. We simply indicate the principal results needed and leave the details as an exercise.

Part (i) follows from the definition of orthogonal projection. Part (ii) follows from part (ii) of Theorem 7.6.26. Parts (iii) and (iv) follow from Theorems 7.1.27 and 7.8.5. ■

7.8.8. Exercise. Prove Theorem 7.8.7.

In Chapter 4 we defined the resolution of the identity operator for Euclidean spaces. We conclude this section with a more general definition.

7.8.9. Definition. Let $\{P_n\}$ be a sequence of linear transformations on X such that $P_n \in B(X, X)$ for each n. If conditions (i), (ii), and (iii) of Theorem 7.8.7 are satisfied, then $\{P_n\}$ is said to be a **resolution of the identity.**

7.9. DIFFERENTIATION OF OPERATORS

In this section we consider differentiation of operators on normed linear spaces. Such operators need not be linear. *Throughout this section, X and Y are normed linear spaces over a field F, where F may be either R, the real numbers, or C, the complex numbers.* We will identify mappings which are, in general, not linear by $f: X \longrightarrow Y$. As usual, $L(X, Y)$ will denote the class of all linear operators from X into Y, while $B(X, Y)$ will denote the class of all bounded linear operators from X into Y.

7.9.1. Definition. Let $x_0 \in X$ be a fixed element, and let $f: X \longrightarrow Y$. If there exists a function $\delta f(x_0, \cdot): X \longrightarrow Y$ such that

$$\lim_{t \to 0} \left\| \frac{f(x_0 + th) - f(x_0)}{t} - \delta f(x_0, h) \right\| = 0 \tag{7.9.2}$$

(where $t \in F$) for all $h \in X$, then f is said to be **Gateaux differentiable** at x_0, and $\delta f(x_0, h)$ is called the **Gateaux differential** of f at x_0 with **increment** h.

The Gateaux differential of f is sometimes also called the **weak differential** of f or the G-**differential** of f. If f is Gateaux differentiable at x_0, then $\delta f(x_0, h)$ need *not* be linear nor continuous as a function of $h \in X$. However, we shall primarily be concerned with functions $f: X \longrightarrow Y$ which have these properties. This gives rise to the following concept.

7.9.3. Definition. Let $x_0 \in X$ be a fixed element, and let $f: X \longrightarrow Y$. If there exists a bounded linear operator $F(x_0) \in B(X, Y)$ such that

$$\lim_{\|h\| \to 0} \frac{1}{\|h\|} \|f(x_0 + h) - f(x_0) - F(x_0)h\| = 0$$

(where $h \in X$), then f is said to be **Fréchet differentiable** at x_0, and $F(x_0)$ is called the **Fréchet derivative** of f at x_0. We define

$$f'(x_0) = F(x_0).$$

If f is Fréchet differentiable for each $x \in D$, where $D \subset X$, then f is said to be Fréchet differentiable on D.

We now show that Fréchet differentiability implies Gateaux differentiability.

7.9.4. Theorem. Let $f: X \longrightarrow Y$, and let $x_0 \in X$ be a fixed element. If f is Fréchet differentiable at x_0, then f is Gateaux differentiable, and furthermore the Gateaux differential is given by

$$\delta f(x_0, h) = f'(x_0)h \text{ for all } h \in X.$$

Proof. Let $F(x_0) = f'(x_0)$, let $\epsilon > 0$, and let $h \in X$. Then there is a $\delta > 0$ such that

$$\frac{1}{||th||}\left\|f(x_0 + th) - f(x_0) - F(x_0)th\right\| < \epsilon \cdot ||h||$$

provided that $||th|| < \delta$ if $th \neq 0$. This implies that

$$\left\|\frac{f(x_0 + th) - f(x_0)}{t} - F(x_0)h\right\| < \epsilon$$

provided that $|t| < \delta/||h||$. Hence, f is Gateaux differentiable at x_0 and $\delta f(x_0, h) = F(x_0)h$. ■

Because of the preceding theorem, if $f: X \longrightarrow Y$ is Frechét differentiable at $x_0 \in X$, the Gateaux differential $\delta f(x_0, h) = f'(x_0)h$ is also called the **Fréchet differential** of f at x_0 with **increment** h.

Let us now consider some examples.

7.9.5. Example. Let X be a Hilbert space, and let f be a functional defined on X; i.e., $f: X \longrightarrow F$. If f has a Fréchet derivative at some $x_0 \in X$, then that derivative must be a bounded linear functional on X; i.e., $f'(x_0) \in X^*$. In view of Theorem 6.14.2, there is an element $y_0 \in X$ such that $f'(x_0)h = (h, y_0)$ for each $h \in X$. Although $f'(x_0) \in X^*$ and $y_0 \in X$, we know by Exercise 6.14.4 that X and X^* are congruent and thus isometric. It is customary to view the corresponding elements of isometric spaces as being one and the same element. With this in mind, we say $f'(x_0) = y_0$ and we call $f'(x_0)$ the **gradient** of f at x_0. ■

As a special case of the preceding example we consider the following specific case.

7.9.6. Example. Let $X = R^n$ and let $|| \cdot ||$ be any norm on X. By Theorem 6.6.5, X is a Banach space. Now let f be a functional defined on X; i.e., $f: X \longrightarrow R$. Let $x = (\xi_1, \ldots, \xi_n) \in X$ and $h = (h_1, \ldots, h_n) \in X$. If f has continuous partial derivatives with respect to ξ_i, $i = 1, \ldots, n$, then the Fréchet differential of f is given by

$$\delta f(x, h) = \frac{\partial f(x)}{\partial \xi_1} h_1 + \ldots + \frac{\partial f(x)}{\partial \xi_n} h_n.$$

For fixed $x_0 \in X$, we define the bounded linear functional $F(x_0)$ on X by

$$F(x_0)h = \sum_{i=1}^{n} \frac{\partial f(x)}{\partial \xi_i} h_i \Bigg|_{x=x_0} \quad \text{for } h \in X.$$

Then $F(x_0)$ is the Fréchet derivative of f at x_0. As in the preceding example, we do not distinguish between X and X^*, and we write

$$F(x_0) = \left(\frac{\partial f(x_0)}{\partial \xi_1}, \ldots, \frac{\partial f(x_0)}{\partial \xi_n}\right).$$

The gradient of f at x is given by

$$f'(x) = \left(\frac{\partial f(x)}{\partial \xi_1}, \ldots, \frac{\partial f(x)}{\partial \xi_n}\right). \quad \blacksquare \tag{7.9.7}$$

In the following, we consider another example of the gradient of a functional.

7.9.8. Example. Let X be a real Hilbert space, let $L: X \longrightarrow X$ be a bounded linear operator, and let $f: X \longrightarrow R$ be given by $f(x) = (x, Lx)$. Then f has a Fréchet derivative which is given by $f'(x) = (L + L^*)x$. To verify this, we let h be an arbitrary element in X and we let $F(x) = (L + L^*)x$. Then

$$\begin{aligned} f(x + h) - f(x) - F(x)h &= (x + h, Lx + Lh) - (x, Lx) - (h, Lx) - (h, L^*x) \\ &= (h, Lh). \end{aligned}$$

From this it follows that

$$\lim_{\|h\| \to 0} \frac{|f(x + h) - f(x) - F(x)h|}{\|h\|} = 0. \quad \blacksquare$$

In the next example we consider a functional which frequently arises in optimization problems.

7.9.9. Example. Let X and Y be real Hilbert spaces, and let L be a bounded linear operator from X into Y; i.e., $L \in B(X, Y)$. Let L^* be the adjoint of L. Let v be a fixed element in Y, and let f be a real-valued functional defined on X by

$$f(x) = \|v - Lx\|^2 \text{ for all } x \in X.$$

Then f has a Fréchet derivative which is given by

$$f'(x) = -2L^*v + 2L^*Lx.$$

To verify this, observe that

$$\begin{aligned} f(x) &= (v - Lx, v - Lx) = (v, v) - 2(v, Lx) + (Lx, Lx) \\ &= (v, v) - 2(L^*v, x) + (x, L^*Lx). \end{aligned}$$

The conclusion now follows from Examples 7.9.5 and 7.9.8. $\blacksquare$

In the next example we introduce the Jacobian matrix of a function $f: R_n \longrightarrow R^m$.

7.9.10. Example. Let $X = R^n$, and let $Y = R^m$. Since X and Y are finite dimensional, we may assume arbitrary norms on each of these spaces and they will both be Banach spaces. Let $f: X \longrightarrow Y$. For $x = (\xi_1, \ldots, \xi_n) \in X$, let us write

$$f(x) = \begin{bmatrix} f_1(x) \\ \cdot \\ \cdot \\ \cdot \\ f_m(x) \end{bmatrix} = \begin{bmatrix} f_1(\xi_1, \ldots, \xi_n) \\ \cdot \\ \cdot \\ \cdot \\ f_m(\xi_1, \ldots, \xi_n) \end{bmatrix}.$$

For $x_0 \in X$, assume that the partial derivatives

$$\left.\frac{\partial f_i(x)}{\partial \xi_j}\right|_{x=x_0} \triangleq \frac{\partial f_i(x_0)}{\partial \xi_j}$$

exist and are continuous for $i = 1, \ldots, m$ and $j = 1, \ldots, n$. The Fréchet differential of f at x_0 with increment $h = (h_1, \ldots, h_n) \in X$ is given by

$$\delta f(x_0, h) = \begin{bmatrix} \frac{\partial f_1(x_0)}{\partial \xi_1} & \cdots & \frac{\partial f_1(x_0)}{\partial \xi_n} \\ \cdot & & \cdot \\ \cdot & & \cdot \\ \cdot & & \cdot \\ \frac{\partial f_m(x_0)}{\partial \xi_1} & \cdots & \frac{\partial f_m(x_0)}{\partial \xi_n} \end{bmatrix} \begin{bmatrix} h_1 \\ \cdot \\ \cdot \\ \cdot \\ h_n \end{bmatrix}.$$

The Fréchet derivative of f at x_0 is given by

$$f'(x_0) = \begin{bmatrix} \frac{\partial f_1(x_0)}{\partial \xi_1} & \cdots & \frac{\partial f_1(x_0)}{\partial \xi_n} \\ \cdot & & \cdot \\ \cdot & & \cdot \\ \cdot & & \cdot \\ \frac{\partial f_m(x_0)}{\partial \xi_1} & \cdots & \frac{\partial f_m(x_0)}{\partial \xi_n} \end{bmatrix}$$

which is also called the **Jacobian matrix** of f at x_0. We sometimes write $f'(x) = \partial f(x)/\partial x$. ■

7.9.11. Example. Let $X = \mathcal{C}[a, b]$, the family of real-valued continuous functions defined on $[a, b]$, and let $\{X; \|\cdot\|_\infty\}$ be the Banach space given in Example 6.1.9. Let $k(s, t)$ be a real-valued function defined and continuous on $[a, b] \times [a, b]$, and let $g(t, x)$ be a real-valued function which is defined and $\partial g(t, x)/\partial x$ is continuous for $t \in [a, b]$ and $x \in R$. Let $f\colon X \longrightarrow X$ be defined by

$$f(x) = \int_a^b k(s, t)g(t, x(t))dt, \; x \in X.$$

For fixed $x_0 \in X$, the Fréchet differential of f at x_0 with increment $h \in X$ is given by

$$\delta f(x_0, h) = \int_a^b k(s, t) \frac{\partial g(t, x_0(t))}{\partial x} h(t)dt. \quad ■$$

7.9.12. Exercise. Verify the assertions made in Examples 7.9.5 to 7.9.11.

We now establish some of the properties of Fréchet differentials.

7.9.13. Theorem. Let $f, g: X \longrightarrow Y$ be Fréchet differentiable at $x_0 \in X$. Then

(i) f is continuous at $x_0 \in X$; and
(ii) for all $\alpha, \beta \in F$, $\alpha f + \beta g$ is Fréchet differentiable at x_0 and $(\alpha f + \beta g)'(x_0) = \alpha f'(x_0) + \beta g'(x_0)$.

Proof. To prove (i), let f be Fréchet differentiable at x_0, and let $F(x_0)$ be the Fréchet derivative of f at x_0. Then

$$f(x_0 + h) - f(x_0) = f(x_0 + h) - f(x_0) - F(x_0)h + F(x_0)h,$$

and

$$||f(x_0 + h) - f(x_0)|| \leq ||f(x_0 + h) - f(x_0) - F(x_0)h|| + ||F(x_0)h||.$$

Since $F(x_0)$ is bounded, there is an $M > 0$ such that $||F(x_0)h|| \leq M||h||$. Furthermore, for given $\epsilon > 0$ there is a $\delta > 0$ such that $||f(x_0 + h) - f(x_0) - F(x_0 h)|| < \epsilon||h||$ provided that $||h|| < \delta$. Hence, $||f(x_0 + h) - f(x_0)|| \leq (M + \epsilon)||h||$ whenever $||h|| < \delta$. This implies that f is continuous at x_0.

The proof of part (ii) is straightforward and is left as an exercise. ■

7.9.14. Exercise. Prove part (ii) of Theorem 7.9.13.

We now show that the chain rule encountered in calculus applies to Fréchet derivatives as well.

7.9.15. Theorem. Let X, Y, and Z be normed linear spaces. Let $g: X \longrightarrow Y$, $f: Y \longrightarrow Z$, and let $\varphi: X \longrightarrow Z$ be the composite function $\varphi = f \circ g$. Let g be Fréchet differentiable on an open set $D \subset X$, and let f be Fréchet differentiable on an open set $E \subset g(D)$. If $x \in D$ is such that $g(x) \in E$, then φ is Fréchet differentiable at x and $\varphi'(x) = f'(g(x))g'(x)$.

Proof. Let $y = g(x)$ and $d = g(x + h) - g(x)$, where $h \in X$ is such that $x + h \in D$. Then

$$\begin{aligned}\varphi(x + h) - \varphi(x) - f'(y)g'(x)h &= f(y + d) - f(y) - f'(y)d \\ &\qquad + f'(y)[d - g'(x)h] \\ &= f(y + d) - f(y) - f'(y)d + f'(y)[g(x + h) - g(x) - g'(x)h].\end{aligned}$$

Thus, given $\epsilon > 0$ there is a $\delta > 0$ such that $||d|| < \delta$ and $||h|| < \delta$ imply

$$||\varphi(x + h) - \varphi(x) - f'(y)g'(x)h|| \leq \epsilon||d|| + ||f'(y)|| \cdot ||h|| \cdot \epsilon.$$

By the continuity of g (see the proof of part (i) of Theorem 7.9.13), it follows that $||d|| \leq M \cdot ||h||$ for some constant M. Hence, there is a constant k

such that

$$||\varphi(x+h)-\varphi(x)-f'(y)g'(x)h|| \leq k\epsilon||h||.$$

This implies that $\varphi'(x)$ exists and $\varphi'(x)=f'(g(x))g'(x)$. ■

We next consider the Fréchet derivative of bounded linear operators.

7.9.16. Theorem. Let T be a linear operator from X into Y. If $f(x)=Tx$ for all $x \in X$, then f is Fréchet differentiable on X if and only if T is a bounded linear operator. In this case, $f'(x)=T$ for all $x \in X$.

Proof. Let T be a bounded linear operator. Then $||f(x+h)-f(x)-Th|| = ||T(x+h)-Tx-Th|| = 0$ for all $x, h \in X$. From this it follows that $f'(x)=T$.

Conversely, suppose T is unbounded. Then, by Theorem 7.9.13, f cannot be Fréchet differentiable. ■

Let us consider a specific case.

7.9.17. Example. Let $X=R^n$ and $Y=R^m$, and let us assume that the natural basis for each of these spaces is being used (see Example 4.1.15). If $A \in B(X, Y)$, then Ax is given in matrix representation by

$$\mathbf{Ax} = \begin{bmatrix} a_{11} & \cdots & a_{1n} \\ \cdot & & \cdot \\ \cdot & & \cdot \\ \cdot & & \cdot \\ a_{m1} & \cdots & a_{mn} \end{bmatrix} \begin{bmatrix} \xi_1 \\ \cdot \\ \cdot \\ \cdot \\ \xi_n \end{bmatrix}.$$

Hence, if $f(x)=Ax$, then $f'(x)=A$, and the matrix representation of $f'(x)=\partial f(x)/\partial x$ is $\mathbf{A}$. ■

The next result is useful in obtaining bounds on Fréchet differentiable functions.

7.9.18. Theorem. Let $f: X \to Y$, let D be an open set in X, and let f be Fréchet differentiable on D. Let $x_0 \in D$, and let $h \in X$ be such that $x_0 + th \in D$ for all t when $0 \leq t \leq 1$. Let $N = \sup_{0<t<1} ||f'(x_0+th)||$. Then $||f(x_0+h)-f(x_0)|| \leq N \cdot ||h||$.

Proof. Let $y=f(x_0+h)-f(x_0)$, and let φ be a bounded linear functional defined on Y (i.e., $\varphi \in Y^*$) such that $\varphi(y)=||\varphi||\cdot||y||$ (see Corollary 6.8.6). Define $g: (0, 1) \to R$ by $g(t)=\varphi(f(x+th))$ for $0<t<1$. By Theorems 7.9.15 and 7.9.16, $g'(t)=\varphi(f'(x+th)h)$. By the mean value theorem of calculus, there is a t_0 such that $0<t_0<1$ and $g(1)-g(0)=g'(t_0)$. Thus,

$$|\varphi(f(x+h))-\varphi(f(x))| \leq ||\varphi|| \cdot \sup_{0<t<1} ||f'(x+th)|| \cdot ||h||.$$

Since

$$|\varphi(f(x+h)) - \varphi(f(x))| = |\varphi(f(x+h) - f(x))| = |\varphi(y)|$$
$$= ||\varphi|| \cdot ||f(x_0 + h) - f(x_0)||,$$

it follows that $||f(x_0 + h) - f(x_0)|| \leq \sup_{0<t<1} ||f'(x + th)|| \cdot ||h||$. ∎

If a function $f\colon X \longrightarrow Y$ is Fréchet differentiable on an open set $D \subset X$, and if $f'(x)$ is Fréchet differentiable at $x \in D$, then f is said to be **twice Fréchet differentiable** at x, and we call the Fréchet derivative of $f'(x)$ the **second derivative** of f. We denote the second derivative of f by f''. Note that f'' is a bounded linear operator defined on X with range in the normed linear space $B(X, Y)$.

We leave the proof of the next result as an exercise.

7.9.19. Theorem. Let $f\colon X \longrightarrow Y$ be twice Fréchet differentiable on an open set $D \subset X$. Let $x_0 \in D$, and $h \in X$ be such that $x_0 + th \in D$ for all t when $0 \leq t \leq 1$. Let $N = \sup_{0<t<1} ||f''(x + th)||$. Then

$$||f(x+h) - f(x) - f'(x)h|| \leq \tfrac{1}{2}N \cdot ||h||^2.$$

7.9.20. Exercise. Prove Theorem 7.9.19.

We conclude the present section by showing that the Gateaux and Fréchet differentials play a role in maximizing and minimizing functionals which is similar to that of the ordinary derivative of functions of real variables.

Let $F = R$, and let f be a functional on X; i.e., $f\colon X \longrightarrow R$. Clearly, for fixed $x_0, h \in X$, we may define a function $g\colon R \longrightarrow R$ by the relation $g(t) = f(x_0 + th)$ for all $t \in R$. In this case, if f is Gateaux differentiable at x_0, we see that $\delta f(x_0, h) = g'(t)|_{t=0}$, where $g'(t)$ is the usual derivative of $g(t)$. We will need this property in proving our next result, Theorem 7.9.22. First, however, we require the following important concept.

7.9.21. Definition. Let f be a real-valued functional defined on a domain $\mathfrak{D} \subset X$; i.e., $f\colon \mathfrak{D} \longrightarrow R$. Let $x_0 \in \mathfrak{D}$. Then f is said to have a **relative minimum (relative maximum) at** x_0 if there exists an open sphere $S(x_0; r) \subset X$ such that for all $x \in S(x_0; r) \cap \mathfrak{D}$ the relation $f(x_0) \leq f(x)$ ($f(x_0) \geq f(x)$) holds. If f has either a relative minimum or a relative maximum at x_0, then f is said to have a **relative extremum** at x_0.

For relative extrema, we have the following result.

7.9.22. Theorem. Let $f\colon X \longrightarrow R$ be Gateaux differentiable at $x_0 \in X$. If f has a relative extremum at x_0, then $\delta f(x_0, h) = 0$ for all $h \in X$.

Proof. As pointed out in the remark preceding Definition 7.9.21, the real-valued function $g(t) = f(x_0 + th)$ must have an extremum at $t = 0$. From the oridnary calculus we must have $g'(t)|_{t=0} = 0$. Hence, $\delta f(x_0, h) = 0$ for all $h \in X$. ■

We leave the proof of the next result as an exercise.

7.9.23. Corollary. Let $f: X \longrightarrow R$ be Fréchet differentiable at $x_0 \in X$. If f has a relative extremum at x_0, then $f'(x_0) = 0$.

7.9.24. Exercise. Prove Corollary 7.9.23.

We conclude this section with the following example.

7.9.25. Example. Consider the real-valued functional f defined in Example 7.9.9; i.e., $f(x) = \|v - Lx\|^2$. For a given $v \in Y$, a necessary condition for f to have a minimum at $x_0 \in X$ is that

$$L^*Lx_0 = L^*v. \quad ■$$

7.10. SOME APPLICATIONS

In this section we consider selected applications of the material of the present chapter. The section consists of three parts. In the first part we consider integral equations, in the second part we give an example in optimal control, while in the third part we address the problem of minimizing functionals by the method of steepest descent.

A. Applications to Integral Equations

Throughout this part, X is a complex Hilbert space while T denotes a completely continuous normal operator defined on X.

We recall that if, e.g., $X = L_2[a, b]$ and T is defined by (see Example 7.3.11 and the comment at the end in Example 7.7.5)

$$Tx(s) = \int_a^b k(s, t)x(t)dt, \tag{7.10.1}$$

then T is a completely continuous operator defined on X. Furthermore, if $k(s, t) = \overline{k(t, s)}$ for all $s, t \in [a, b]$, then T is hermitian (see Exercise 7.4.20) and, hence, normal.

In the following, we shall focus our attention on equations of the form

$$Tx - \lambda x = y, \tag{7.10.2}$$

where $\lambda \in C$ and $x, y \in X$. If, in particular, T is defined by Eq. (7.10.1), then Eq. (7.10.2) includes a large class of integral equations. Indeed, it was the study of such equations which gave rise to much of the development of functional analysis.

We now prove the following existence and uniqueness result.

7.10.3. Theorem. If $\lambda \neq 0$ and if λ is not an eigenvalue of T, then Eq. (7.10.2) has a unique solution, which is given by

$$x = \frac{P_0 y}{-\lambda} + \sum_{n=1}^{\infty} \frac{P_n y}{\lambda_n - \lambda}, \tag{7.10.4}$$

where $\{\lambda_n\}$ are the non-zero distinct eigenvalues of T, P_n is the projection of X onto $\mathfrak{N}_n = \mathfrak{N}(T - \lambda_n I)$ along $\mathfrak{N}_n^{\perp}$ for $n = 1, 2, \ldots$, and $P_0 x$ is the projection of x onto $\mathfrak{N}(T)$.

Proof. We first prove that the infinite series on the right-hand side of Eq. (7.10.4) is convergent. Since $\lambda \neq 0$, it cannot be an accumulation point of $\{\lambda_n\}$. Thus, we can find a $d > 0$ such that $|\lambda| \geq d$ and $|\lambda - \lambda_k| \geq d$ for $k = 1, 2, \ldots$. We note from Theorem 7.8.7 that $P_i P_j = 0$ for $i \neq j$. Now for $N < \infty$, we have by the Pythagorean theorem,

$$\begin{aligned}
\left\| -\frac{P_0 y}{\lambda} + \sum_{k=1}^{N} \frac{P_k y}{\lambda_k - \lambda} \right\|^2 &= \frac{1}{|\lambda|^2} \|P_0 y\|^2 + \sum_{k=1}^{N} \frac{1}{|\lambda - \lambda_k|^2} \|P_k y\|^2 \\
&\leq \frac{1}{d^2} \|P_0 y\|^2 + \frac{1}{d^2} \sum_{k=1}^{N} \|P_k y\|^2 \\
&= \frac{1}{d^2}\left[\|P_0 y\|^2 + \sum_{k=1}^{N} \|P_k y\|^2\right] \\
&= \frac{1}{d^2}\left\| P_0 y + \sum_{k=1}^{N} P_k y \right\|^2 \\
&\leq \frac{1}{d^2}\left\| P_0 y + \sum_{k=1}^{\infty} P_k y \right\|^2 \\
&= \frac{1}{d^2} \|y\|^2.
\end{aligned}$$

This implies that $\sum_{k=1}^{\infty} \frac{1}{|\lambda - \lambda_k|^2} \|P_k y\|^2$ is convergent, and so it follows from Theorem 6.13.3 that $\sum_{n=1}^{\infty} \frac{P_n y}{\lambda_n - \lambda}$ is convergent to an element in X.

Let j be a positive integer. By Theorem 7.5.12, P_j is continuous, and so by Theorem 7.1.27, $P_j \left(\sum_{n=1}^{\infty} \frac{P_n y}{\lambda_n - \lambda}\right) = \sum_{n=1}^{\infty} \frac{P_j P_n y}{\lambda_n - \lambda}$. Now let x be given by Eq. (7.10.4) for arbitrary $y \in X$. We want to show that $Tx - \lambda x = y$. From

Eq. (7.10.4) we have

$$P_0 x = -\frac{1}{\lambda} P_0 y$$

and

$$P_j x = \frac{1}{\lambda_j - \lambda} P_j y \text{ for } j = 1, 2, \ldots.$$

Thus, $P_0 y = -\lambda P_0 x$ and $P_j y = \lambda_j P_j x - \lambda P_j x$. Now from the spectral theorem (Theorem 7.8.7), we have $y = P_0 y + \sum_{j=1}^{\infty} P_j y$, $Tx = \sum_{j=1}^{\infty} \lambda_j P_j x$, and $\lambda x = \lambda P_0 x + \sum_{j=1}^{\infty} \lambda P_j x$. Hence, $y = Tx - \lambda x$.

Finally, to show that x given by Eq. (7.10.4) is unique, let x and z be such that $Tx - \lambda x = Tz - \lambda z = y$. Then it follows that $T(x - z) - \lambda(x - z) = y - y = 0$. Hence, $T(x - z) = \lambda(x - z)$. Since λ is by assumption not an eigenvalue of T, we must have $x - z = 0$. This completes the proof. ■

In the next result we consider the case where λ is a non-zero eigenvalue of T.

7.10.5. Theorem. Let $\{\lambda_n\}$ denote the non-zero distinct eigenvalues of T, and let $\lambda = \lambda_j$ for some positive integer j. Then there is a (non-unique) $x \in X$ satisfying Eq. (7.10.2) if and only if $P_j y = 0$, where P_j is the orthogonal projection of X onto $\mathfrak{N}_j = \{x : (T - \lambda_j I)x = 0\}$. If $P_j y = 0$, then a solution to Eq. (7.10.2) is given by

$$x = x_0 - \frac{P_0 y}{\lambda} + \sum_{\substack{k=1 \\ k \neq j}}^{\infty} \frac{P_k y}{\lambda_k - \lambda}, \tag{7.10.6}$$

where P_0 is the orthogonal projection of X onto $\mathfrak{N}(T)$ and x_0 is any element in $\mathfrak{N}_j$.

Proof. We first observe that $\mathfrak{N}_j$ reduces T by part (iii) of Theorem 7.6.26. It therefore follows from part (ii) of Theorem 7.5.22 that $TP_j = P_j T$. Now suppose that y is such that Eq. (7.10.2) is satisfied for some $x \in X$. Then it follows that $P_j y = P_j(Tx - \lambda_j x) = TP_j x - \lambda_j P_j x = \lambda_j P_j x - \lambda_j P_j x = 0$. In the preceding, we used the fact that $Tx = \lambda_j x$ for $x \in \mathfrak{N}_j$ and $P_j x \in \mathfrak{N}_j$ for all $x \in X$. Hence, $P_j y = 0$.

Conversely, suppose that $P_j y = 0$, and let x be given by Eq. (7.10.6). The proof that x satisfies Eq. (7.10.2) follows along the same lines as the proof of Theorem 7.10.3, and the details are left as an exercise. The non-uniqueness of the solution is apparent, since $(T - \lambda_j I)x_0 = 0$ for any $x_0 \in \mathfrak{N}_j$. ■

7.10.7. Exercise. Complete the proof of Theorem 7.10.5.

B. An Example from Optimal Control

In this example we consider systems which can appropriately be described by the system of first-order ordinary differential equations

$$\dot{\mathbf{x}}(t) = \mathbf{A}\mathbf{x}(t) + \mathbf{B}\mathbf{u}(t), \tag{7.10.8}$$

where $\mathbf{x}(0) \triangleq \mathbf{x}_0$ is given. Here $\mathbf{x}(t) \in R^n$ and $\mathbf{u}(t) \in R^m$ for every t such that $0 \leq t \leq T$ for some $T > 0$, and $\mathbf{A}$ is an $n \times n$ matrix, and $\mathbf{B}$ is an $n \times m$ matrix. As we saw in part (vi) of Theorem 4.11.45, if each element of the vector $\mathbf{u}(t)$ is a continuous function of t, then the unique solution to Eq. (7.10.8) at time t is given by

$$\mathbf{x}(t) = \mathbf{\Phi}(t, 0)\mathbf{x}(0) + \int_0^t \mathbf{\Phi}(t, \tau)\mathbf{B}\mathbf{u}(\tau)d\tau, \tag{7.10.9}$$

where $\mathbf{\Phi}(t, \tau)$ is the state transition matrix for the system of equations given in Eq. (7.10.8).

Let us now define the class of vector valued functions $L_2^m[0, T]$ by

$$L_2^m[0, T] = \{\mathbf{u}\colon \mathbf{u}^T = (u_1, \ldots, u_m), \quad \text{where } u_i \in L_2[0, T], i = 1, \ldots, m\}.$$

If we define the inner product by

$$(\mathbf{u}, \mathbf{v}) = \int_0^T \mathbf{u}^T(t)\mathbf{v}(t)dt$$

for $\mathbf{u}, \mathbf{v} \in L_2^m[0, T]$, then it follows that $L_2^m[0, T]$ is a Hilbert space (see Example 6.11.11). Next, let us define the linear operator $L\colon L_2^m[0, T] \longrightarrow L_2^n[0, T]$ by

$$[L\mathbf{u}](t) = \int_0^t \mathbf{\Phi}(t, \tau)\mathbf{B}\mathbf{u}(\tau)d\tau \tag{7.10.10}$$

for all $\mathbf{u} \in L_2^m[0, T]$. Since the elements of $\mathbf{\Phi}(t, \tau)$ are continuous functions on $[0, T] \times [0, T]$, it follows that L is completely continuous.

Now recall from Exercise 5.10.59 that Eq. (7.10.9) is the unique solution to Eq. (7.10.8) when the elements of the vector $\mathbf{u}(t)$ are continuous functions of t. It can be shown that the solution of Eq. (7.10.8) exists in an extended sense if we permit $\mathbf{u} \in L_2^m[0, T]$. Allowing for this generalization, we can now consider the following **optimal control problem.** Let $\gamma \in R$ be such that $\gamma > 0$, and let f be the real-valued functional defined on $L_2^n[0, T]$ given by

$$f(\mathbf{u}) = \int_0^T \mathbf{x}^T(t)\mathbf{x}(t)dt + \gamma \int_0^T \mathbf{u}^T(t)\mathbf{u}(t)dt, \tag{7.10.11}$$

where $\mathbf{x}(t)$ is given by Eq. (7.10.9) for $\mathbf{u} \in L_2^m[0, T]$. The **linear quadratic cost control problem** is to find $\mathbf{u} \in L_2^m[0, T]$ such that $f(\mathbf{u})$ in Eq. (7.10.11) is minimum, where $\mathbf{x}(t)$ is the solution to the set of ordinary differential equations (7.10.8). This problem can be cast into a minimization problem in a Hilbert space as follows.

Let

$$\mathbf{v}(t) = -\mathbf{\Phi}(t, 0)\mathbf{x}_0 \text{ for } 0 \leq t \leq T.$$

Then we can rewrite Eq. (7.10.9) as

$$\mathbf{x} = L\mathbf{u} - \mathbf{v},$$

and Eq. (7.10.11) assumes the form

$$f(\mathbf{u}) = ||L\mathbf{u} - \mathbf{v}||^2 + \gamma||\mathbf{u}||^2.$$

We can find the desired minimizing $\mathbf{u}$ in the more general context of arbitrary real Hilbert spaces by means of the following result.

7.10.12. Theorem. Let X and Y be real Hilbert spaces, let $L: X \longrightarrow Y$ be a completely continuous operator, and let L^* denote the adjoint of L. Let v be a given fixed element in Y, let $\gamma \in R$, and define the functional $f: X \longrightarrow R$ by

$$f(u) = ||Lu - v||^2 + \gamma||u||^2 \tag{7.10.13}$$

for $u \in X$. (In Eq. (7.10.13) we use the norm induced by the inner product and note that $||u||$ is the norm of $u \in X$, while $||Lu - v||$ is the norm of $(Lu - v) \in Y$.) If in Eq. (7.10.13), $\gamma > 0$, then there exists a unique $u_0 \in X$ such that $f(u_0) \leq f(u)$ for all $u \in X$. Furthermore, u_0 is the solution to the equation

$$L^*Lu_0 + \gamma u_0 = L^*v. \tag{7.10.14}$$

Proof. Let us first examine Eq. (7.10.14). Since L is a completely continuous operator, by Corollary 7.7.12, so is L^*L. Furthermore, the eigenvalues of L^*L cannot be negative, and so $-\gamma$ cannot be an eigenvalue of L^*L. Making the association $T = L^*L$, $\lambda = -\gamma$, and $y = L^*v$ in Eq. (7.10.2), it is clear that T is normal and it follows from Theorem 7.10.3 that Eq. (7.10.14) has a unique solution. In fact, this solution is given by Eq. (7.10.4), using the above definitions of symbols.

Next, let us assume that u_0 is the unique element in X satisfying Eq. (7.10.14), and let $h \in X$ be arbitrary. It follows from Eq. (7.10.13) that

$$\begin{aligned} f(u_0 + h) &= (Lu_0 + Lh - v, Lu_0 + Lh - v) + \gamma(u_0 + h, u_0 + h) \\ &= (Lu_0 - v, Lu_0 - v) + 2(Lh, Lu_0 - v) \\ &\quad + (v, v) + \gamma(u_0, u_0) + 2\gamma(u_0, h) + \gamma(h, h) \\ &= (Lu_0 - v, Lu_0 - v) + (v, v) + \gamma(u_0, u_0) \\ &\quad + 2(h, L^*Lu_0 + \gamma u_0 - L^*v) + \gamma(h, h) \\ &= ||Lu_0 - v||^2 + ||v||^2 + \gamma||u_0||^2 + \gamma||h||^2. \end{aligned}$$

Therefore, $f(u_0 + h)$ is minimum if and only if $h = 0$. ■

The solution to Eq. (7.10.14) can be obtained from Eq. (7.10.4); however, a more convenient method is available for the finding of the solution when L is given by Eq. (7.10.10). This is summarized in the following result.

7.10.15. Theorem. Let $\gamma > 0$, and let $f(u)$ be defined by Eq. (7.10.11), where $\mathbf{x}(t)$ is the solution to Eq. (7.10.8). If

$$\mathbf{u}(t) = -\frac{1}{\gamma}\mathbf{B}^T\mathbf{P}(t)\mathbf{x}(t)$$

for all t such that $0 \leq t \leq T$, where $\mathbf{P}(t)$ is the solution to the matrix differential equation

$$\dot{\mathbf{P}}(t) = -\mathbf{A}^T\mathbf{P}(t) - \mathbf{P}(t)\mathbf{A} + \frac{1}{\gamma}\mathbf{P}(t)\mathbf{B}\mathbf{B}^T\mathbf{P}(t) - \mathbf{I} \qquad (7.10.16)$$

with $\mathbf{P}(T) = \mathbf{0}$, then $\mathbf{u}$ minimizes $f(\mathbf{u})$.

Proof. We want to show that $\mathbf{u}$ satisfies Eq. (7.10.14), where $L\mathbf{u}$ is given by Eq. (7.10.10). We note that if $\mathbf{u}$ satisfies Eq. (7.10.14), then $\mathbf{u} = -\frac{1}{\gamma}L^*(L\mathbf{u} - \mathbf{v}) = -\frac{1}{\gamma}L^*\mathbf{x}$. We now find the expression for evaluating $L^*\mathbf{w}$ for arbitrary $\mathbf{w} \in L_2[0, T]$. We compute

$$\begin{aligned}(\mathbf{w}, L\mathbf{u}) &= \int_0^T \left[\int_0^s \mathbf{\Phi}(s, t)\mathbf{B}\mathbf{u}(t)dt\right]^T \mathbf{w}(s)ds \\ &= \int_0^T \int_0^s \mathbf{u}^T(t)\mathbf{B}^T\mathbf{\Phi}^T(s, t)\mathbf{w}(s)dtds \\ &= \int_0^T \mathbf{u}^T(t)\left[\int_t^T \mathbf{B}^T\mathbf{\Phi}^T(s, t)\mathbf{w}(s)ds\right]dt.\end{aligned}$$

In order for this last expression to equal $(L^*\mathbf{w}, \mathbf{u})$, we must have

$$[L^*\mathbf{w}](t) = \int_t^T \mathbf{B}^T\mathbf{\Phi}^T(s, t)\mathbf{w}(s)ds.$$

Thus, $\mathbf{u}$ must satisfy

$$\mathbf{u}(t) = -\frac{1}{\gamma}\mathbf{B}^T \int_t^T \mathbf{\Phi}^T(s, t)\mathbf{x}(s)ds$$

for all t such that $0 \leq t \leq T$. Now assume there exists a matrix $\mathbf{P}$ such that

$$\mathbf{P}(t)\mathbf{x}(t) = \int_t^T \mathbf{\Phi}^T(s, t)\mathbf{x}(s)ds. \qquad (7.10.17)$$

We now find conditions for such a matrix $\mathbf{P}(t)$ to exist. First, we see that $\mathbf{P}(T) = \mathbf{0}$. Next, differentiating both sides of Eq. (7.10.17) with respect to t, and noting that $\dot{\mathbf{\Phi}}^T(s, t) = -\mathbf{A}^T\mathbf{\Phi}(s, t)$, we have

$$\begin{aligned}\dot{\mathbf{P}}(t)\mathbf{x}(t) + \mathbf{P}(t)\dot{\mathbf{x}}(t) &= -\mathbf{x}(t) - \mathbf{A}^T \int_t^T \mathbf{\Phi}^T(s, t)\mathbf{x}(s)ds \\ &= -\mathbf{x}(t) - \mathbf{A}^T\mathbf{P}(t)\mathbf{x}(t).\end{aligned}$$

Therefore,

$$\dot{\mathbf{P}}(t)\mathbf{x}(t) + \mathbf{P}(t)[\mathbf{A}\mathbf{x}(t) + \mathbf{B}\mathbf{u}(t)] = -\mathbf{x}(t) - \mathbf{A}^T\mathbf{P}(t)\mathbf{x}(t).$$

But

$$\mathbf{u}(t) = -\frac{1}{\gamma}L^*\mathbf{x}(t) = -\frac{1}{\gamma}\mathbf{B}^T\mathbf{P}(t)\mathbf{x}(t)$$

so that

$$\dot{\mathbf{P}}(t)\mathbf{x}(t) + \mathbf{P}(t)\mathbf{A}\mathbf{x}(t) - \frac{1}{\gamma}\mathbf{P}(t)\mathbf{B}\mathbf{B}^T\mathbf{P}(t)\mathbf{x}(t) = -\mathbf{x}(t) - \mathbf{A}^T\mathbf{P}(t)\mathbf{x}(t).$$

Hence, $\mathbf{P}(t)$ must satisfy

$$\dot{\mathbf{P}}(t) = -\mathbf{A}^T\mathbf{P}(t) - \mathbf{P}(t)\mathbf{A}^T + \frac{1}{\gamma}\mathbf{P}(t)\mathbf{B}\mathbf{B}^T\mathbf{P}(t) - \mathbf{I}$$

with $\mathbf{P}(T) = \mathbf{0}$.

If

$$\mathbf{u}(t) = -\frac{1}{\gamma}\mathbf{B}^T\mathbf{P}(t)\mathbf{x}(t),$$

it follows that $\mathbf{u}$ satisfies

$$L^*L\mathbf{u} + \gamma\mathbf{u} = L^*\mathbf{v},$$

where $\mathbf{v} = -\mathbf{\Phi}(t, 0)\mathbf{x}_0$ and so, by Theorem 7.10.12, $\mathbf{u}$ minimizes f given by Eq. (7.10.11). This completes the proof of the theorem. ■

The differential equation for $\mathbf{P}(t)$ in Eq. (7.10.16) is called a **matrix Riccati equation** and can be shown to have a unique solution for all $t < T$.

C. *Minimization of Functionals: Method of Steepest Descent*

The problem of finding the minimum (or maximum) of functionals arises frequently in many diverse areas in applications. In this part we turn our attention to an iterative method of obtaining the minimum of a functional f defined on a real Hilbert space X.

Consider a functional $f: X \longrightarrow R$ of the form

$$f(x) = (x, Mx) - 2(w, x) + \beta, \tag{7.10.18}$$

where w is a fixed vector in X, where $\beta \in R$, and where M is a linear self-adjoint operator having the property

$$c_1 \|x\|^2 \leq (x, Mx) \leq c_2 \|x\|^2 \tag{7.10.19}$$

for all $x \in X$ and some constants $c_2 \geq c_1 > 0$. The reader can readily verify that the functional given in Eq. (7.10.13) is a special case of f, given in Eq. (7.10.18), where we make the association $M = L^*L + \gamma I$ (provided $\gamma > 0$), $w = L^*v$, and $\beta = (v, v)$.

Under the above conditions, the equation

$$Mx = w \tag{7.10.20}$$

has a unique solution, say x_0, and x_0 minimizes $f(x)$. Iterative methods are based on beginning with an initial guess to the solution of Eq. (7.10.20) and then successively attempting to improve the estimate according to a recursive relationship of the form

$$x_{n+1} = x_n + \alpha_n r_n, \quad n = 1, 2, \ldots, \tag{7.10.21}$$

where $\alpha_n \in R$ and $r_n \in X$. Different methods of selecting α_n and r_n give rise to various algorithms of minimizing $f(x)$ given in Eq. (7.10.18) or, equivalently, finding the solution to Eq. (7.10.20). In this part we shall in particular consider the **method of steepest descent**. In doing so we let

$$r_n = w - Mx_n, \quad n = 1, 2, \ldots. \tag{7.10.22}$$

The term r_n defined by Eq. (7.10.22) is called the **residual** of the approximation x_n. If, in particular, x_n satisfies Eq. (7.10.20), we see that the residual is zero. For $f(x)$ given in Eq. (7.10.18), we see that

$$f'(x_n) = -2r_n,$$

where $f'(x_n)$ denotes the gradient of $f(x_n)$. That is, the residual, r_n, is "pointing" into the direction of the negative of the gradient, or in the direction of steepest descent. Equation (7.10.21) indicates that the correction term $\alpha_n r_n$ is to be a scalar multiple of the gradient, and thus the steepest descent method constitutes an example of one of the so-called "gradient methods." With r_n given by Eq. (7.10.22), α_n is chosen so that $f(x_n + \alpha_n r_n)$ is minimum. Substituting $x_n + \alpha_n r_n$ into Eq. (7.10.18), it is readily shown that

$$\alpha_n = \frac{(r_n, r_n)}{(r_n, Mr_n)}$$

is the minimizing value. This method is illustrated pictorially in Figure B.

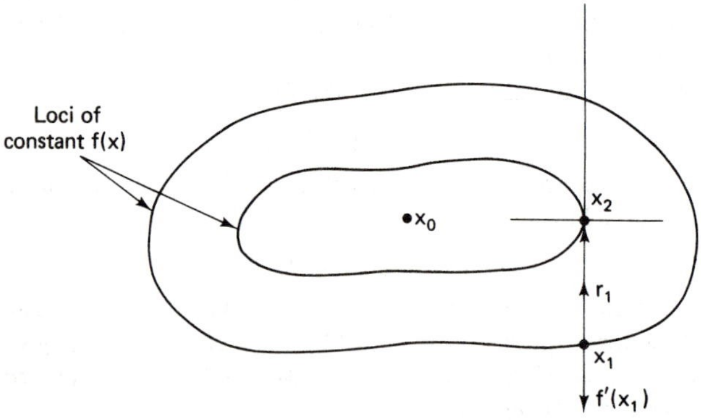

7.10.23. **Figure B.** Illustration of the method of steepest descent.

In the following result we show that under appropriate conditions the sequence $\{x_n\}$ generated in the heuristic discussion above converges to the unique minimizing element x_0 satisfying Eq. (7.10.20).

7.10.24. Theorem. Let $M \in B(X, X)$ be a self-adjoint operator such that for some pair of positive real numbers γ and μ we have $\gamma\|x\|^2 \leq (x, Mx) \leq \mu\|x\|^2$ for all $x \in X$. Let $x_1 \in X$ be arbitrary, let $w \in X$, and let $r_n = w - Mx_n$, where $x_{n+1} = x_n + \alpha_n r_n$ for $n = 1, 2, \ldots$, and $\alpha_n = (r_n, r_n)/(r_n, Mr_n)$. Then the sequence $\{x_n\}$ converges to x_0, where x_0 is the unique solution to Eq. (7.10.20).

Proof. In view of the Schwarz inequality we have $(x, Mx) \leq \|Mx\|\|x\|$. This implies that $\gamma\|x\| \leq \|Mx\|$ for all $x \in X$, and so M is a bijective mapping by Theorem 7.4.21, with $M^{-1} \in B(X, X)$ and $\|M^{-1}\| \leq 1/\gamma$. By Theorem 7.4.10, M^{-1} is also self-adjoint. Let x_0 be the unique solution to Eq. (7.10.20), and define $F: X \longrightarrow R$ by

$$F(x) = (x - x_0, M(x - x_0)) \text{ for } x \in X.$$

We see that F is minimized uniquely by $x = x_0$, and furthermore $F(x_0) = 0$. We now show that $\lim_n F(x_n) = 0$. If for some n, $F(x_n) = 0$, the process terminates and we are done. So assume in the following that $F(x_n) \neq 0$. Note also that since M is positive, we have $F(x) \geq 0$ for all $x \in X$.

We begin with the fact that

$$F(x_{n+1}) = F(x_n) - 2\alpha_n(r_n, My_n) + \alpha_n^2(r_n, Mr_n),$$

where we have let $y_n = x_0 - x_n$. Noting that $r_n = My_n$, so that $F(x_n) = (y_n, My_n) = (M^{-1}r_n, r_n)$, we have

$$\frac{F(x_n) - F(x_{n+1})}{F(x_n)} = \frac{(r_n, r_n)^2}{(r_n, Mr_n)(M^{-1}r_n, r_n)} \geq \frac{\gamma}{\mu}.$$

Hence, $F(x_{n+1}) \leq \left(1 - \frac{\gamma}{\mu}\right)F(x_n) \leq \left(1 - \frac{\gamma}{\mu}\right)^n F(x_1)$. Thus, $\lim_n F(x_n) = 0$ and so $x_n \longrightarrow x_0$, which was to be proven. ■

7.11. REFERENCES AND NOTES

Many of the excellent sources dealing with linear operators on Banach and Hilbert spaces include Balakrishnan [7.2], Dunford and Schwarz [7.5], Kantorovich and Akilov [7.6], Kolmogorov and Fomin [7.7], Liusternik and Sobolev [7.8], Naylor and Sell [7.11], and Taylor [7.12]. The exposition by Naylor and Sell is especially well suited from the viewpoint of applications in science and engineering.

For applications of the type considered in Section 7.10, as well as additional applications, refer to Antosiewicz and Rheinboldt [7.1], Balakrishnan [7.2], Byron and Fuller [7.3], Curtain and Pritchard [7.4], Kantarovich and Akilov [7.6], Lovitt [7.9], and Luenberger [7.10]. Applications to integral equations (see Section 7.10A) are treated in [7.3] and [7.9]. Optimal control problems (see Section 7.10B) in a Banach and Hilbert space setting are presented in [7.2], [7.4], and [7.10]. Methods for minimization of functionals (see Section 7.10C) are developed in [7.1], [7.6], and [7.10].

REFERENCES

[7.1] H. A. ANTOSIEWICZ and W. C. RHEINBOLDT, "Numerical Analysis and Functional Analysis," Chapter 14 in *Survey of Numerical Analysis*, ed. by J. TODD. New York: McGraw-Hill Book Company, 1962.

[7.2] A. V. BALAKRISHNAN, *Applied Functional Analysis*. New York: Springer-Verlag, 1976.

[7.3] F. W. BYRON and R. W. FULLER, *Mathematics of Classical and Quantum Physics*. Vols. I, II. Reading, Mass.: Addison-Wesley Publishing Co., Inc., 1969 and 1970.*

[7.4] R. F. CURTAIN and A. J. PRITCHARD, *Functional Analysis in Modern Applied Mathematics*. London: Academic Press, Inc., 1977.

[7.5] N. DUNFORD and J. SCHWARZ, *Linear Operators*, Parts I and II. New York: Interscience Publishers, 1958 and 1964.

[7.6] L. V. KANTOROVICH and G. P. AKILOV, *Functional Analysis in Normed Spaces*. New York: The Macmillan Company, 1964.

[7.7] A. N. KOLMOGOROV and S. V. FOMIN, *Elements of the Theory of Functions and Functional Analysis*. Vols. I, II. Albany, N.Y.: Graylock Press, 1957 and 1961.

[7.8] L. A. LIUSTERNIK and V. J. SOBOLEV, *Elements of Functional Analysis*. New York: Frederick Ungar Publishing Company, 1961.

[7.9] W. V. LOVITT, *Linear Integral Equations*. New York: Dover Publications, Inc., 1950.

[7.10] D. G. LUENBERGER, *Optimization by Vector Space Methods*. New York: John Wiley & Sons, Inc., 1969.

[7.11] A. W. NAYLOR and G. R. SELL, *Linear Operator Theory*. New York: Holt, Rinehart and Winston, 1971.

[7.12] A. E. TAYLOR, *Introduction to Functional Analysis*. New York: John Wiley & Sons, Inc., 1958.

*Reprinted in one volume by Dover Publications, Inc., New York, 1992.

INDEX

A

B

C

G

H

I

J

K

L

P

Q

R

S

T

U

V

W

X Y Z

A CATALOG OF SELECTED

DOVER BOOKS

IN SCIENCE AND MATHEMATICS

THE FOUR-COLOR PROBLEM: Assaults and Conquest, Thomas L. Saaty and Paul G. Kainen. Engrossing, comprehensive account of the century-old combinatorial topological problem, its history and solution. Bibliographies. Index. 110 figures. 228pp. 5⅜ × 8½. 65092-8 Pa. $6.95

CATALYSIS IN CHEMISTRY AND ENZYMOLOGY, William P. Jencks. Exceptionally clear coverage of mechanisms for catalysis, forces in aqueous solution, carbonyl- and acyl-group reactions, practical kinetics, more. 864pp. 5⅜ × 8½. 65460-5 Pa. $19.95

PROBABILITY: An Introduction, Samuel Goldberg. Excellent basic text covers set theory, probability theory for finite sample spaces, binomial theorem, much more. 360 problems. Bibliographies. 322pp. 5⅜ × 8½. 65252-1 Pa. $8.95

LIGHTNING, Martin A. Uman. Revised, updated edition of classic work on the physics of lightning. Phenomena, terminology, measurement, photography, spectroscopy, thunder, more. Reviews recent research. Bibliography. Indices. 320pp. 5⅜ × 8¼. 64575-4 Pa. $8.95

PROBABILITY THEORY: A Concise Course, Y.A. Rozanov. Highly readable, self-contained introduction covers combination of events, dependent events, Bernoulli trials, etc. Translation by Richard Silverman. 148pp. 5⅜ × 8¼. 63544-9 Pa. $5.95

THE CEASELESS WIND: An Introduction to the Theory of Atmospheric Motion, John A. Dutton. Acclaimed text integrates disciplines of mathematics and physics for full understanding of dynamics of atmospheric motion. Over 400 problems. Index. 97 illustrations. 640pp. 6 × 9. 65096-0 Pa. $17.95

STATISTICS MANUAL, Edwin L. Crow, et al. Comprehensive, practical collection of classical and modern methods prepared by U.S. Naval Ordnance Test Station. Stress on use. Basics of statistics assumed. 288pp. 5⅜ × 8½. 60599-X Pa. $6.95

DICTIONARY/OUTLINE OF BASIC STATISTICS, John E. Freund and Frank J. Williams. A clear concise dictionary of over 1,000 statistical terms and an outline of statistical formulas covering probability, nonparametric tests, much more. 208pp. 5⅜ × 8½. 66796-0 Pa. $6.95

STATISTICAL METHOD FROM THE VIEWPOINT OF QUALITY CONTROL, Walter A. Shewhart. Important text explains regulation of variables, uses of statistical control to achieve quality control in industry, agriculture, other areas. 192pp. 5⅜ × 8½. 65232-7 Pa. $6.95

THE INTERPRETATION OF GEOLOGICAL PHASE DIAGRAMS, Ernest G. Ehlers. Clear, concise text emphasizes diagrams of systems under fluid or containing pressure; also coverage of complex binary systems, hydrothermal melting, more. 288pp. 6½ × 9¼. 65389-7 Pa. $10.95

STATISTICAL ADJUSTMENT OF DATA, W. Edwards Deming. Introduction to basic concepts of statistics, curve fitting, least squares solution, conditions without parameter, conditions containing parameters. 26 exercises worked out. 271pp. 5⅜ × 8½. 64685-8 Pa. $7.95